ORGANIC CHEMISTRY II AS A SECOND LANGUAGE

DR. DAVID R. KLEIN

Johns Hopkins University

JOHN WILEY & SONS, INC.

SENIOR ACQUISITIONS EDITOR	Kevin Molloy
EXECUTIVE PUBLISHER	Kaye Pace
MARKETING MANAGER	Amanda Wygal
PRODUCTION MANGER	Pam Kennedy
PRODUCTION EDITOR	Sarah Wolfman-Robichaud
CREATIVE DIRECTOR	Harry Nolan
SENIOR DESIGNER	Kevin Murphy
SENIOR EDITORIAL ASSISTANT	Cathy Donovan
PROJECT MANAGEMENT SERVICES	Pam Lininger/Matrix Publishing Services
COPYEDITOR	Betty Pessagno

This book was set in 10/12 Times Roman by Matrix Publishing and printed and bound by Courier Westford. The cover was printed by Courier Westford.

This book is printed on acid-free paper. ∞

To order books or for customer service please, call 1-800-CALL WILEY (225-5945).

ISBN   978-0-471-73808-4

Printed in the United States of America

15 14 13 12 11

TABLE OF CONTENTS

CHAPTER 1

THE SKILLS YOU NEED

1.1 THE GOAL OF THIS BOOK

There are a lot of reactions that you will learn this year. Perhaps you could try to memorize them. Some people are good at that. But rather than memorizing, you would actually be better off if you tried to focus on building your *skills*. You will need certain skills in order to do well in this course. In this book, we will focus on those skills.

Specifically, you will learn the skills that you need to do three very important types of problems:

1. Proposing a mechanism
2. Predicting products
3. Proposing a synthesis

As you progress through the course, you will soon realize that it is not enough just to learn the reactions. To do well in this course, you MUST learn how to approach and solve these three types of problems. You must become a master of very specific skills. These skills will guide you in solving problems. These three types of problems represent the core of an organic chemistry course. If you master the skills you need, you will do very well.

Each chapter in this book will focus on the skills that you need in order to master a particular topic. The chapters in this book are designed to map out fairly well onto the chapters in your textbook. For instance, when you are learning about carboxylic acid derivatives, there will be a chapter in this book with the same title.

We will not have enough space to cover every topic in your textbook. This supplement is not designed to replace your textbook or your instructor. Rather, it is meant to provide you with the core skills that will allow you to study more efficiently.

Although we will focus on three major types of problems, we must place the major emphasis on mechanisms. Mechanisms are your keys to success in organic chemistry. If you master the mechanisms, you will do very well in the class; if you don't master them, you will do poorly. It is hard to talk about synthesis problems if you don't know the reactions well enough (and the same is true for predicting products). That is why Chapter 2 is devoted to laying the foundation you need to master mechanisms. That chapter is important. So, even though it won't correspond to a specific chapter in your textbook, make sure to go through Chapter 2 anyway.

In Chapter 2, we will see that mechanisms follow a small number of basic themes and ideas. By focusing on these basic themes, you will see the common threads between mechanisms that would otherwise appear to be very different. This approach will minimize the need for memorization. In fact, we will soon argue that students who focus on memorization will miss problems that are trivial when you understand the basic concepts. This book will provide you with the fundamental language and tools that you need in order to master mechanisms. And while we are at it, we will work on the skills you need to solve synthesis problems and predicting products as well.

1.2 MECHANISMS ARE YOUR KEYS TO SUCCESS

What are mechanisms, and why are they so important?

To understand the important role that mechanisms play, let's consider an analogy. I recall the time that I had to teach my children how to put on their shoes. It is amazing how many steps are involved when tying your shoelaces. Next time you tie your shoelaces, think about how many individual steps your hands are doing, and think about how you would explain each of these steps to someone who never learned how to tie their own shoes. It is a difficult thing to teach. I am embarrassed to say that my first response was to take the easy road—I bought them shoes with Velcro straps. I had temporarily solved my problem because the new shoes required so few steps to put on.

The end result (a child wearing shoes) looks the same regardless of the type of shoe used. But the process of getting the shoes on is very much dependent on the type of shoe. In particular, the number of steps involved to compete the process is vastly different from one type of shoe to the other. The same is true with reactions. All reactions involve steps to get from starting materials to products. Some reactions proceed through a lot of steps, while others go through just a few steps. A detailed list of steps that a reaction follows is called a mechanism.

When two compounds react with each other to form new and different products, we try to understand *how* the reaction occurred—what are the individual steps in the process? Every step involves the flow of electron density; electrons move to break bonds or to form new bonds. Mechanisms illustrate how the electrons move during each step of a reaction. The flow of electrons is shown with curved arrows, for example:

The curved arrows in each of the three steps above show us *how* the reaction took place. You should think of a mechanism as the "book-keeping of electrons." Just like an accountant will do the book-keeping of a company's cash flow (money coming in and money going out), so, too, a reaction mechanism is the book-keeping of the electron-flow in each step of a reaction.

In my previous book (*Organic Chemistry as a Second Language: Translating the Basic Concepts*), we saw that bond-line drawings (the way we draw molecules in organic chemistry—see the reaction above) are the "hieroglyphics" of organic chemistry. We saw that these drawings focus on the electrons (the atoms themselves are generally not drawn, but are implied). Each line in the drawing represents a bond, which shows you where the electrons are. Look at the mechanism above, and in your mind, think of it as a sentence. The drawings of the molecules (the bond-line drawings) represent the nouns of the sentence. The curved arrows are the verbs of the sentence. So a mechanism is essentially a sentence. It is truly a language, and you need to learn how to combine the nouns and verbs, just as you do in any other language. In my previous book, we focused on the nouns of the sentences—we focused on molecules. In this book, we focus on the verbs. Imagine learning French without learning any verbs. You wouldn't be able to get out a single sentence!!!!

So, if you want to master mechanisms, then you really need to master curved arrows, which are the verbs of our language. You need to know when and where to draw curved arrows. In other words, you need to learn to predict how electrons flow.

Once you appreciate the idea that electrons move in predictable ways, you will begin to see similarities in all of the mechanism in this course. As an analogy, consider the way water flows. You know that, because of gravity, water flows to the lowest spot. That is true on a mountain, on a bumpy road, on a roof, and so on. In each situation, you could probably predict how water would flow on the terrain by simply inspecting it and looking for grooves that lead to the lowest point. Simple, right? Well, imagine if you had a friend who did not understand that water always flows to the lowest spot. Your friend would be absolutely amazed at your ability to predict how the water would flow in every (seemingly different) situation. After all, a roof is certainly a different situation from a bumpy road. But in fact, the situations are not all that different when you understand the one simple concept that water flows to the lowest point.

Similarly, the flow of electrons can be predicted by understanding just a few simple concepts. If you really want to make your life miserable, then you could memorize every single mechanism in the course. But that would be silly—that would be like your friend (who does not know how water flows) trying to memorize every different situation (the roof, the mountain, the road, etc.). Your friend might have a good memory, but if he does not understand how water flows, he will be completely stumped when he sees a new terrain that he has not memorized. Without knowing the one guiding principle, he would be lost. But when you know the guiding principle, it is trivial to make the prediction of how the water will flow on ANY new terrain.

When you understand a mechanism, you will understand why the reaction took place, why the stereocenters turned out the way they did, and the like. If you do not understand the mechanism, then you will find yourself memorizing the exact details of every single reaction. Unless you have a photographic memory, that will be a very difficult challenge, and as we have just seen, you will not be able to extrapolate to situations that you have never seen. By understanding mechanisms, you will be able to make more sense of the course content, you will be able to better organize all of the reactions in your mind, and you will be able to propose mechanisms in new situations.

At this point, you might be wondering why everyone always says that organic chemistry is all about memorization. Well, the truth is, they are all WRONG. Before you can master organic chemistry, you must let go of the myth that so many former students have engrained in your psyche. There is actually very little memorization in organic chemistry. And if you try to replace true understanding with rote memorization, you will not do as well.

Organic chemistry is all about taking the principles you learn and applying them to new situations. This is easy to do if you understand the rules. It is VERY hard to do if you try to memorize 200 mechanisms. So, don't memorize mechanisms. Instead, focus on *understanding* them with an emphasis on the guiding principles of how electrons flow. That way, you will be able to predict reactions that you have never seen. And when you can do that, you will feel really good about organic chemistry. It might seem like a lofty goal right now, but be patient; this book will guide you through the process, step by step.

Every year, my students ask me how many mechanisms they need to know. I always tell them that it depends how you count the mechanisms. If your strategy is memorization, then you will need to know about 200 mechanisms. However, if you focus on just a few simple principles and rules, you will see that there are only about a dozen unique mechanisms. In fact, those dozen mechanisms are just different combinations of four basic moves.

So, we will begin our step-by-step process by going over the basic moves. We will learn them and practice them. Then we will explore the various combinations of these basic moves, once again in a step-by-step fashion.

In the end, you will see that proposing a mechanism is just as simple as predicting how water will flow down your roof.

CHAPTER **2**

INTRO TO IONIC MECHANISMS

Ionic reactions (those reactions that involve either full charges or partial charges) represent most (95%) of the mechanisms you will see this semester. The other two major categories, **radical** mechanisms and **pericyclic** mechanisms, occupy a much smaller focus in the typical undergraduate organic chemistry course. Accordingly, we will devote most of our attention to ionic mechanisms.

In this chapter, we will learn about the basic steps involved in all ionic mechanisms. There are only four basic moves. Let's begin with a quick review (from last semester) of curved arrows.

2.1 CURVED ARROWS

Before you can understand the four basic moves, you must first become a master of drawing curved arrows. These are the tools of drawing mechanisms. In the previous book, *Organic Chemistry as a Second Language: Translating the Basic Concepts*, there was an entire chapter devoted to mechanisms. The first nine pages of that chapter were devoted to building skills for drawing proper curved arrows. If you have a copy of that book, I highly recommend that you review those nine pages before continuing in this book. If you do not have a copy of that book, you should look in your textbook to see if there is any introduction to curved arrows and mechanisms.

Even for those students who feel comfortable with curved arrows, a short and quick review cannot hurt. To quickly summarize, every curved arrow has a *head* and a *tail*. It is **essential** that the head and tail of every arrow be drawn in precisely the proper place. ***The tail shows where the electrons are coming from, and the head shows where the electrons are going***:

Tail ⌒⟶ Head

Some students confuse the meaning of an arrow. They think the arrow shows where *atoms* are moving. But this is wrong. Curved arrows actually show the motion of **electrons**. As an example, consider a simple acid-base reaction:

We see that one curved arrow comes from the base (OH^-), showing the *electrons* of the base grabbing the proton. The next curved arrow shows what happens to the electrons that were originally holding onto the proton. In the end, a proton has been transferred from one place to another. So, we call this a proton transfer. But don't let the name fool you into thinking that the mechanism is like this:

WRONG

This is wrong because a mechanism does NOT show how a *proton* moves. Rather, the mechanism shows how the *electrons* move. So, for every curved arrow that you draw, you must always make sure that it is going in the right direction. Otherwise, your arrow (and therefore your mechanism) will be wrong.

There is one more thing to clarify before we can move on. Notice that the tail of the curved arrow (coming from the oxygen) is placed on a lone pair:

This makes sense because the tail of the arrow represents where the electrons are coming from. And electrons can either come from lone pairs or from bonds (we will see many examples very soon). In the example above, a *lone pair* is grabbing the proton. However, it is very common for organic chemists to draw compounds *without* drawing the lone pairs because it is faster that way. For example:

and here is another, more striking, example:

Lone pairs do not *have to* be drawn because they are implied (we saw that in the first semester). In other words, you are supposed to know that the lone pairs are really there, despite their absence from our drawing. As a result, you will often see something like this:

Notice that the tail of the curved arrow is placed on the negative charge. This is the way we must draw the curved arrow in situations where we omit the lone pairs from our drawings. But don't be confused; it really isn't the negative charge that is grabbing the proton. Rather, it is a lone pair that is grabbing the proton.

There are two reasons why chemists will often omit the lone pairs when drawing mechanisms. First of all, it is faster to draw mechanisms if we omit most of the lone pairs. But more importantly, it is a lot easier to follow a mechanism when it is not cluttered with lone pairs. Compare for yourself:

Clearly, the second way of drawing it is less cluttered and easier to follow.

Your textbook will most likely draw all lone pairs (at least the ones participating in a reaction), but your instructor might draw some mechanisms without lone pairs during the lecture (to save time). Both ways are correct. For purposes of clarity and simplicity, most drawings in this book will leave out the lone pairs. I will only show the lone pairs when their presence does not compromise the clarity of the presentation.

Since we will be leaving out the lone pairs very often, it is important that you get accustomed to "seeing" the lone pairs even though they are not drawn. Oxygen is the most common element that you will see in the course (that possesses lone pairs), so let's start there. Each oxygen atom with no charge will have two lone pairs:

is the same as

An oxygen atom with one bond and a negative charge will always have three lone pairs:

is the same as

And an oxygen atom with three bonds and a positive charge will always have one lone pair:

is the same as

and

is the same as

You will see many other elements in this course (nitrogen, sulfur, phosphorus, etc.), but oxygen is the most common of them all. So, you would do yourself a big favor if you would remember the rules above.

One other very common example is a carbon atom with a negative charge:

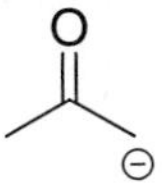

is the same as

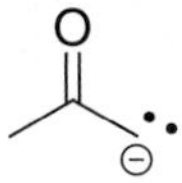

It is important that you can see the lone pairs even when they are not drawn, so let's get some quick practice to make sure that you can see the lone pairs:

EXERCISE 2.1.   In the following intermediate, the lone pairs are not drawn. Draw them.

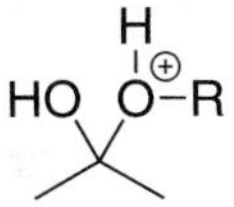

Answer: Oxygen atoms with no charge will have two lone pairs:

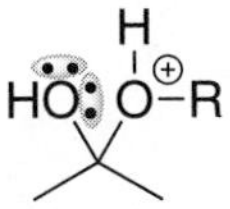

And oxygen atoms with three bonds and a positive charge, will always have one lone pair:

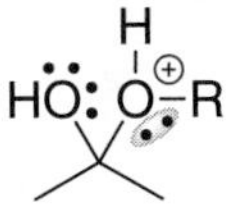

For each of the following, draw in the lone pairs that are not shown:

2.2. 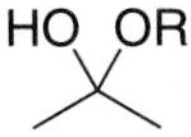   **2.3.** 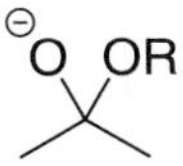   **2.4.**

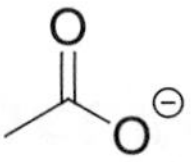

2.5. 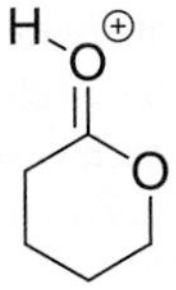   **2.6.** 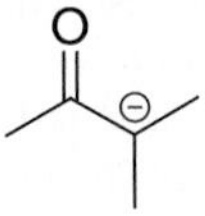   **2.7.**

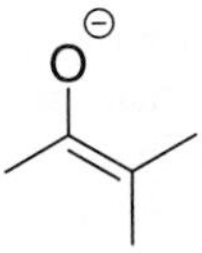

2.8. 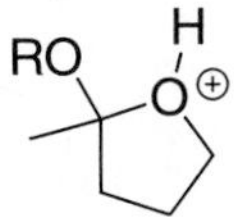   **2.9.** 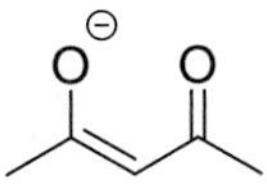   **2.10.**

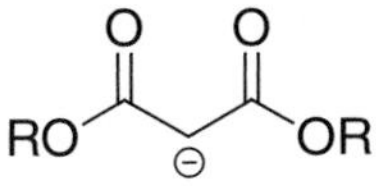

When drawing curved arrows, you must focus on two things:

1. The *tail* of every curved arrow needs to be in the right place, and
2. The *head* of every curved arrow needs to be in the right place.

Now let's make sure that you are comfortable with curved arrows. For each of the reactions below, try to draw the curved arrows *without* flipping to the answers in the back of the book. When you are finished, look up the answers in the back of the book to make sure that you drew the arrows correctly. When looking at the answers, make sure to focus on the position of every head and every tail of the arrows in your answers. If any heads or tails were not drawn in exactly the correct place, then you should go back and review the first nine pages of the mechanisms chapter from the previous book.

PROBLEMS. For each transformation below, complete the mechanism by drawing the proper arrows (most of these reactions are from the first semester of organic chemistry):

2.11.

2.12.

2.13.

2.14.

2.15.

2.2 THE BASIC MOVES

To truly become the master of arrow pushing, you must master the "basic moves" that are at your disposal when proposing a mechanism. These basic moves are your tools for proposing mechanisms. That's what this chapter is all about: mastering the most basic moves of any ionic mechanism. It is comforting to realize that there are only four basic moves. (That will cover you for any ionic mechanism that you will see in this course.) Let's go through them, one by one:

1. *Nucleophile attacks electrophile*. For example:

Remember that a nucleophile is a compound with a region of high electron density. In some cases, the nucleophile will actually have a negative charge (like the bromide ion in the example above). In many other cases, the nucleophile will not have a charge—you must keep in mind that lone pairs and π bonds can also be nucleophilic:

Here is one example of a *lone pair* acting as a nucleophile:

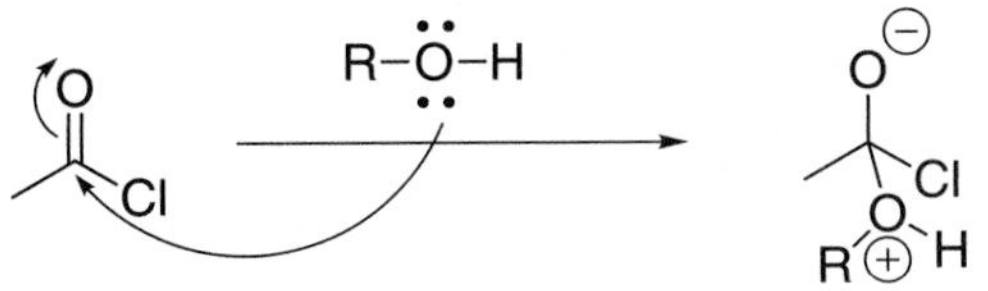

The nucleophile here is the lone pair on the oxygen of the alcohol above (ROH). This oxygen atom is using one of its lone pairs to attack the C=O bond. Notice that there is no negative charge on the attacking oxygen atom. ROH does not bear a negative charge. That's OK, because a nucleophile is just a region in space of high-electron density. And the lone pairs of the oxygen atom are regions in space of high-electron density.

Also notice that there is more than one curved arrow on the step above. The first curved arrow shows the nucleophile (ROH) attacking the electrophile. The second curved arrow (pushing the electron density up onto the oxygen) is part of the same basic move—nucleophile attacking an electrophile. To see how this second arrow is part of the same move, consider the following: what if I were to draw a resonance structure of the starting compound BEFORE it gets attacked:

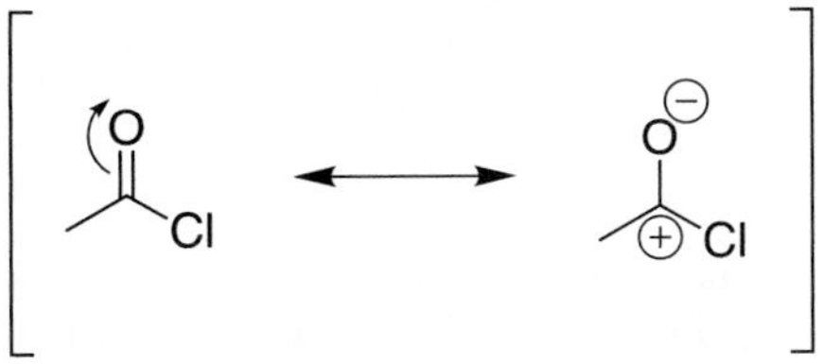

Now, I can draw the attack taking place at the second resonance structure, like this:

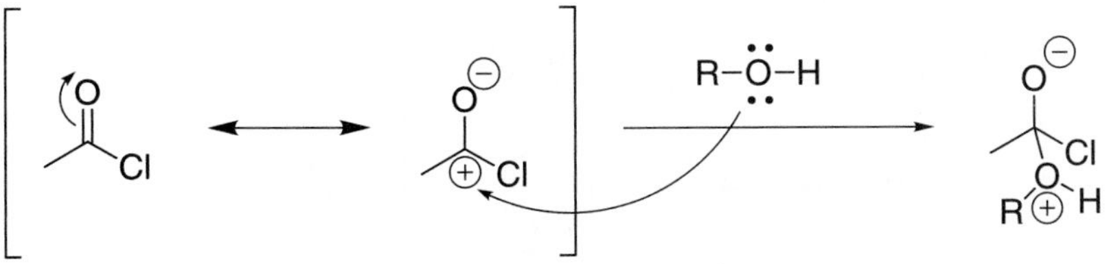

When we think of it like this, we realize that only one of the arrows is actually showing the attack:

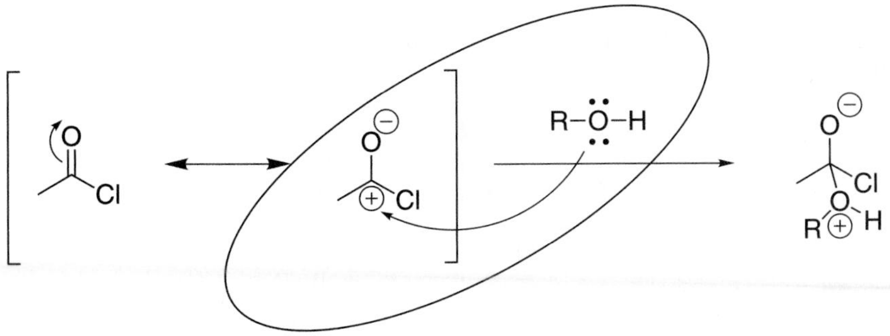

The second arrow can be thought of as resonance, or alternatively, you can think of the second arrow as an actual flow of electron density that goes up onto the oxygen when the nucleophile attacks:

Electron density
is pushed up
onto the oxygen

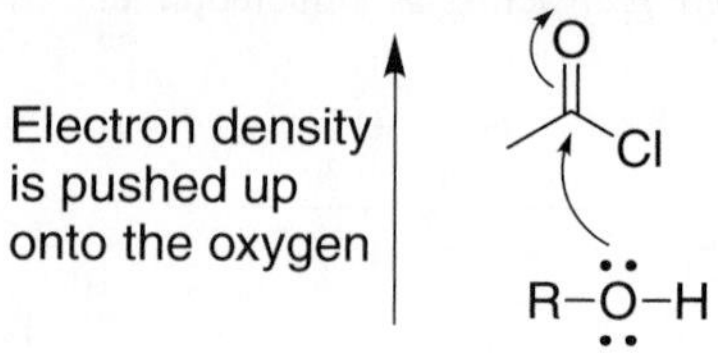

There is a subtle difference between these two ways of looking at the second arrow (either thinking of it as resonance or thinking of it as an actual flow of electron density). It is probably more accurate to think of it as an actual flow of electron density up onto the oxygen because that is the way most organic chemists think about it. However you wish to view this, you should realize that both curved arrows are used to show only one basic move: a nucleophile is attacking an electrophile.

In fact, it is common to see even *more* than two arrows when a nucleophile attacks an electrophile. For example, consider the following:

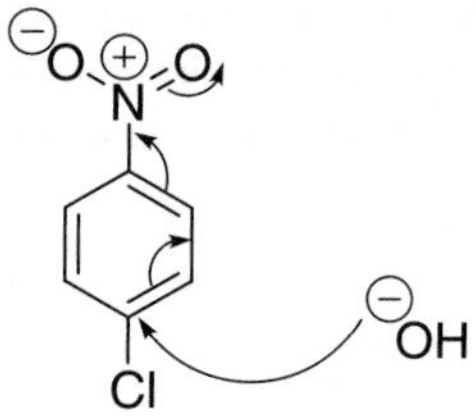

All of these arrows show *one* thing happening: a nucleophile (OH⁻) is attacking an electrophile. The first curved arrow (coming from OH⁻) shows this attack. The other curved arrows can be viewed as resonance arrows before the attack:

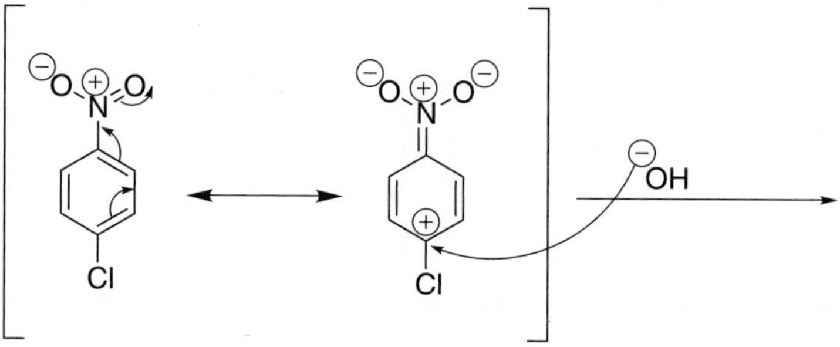

Or they can be viewed as showing the flow of electron density that takes place when the nucleophile attacks:

Electron density
is pushed up
onto the oxygen

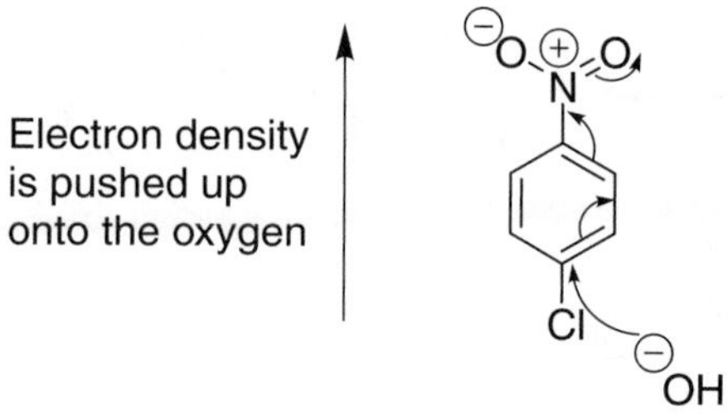

To be consistent with the way organic chemists speak, you will probably be better off if you think of it as a flow of electron density (rather than just one attacking arrow and the remaining arrows being considered as resonance arrows). I only gave the resonance argument so that we can justify in our minds that all of these arrows are really just showing one basic thing happening.

In all of the examples we have seen, the nucleophile had a lone pair that was responsible for attacking the electrophile. But we said that a π bond can also serve as a nucleophile. For example:

Once again, there is more than one curved arrow here. There are two curved arrows. The first arrow (coming from the ring to attack SO_3) is the arrow showing the attack. The second arrow can be thought of in two ways. You can think of it as a resonance arrow:

Or you can think of it as the flow of electron density when the nucleophile attacks:

Electron density
is pushed up
onto the oxygen

Either way, you should realize that this is just one basic move: a nucleophile attacking an electrophile.

As we move through this course, we will see many more examples of nucleo-philes. For now, let's just appreciate the basic feature of a nucleophile—it is a region of high-electron density (either a lone pair or a π bond). And nucleophiles always do what they are supposed to do—they attack electrophiles.

An electrophile is a compound with a region of low-electron density. In some cases, the electrophile will actually have a positive charge (as in the example we saw earlier where the bromide ion is attacking a compound with a positive charge). In many other cases, the electrophile will not have a charge. For example, a ketone is an excellent electrophile, which we can understand when we draw the resonance structures:

We can see from the second resonance structure that the carbon of the C=O double bond is electron-poor (a site of low-electron density). When you normally draw a ketone, you don't see a positive charge on the carbon atom:

But even though there is no net charge, there is still an electrophilic center (that is waiting to be attacked by a nucleophile). Later in this semester, we will see a lot of reactions that deal with the chemistry of compounds containing C=O bonds. Most of this chemistry revolves around the electrophilicity of the C=O group.

As we move through the course, we will see more examples of electrophiles. Now that we have identified the first basic move (nucleophiles attacking electrophiles), let's continue with a summary of the second basic move.

2. ***Loss of a leaving group***. For example:

This move can be thought of as the reverse process of the first basic move that we saw. In the reaction we are showing here, a leaving group leaves to form a carbocation. If you were to imagine taking a video of this process, and then playing the video backwards (rewinding it so that you can see it while it is rewinding), you would see a bromide ion (a nucleophile) attacking a carbocation (an electrophile) to form the compound on the left above. So, in fact, the basic move we are discussing right now (loss of a leaving group) is just the reverse process of the first basic move that we saw (nucleophile attacking an electrophile).

Sometimes, you will see more than one arrow being used for the loss of a leaving group. For example, consider the following:

There is one curved arrow that actually shows the Cl leaving. The remaining arrows can be thought of in two different ways (just as we saw earlier). You can think of the other arrows as being resonance arrows after the leaving group leaves:

Or you can think of this as a flow of electron density that pushes out the leaving group:

Electron density
is pushed down
to kick off a
leaving group

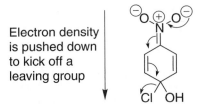

You are probably better off if you think of this as a flow of electron density that pushes out the leaving group. But however you think of it, you should realize that all of these arrows are just showing one basic move: a leaving group is leaving.

In the first semester of organic chemistry, we saw tips for identifying good leaving groups and bad leaving groups. If you are rusty on leaving groups, I recommend that you go back and review that material. You will need to be able to identify good leaving groups in order to propose mechanisms.

So far, we have seen two basic moves: a nucleophile attacking an electrophile and a leaving group leaving. Now let's take a look at our third basic move:

3. ***Proton Transfers***. For example:

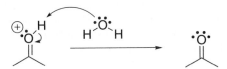

Proton transfers are just acid-base reactions. We talked about acidity (proton transfers) at length in the first semester of organic chemistry. For now, let's just focus on the fact that this basic move always requires at least two curved arrows. As an example, look at the proton transfer we just saw. The ketone is grabbing a proton, and we need *two* curved arrows to show this. One arrow goes from the ketone to the proton, and the second arrow shows what happens to the electrons that were previously holding the proton.

There are always at least two arrows, whether the compound is grabbing a proton (like the case above), or whether a compound is losing a proton, like this:

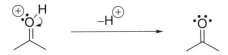

Notice that, once again, we need two arrows to show the proton transfer. When a compound is losing a proton, some textbooks and instructors will skip showing what grabs the proton. And they will only use one arrow, like this:

Other instructors are particular that you show both arrows. In other words, you need to show what grabs the proton. Certainly, it can't hurt to get into the habit of always showing what grabs the proton, right? So get in the habit of using two arrows.

Sometimes, you will see *more* than two arrows. For example, consider this case which shows three arrows:

To explain this, we will use the same logic we used when describing the other basic moves. We can understand this in either of two ways. We can say that there are only two arrows needed for the reaction, and the remaining arrows are just resonance arrows:

Or we could argue that all of the arrows together show the flow of electron density that takes place during the proton transfer:

Electron density flows up onto the oxygen of the ketone.

For proton transfers, both arguments are equally good, and you should think of it both ways. And most importantly, you should realize that there is just one basic move happening here: a proton is being transferred.

Now we are ready to see the last basic move:

4. *Rearrangements*. For example:

When a compound has a positively charged carbon atom (a carbocation), then it is possible for the compound to undergo a rearrangement. There are other rare instances when rearrangements can take place (other than carbocations). We will point them out later in this book, when we get to them. For now, let's focus on carbocation rearrangements.

Believe it or not, that's it!!! Just four basic moves to master:

1. Nucleophiles attacking electrophiles
2. Leaving groups leaving
3. Proton transfers
4. Rearrangements

In fact, we can summarize it even better, like this:

1. Attack
2. Leave
3. Protonate (or deprotonate)
4. Rearrange

Each of these four basic moves has subtle details that you will have to learn. For example, when should you use H_2O instead of OH^- to grab a proton? As another example, if you have a reagent like OH^- that can function as a nucleophile or a base, which basic move should you use? (Should you use the reagent as a nucleophile to attack something, or should you use it as a base to grab a proton?) These questions, and many others, are the fine details of these four basic moves. As we move through the course, I will point out these details for you. Make sure to appreciate how important these details are. When you learn a new detail—for example, when you learn that proton transfers are generally faster than nucleophilic attacks—you should realize how important it is to know that. It will help you with ALL mechanisms throughout the course. You will only be able to master the mechanisms if you master the details.

The goal of this chapter was to introduce you to the four basic moves. So let's make sure that you can identify each of the four moves when you see them.

EXERCISE 2.16.   Consider the following step:

This step is one of the four basic moves. Which basic move is this step?

Answer: You might be tempted to call this a rearrangement because it looks like the location of C+ is moving. But take a closer look. This is not a hydride shift, nor is it a methyl shift. If we had to describe what is happening, we would say that the double bond (the π bond) is attacking the C+. The π bond is acting as a nucleophile to attack an electrophile. It is an interesting case because the nucleophilic site and the electrophilic site are both contained in the same compound. So this step is called *intramolecular* because the attack takes place entirely within one molecule. This step is definitely the first basic move—a nucleophile attacking an electrophile.

PROBLEMS.   For each of the following, identify the basic move that you see. Your choices are: (1) a nucleophile attacking an electrophile, (2) a leaving group leaving, (3) a proton being transferred, or (4) a rearrangement.

2.17.

Answer _____

2.18.

Answer _____

2.19.

Answer _____

2.20.

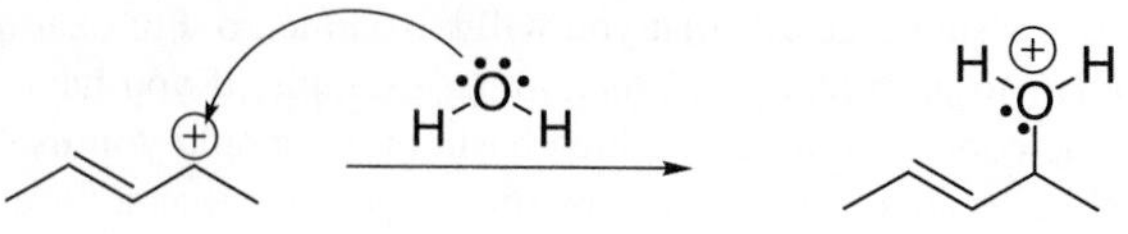

Answer ——————

2.21.

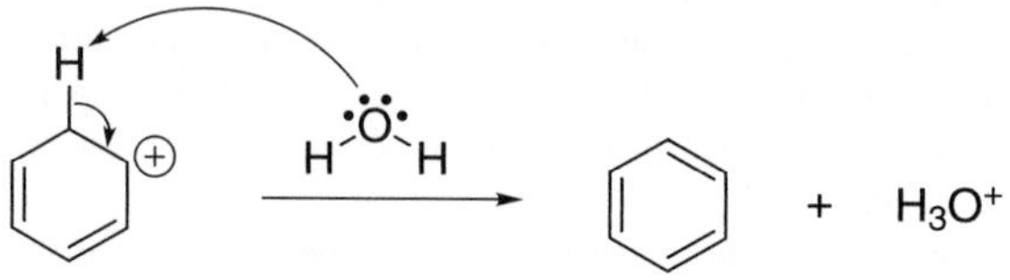

Answer ——————

2.22.

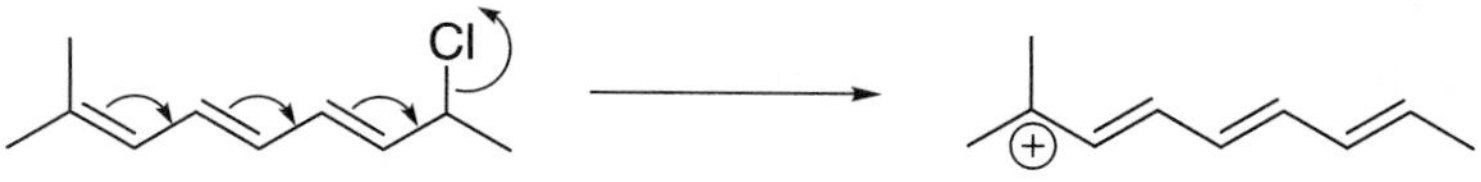

Answer ——————

2.23.

Answer ——————

2.24.

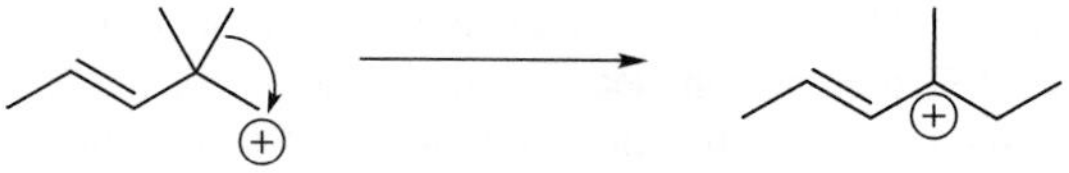

Answer ——————

2.25.

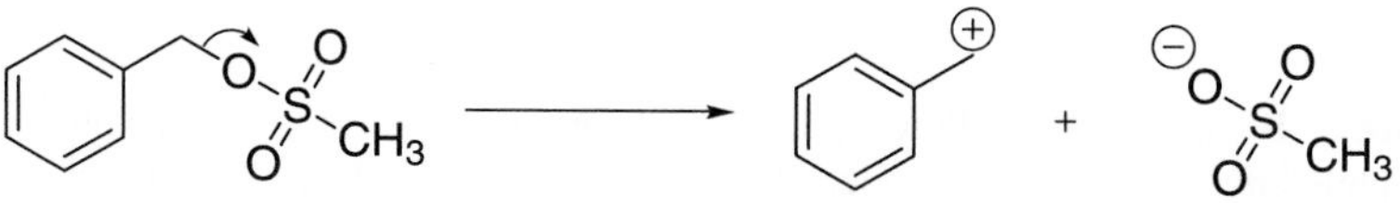

Answer ——————

2.26.

Answer ——————

2.3 COMBINING THE BASIC MOVES

In order to conquer mechanism problems, you must have these four basic moves at your fingertips. To understand why, you must realize that all ionic mechanisms, regardless of how complex, are just different combinations of these four basic moves. Let's see a few examples.

Let's start by looking at some mechanisms that we saw in the first semester of organic chemistry (and then we can move on to a preview of reactions we will see this semester). For example, take a look at this S_N1 reaction:

Notice that the first step is *loss of a leaving group*, and the second step is *nucleo-phile attacks electrophile*. So, we see that the S_N1 mechanism is just two of the four basic moves (one after the other). Now, let's look at an S_N2 reaction:

In this mechanism, we are still doing the same two basic moves as we did in the previous mechanism (*loss of a leaving group* and *nucleophile attacks electrophile*). The only difference is that we are now doing both *at the same time*.

What about an elimination reaction? Consider the following E1 reaction:

The first step is *loss of a leaving group*, and the second step is just a *proton transfer* (which always has two curved arrows—one curved arrow is showing the base pulling off the proton, and the second curved arrow is showing where to place the electrons that used to hold the proton). If we use a stronger base, and we do all of these steps at the same time, we get an E2 mechanism:

Let's do one more example. Consider the following addition reaction from last semester:

Notice that the first step is a *proton transfer*, and the second step is a *nucleophile attacking an electrophile*.

We have now seen many examples from last semester. These examples demonstrate that there are only four basic moves, and these four moves are your tools for solving mechanism problems. By mastering these tools, you will then be able to master how to use them together (one after another) to build up a mechanism. The trick is to recognize the various patterns that can arise when you combine these basic moves in specific ways. And that is what organic chemistry mechanisms are all about. For example, consider the following reaction that we will see later in this semester:

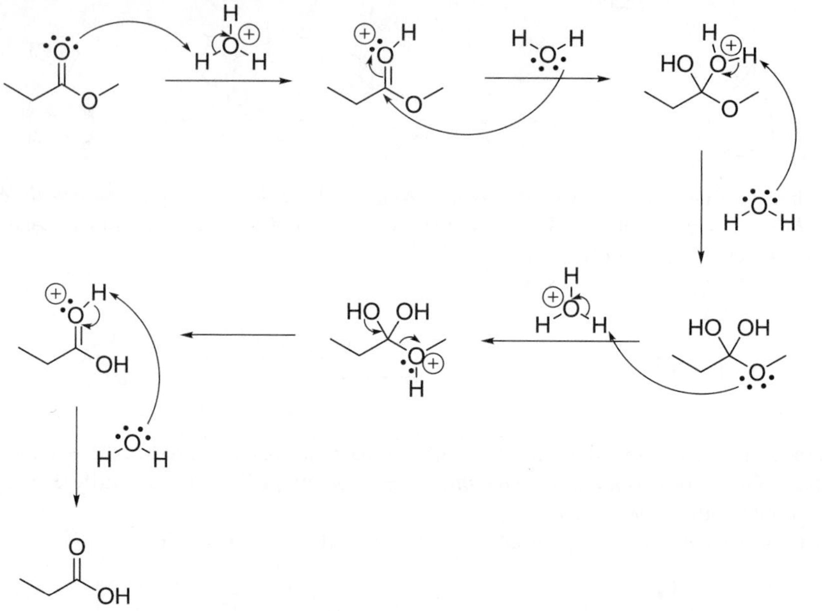

This is certainly a longer mechanism than the reactions you learned in the first semester of organic chemistry. But let's try to make sense of this long mechanism by taking a close look at the steps. Let's see if we can convince ourselves that this mechanism is just a combination of the basic moves. The first step is a proton transfer:

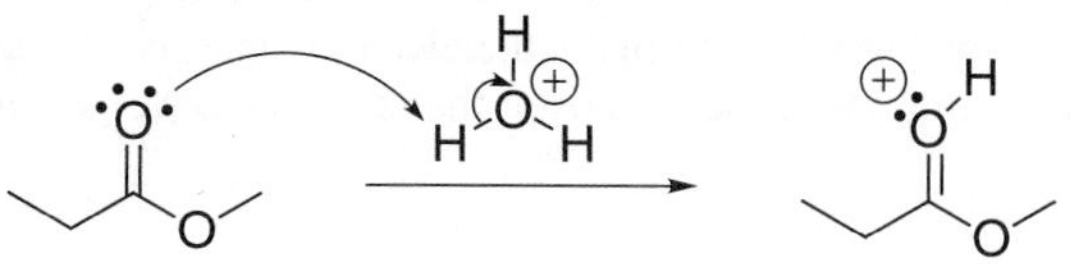

The second step is a nucleophile attacking an electrophile.

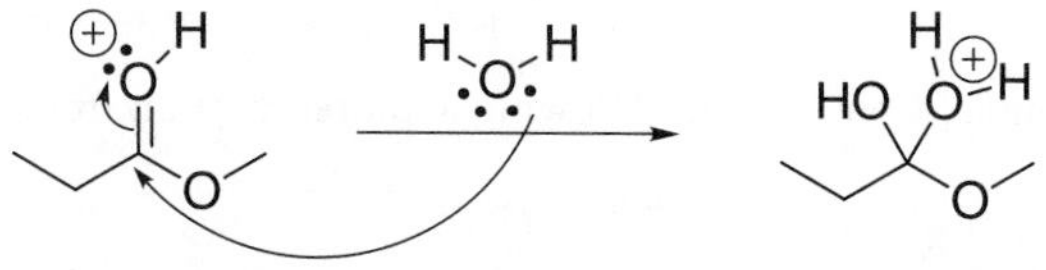

The next two steps are both just proton transfers:

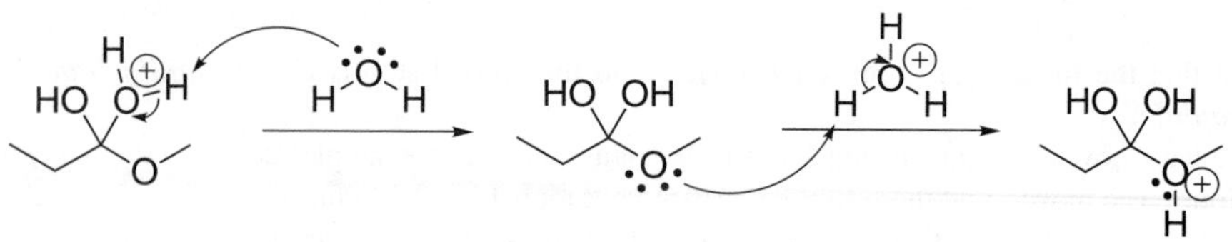

The second-to-last step is just the loss of a leaving group:

And the final step of the mechanism is just a proton transfer:

So, after analyzing the entire mechanism, we see that the entire mechanism is just made up of a series of individual steps, and each one of those steps is one of the four basic moves.

If you go through that mechanism carefully, you will find that only two steps are not proton transfers. There is one step toward the beginning (the second step), which is a nucleophile attacking an electrophile. Then there is one step toward the end (the second-to-last step), which is the loss of a leaving group. *Everything else* was just proton transfers. In fact, if you think about it, this reaction is not much different from a regular substitution reaction. A few pages ago, we saw that a regular substitution reaction (S$_N$2 for example) is just two major steps: nucleophile attacking an electrophile, and the loss of a leaving group. Here, in our long mechanism, we have the same two steps: nucleophile attacks electrophile, followed by loss of a leaving group. The major difference here is that we have many more steps that are all proton transfers.

This is a common theme that you will see when you analyze "long" mech-anisms. You will find that usually only two or three critical steps define the mechanism, and the rest of the steps are all just proton transfers that facilitate the reaction. But don't worry, you won't have to memorize when to do all of the proton transfers. There are some simple rules that will tell you when to use a proton transfer in a mechanism. We will see those rules as we move through the course. For now, I just want you to appreciate two things:

1. Even long mechanisms are just a special sequence of the four basic moves. These four basic moves are your tools for understanding and (eventually) proposing mechanisms.
2. Long mechanisms usually have only two or three critical steps that define the reaction. The rest of the steps are usually just proton transfers.

It will help you to think about mechanisms in this way because it will help you to see the patterns and similarities between "apparently" different mechanisms. When you view mechanisms in this way, you will appreciate just how similar so many mechanisms are. In fact, so many mechanisms are identical, with the exception of a proton transfer here or there. Once you master these tools (when and how to use them), you will be able to actually propose a mechanism for a reaction that you have never seen. At each step of the mechanism, you will be able to logically argue what *should* happen next. When you get to that point, you will realize that there is no memorization here. It all makes sense, and it is all predictable, once you have mastered the basic moves.

It's all about recognizing patterns and getting into the habit of using a very small number of basic moves. It is my intention to greatly simplify organic chemistry for you by bundling the material into little bite-sized packages that are very easy to master.

Now let's just get a bit of practice.

EXERCISE 2.27.   Consider the mechanisms of the two different reactions shown below:

REACTION 1

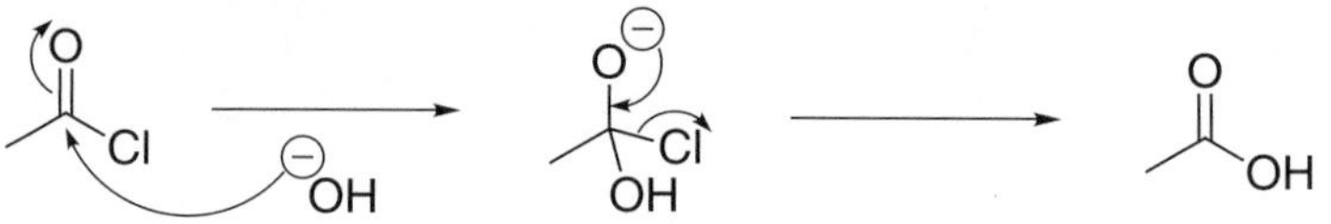

REACTION 2

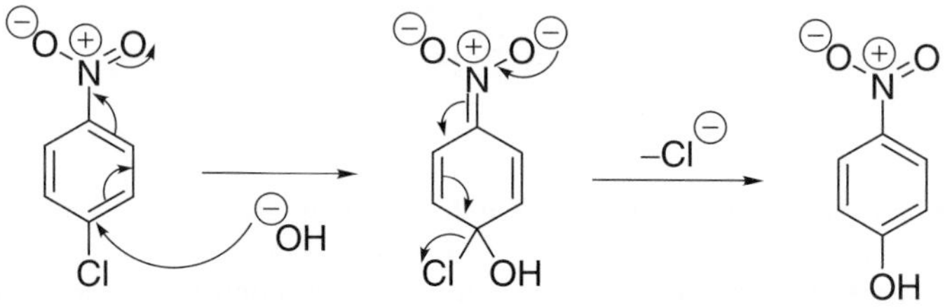

For each reaction, identify the sequence of basic moves. Then compare these two mechanisms in terms of their sequences.

Answer: The first reaction has two steps: (1) nucleophile attacks electrophile, followed by (2) loss of a leaving group. In the second reaction, we see the same two basic moves, in the same order: (1) nucleophile attacks electrophile, followed by (2) loss of a leaving group.

When you think of it this way, you can begin to see the striking similarity between these two seemingly different reactions. In the first reaction, a hydroxide ion attacks, kicking up electron density up onto an oxygen atom. Then, in step 2, that electron density comes back down to kick off chloride as a leaving group. In the second reaction, we see the exact same order of events: a hydroxide ion attacks, kicking up electron density up onto an oxygen atom, and the electron density then comes back down to kick off chloride as a leaving group.

PROBLEMS.   For each mechanism below, "read" the mechanism, and identify the sequence of basic moves that is shown.

2.28.

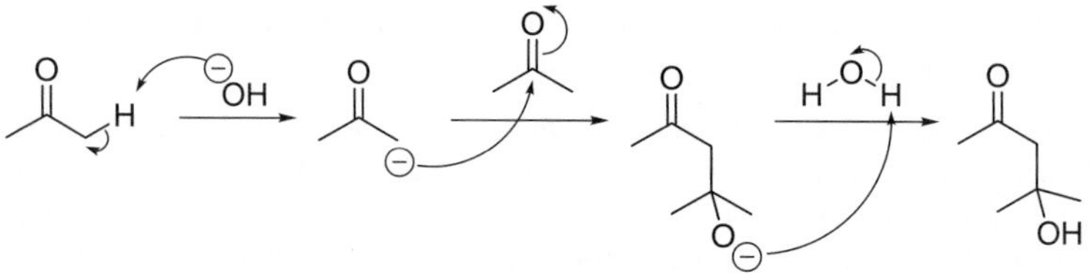

Sequence of moves is: _____

2.29.

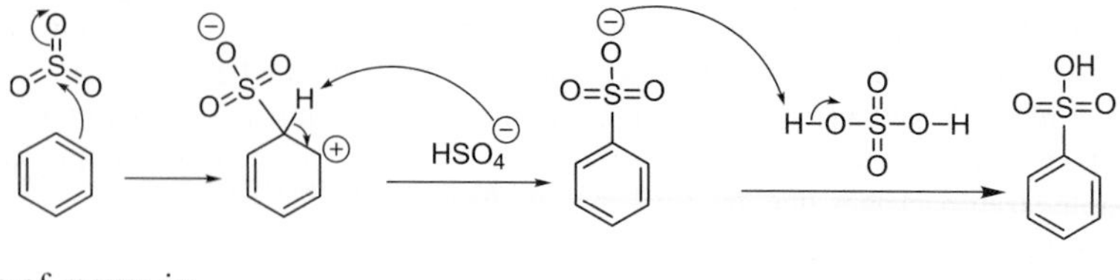

Sequence of moves is: _____

2.30.

Sequence of moves is: _____

2.31.

Sequence of moves is: _____

2.32.

Sequence of moves is: _____

2.33.

Sequence of moves is: _____

In summary, we have seen four basic moves.

Let's give them shorter names, and let's turn them all into verbs:

1. Attack
2. Leave
3. Protonate (or deprotonate)
4. Rearrange

As you study, you should view every mechanism as a particular sequence of these basic moves. That sequence will represent a short summary of the mechanism. As an example, in the next chapter, we will learn about electrophilic aromatic substitution reactions. We will see that all of them follow the same basic sequence: (1) attack, (2) deprotonate. That's it.

We will also spend a lot of time on the chemistry of compounds containing a C=O bond. Once again, all of those mechanisms will follow a similar sequence of basic moves. When you see it that way, it helps you remember the mechanism. But more importantly, it will help you to see the similarities between organic reactions. There might be over 100 mechanisms in an entire organic chemistry course, but there are less than a dozen different patterns (or sequences of basic moves). And as the course moves on, you will begin to master the fine details of the basic moves, which will empower you to begin proposing mechanisms for reactions you have never seen. One goal of this book is to get you to that point. And our first stop is electrophilic aromatic substitution. . . .

CHAPTER *3*

ELECTROPHILIC AROMATIC SUBSTITUTION

We must begin this chapter with a review of a reaction from the first semester of organic chemistry. Recall the addition of bromine (Br_2) across a double bond:

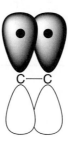

When we learned this reaction in the first semester, we saw that this reaction involves a nucleophile attacking an electrophile. The nucleophile is the double bond, and the electrophile is Br_2. To understand how a double bond can function as a nucleophile, recall that a double bond is the overlap of two neighboring p orbitals:

Therefore, a double bond is a region in space of high-electron density. Even though there is no full negative charge anywhere, the double bond can function as a nucleophile and can attack an electrophile. But the obvious question is: why is Br_2 an electrophile? After all, the bond between the two bromine atoms is covalent, and therefore, we cannot say that one of the bromine atoms has any more electron density than the other. (There is no induction here because both atoms, Br and Br, have the same electronegativity.)

There is a simple reason why bromine can act as an electrophile here. We need to consider what happens when a molecule of Br_2 approaches an alkene. To help us see this, think of Br_2 in terms of the electron cloud surrounding it:

As a molecule of Br_2 gets close to an alkene, the electron density of the alkene begins to repel the electron cloud around Br_2. This effect gives the Br_2 molecule an induced dipole moment (this is a temporary interaction—it only happens while the bromine molecule is near the alkene):

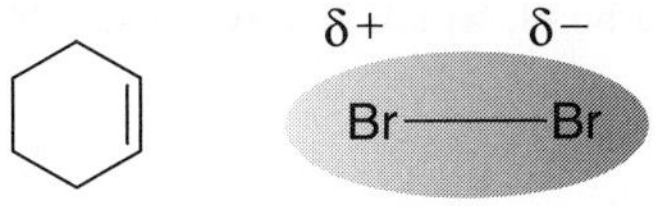

$$\delta+ \qquad \delta-$$

So, we have an electron-rich alkene, which can attack the nearby, electron-poor bromine atom. This gives us the reaction that we saw in the first semester of organic chemistry:

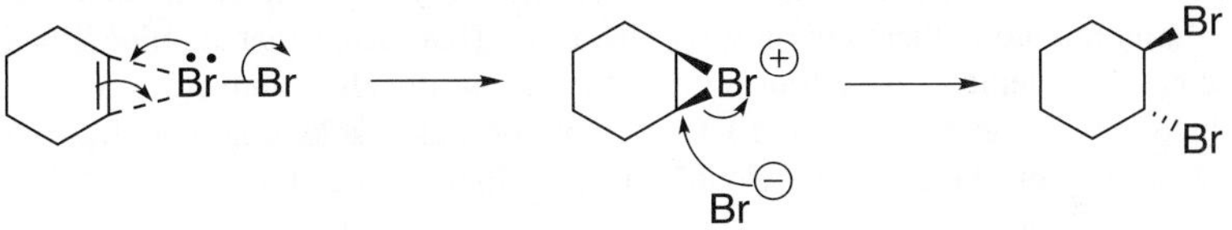

Now let's consider what happens if we try to do this exact same reaction with benzene as our nucleophile. So we are trying to do this reaction:

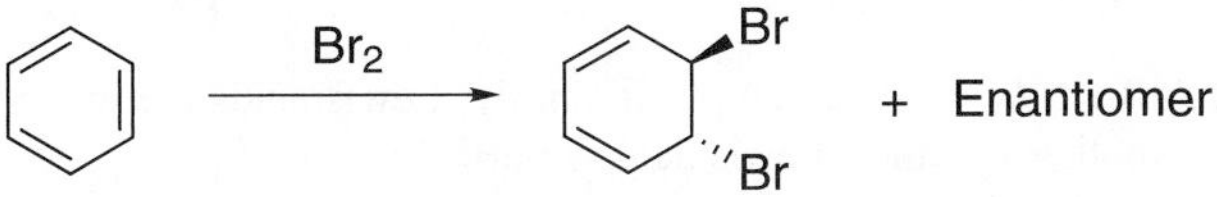

$$\xrightarrow{Br_2} \quad + \quad \text{Enantiomer}$$

But when we heat up benzene in the presence of Br_2, we find that there is no reaction:

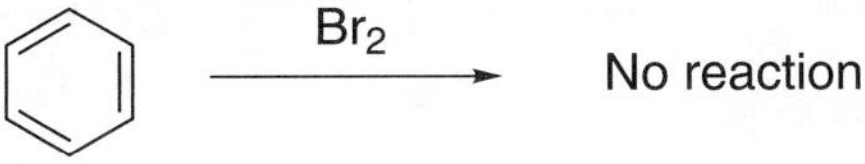

$$\xrightarrow{Br_2} \quad \text{No reaction}$$

We can understand this because benzene is an aromatic compound. It has a special stability due to its aromaticity. If we add Br_2 across benzene, we will lose aromaticity. And that is why the reaction does not take place—it would be going "uphill" in energy. But is it possible to try to "force" the reaction to happen?

This brings us to a simple and important concept in organic chemistry. One of the driving forces for any reaction between a nucleophile and an electrophile is the difference in the electron density between the two compounds. The nucleophile is electron-rich, and the electrophile is electron-poor. Therefore, they are attracted to each other in space (opposite charges attract). So, if the reaction is not proceeding, we can try to force it along by making the attraction even stronger between the nucleophile and the electrophile. We can accomplish this in one of two ways. We can either make the nucleophile even more electron-rich (more nucleophilic), or we can make the electrophile even more electron-poor (more electrophilic).

In this chapter, we will explore both of these scenarios. For now, let us start by trying to make the electrophile a better electrophile. How do we make Br_2 a better electrophile? Let's remember why Br_2 is an electrophile in the first place. Just a few moments ago, we saw that an induced dipole moment is formed when Br_2 gets close to an alkene. This creates a partial positive charge on one of the bromine atoms. Clearly, if we had Br^+ instead of Br_2, then that would be an even better electrophile. We would not have to wait around for Br_2 to get slightly polarized. Instead, we would have Br^+ floating around in solution.

But how do we form Br^+? That is where Lewis acids come into the picture.

3.1 HALOGENATION AND THE ROLE OF LEWIS ACIDS

Consider the compound $AlBr_3$. The central atom in this structure is aluminum. Aluminum is in the third column of the periodic table, and therefore, it has three valence electrons. It uses each of these electrons to form a bond, which is why we see three bonds to the aluminum atom in $AlBr_3$

$$Br\overset{\overset{\displaystyle Br}{|}}{\underset{}{Al}}\diagdown Br$$

But you should notice that the aluminum atom does not have an octet. If you count the electrons around the aluminum atom, there are only six electrons. That means that aluminum has one empty orbital. (Second-row elements have four orbitals to use, but the aluminum atom is only using three of them, which leaves one empty orbital.) That empty orbital is able to accept electrons. In fact, it *wants* to accept electrons because that would give aluminum its octet:

$$Br\overset{\overset{\displaystyle Br}{|}}{\underset{\displaystyle Br}{Al}} \quad :X \longrightarrow Br\overset{\ominus}{\overset{\overset{\displaystyle Br}{|}}{\underset{\displaystyle Br}{Al}}}-X^{\oplus}$$

Therefore, we call $AlBr_3$ a *Lewis acid*. To put it simply, Lewis acids are just compounds that can *accept an electron*. Another common Lewis acid is $FeBr_3$:

$$Br\overset{\overset{\displaystyle Br}{|}}{\underset{\displaystyle Br}{Fe}} \quad :X \longrightarrow Br\overset{\ominus}{\overset{\overset{\displaystyle Br}{|}}{\underset{\displaystyle Br}{Fe}}}-X^{\oplus}$$

Lewis acid

Now let's consider what happens when we mix Br_2 and a Lewis acid. The Lewis acid can accept electrons from Br_2:

$$Br\overset{\overset{\displaystyle Br}{|}}{\underset{\displaystyle Br}{Al}} \quad :\ddot{B}r-\ddot{B}r: \longrightarrow Br\overset{\ominus}{\overset{\overset{\displaystyle Br}{|}}{\underset{\displaystyle Br}{Al}}}-\overset{\oplus}{\ddot{B}r}-\ddot{B}r:$$

This intermediate can then liberate Br^+ to give the following complex:

$$Br\overset{\ominus}{\overset{\overset{\displaystyle Br}{|}}{\underset{\displaystyle Br}{Al}}}-\overset{\oplus}{\ddot{B}r}-\ddot{B}r: \longrightarrow Br\overset{\ominus}{\overset{\overset{\displaystyle Br}{|}}{\underset{\displaystyle Br}{Al}}}-\ddot{B}r: \quad \boxed{\overset{\oplus}{\ddot{B}r:}}$$

In essence, the Lewis acid approached a molecule of Br_2 and pulled off a bromine atom to leave behind Br^+. It is probably not accurate to think of this as a free Br^+ that can float around the solution by itself. Rather, it most likely exists as a complex with $AlBr_4^-$:

$$\left[Br\overset{\ominus}{\overset{\overset{\displaystyle Br}{|}}{\underset{\displaystyle Br}{Al}}}-\ddot{B}r: \quad \overset{\oplus}{\ddot{B}r:} \right]$$

But the important point is that we have now formed Br^+, and that is what we needed in order to try to force our reaction to go. So, now let's try our reaction again. Let's try to react benzene with bromine, but this time, we will add $AlBr_3$ as a Lewis acid. When we try to run this reaction, we do in fact get a reaction. BUT it is not the reaction that we expected. Look closely at the product:

We did NOT get an addition reaction. Rather, we got a *substitution* reaction. Br^+ *substituted* for one of the protons on the *aromatic* ring. So we call this an *aromatic substitution* reaction. Since the reagent reacting with the ring is an electrophile (Br^+), we call this reaction an *electrophilic aromatic substitution*.

To see how this reaction happens, let's take a close look at the mechanism. It is absolutely critical that you fully understand the mechanism of this reaction. We will soon see that ALL electrophilic aromatic substitution reactions follow the same mechanism. The first step shows the ring acting as a nucleophile to attack Br^+:

This creates an intermediate that does not have aromaticity. So, it is true that we have destroyed aromaticity *temporarily*, but we will soon reestablish aromaticity in the final product. In this first step of the mechanism (shown above), the ring attacks Br^+ to form an intermediate that has three important resonance structures:

It is important to remember what resonance structures are. Recall from the first semester that resonance is NOT a molecule flipping back and forth between different states. Rather, resonance is the way we deal with the inadequacy of our drawings. There is no one single drawing that adequately captures the essence of the intermediate, so we draw three drawings, and we meld them all together in our mind to understand what the intermediate looks like.

Attempts have been made to draw a single drawing for this intermediate:

You might even see this in your textbook. I usually try to avoid using this drawing because it implies that the positive charge is spread out over five atoms in the ring. This is not the case. The majority of the positive charge is actually only spread out over three atoms of the ring (which we can clearly see when we look at all three resonance structures above).

This intermediate has some special names. We often call it a sigma complex, or sometimes, we call it an Arrhenium ion. These are just two different names for the same intermediate. From now on, in this book, we will call it a sigma complex:

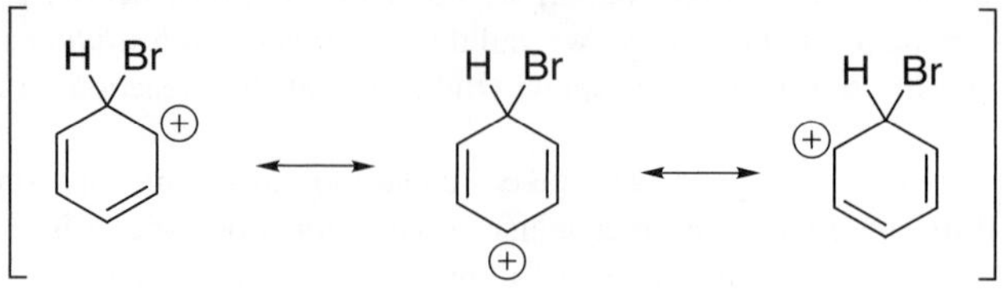

SIGMA COMPLEX

The last step of the mechanism is the loss of a proton to reform aromaticity:

Notice that we are using something to grab the proton (we are using some base). Technically, it is not correct to just let a proton fall off into space by itself, like this:

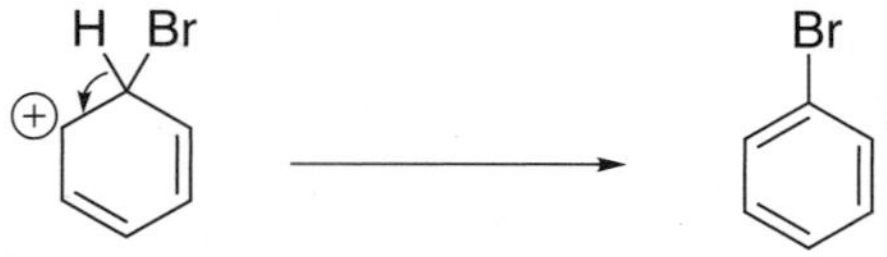

Whenever you are drawing a proton transfer, you should show what is grabbing the proton. In this particular case, it might be tempting to use Br⁻ to pull off the proton. But Br⁻ is not a good base. (In the first semester, we learned the difference between basicity and nucleophilicity, and we saw that Br⁻ is a very good nucleophile but a very poor base.) So, instead, we must use aluminum tetrabromide to pull off the proton. Notice that aluminum tetrabromide functions as a "delivery agent" of Br⁻.

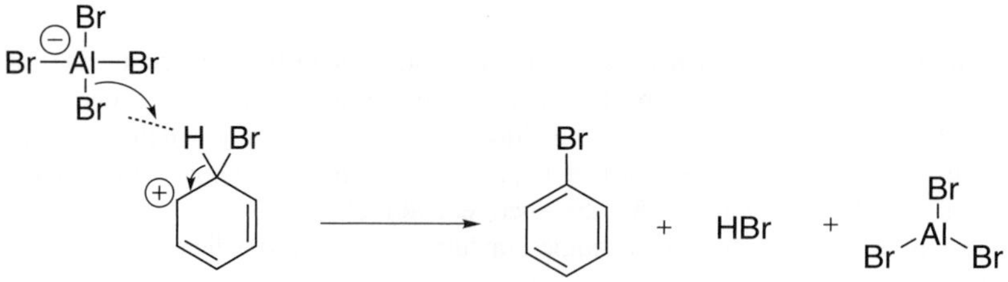

Notice that, in the end, we regenerate our Lewis acid ($AlBr_3$). So the Lewis acid is actually not being consumed by the reaction. It is only there to help the reaction along, which is why we call it a *catalyst* in this case. That is why we don't need to add a lot of the Lewis acid—a pinch of it will suffice.

Now that we have seen all of the individual steps of the mechanism, let's take a close look at the entire mechanism all at once:

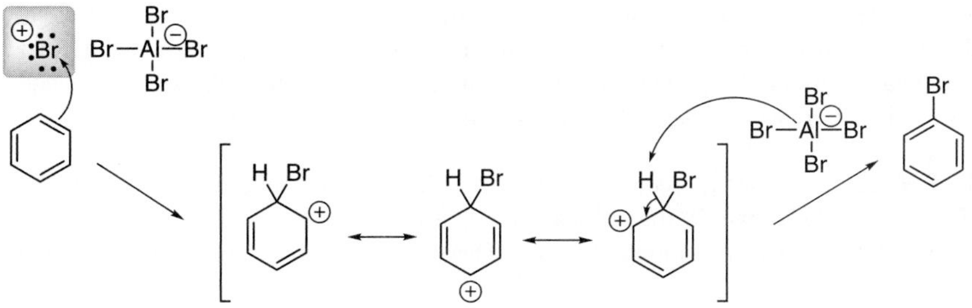

SIGMA COMPLEX

On the surface, it would seem like a lot of steps. However, remember that resonance structures are not actually steps. Those three resonance structures (in the center of the mechanism) are necessary so that we can understand the nature of the one and only intermediate (the sigma complex) of this reaction. So, when we really take a close look at the reaction, we see that there are only two steps. In step 1, benzene acts as a nucleophile attacking Br^+ to form the sigma complex, and in step 2, H^+ is pulled off of the ring to re-form aromaticity. So, the sequence of basic moves (from the previous chapter) is: attack, then deprotonate. To put it in other terms: Br^+ comes on, and then H^+ comes off. That's all there is to it. So we see that the mechanism of this reaction is actually quite simple.

PROBLEM 3.1. It is absolutely critical that you MASTER this mechanism. Without looking at the mechanism above, try to redraw the entire mechanism on a separate sheet of paper. Don't look above—you can figure it out. Just remember that there are two steps: E^+ on and then H^+ off. Don't forget to draw all three resonance structures of the intermediate sigma complex.

This reaction can also be used to put a chlorine atom on the ring. We would just use the following reagents:

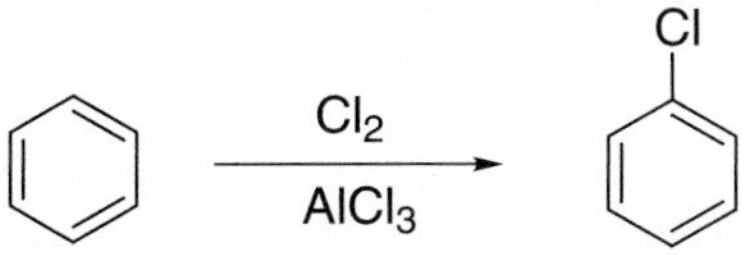

The mechanism is exactly the same. Cl_2 reacts with $AlCl_3$ to form Cl^+, and that gives us the electrophile we need (Cl^+). Then, we get our reaction, which has two steps: Cl^+ on, and then H^+ off.

PROBLEM 3.2. Draw the mechanism for formation of Cl^+ from $AlCl_3$ and Cl_2.

PROBLEM 3.3. Draw the mechanism of the reaction between benzene and Cl^+ to form chlorobenzene. The mechanism is exactly the same as putting a Br on the ring. But PLEASE, do not look back at that mechanism to copy it. Try to do it *without* looking back. Then, when you are finished, compare your answer to the answer in the back of the book (and compare every arrow to make sure that all of your arrows were drawn correctly):

We can also use a similar reaction to place I^+ on the ring. There are many ways to form I^+; you should look in your textbook and in your lecture notes to see if you are responsible for knowing how to iodinate benzene. If so, be aware that the mechanism is exactly the same as what we have seen. The only difference will be in the mechanism of how I^+ is formed.

PROBLEM 3.4. Draw the mechanism of the reaction between benzene and I^+ to form iodobenzene. The mechanism is exactly the same as putting a Br or a Cl on the ring. Once again, try to do it *without* looking back at your previous work. The whole point is to master this mechanism.

3.2 NITRATION

In the previous section, we saw the mechanism of an electrophilic aromatic substitution reaction. We saw that the mechanism is the same, whether you are putting Br^+ on the ring, Cl^+ on the ring, or I^+ on the ring. We also said that we will use this same mechanism to explain how we put *any* electrophile (E^+) on the ring. For example, let's say we wanted to convert benzene into nitrobenzene:

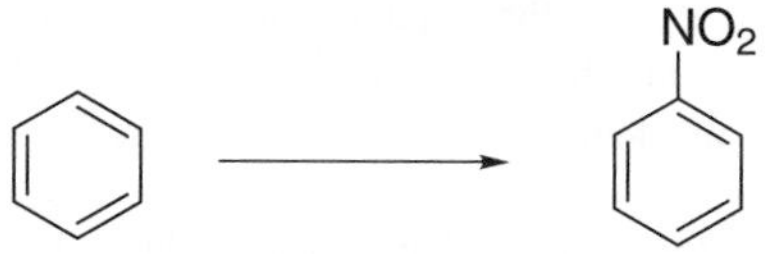

In order to form nitrobenzene, we will need NO_2^+ as our electrophile. But how do we make NO_2^+? If we look at how we made Br^+ or Cl^+ in the previous section, we might be tempted to use NO_2Br and $AlBr_3$, to get the following reaction:

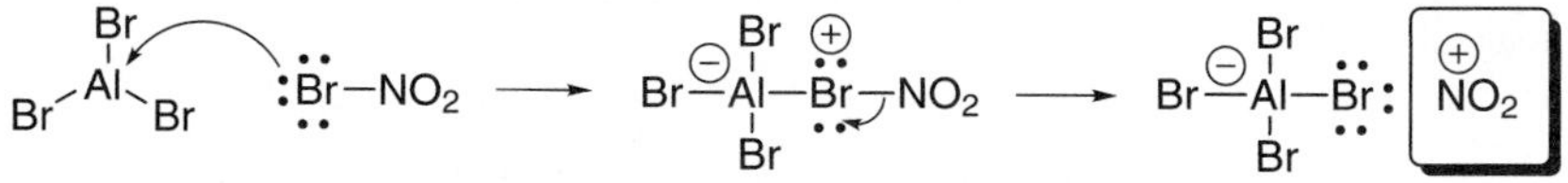

In essence, we would be using the Lewis acid to pull a Br off of NO_2Br to leave behind NO_2^+. This would be directly analogous to the method we have already used for forming Br^+ or Cl^+. The problem is that NO_2Br is nasty stuff, and you probably would not want to work with it in a lab, especially since there is a much simpler way to make NO_2^+. We can form NO_2^+ by mixing sulfuric acid with nitric acid:

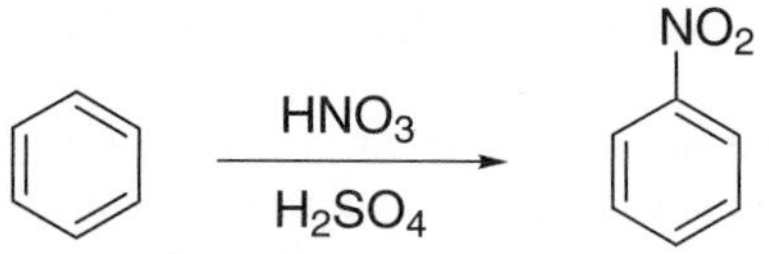

We need to take a close look at how NO_2^+ is formed under these conditions. Let's begin by drawing the structures of sulfuric acid and nitric acid:

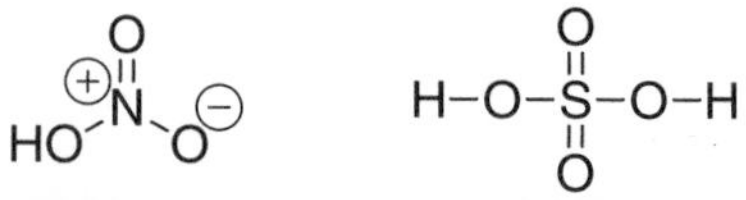

Nitric acid   Sulfuric acid

Notice that nitric acid has charge separation. It might be tempting to get rid of the charges and draw it like this:

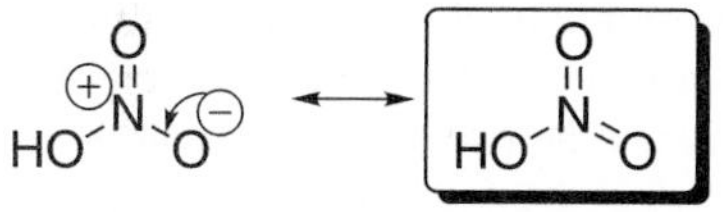

DON'T DRAW THIS

But you cannot do that because this would give the central nitrogen atom five bonds. It cannot EVER have five bonds because it only has four orbitals that it can use to form bonds. So we must draw nitric acid with charge separation.

Now that we have seen the structures of our two acids, we must remember that the term *acidic* is a relative term. It is true that nitric acid is acidic, and it is also true that sulfuric acid is acidic. But between the two of them, sulfuric acid is a *better* acid. In fact, it is so much stronger as an acid that it is able to give a proton to nitric acid:

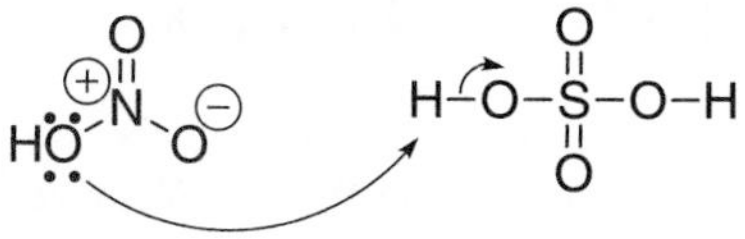

That's right—it might seem weird because nitric acid is essentially functioning *as a base* to pull a proton off of sulfuric acid. And it might make us uncomfortable to use nitric acid as a base, but that is exactly what is happening. Why? Because *relative* to sulfuric acid, nitric acid is a base. It's all relative.

OK, so nitric acid grabs a proton from sulfuric acid. But the obvious question is: why does the uncharged oxygen grab the proton? Wouldn't it make more sense for the negatively charged oxygen to grab the proton? Like this:

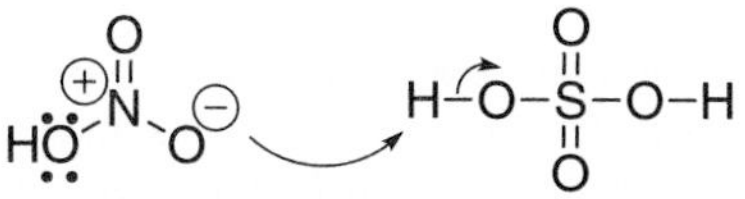

The answer is: yes, this probably would make more sense. And it probably happens like this a lot more often. The oxygen with the negative charge probably pulls off the proton much more readily than the uncharged oxygen atom does. However, proton transfers are reversible. Protons are being transferred back and forth all of the time. And all of this is happening very quickly. So, it is true that the negatively charged oxygen pulls off the proton more often—but when that happens, the only thing that can happen next is for the proton to be given right back to reform nitric acid.

Every once in a while, however, the uncharged oxygen can grab the proton. And when that happens, something new can happen next: water can leave:

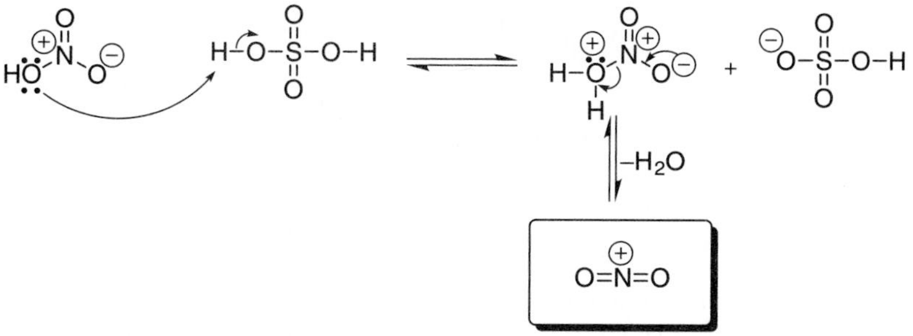

And when this happens, we get NO_2^+. So when we mix sulfuric acid and nitric acid, we get a little bit of NO_2^+ in the equilibrium mixture, and that NO_2^+ functions as the electrophile we need in order to put a nitro group on a benzene ring.

Once again, this mechanism is essentially the same mechanism that we saw in the previous section: NO_2^+ on and then H^+ off. There are only two very subtle differences. Let's take a look at the mechanism, and then we will focus on the two subtle differences:

SIGMA COMPLEX

In the first step, when we attack the electrophile, we need to use *two* arrows (whereas we only used one arrow to attack E^+ in the previous reactions). We have the second arrow here so that we won't give five bonds to nitrogen. (This is similar to the argument we gave when we looked at the structure of nitric acid—NEVER give nitrogen five bonds because it only has four orbitals with which to form bonds.)

The only other subtle difference is in the last step of the mechanism. In this case, we are using HSO_4^- to pull off the proton (instead of $AlBr_4^-$), which should make sense because we don't have any $AlBr_4^-$ in this reaction.

Other than those two subtle differences, the mechanism is *identical* to what we have already seen in the previous reactions.

So far, we have seen how to put a halogen on the ring (Cl, Br, or I), and we have seen how to put a nitro group onto the ring. Before we move on, let's just make sure that you are familiar with the reagents necessary to do these reactions. For each of the following problems, fill in the reagents that you would need in order to carry out the transformation:

3.5.

3.6.

3.7.

3.8.   *Without looking back* at the previous section, try to draw the mechanism for the nitration of benzene. You will need a separate piece of paper to record your answer. Make sure to start by drawing the mechanism for formation of NO_2^+, and then show the reaction of benzene with NO_2^+.

3.3 FRIEDEL-CRAFTS ALKYLATION AND ACYLATION

In the previous sections, we saw how to put several different groups (Br, Cl, I, or NO_2) onto a benzene ring, using an electrophilic aromatic substitution. In each case, the mechanism was the same: E^+ **on** the ring, and then H^+ **off**. In this section, we will learn how to put an alkyl group on the ring.

Let's start with the simplest of all alkyl groups: a methyl group. So, the question is: what reagents would we need to do the following transformation:

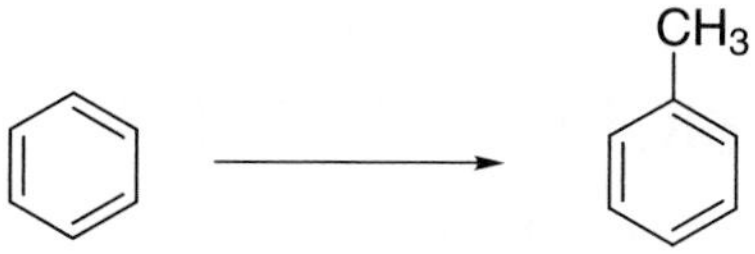

Using the logic that we have developed in this chapter, we would want to use CH_3^+ as our electrophile. But you should probably cringe when you see CH_3^+. After all, you probably remember what you learned about the stability of carbocations—that tertiary carbocations are more stable than secondary carbocations, and so on. Certainly a methyl carbocation would not be very stable at all. In fact, we deliberately avoid using methyl or primary carbocations in our mechanisms. But here we are, trying to make a methyl carbocation. Is it even possible? The answer is: yes. In fact, we will make it using the same method we have used in the previous sections.

If we take methyl chloride and we mix it with a pinch of $AlCl_3$, we will have a source of CH_3^+:

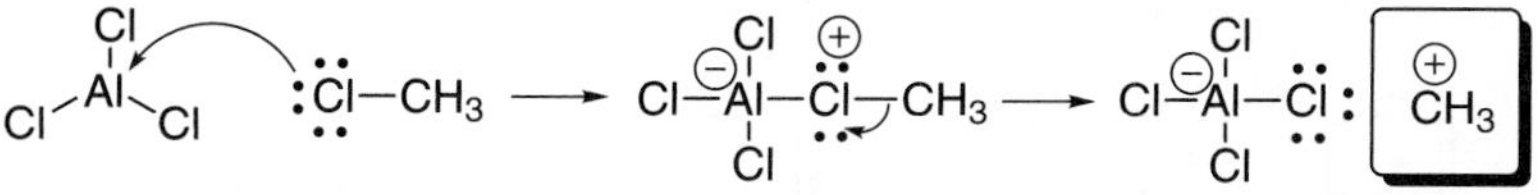

Does not really exist as free CH_3^+

The truth is that we are NOT really forming a free methyl carbocation that can float off into solution. CH_3^+ would be too unstable by itself (think back to carbocation stability). So, instead, we must view this as a complex that can serve as a "source" of CH_3^+.

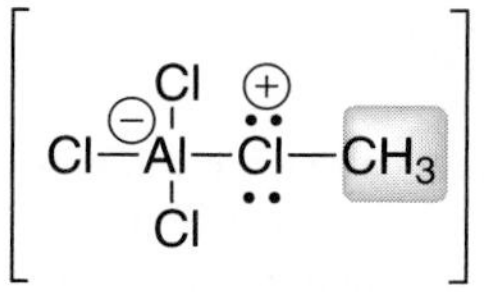

Source of CH_3^+

This gives us a way to methylate a benzene ring:

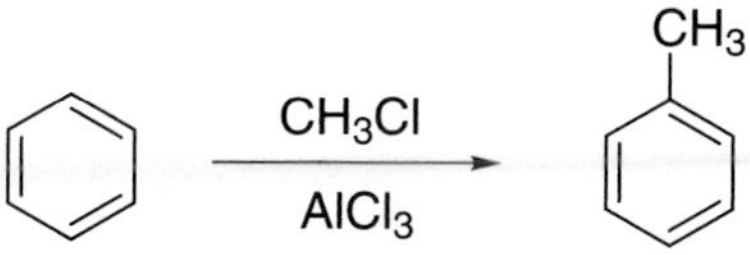

And the mechanism is, once again, the same mechanism that we have seen over and over again. It is an electrophilic aromatic substitution: CH_3^+ *on* the ring and then H^+ *off*:

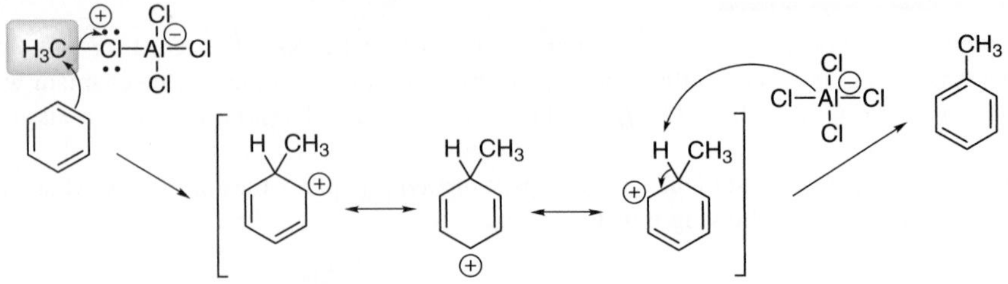

SIGMA COMPLEX

We can use the exact same process to put an ethyl group on a ring:

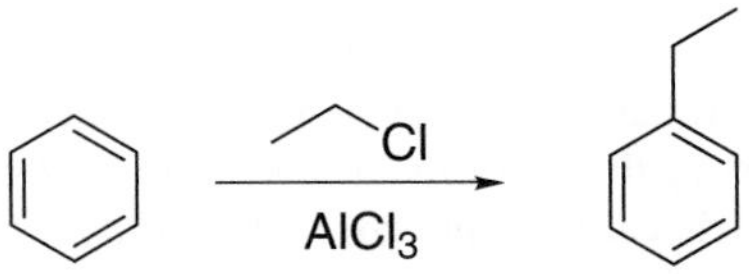

This process (putting an alkyl group onto a ring) is called a Friedel-Crafts Alkylation. It works very well for putting a methyl group or an ethyl group on the ring. BUT we run into problems when we try to put a propyl group onto the ring. We actually get a mixture of products when we try to put a propyl group on the ring:

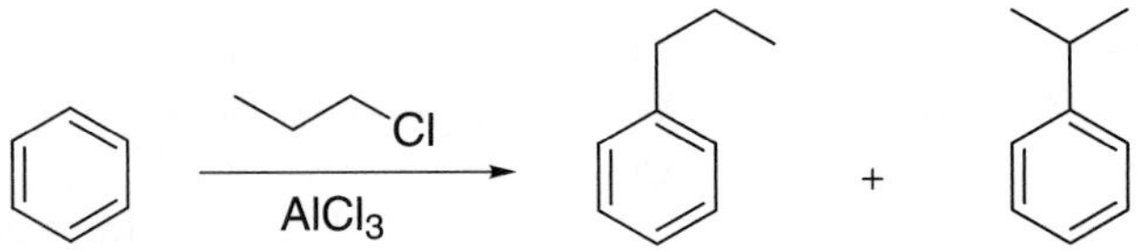

The reason for this is simple. Since we are forming a carbocation, it is possible to get a carbocation rearrangement. It was not possible for a methyl carbocation to rearrange. Similarly, an ethyl carbocation cannot rearrange to become any more stable. But a propyl carbocation CAN rearrange (via a hydride shift):

And since we are forming a propyl carbocation, we can expect that sometimes it will rearrange before reacting with benzene (while other times, it will not get a chance to rearrange before it reacts with benzene). And that is why we get a mixture of products. So you need to be careful when using a Friedel-Crafts Alkylation to look out for unwanted carbocation rearrangements.

Now, if we wanted to make isopropyl benzene, we could avoid this whole issue by just using isopropyl chloride as our reagent:

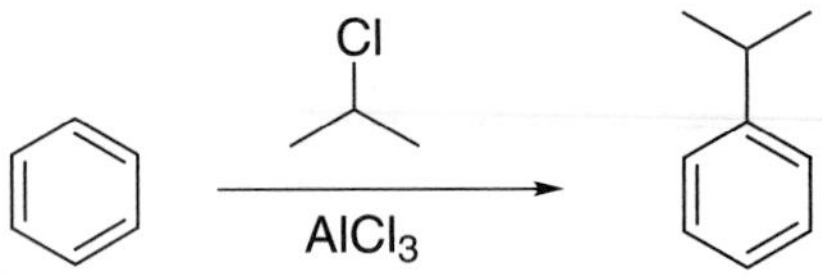

But what if we want to make propylbenzene?

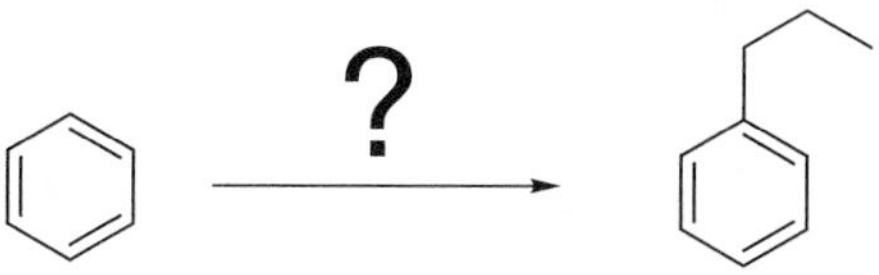

How would we do that? If we use propyl chloride, we have already seen that we will get some re-arrangement, and we will not get a good yield of the desired product. In fact, we can generalize the question like this: how do we attach *any* alkyl group and avoid a potential carbocation rearrangement. For example, how could we do the following transformation, *without* a carbocation rearrangement?

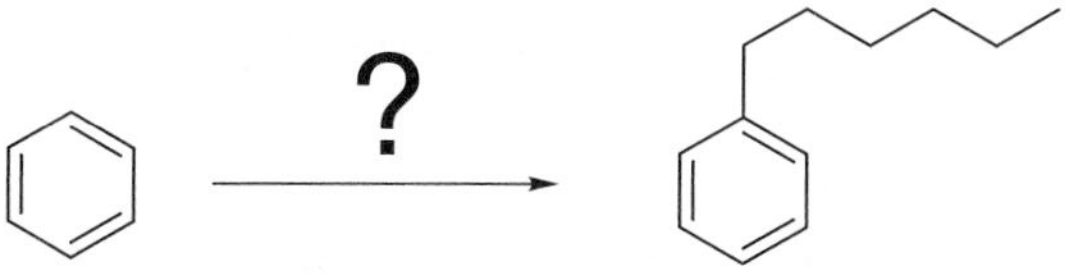

If we just use chlorohexane (and AlCl$_3$), we will likely get a mixture of products.

Clearly, we need a trick. And there is a trick. To see how it works, we need to take a close look at a similar reaction that also bears the name Friedel-Crafts. But this reaction is not an *alkyl*ation. Rather, it is called an *acyl*ation. To see the difference, let's quickly compare an alkyl group with an acyl group.

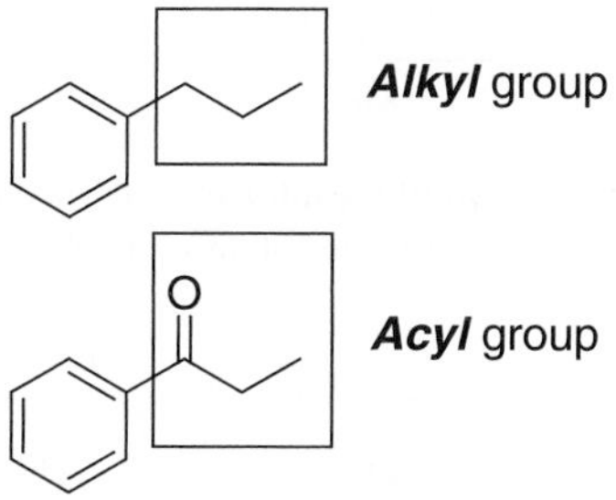

Alkyl group

Acyl group

We can put an *acyl* group onto a benzene ring in exactly the same way that we put an alkyl group on the ring. We just need to use the following reagents:

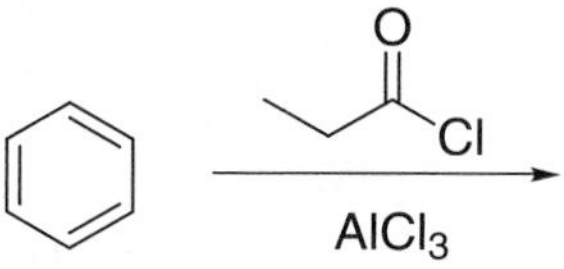

The first reagent is called an acyl chloride (or acid chloride), and we are already familiar with the role of AlCl$_3$ (the Lewis acid). The Lewis acid is used to pull a Cl atom off of the acyl chloride, like this:

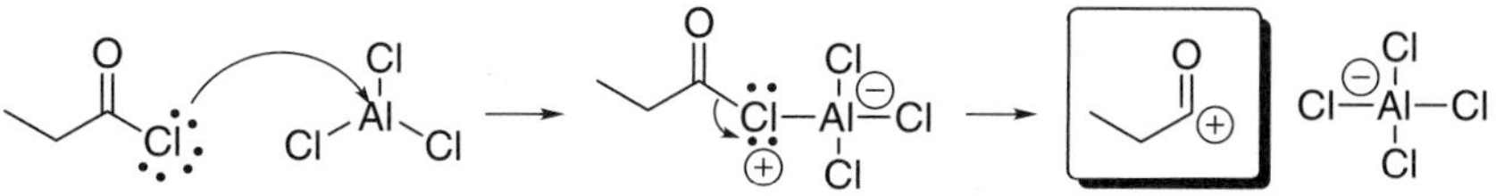

The end result is that we have a new kind of electrophile that we can use to react with benzene. This electrophile is called an **acylium** ion (that should make sense—"acyl" because we can see that this electrophile has an acyl group; and "*ium*" because there is a positive charge). This electrophile actually has an important resonance structure:

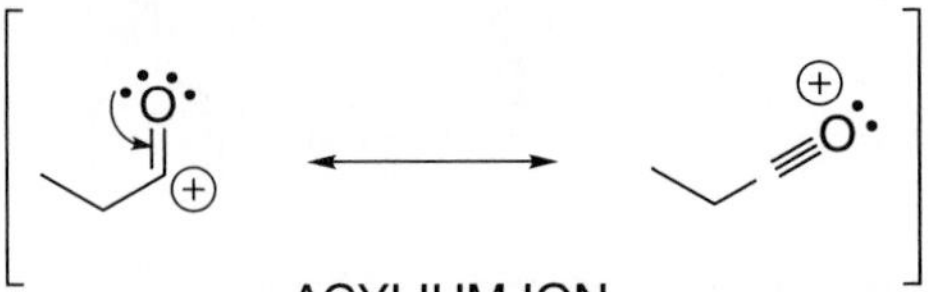

ACYLIUM ION

These resonance structures are important because they tell us that an acylium ion is *stabilized* by resonance, and therefore, it will NOT undergo a carbocation rearrangement. (If it did, it would lose this resonance stabilization.) Compare the following two cases:

CAN REARRANGE

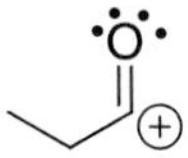

CANNOT REARRANGE

So, we see that if we do a Friedel-Crafts *Acyl*ation, we can cleanly hook an acyl group onto a benzene ring (without any rearrangements):

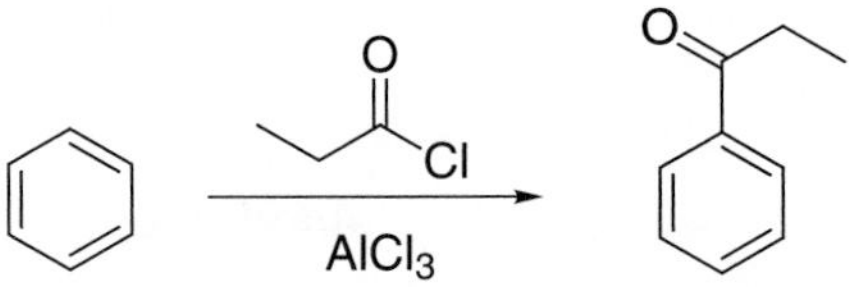

And we don't get any side products that would result from a carbocation rearrangement. Once again, compare a Friedel-Crafts *Alkyl*ation with a Friedel-Crafts *Acyl*ation:

ALKYLATION

Mixture of products

ACYLATION

One product

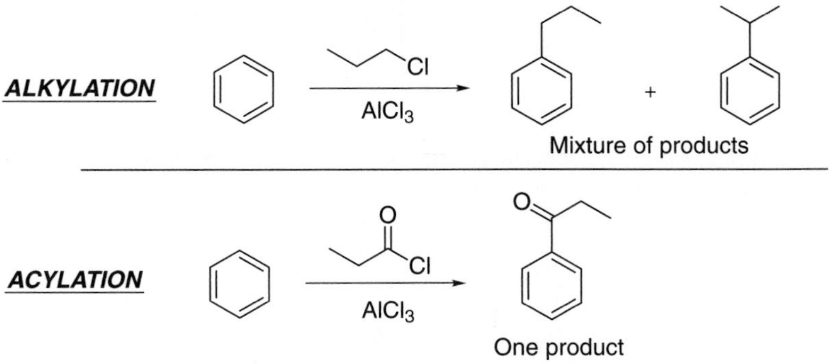

Now take a close look at the acylation above, and let's point out a very important feature. Notice that we have hooked on a three-carbon chain onto the ring, but this chain is attached by the *first* carbon of the chain, as opposed to the middle carbon of the chain:

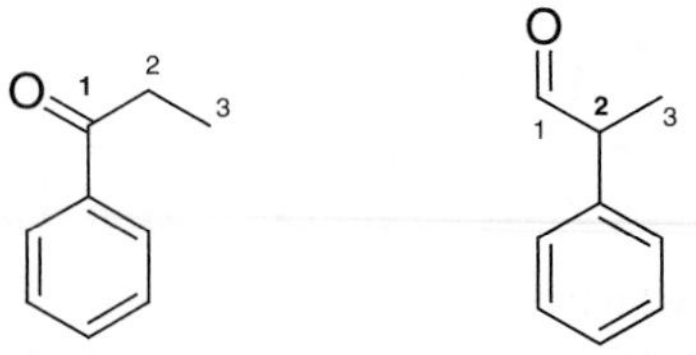

We get this   We *don't* get this

Once again, it hooks on by the first carbon because we don't get a rearrangement (the acylium ion is resonance stabilized, and does not rearrange). Now, all we need is a way to remove the oxygen, and then we will have a two-step synthesis for attaching a propyl group to a benzene ring:

And luckily, there is a simple way to remove the oxygen; in fact, there are *three* very common ways to remove the oxygen. We will just focus on one method right now (which uses acidic conditions), but keep in mind that we will see two other ways of doing this in the upcoming chapters (one way uses basic conditions and the other way uses neutral conditions). When we reduce a ketone under acidic conditions, we call the reaction a Clemmensen reduction:

Let's quickly look at the reagents we used here. In the first reagent, Zn[Hg], the brackets around the mercury indicate that we are using an *amalgam* of zinc and mercury. An amalgam is what you get when you take two metals, heat them up until they are liquids, mix those liquids, and then let them cool back down to give you one solid that is composed of both metals. In addition, we also need HCl and heat to have the conditions for a Clemmensen reduction.

Chemists have not reached a clear consensus about the mechanism of a Clemmensen reduction, so most textbooks do not go into the mechanism of this reaction. The important thing to know here is that you can use a Clemmensen reduction as the second step of a two-step synthesis that places an alkyl group onto a ring *without* any rearrangement taking place:

Before we move on, there is one subtle point to mention about a Friedel-Crafts Acylation (step 1 of the synthesis above). Remember that this reaction takes place in the presence of a Lewis acid ($AlCl_3$), which is constantly on the lookout for electrons that it can latch onto. Well, the product of the acylation step is a ketone. And a ketone has electrons that the Lewis acid can latch onto:

Therefore, whenever we do an acylation reaction, we need to pull the Lewis acid off our product in the end. And there is a simple way to do this. We just give the Lewis acid some other source of electrons to latch onto instead. The easiest thing to use is water (H_2O). So, whenever you use a

Friedel-Crafts Acylation in a synthesis problem, you should make sure to include water in your list of reagents (immediately after the acylation):

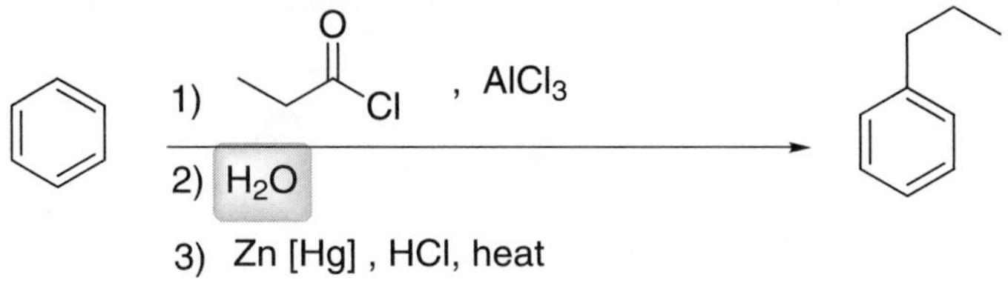

To summarize, we have seen that a Friedel-Crafts Acylation can be followed up by a Clemmensen reduction, as a clever way of putting an alkyl group on a ring (without rearrangements). But there are actually times when you will want to put an acyl group on the ring, and you won't want to do a Clemmensen reduction afterward. For example:

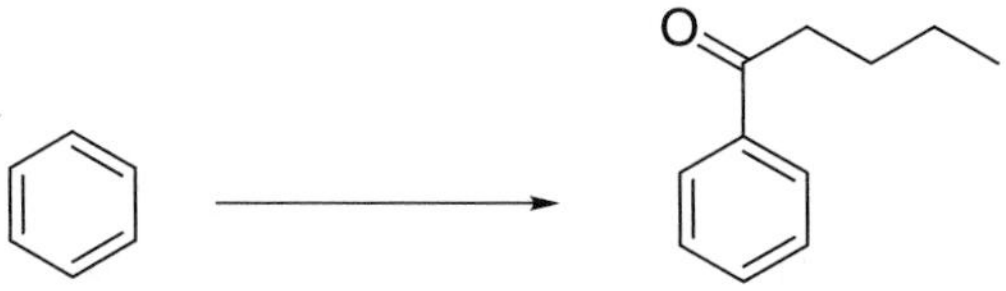

To do this transformation, you would just use a Friedel-Crafts Acylation (including the step where you use water), and that's it. There is no need for a Clemmensen reduction, because we don't want to pull off the oxygen in the end.

EXERCISE 3.9. Show the reagents you would use to carry out the following synthesis:

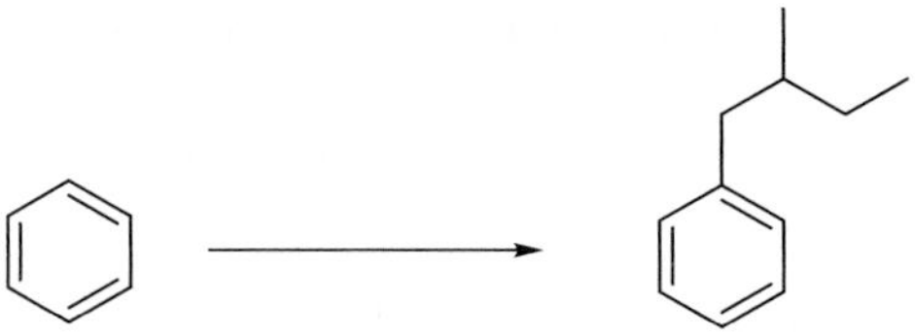

Answer: In this problem, we need to attach an alkyl group to a benzene ring. So, we first look to see if we could do this one step, using a Friedel-Crafts Alkylation. In this case, we cannot do it in one step because we have to worry about a carbocation rearrangement. If we think about the electrophile that we would need to make, we will see that it could rearrange:

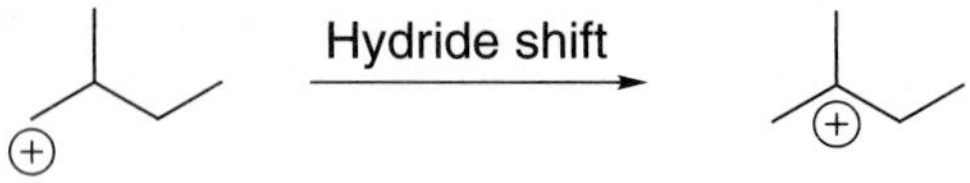

And this would give us a mixture of products:

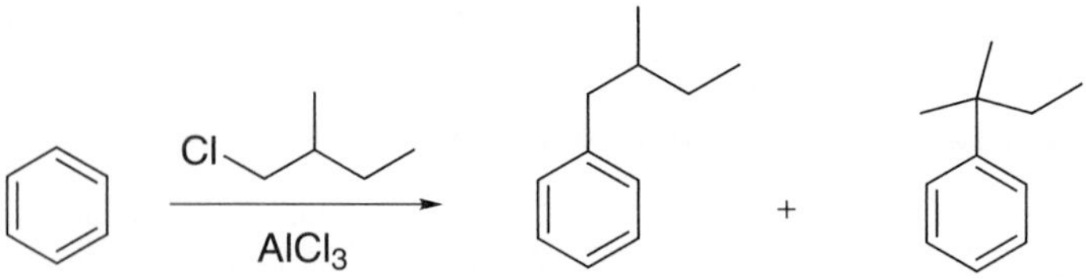

So, instead we will have to use a Friedel-Crafts Acylation (don't forget the water) followed by a Clemmensen reduction:

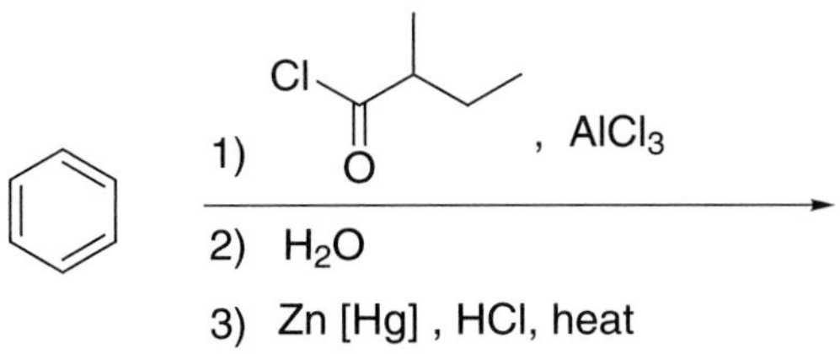

For each of the following problems, show what reagents you would use to accomplish the transformation. In some situations, you will want to use a Friedel-Crafts Alkylation, while in other situations, you will want to use a Friedel-Crafts Acylation.

3.10.

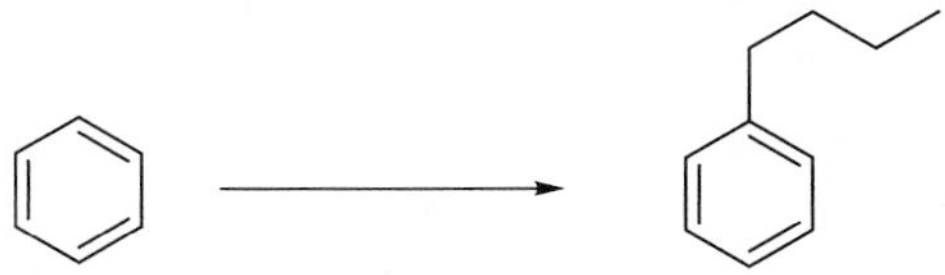

3.11.

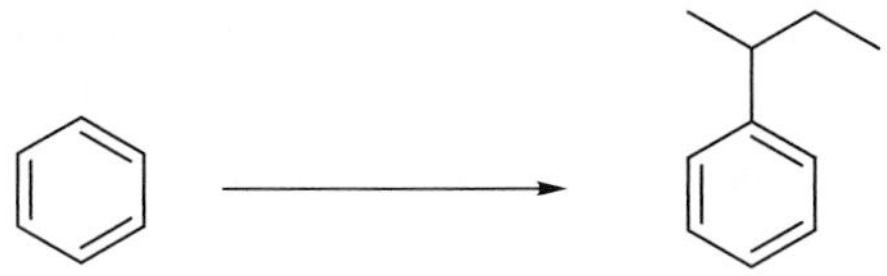

3.12.

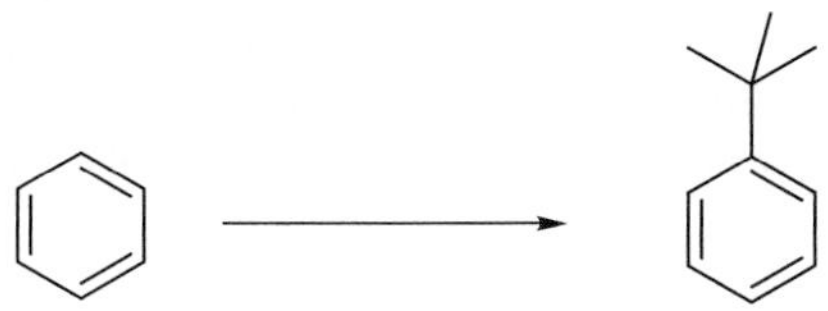

3.13.

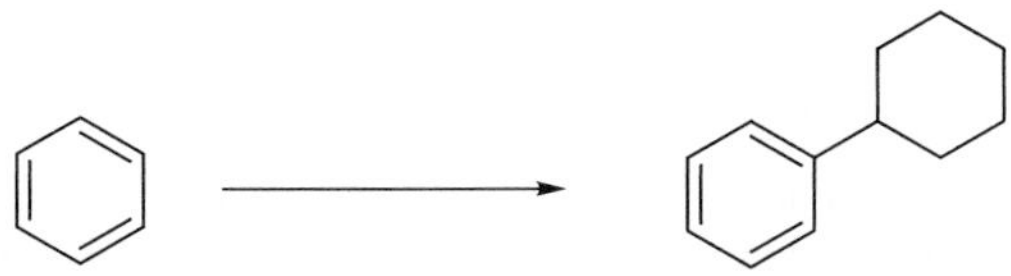

3.14.

3.15. Predict the products that would form from the following reaction.

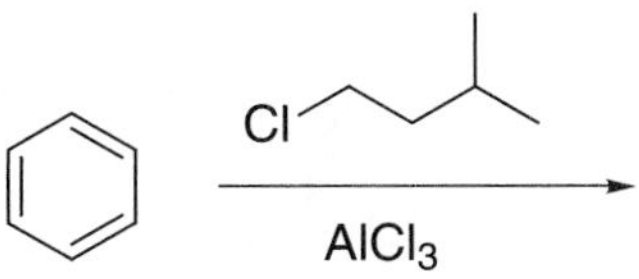

(*Hint*: There should be a mixture of *three* products in this case. Be sure to consider all of the possible rearrangements that can take place. If you are rusty on carbocation rearrangements, then you should take a break to go back and review them now.)

3.16. On a separate piece of paper, draw the mechanism of formation for each one of the three products from the previous problem.

3.17. On a separate piece of paper, draw the mechanism of the following reaction. Make sure to show the mechanism of formation of the acylium ion before you use it to react with the ring:

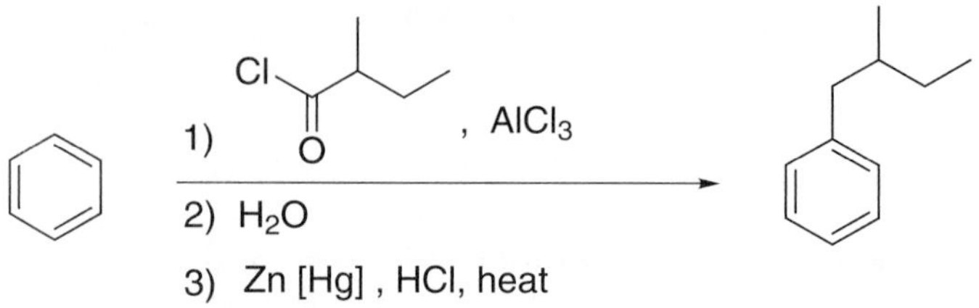

Using Friedel-Crafts reactions involves a few limitations. You should take a moment to read about them in your textbook. The two most important limitations are as follows:

1. When doing a Friedel-Crafts *Alkylation*, it is often difficult to put on just one alkyl group. Each alkyl group makes the ring *more* reactive toward a subsequent attack on the same ring.

2. When doing a Friedel-Crafts *Acylation*, it is generally not possible to place more than one acyl group onto a ring. The presence of one acyl group makes the ring *less* reactive toward a second acylation.

We need to understand WHY an alkyl group makes the ring more reactive, and WHY an acyl group makes the ring less reactive. We will explain this in greater detail when we turn our attention to the *nucleophile* of an electrophilic aromatic substitution reaction. (Remember that the benzene ring is the nucleophile of the reaction, and it attacks some electrophile, E^+.) Until now, we have focused on the electrophile of the reaction. So far, we have seen how to use Br^+, Cl^+, I^+, NO_2^+, alkyl$^+$, and acyl$^+$. Before we can shift our attention to the nucleophile of the reaction, we have one more electrophile to discuss.

3.4   SULFONATION

The reaction we will discuss now is probably one of the most important reactions for you to have at your fingertips. This reaction will be used extensively in synthesis problems later in this chapter. If you do not keep this reaction in mind while you are solving synthesis problems, then you will be at a severe loss. We will explain why this reaction is so important in the upcoming sections. For now, just take my word for it, and let's just master the reaction.

All of the electrophiles we have seen so far have always been positively charged. But now we will deal with one electrophile that is not positively charged. The electrophile is SO_3. Let's take a close look at the structure:

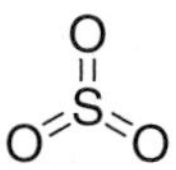

Notice that there are three $S{=}O$ double bonds here. But these double bonds are not such great double bonds. Recall that a double bond is formed from the overlap of two p orbitals:

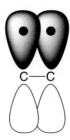

When we are talking about a carbon-carbon double bond, the overlap of the p orbitals is pretty good because the p orbitals are the same size. But what happens when you try to overlap the p orbital of an oxygen atom with the p orbital of a sulfur atom? The p orbitals are different sizes (oxygen is in the second row of the periodic table, which means that it is using a p orbital from the second energy level; but sulfur is in the third row of the periodic table, so it is using a p orbital from the third energy level):

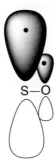

Therefore, the overlap is not so good, and it is misleading to think of $S{=}O$ as being a double bond. It is probably much closer in nature to being like this:

$$\overset{\oplus}{S}{-}\overset{\ominus}{O} \quad \text{rather than} \quad S{=}O$$

When we do this analysis for each of the three double bonds in SO_3, we begin to see that the sulfur atom is VERY electron-poor:

In fact, the sulfur atom is so electron-poor that it is an *excellent* electrophile, even though the compound is overall neutral (no net charge). Now we are going to see a reaction that uses SO_3 as an electrophile. But first, let's see where SO_3 comes from.

Sulfuric acid is constantly in equilibrium with SO_3 and water:

$$H_2SO_4 \rightleftharpoons SO_3 + H_2O$$

That means that any bottle of sulfuric acid will have some SO_3 in it. At room temperature, SO_3 is a gas, and it is possible to add extra SO_3 gas to sulfuric acid (which shifts the equilibrium a bit).

When we do this, we call the mixture *fuming* sulfuric acid. So, from now on, whenever you see concentrated, fuming sulfuric acid, you should realize that we are talking about SO_3 as the reagent. And here is the reaction:

Notice that, in the end, we have placed the SO_3H group on the ring. The obvious question is: why is the H attached to the SO_3 in the end? To see why, let's take a closer look at the mechanism. Remember our two steps for any electrophilic aromatic substitution: E^+ goes on the ring, and then H^+ comes off. But wait a second. . . . In this case, we are not using an electrophile with a net positive charge. The electrophile in this case has no net charge. In all of the reactions we have seen so far, we put something positively charged onto the ring, and then we took something positively charged off of the ring. So in the end, our ring never gained or lost any charges. But in this case, we are putting something neutral (SO_3) onto the ring, and then we are removing something positively charged (H^+). That should leave our product with an overall negative charge, which is exactly what happens:

SIGMA COMPLEX

So, we need to add one more step to our mechanism. The negative charge on our product grabs a proton from sulfuric acid (remember that there is plenty of sulfuric acid around because we used fuming sulfuric acid as our source of SO_3):

Although this reaction has this one extra step at the end of the mechanism, keep in mind that this extra step is just a proton transfer. The core reaction is still the same as we have seen in all of the previous reactions: the electrophile comes on the ring, and then H^+ comes off of the ring.

An important feature of this reaction (and this is the feature that will make this reaction so important for synthesis problems) is how easily the reaction can be reversed. The amount of product that you get is equilibrium controlled, and it is very sensitive to the conditions. So, if you use dilute sulfuric acid instead, the equilibrium leans the other way (look closely at the equilibrium arrows below):

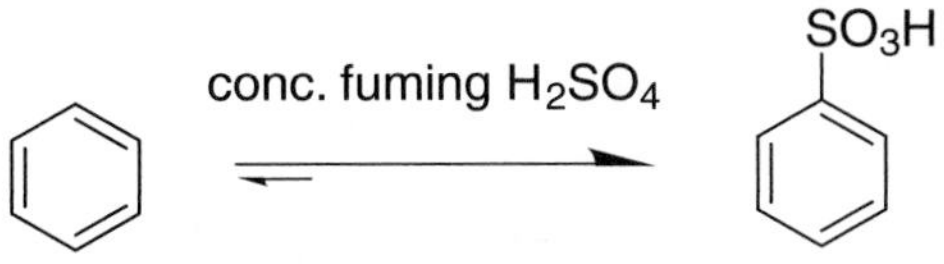

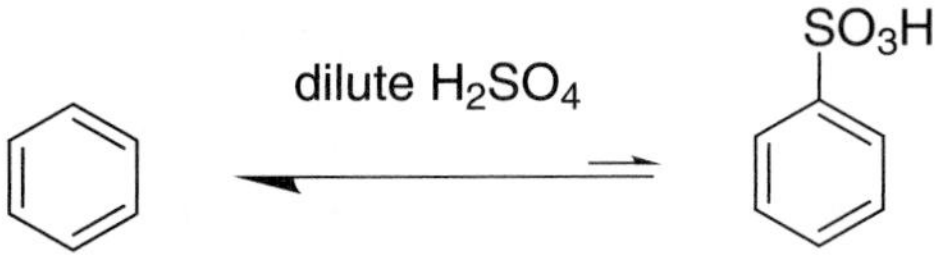

We can use this to our advantage because this gives us a way to remove the SO₃H group whenever we want. We would just use dilute sulfuric acid to pull it off:

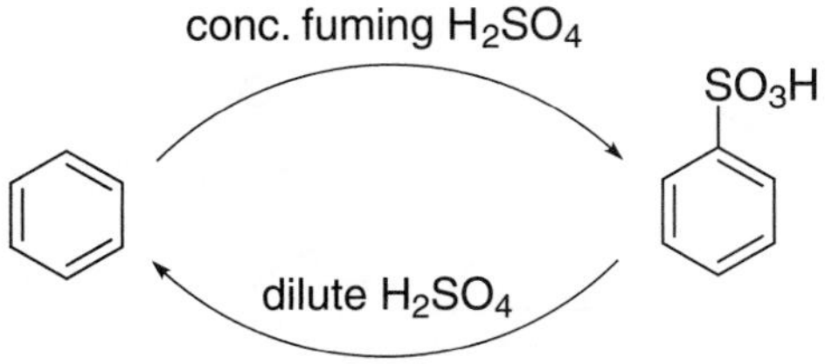

So we now have the ability to put the SO₃H group onto the ring whenever we want, AND we can take it off whenever we want as well. You might wonder why you would want to put a group on in order just to take it off later. On the surface, that would seem like a waste of time. But in the upcoming chapters, we will see that this will become very important in synthesis problems.

For now, let's make sure that we are comfortable with the reagents.

EXERCISE 3.18. Identify the reagents that you would use to carry out the transformation shown below:

Answer: We know that we use fuming sulfuric acid to put an SO₃H group onto the ring, and we use dilute sulfuric acid to take the group off. In this case, we are taking the group off, so we need to use dilute sulfuric acid:

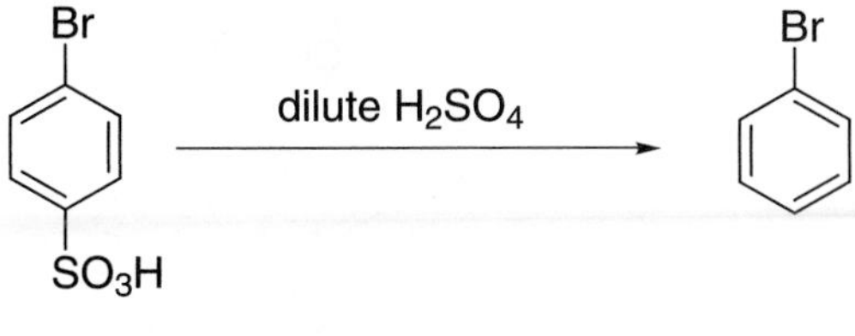

For each of the following problems, identify the reagents you would use to carry out the transformation shown:

3.19.

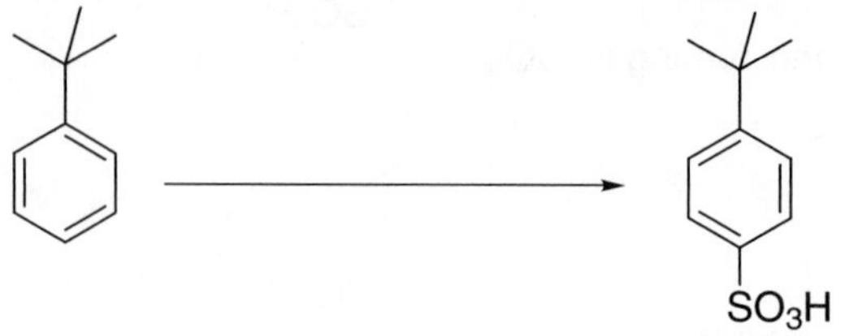

3.20.

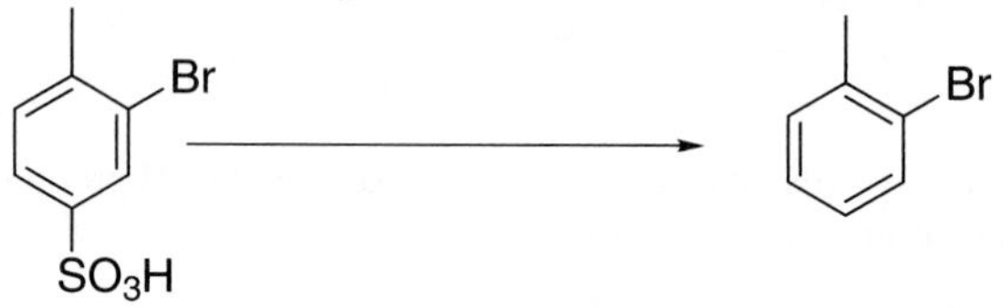

3.21.

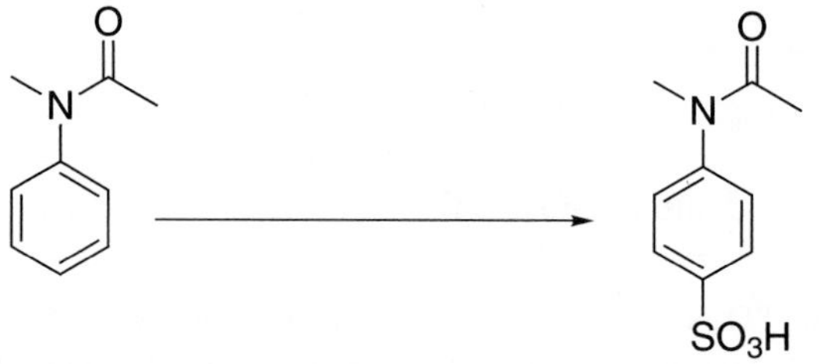

3.22.

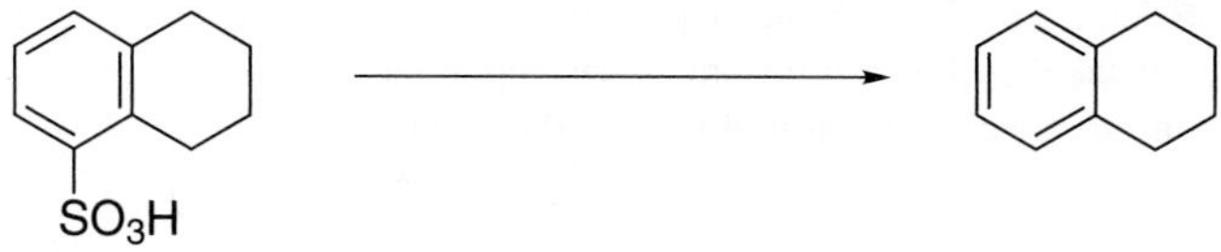

And to make sure that you are not getting rusty on the other reactions we have learned in this chapter so far, fill in the reagents you would use for the following transformations:

3.23.

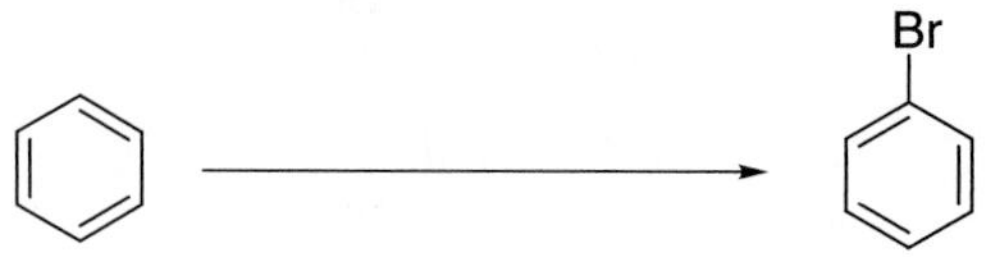

3.24.

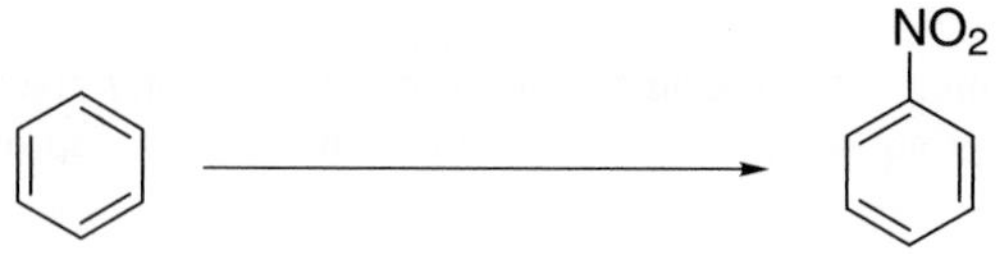

3.25.

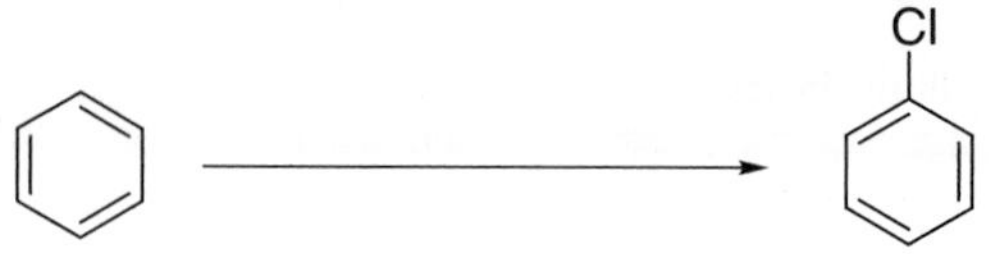

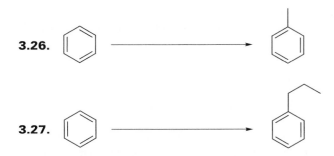

3.28.   Now, let's just make sure that you can draw the mechanism of a sulfonation reaction. (That's just the fancy name we give to the reaction where we put an SO_3H group onto the ring.) On a separate piece of paper, take a moment and try to draw the mechanism for the sulfonation of benzene. Remember that there will be three steps: (1) electrophile comes on the ring, (2) H^+ comes off, and then (3) proton transfer gets rid of the negative charge. Don't forget to draw the resonance structures of the intermediate sigma complex. Try to draw the mechanism without looking back to where it is shown earlier in this section.

3.29.   And now, for a challenging problem, try to draw the mechanism of a desulfonation reaction (a reaction where we take the SO_3H group off of the ring). The process will be exactly the reverse of what you just drew in the previous problem. There will be three steps: (1) remove the proton from the SO_3H group, (2) H^+ comes on the ring, and then (3) SO_3 comes off of the ring. The truth is that there are only two core steps here: H^+ comes on the ring, and then SO_3 comes off of the ring. You can actually pull the proton off of the SO_3H group at the same time that SO_3 comes off of the ring. Try to do it yourself, and if you get stuck, you can look at the answer in the back of the book. But try to do the problem without looking at the answer.

3.5   MODIFYING THE NUCLEOPHILICITY OF THE NUCLEOPHILE

When we began this chapter, we saw that benzene and bromine do not react with each other (without a Lewis acid):

$$\text{benzene} \xrightarrow{\text{Br}_2} \text{No reaction}$$

So, we began to investigate how we could force the reaction to take place. We said that the benzene ring is the nucleophile and Br_2 is the electrophile. And if we want to force the reaction to go, then we have two choices: we can either use a better electrophile (something more electrophilic than Br_2) or we can use a better nucleophile (something more nucleophilic than benzene). We then focused all of our attention on making the electrophile more electrophilic. We saw how to use a Lewis acid to make Br^+. We then saw many other positively charged electrophiles (Cl^+, I^+, NO_2^+, alkyl$^+$, and acyl$^+$), and we even saw one electrophile that is an excellent electrophile, even though it does not have a net positive charge (SO_3). Now, we will turn our attention to the nucleophile—the aromatic ring. How do we make an aromatic ring more nucleophilic?

To answer this question, we need to focus on the groups that are already connected to the ring. Benzene itself (C_6H_6) has no groups on it. But consider the structure of methyl benzene (commonly called toluene):

Methylbenzene

(Common name = Toluene)

Here, we have an aromatic ring with a methyl group on it, and the question is: what effect does that methyl group have on the nucleophilicity of the aromatic ring? Is this compound a better nucleophile than benzene? Also consider the structure of hydoxybenzene (commonly called phenol):

OH

Hydroxybenzene

(Common name = Phenol)

What is the effect of the OH group? Does it make the ring a better nucleophile?

Let's begin by showing a trend. Consider the following three reactions:

Br_2 / heat → No reaction

Br_2 / heat →

OH / Br_2 →

The first reaction shows us that benzene and bromine do not react with each other (without a Lewis acid to make the electrophile more electrophilic). Then in the second reaction, we see that toluene WILL react with Br_2. So, clearly, toluene is more nucleophilic than benzene. Then, in the third reaction, we see that phenol will react with bromine several times to give a tri-brominated product. So, we can see that phenol is even more nucleophilic than toluene. But a number of obvious questions follow from this:

1. WHY does a methyl group make the ring more nucleophilic?

2. WHY does an OH group make the ring even more nucleophilic than the methyl group does?

3. When bromine does go on the ring, why is it going in specific locations? Why don't we get a penta-brominated product? To better understand this question, let's make sure we are us-

ing the correct terminology. When you have an aromatic ring with one group on it, we describe the other positions with the following terms:

The two positions that are closest to the R group are called *ortho* positions. Then we have the *meta* positions. And finally, the farthest position from the group is called the *para* position. Using this terminology, we see in our reactions above that the substitution reactions always seem to take place at the ortho and para positions. Why? Why don't we get a substitution reaction at the meta position?

In order to answer these questions, we will need to take a close look at the factors that determine the electronics of any compound. If you ever want to know whether a compound is nucleophilic (or how nucleophilic it is), you are really asking about the electronics of that compound. And it is comforting to know that there are only two factors that you ever have to consider: *induction* and *resonance*.

Let's use phenol as our example, and let's explore the electronics of that compound. Let's begin by looking at the first factor: induction. Recall from the first semester that induction is fairly simple to assess. You just look at the relative electronegativity of the atoms. If you want to know what inductive effect the OH group will have on the aromatic ring, you look at the bond connecting the OH group to the ring:

Oxygen is more electronegative than carbon, so there is an inductive effect (shown by the arrow above). The oxygen is withdrawing electron density from the ring. Remember that the ring is only a nucleophile because it is electron-rich (from all of those pi electrons), so if we take away electron density from the ring, then we are decreasing the nucleophilicity of the ring. So the inductive effect of the OH group is to make the ring *less* nucleophilic. But we're not done yet. We need to consider the other factor that determines the electronics of a compound: resonance.

In order to see how resonance affects the electronics of our compound, we need to draw the resonance structures (if you did not master the steps for drawing resonance structures in the first semester, then you should go back and do so now—resonance will reappear in almost every topic in the second semester):

Notice that there is a negative charge spread throughout the ring. When we meld all of these resonance structures together in our minds, we get the following picture for the molecule:

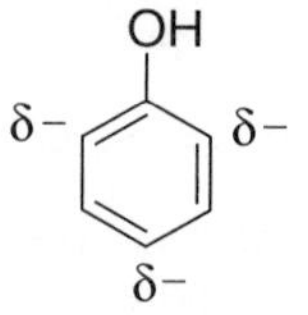

The δ- shows that there is a partial negative charge spread throughout the ring. Therefore, the effect of resonance is to *give* electron density to the ring. So, now we have a competition. By induction, the OH group is *electron-withdrawing*, which makes the ring *less* nucleophilic. But by resonance, the OH group is *electron-donating*, which makes the ring *more* nucleophilic. So the question is: which effect is stronger? Resonance or induction? This is a common scenario in organic chemistry (where induction and resonance are in competition), and the general rule is: resonance usually wins. There are some important exceptions. We will soon see one of these exceptions, but in general, resonance is much stronger than induction.

Now let's apply this general rule to our case of phenol. If we say that resonance wins, that means that the net effect of the OH group is to *donate* electron density to the ring. Therefore, the net effect of the OH group is to make the ring *more nucleophilic* (as compared with benzene).

A simple analogy sums this up. Imagine that you have $100 in your hand. You come to me, and I do two things. First, I take some money away from you. Then, I give you some money back. And the question is: do you have more money than you came with, or less? The only way to answer the question is to know whether I took more than I gave you back, or whether I gave back more than I took. The case of phenol is like a case where I take $10 away from you (which brings you down to $90), but then I give you back $200. So, in the end, you end up with $290, which is a lot more than you had originally. Is it true that I first took some money away from you? Yes, it is. But it is almost insignificant compared with the major gift of $200 that I gave you.

This analogy is overly simplistic, and it is probably completely unnecessary in this case because it is easy enough to understand without the analogy. I am using the analogy, however, because we will soon see arguments that are somewhat difficult to understand. Use of this money analogy throughout our discussion will help you grasp the more difficult concepts that we will see in the upcoming section.

We took a very close look at phenol, and we were able to see that the effect of the OH group is to make the ring more nucleophilic. We call this "activation." In other words, the OH group is activating the ring (making it more nucleophilic). So, the OH group is called an *activator*. The methyl group is also an activator (recall from the first semester that alkyl groups are electron donating). Some groups, however, actually *withdraw* electron density from the ring; we call those groups *deactivators* because they deactivate the ring (make the ring *less* nucleophilic). In the next section, we will see an example of a deactivator.

But we have not answered one of the questions we asked earlier in this section. We saw two reactions (bromination of toluene and bromination of phenol) where the substitution took place at the ortho and para positions only. We did *not* get any meta substitution in those cases. Why not? Are there cases where you do get meta substitution? The answers to these questions are critical for solving synthesis problems. Let's explore these questions in more detail now.

3.6   PREDICTING DIRECTING EFFECTS

When we talk about the preference for ortho and para substitution, we are talking about an issue of *regiochemistry*. In other words, in what region of the aromatic ring does the reaction take place? Let's review what we saw about the case of phenol.

In the previous section, we saw that the OH has two effects on the aromatic ring. It is electron-withdrawing by induction, and it is electron-donating by resonance. We saw that resonance was stronger, and therefore, the net effect of the OH group was to donate electron density to the ring (which activates the ring). When we drew the resonance structures for phenol and then melded them together in our minds, we got the following picture:

When we draw it like this, we clearly see that the OH group is *donating* electron density to the ring. But take a close look at *where* the electron density is being donated. It is not in all positions on the ring. It is only on the ortho and para positions. So it is true that the OH group makes the ring more nucleophilic, but it is only more nucleophilic at the ortho and para positions. Thus, when this ring reacts with an electrophile, the reaction takes place at the ortho and para positions:

Notice that the ring is SO incredibly activated that the reaction takes place at all three spots (the two ortho positions AND the para position). And we don't even need a Lewis acid to form Br^+. The ring is so activated by the OH group that even a poor electrophile (Br_2) will do the job (three times).

We have just given a specific explanation for the preference for ortho-para direction in the case of phenol. Our explanation was based on the electronics *of the starting material*. But another explanation can be given and is based on *the stability of the intermediate*. Luckily, this second explanation gives the same result (that there should be a preference for ortho-para direction). Most textbooks give this second explanation, probably because it is the only way to truly explain why alkyl groups are ortho-para directing (the first explanation, which we just gave, cannot be used to explain the directing effects of alkyl groups). You should look in your textbook to review that second explanation for ortho-para direction, but for now, we will just give a quick summary of the explanation.

You start by drawing out three mechanisms: You draw one mechanism for a substitution taking place at the *ortho* position. Then you draw another mechanism for a substitution at the *meta* position. Next you draw one last mechanism for a substitution at the *para* position. Finally, you compare the sigma complexes in each of the three mechanisms. You will find that the sigma complexes for ortho-substitution and para-substitution are more stable (than the sigma complex of a meta-substitution) because each has an extra resonance structure that is not present in the sigma complex of a meta-substitution.

Whether or not you fully understand that second explanation, the take-home message is that the OH group and the methyl group are both ortho-para directors. In fact, this is true not only for these two groups, but for ALL activators. All activators are ortho-para directors. In the next section, we will learn how to predict whether a group is an activator. But before we are ready for that, we need to explore what happens when we have a group on the ring that is a *de*activator.

The best example is the nitro group. Consider the structure of nitrobenzene:

If we want to understand the electronics of this compound, we will need to look at the two factors that determine electronics: induction and resonance. Induction is simple. Nitrogen is more electronegative than carbon, so the nitro group withdraws electron density from the ring by induction. The real question is: what is the effect of resonance? To see that, we will have to draw out the resonance structures:

Notice that we now have a *positive* charge spread throughout the ring. In the case of phenol, we had a *negative* charge spread throughout the ring. In that case, we argued that the OH group *gives* electron density to the ring, making the ring more nucleophilic. But here, we have a positive charge on the ring. So, the nitro group is *withdrawing* electron density from the ring, making the ring *less* nucleophilic:

So, the nitro group is electron-withdrawing by resonance and by induction. Therefore, in this case, there is no competition between resonance and induction. Both factors tell us that a nitro group should *deactivate* the ring. In other words, it should be difficult to do an electrophilic aromatic substitution on nitrobenzene.

So, let's say we take nitrobenzene and try to do an electrophilic aromatic substitution. For example, let's say we try to brominate it. We know that benzene will **not** react with Br_2. So it goes without saying that nitrobenzene, which is *less* nucleophilic than benzene, will definitely not react with Br_2. But what happens if we try to force the reaction to go? Recall how we forced benzene to react—we used Br_2 *together with a Lewis acid* to form Br^+. So, what happens if we try the same trick here? We do, in fact, get a reaction:

But you should immediately notice that the bromine went into the *meta* position only. To understand this, we need to take a close look at the electronics of nitrobenzene. We just saw that resonance allows us to predict that the nitro group is withdrawing a significant amount of electron density from the ring:

$$NO_2$$

$$\delta+ \qquad \delta+$$

$$\delta+$$

But notice that the electron density has not been withdrawn from the entire ring. Rather, it is the ortho and para positions that have been the most affected. So what happens when we force this reaction to go by using Br^+? It certainly cannot react at the ortho and para positions. Remember that the ring is acting as our nucleophile. And the ortho and para positions just don't have the electron density that it would take to attack Br^+. So, if we force the reaction to go, it will have to go in the meta position BY DEFAULT. The meta position has NOT been activated. But rather, the ortho and para positions have been deactivated. So, by default, the reaction must go meta.

The explanation we just gave can be used to show why a nitro group (which is a deactivator) is a meta-director. Our explanation was based on the electronics *of the starting material*. But once again, a second explanation can be given, which is based on *the stability of the intermediate*. Luckily, this second explanation gives the same result (that there should be a preference for meta direction). Most textbooks give this second explanation. You should look in your textbook to review that second explanation for meta direction. Most importantly, you should know the bottom line: *deactivators are meta-directors.*

So, we can now summarize the two important concepts that we have learned so far:

- Activators are ortho-para directors.
- Deactivators are meta-directors.

And now for the obvious question: are there any exceptions to these general statements? The answer is: yes. There is one very important exception. Halogens (F, Br, Cl, or I) are deactivators, so we might expect them to be meta-directors. But instead, they are actually ortho-para directors. Let's try to understand why halogens are the exception.

Before we explain this, let me just say upfront that this explanation is perhaps one of the most difficult concepts in all of organic chemistry. If you get lost and you have trouble understanding WHY halogens are the exception, then don't feel bad. Most students have a hard time with this. It takes time and patience—you should know that before we start. But if you find that you completely understand this explanation, and it makes perfect sense to you, then you should feel very good about yourself because organic chemistry doesn't get much tougher than this. And now, with that disclaimer in mind, let's jump into it.

If we want to understand why halogens are the exception, we need to remember a rule that we saw earlier in this chapter. When we were analyzing the effect of the OH group on an aromatic ring, we saw that there were two competing effects: induction and resonance. We saw that induction *withdraws* electron density from the ring but that resonance *donates* electron density to the ring. In order to know which factor dominates, we gave a general rule: **resonance usually beats induction**. We also said that there was an important exception to this general rule that we would see later. Well, it is now later. Halogens are the exception. We will look at this very closely over the next couple of pages, but for starters, here is the one-paragraph summary:

We have a general rule that resonance usually beats induction. But halogens are the exception to this rule. So, in the case of halogens, induction actually beats resonance. We will then use this concept to explain why halogens violate our rule for directing effects (that all activators are ortho-para directors and all deactivators are meta directors). We will explain why halogens are deactivators but ortho-para directors. And we will see that the answer comes directly from the fact that, in the case of halogens, induction actually beats resonance.

Now that we have seen the summary, let's take a closer look at the explanation. Let's use the example of chlorobenzene:

If we want to understand the electronics of this compound, we will need to look at two factors: induction and resonance. Let's start with induction. The inductive effect is similar to the effect we saw with an OH group on the ring. Just like an OH group, the Cl group is also electron-withdrawing by induction:

However, we also need to look at resonance effects. So we draw the resonance structures:

And once again, we see that the Cl group is very similar to the OH group. It is donating electron density by resonance:

So we have a similar analysis that we had with the case of the OH group. Once again, we have a competition between resonance and induction. The Cl group is *withdrawing* electron density by induction, and *donating* electron density by resonance. But in the case of the OH group, we used the argument that resonance beats induction (we said that was a general rule that holds true most of the time). Therefore, the net effect of the OH group was to donate electron density to the ring (thus, the OH group was an activator). But with the Cl group, resonance does NOT beat induction. This is one of the rare cases where induction actually beats resonance.

If we want to understand why, we must take another look at the resonance structures for chlorobenzene (see above). Notice that these resonance structures show a positive charge on Cl. That is very bad. Halogens do not like to bear positive charges (even more so than oxygen). So these resonance structures do not contribute very much character to the overall electronics of the molecule. Therefore, very little electron density is given to the ortho and para positions:

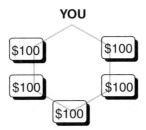

Only a small amount of electron density is given to these three positions. The $\delta-$ in each position is therefore very small.

The effect of resonance in this case is very small, and as a result, induction actually beats resonance in this case. Therefore, the net effect of the Cl group is to *withdraw* electron density from the ring. This explains why the Cl group is a *deactivator*. This also explains how the Cl group can be an ortho-para director (even though it is a deactivator). It is true that resonance is a weak effect in this case, BUT the effect of resonance is not completely negligible. Granted, resonance is so weak that induction actually wins the competition here, but resonance does still give a "tiny" amount of electron density back to the ortho and para positions.

To see this more clearly, let's use our money analogy again. Imagine that five people surround you, each carrying $100:

You begin by taking $30 from every person (which leaves each person with $70). Then, you give back $20 to each of three people (the *ortho* and *para* people):

The two *meta* people each have $70. The ortho and para people each have $90. Now think about what you have done. Overall, you have taken more money than you have given back. But when you compare how much money each person has, you find that the ortho and para people have *more* money than the meta people.

Similarly, the net effect of the Cl group is to *take away* electron density from the entire ring. Therefore, it is a deactivator. However, the Cl group does give a tiny bit of electron density back to the ortho and para positions. If we force the reaction to go, then, it will have to go ortho or para:

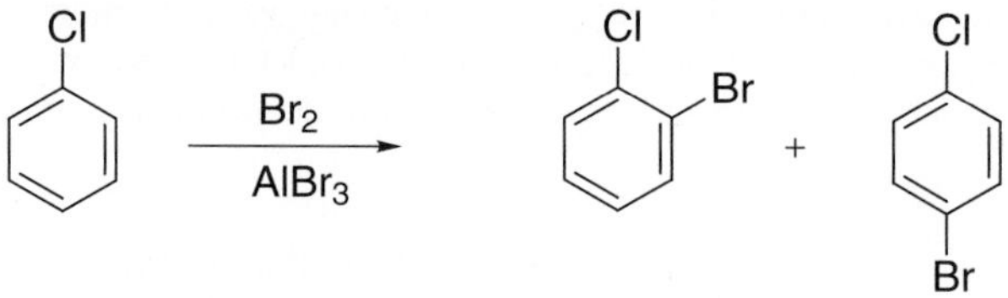

Now we are ready to modify the rules we gave earlier when we said that all activators are ortho-para directors and all deactivators are meta-directors. Here is our new-and-improved formula:

- All activators are ortho-para directors.
- All deactivators are meta-directors, *except for halogens (which are deactivators, but nevertheless, they are <u>ortho-para</u> directors).*

With that in mind, let's try to predict some directing effects.

EXERCISE 3.30. Look closely at the following monosubstituted benzene ring.

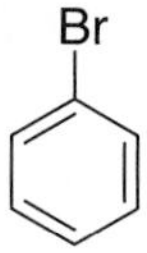

If we tried to do an electrophilic aromatic substitution on this compound, identify where the substitution reaction would take place.

Answer: Br is a halogen (remember that the halogens are F, Cl, Br, and I). We have seen that halogens are the one exception to the general rules (halogens are deactivators, but nevertheless they are ortho-para directors). Therefore, if we use this compound in an electrophilic aromatic substitution, we expect the substitution to take place at the ortho and para positions:

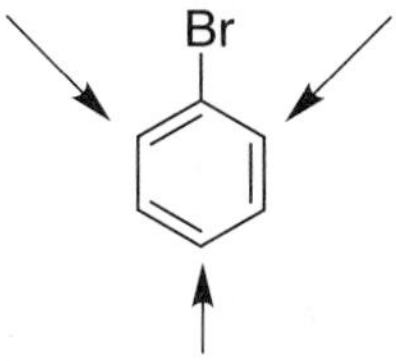

For each of the following problems, predict the directing effects.

3.31. 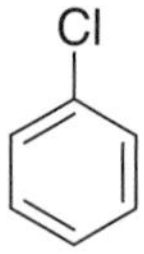 **3.32.** 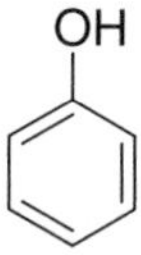 **3.33.**

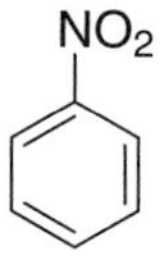

3.34.

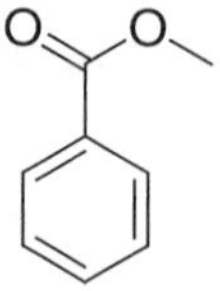

This group is a deactivator.

3.35.

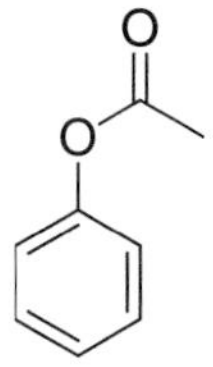

This group is an activator.

3.36.

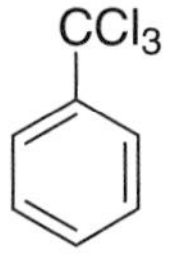

This group is a deactivator.

3.37.

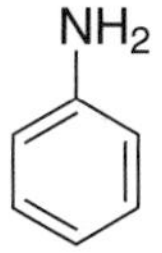

This group is an activator.

Clearly, you can predict where the substitution will take place only if you know whether the group is an activator or a deactivator. In the next section, we will learn how to predict whether a group is an activator or a deactivator. But for now, let's get some practice on some real reactions.

EXERCISE 3.38. Predict the products of the following reaction:

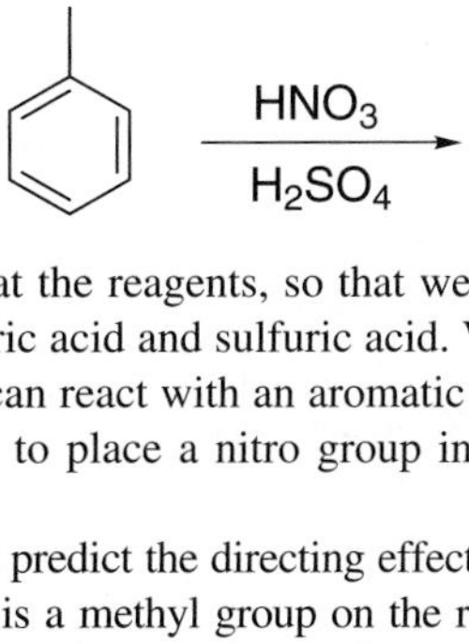

Answer: We begin by looking at the reagents, so that we can determine what kind of reaction is taking place. The reagents are nitric acid and sulfuric acid. We have seen that these reagents give us NO_2^+ as an electrophile, which can react with an aromatic ring in an electrophilic aromatic substitution reaction. The end result is to place a nitro group into the ring. So, now the question is: where do we put the nitro group?

To answer this question, we must predict the directing effects of the group already on the ring (before the reaction takes place). There is a methyl group on the ring, and we have seen that the methyl group is an activator. Therefore, we predict that the reaction will take place at the ortho and para positions, relative to the methyl group:

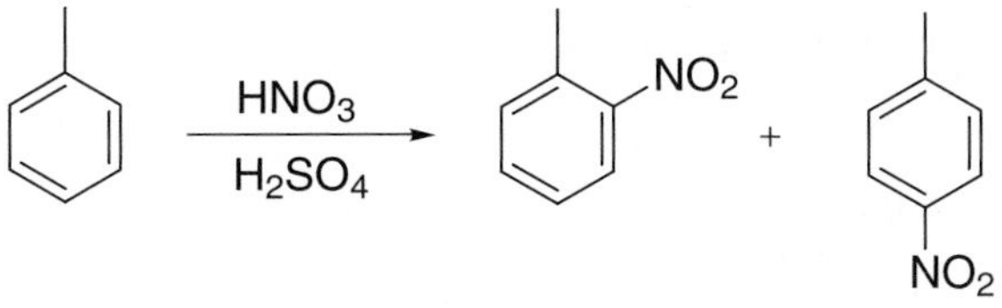

Notice that in this case, we don't draw a substitution at both ortho positions because we get the same product either way:

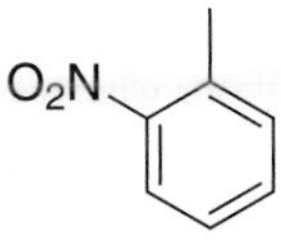

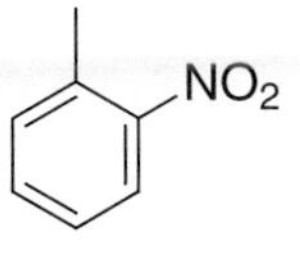

Predict the products of the following reactions:

3.39.

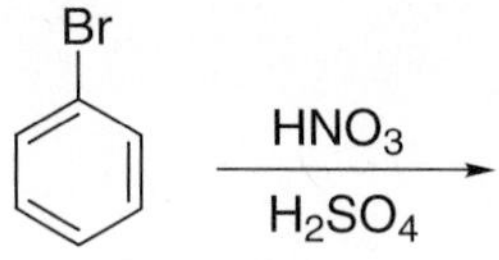

$$\xrightarrow[\text{H}_2\text{SO}_4]{\text{HNO}_3}$$

3.40.

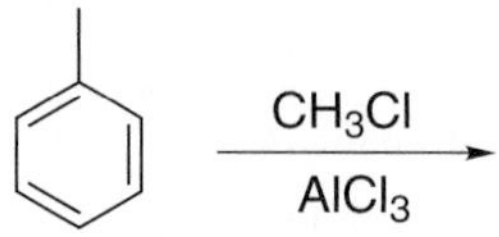

$$\xrightarrow[\text{AlCl}_3]{\text{CH}_3\text{Cl}}$$

3.41.

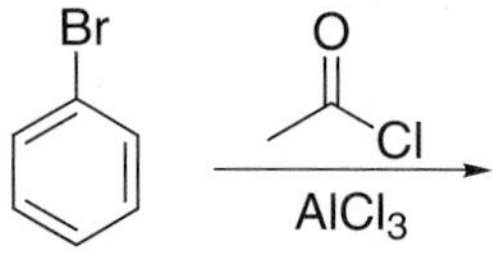

$$\xrightarrow{\text{AlCl}_3}$$

3.42.

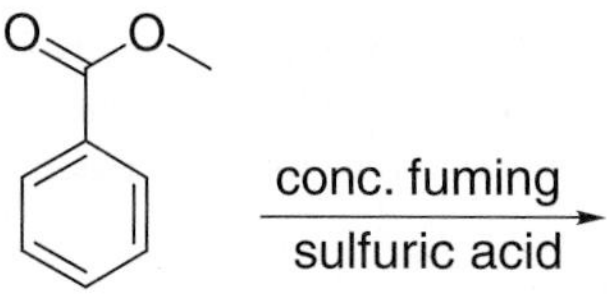

$$\xrightarrow{\substack{\text{conc. fuming} \\ \text{sulfuric acid}}}$$

Hint: The group on the ring is a deactivator.

3.43.

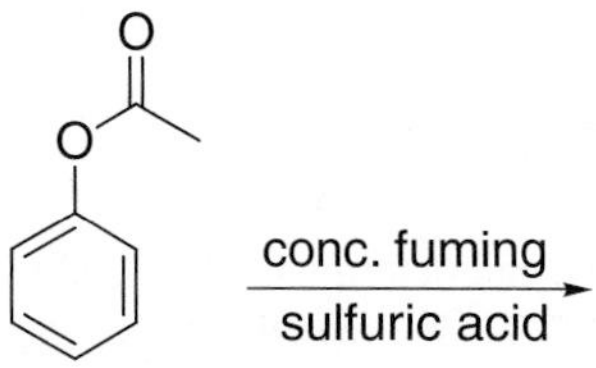

$$\xrightarrow{\substack{\text{conc. fuming} \\ \text{sulfuric acid}}}$$

Hint: The group on the ring is an activator.

3.44.

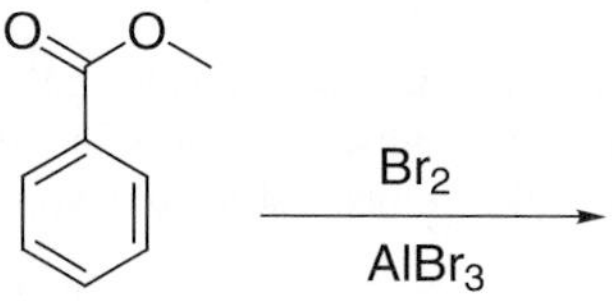

$$\xrightarrow[\text{AlBr}_3]{\text{Br}_2}$$

Hint: The group on the ring is a deactivator.

3.45.

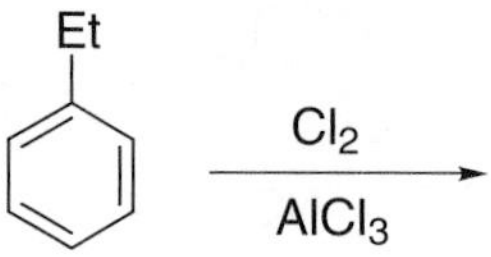

$$\xrightarrow[\text{AlCl}_3]{\text{Cl}_2}$$

Hint: The group on the ring is an activator.

So far, we have focused on the directing effects when you have *only one group* on a ring. And we have seen that activators direct toward the ortho and para position, whereas deactivators direct toward the meta positions:

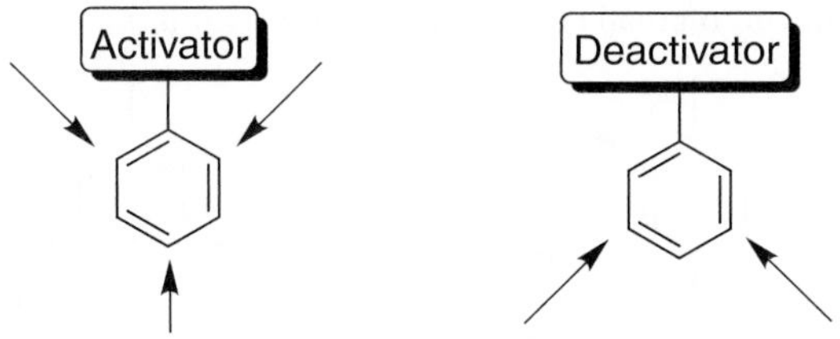

and we saw only one exception to these general rules (the halogens).

But how do you predict the directing effects when you have *more than one group* on a ring? For example, consider the following compound:

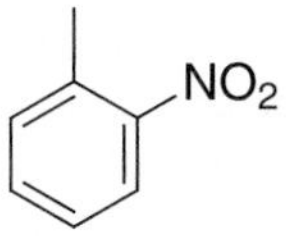

What if we used this compound in an electrophilic aromatic substitution reaction. For example, let's say we try to brominate this compound. Where would the bromine go?

Let's first consider the effect of the methyl group. We mentioned before that a methyl group is an activator, so we predict that it will direct toward the ortho and para positions:

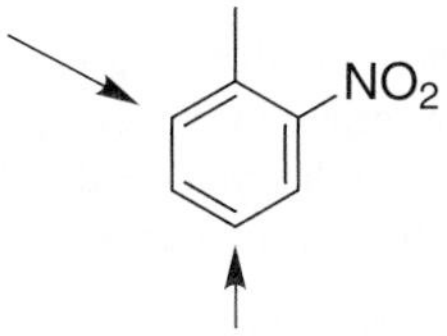

Notice that we do **not** point to the ortho position that already bears the nitro group (we only look at positions where there are currently no groups—remember that in an electrophilic aromatic substitution, E+ comes on the ring and **H+** comes off). So, the methyl group is directing toward *two* spots, as shown above.

Now let's consider the effect of the nitro group. We mentioned before that the nitro group is a powerful deactivator. Therefore, we predict that it should direct to the positions that are *meta **to the nitro group***:

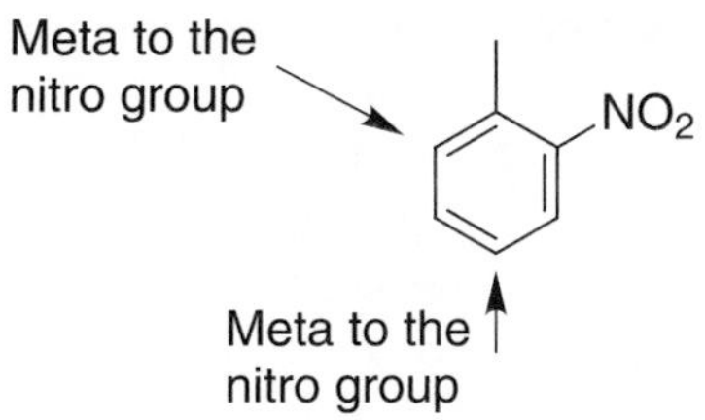

Meta to the
nitro group

Meta to the
nitro group

So, we see that the nitro group and the methyl group are directing toward the same two spots. So, in this case there is no conflict between the directing effects of the nitro group and the methyl group.

But consider this case:

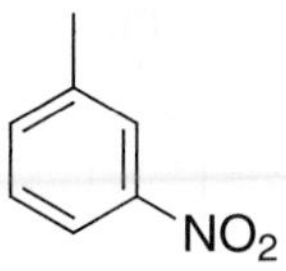

The methyl group and the nitro group are now directing toward different positions:

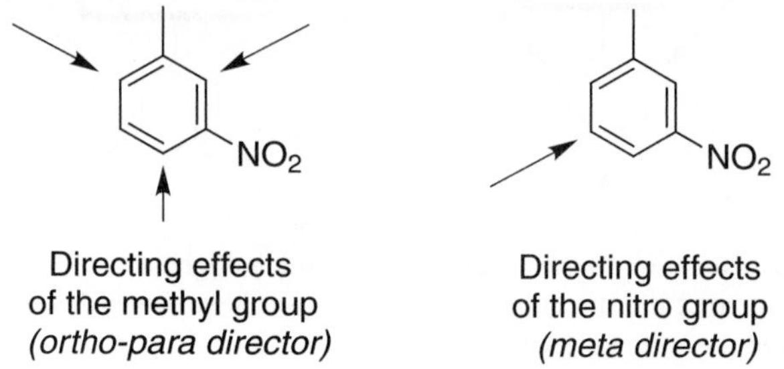

Directing effects
of the methyl group
(ortho-para director)

Directing effects
of the nitro group
(meta director)

So, the big question is: which group wins? It turns out that the directing effects of the methyl group are stronger than the directing effects of the nitro group. So, if we brominate this ring, we will get the following products (where the Br goes ortho or para to the methyl group):

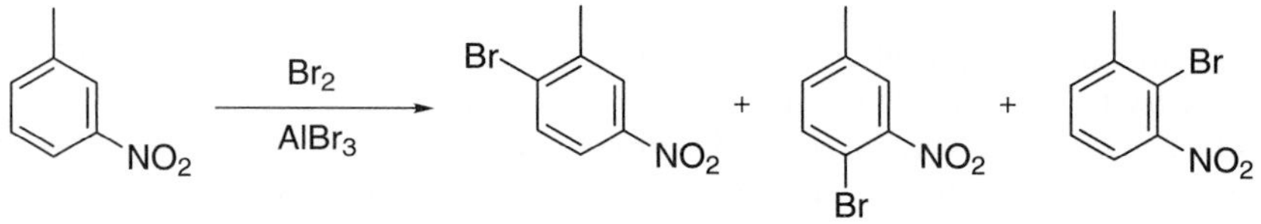

It is common to see a situation where the directing effects of two groups are competing with each other (like the methyl and the nitro group in the above example). So we clearly need rules for determining which group wins. It turns out that you need to know just two simple rules in order to determine which group will dominate the directing effects:

1. *Ortho-para directors always beat meta-directors.* The example we just saw is a perfect illustration of this rule. The methyl group is an activator (an ortho-para director), and the nitro group is a deactivator (a meta-director), so the methyl group wins. This rule should make sense, when we consider how directing effects work. Recall that meta directors do not actually do anything good for the meta positions. Instead, they simply *deactivate the ortho and para positions,* so if we force a reaction to go, it must go meta BY DEFAULT. But ortho-para directors are actually doing something good for the ortho and para positions. They *activate* the ortho and para positions. Thus, ortho-para directors will always beat meta-directors.

2. **Strong** *activators always beat* **weak** *activators.* For example, consider the following case:

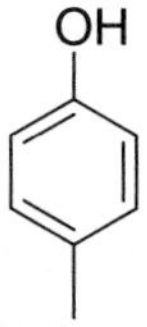

The OH group is a *strong* activator, and the methyl group is a *weak* activator. (We will learn in the next section how to predict which groups are strong and which are weak—for now, just take my word for it.) So, the OH group will win, and the directing effects are ortho and para to the OH group:

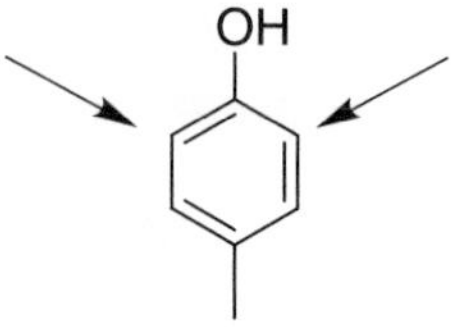

So, we have seen two rules:

- *Ortho-para directors always beat meta-directors.*
- ***Strong*** *activators always beat **weak** activators.*

Keep in mind that the first rule always trumps the second rule. So if you have a weak activator against a strong deactivator, the weak activator wins. Even though the activator is weak, it still beats a strong deactivator because activators (ortho-para directors) always beat deactivators (meta-directors). To demonstrate this, let's consider the following example:

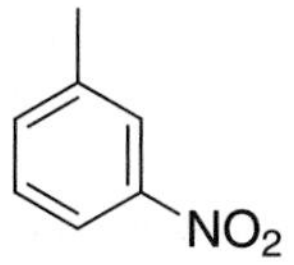

The methyl group is a weak activator, and the nitro group is a strong deactivator. So, in this case, the methyl group wins (and the directing effects are ortho and para to the methyl group; and ***not*** meta to the nitro group):

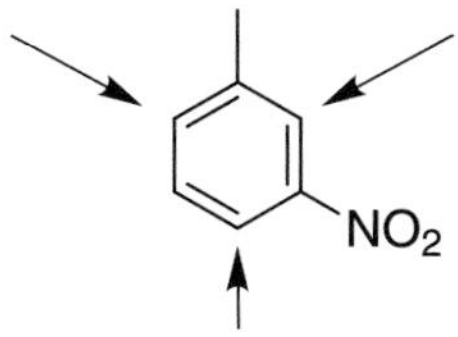

EXERCISE 3.46. Predict the directing effects in the following scenario.

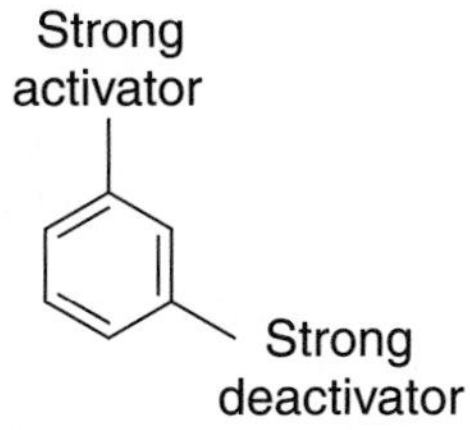

For this problem, you should assume that the deactivator is *not* a halogen.

Answer: We have two groups. The activator will direct toward the positions that are *ortho or para* to itself, and the deactivator will direct toward the positions that are *meta* to itself:

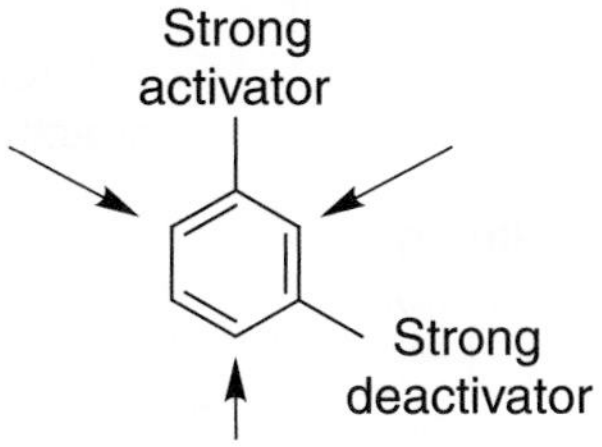

**Directing effects
of the activator**

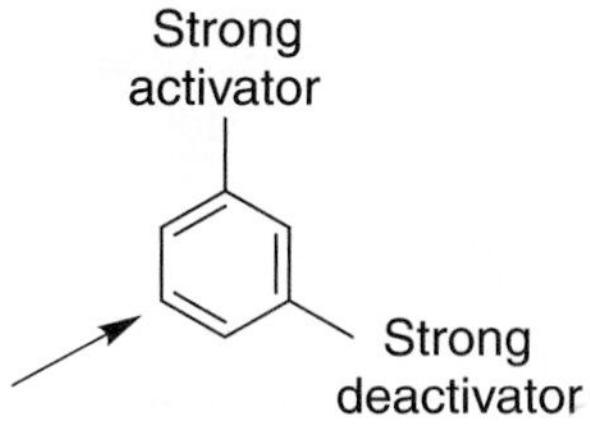

**Directing effects
of the deactivator**

So, there is a competition in the directing effects. Between the two groups, the strong activator beats the strong deactivator because the strong activator is an ortho-para director. So the directing effects are:

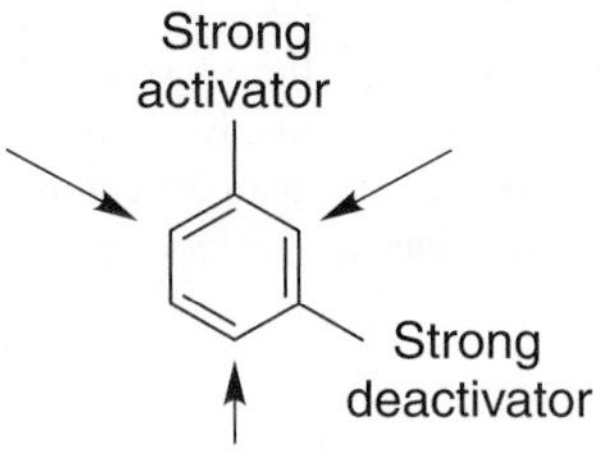

If we did an electrophilic aromatic substitution on a compound of this type, we would expect three products (because the directing effects are toward three positions, shown above). Here is a specific example of a reaction like this

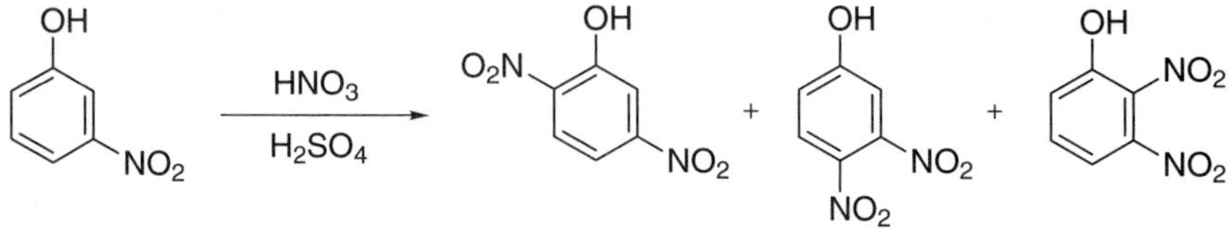

The ring has two groups on it (before we do the reaction). The OH group is a strong activator, and the nitro group is a strong deactivator.

For each of the following problems, predict the directing effects. Unless otherwise indicated, assume that anything labeled as a deactivator is not a halogen (unless it is specifically indicated as a halogen).

3.47.

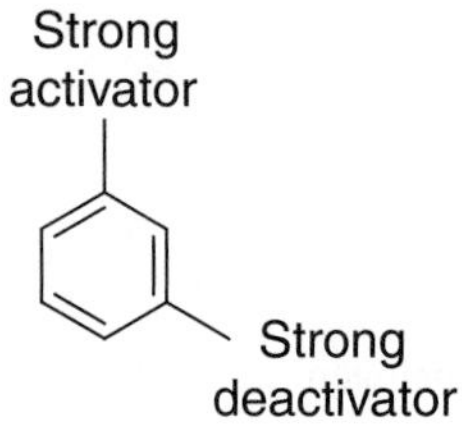

3.48.

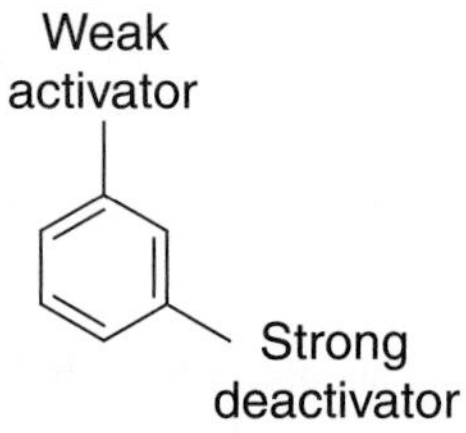

3.49.

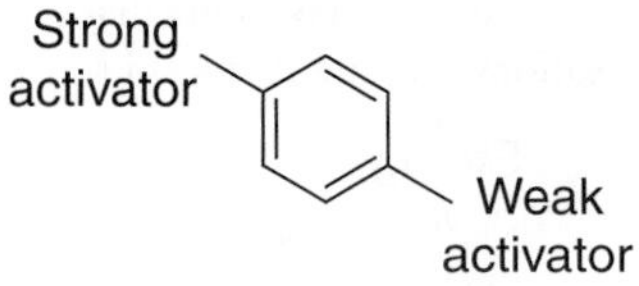

3.50.

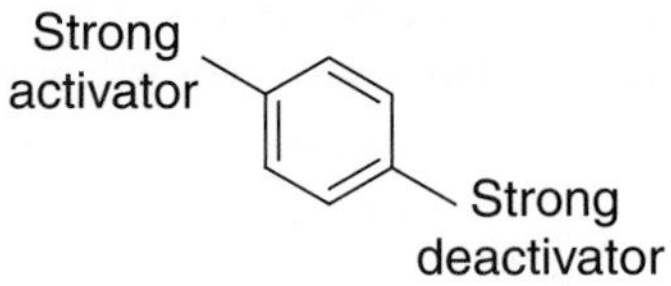

3.51.

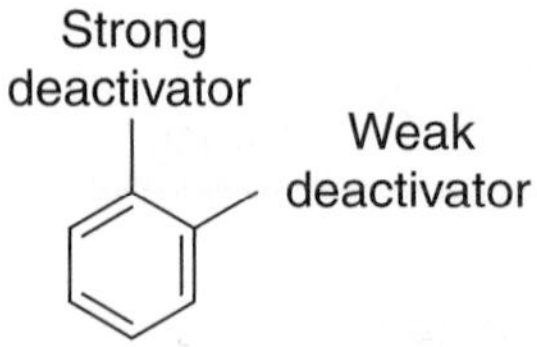

3.52.

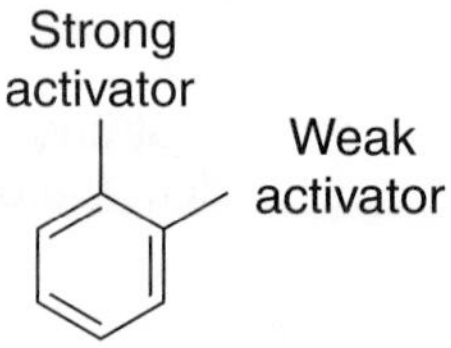

3.53. Weak activator   Strong deactivator

3.54. Strong activator / Weak activator

3.55. Strong deactivator / Me

3.56. Strong activator / Br

3.7 IDENTIFYING ACTIVATORS AND DEACTIVATORS

In the previous section, we learned how to predict the directing effects in a situation where you have more than one group on the ring. But in all of the cases in the previous section, I had to tell you whether each group was an activator or a deactivator and whether it was strong or weak. In this section, we will learn how to predict this, so that you won't have to memorize the characteristics of every possible group. In fact, very little memorization is actually involved here. We will see a few concepts that should make sense. And with those concepts, you should be able to identify the nature of any group, even if you have never seen it before.

We will go through this methodically, starting with strong activators.

Strong activators are groups that have a lone pair next to the aromatic ring. We have already seen an example of this. When an OH group is connected to the ring, there is a lone pair next to the ring, which gives rise to the following resonance structures:

We concluded in the previous section that this resonance effect is very strong and that the OH group is therefore donating a lot of electron density to the ring:

This is true, not only for the OH group, but for other groups that have a lone pair next to the ring. The same kind of resonance structures can be drawn for an amino group connected to a ring:

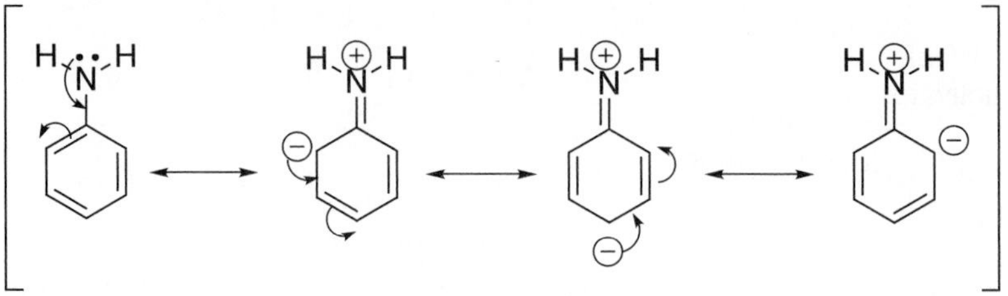

Here are a few examples of strong activators. Make sure that you can easily see the common feature (the lone pair next to the ring):

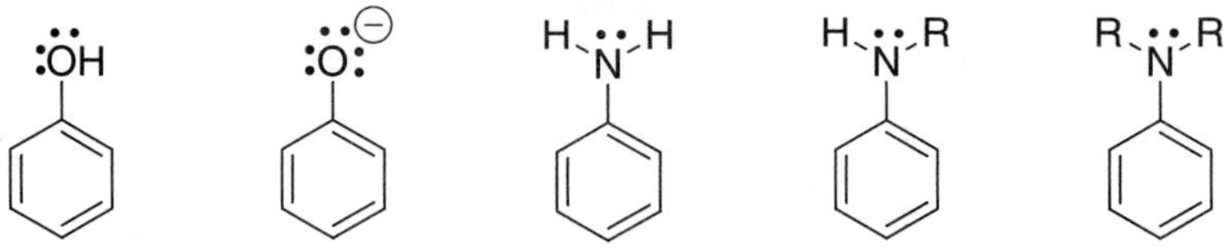

Next, we move on to the *moderate* activators. Moderate activators are groups that have a lone pair next to the ring, BUT that lone pair is already partially tied up in resonance. For example, consider the following group:

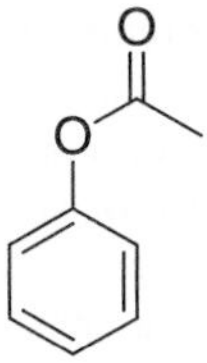

This compound has all of the resonance structures that place electron density into the ring (just like an OH group does):

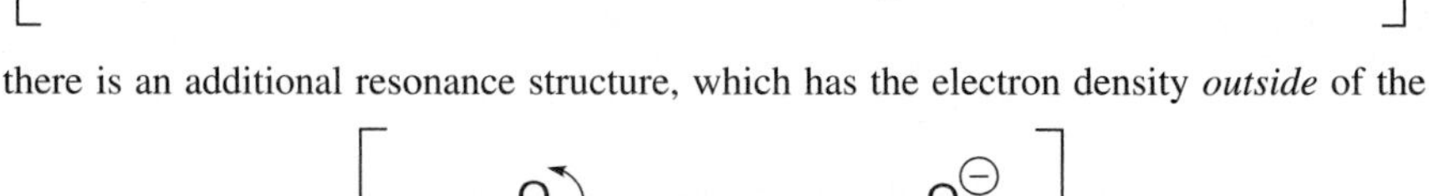

BUT there is an additional resonance structure, which has the electron density *outside* of the ring:

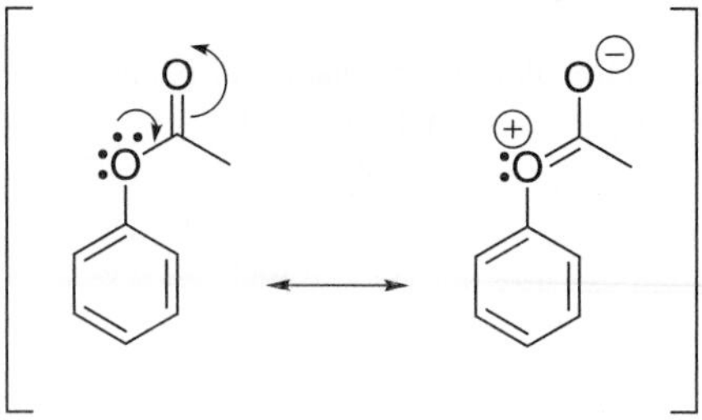

Therefore, the electron density is more spread out (with some in the ring and some out of the ring). This group is therefore not a *strong* activator. Rather, we call it a *moderate* activator. (Some textbooks do not point out this subtle distinction between strong activators and moderate activators.) Here are several examples of moderate activators:

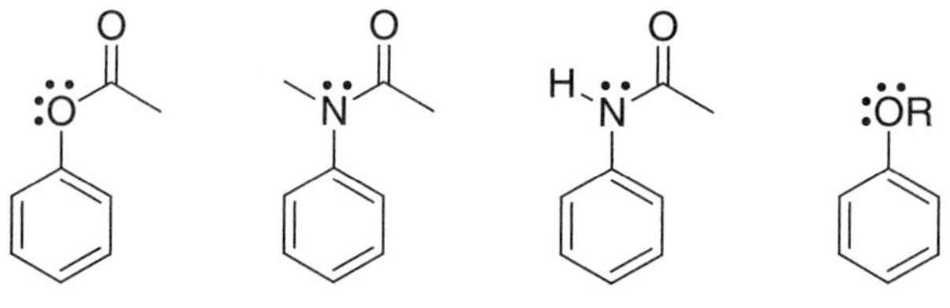

Look closely at the examples above. They all have a lone pair that is tied up in resonance outside of the ring. BUT WAIT A SECOND. What about the last group on this list (the OR group)? This group has a lone pair that is NOT tied up in resonance outside of the ring. We should predict that this group should belong in the first category (*strong* activators), but for some reason, it isn't in that category. It is actually just a *moderate* activator. This is one of the rare examples that departs from the logical explanations that we have given so far. I have spent quite a bit of time trying to figure out why the OR group is a moderate activator (rather than a strong activator). I have come up with several answers over the years, but I am not going to spend several pages dedicated to an esoteric topic that you will certainly not need for your exams (perhaps you might think of it as a brain teaser—something to think about . . .). For now, you will just have to remember that the OR group doesn't follow the trends we have seen. It is a moderate activator.

Now let's turn our attention to *weak* activators. Weak activators donate electron density to the ring through a weak effect, called *hyperconjugation*.

In the first semester of organic chemistry, we saw that ***alkyl groups are electron-donating***. That was important when we learned about carbocation stability (we saw that tertiary carbocations are more stable than secondary carbocations, which are more stable than primary carbocations—because ***alkyl groups are electron-donating***, which stabilizes a carbocation). There is a simple reason why alkyl groups are electron-donating. It is due to a phenomenon called hyperconjugation. If you do not remember this term from the first semester, you can go back and review it if you like. Whether or not you go back, make sure that you remember that ***alkyl groups are electron-donating***. I keep stressing this because there are many more concepts in organic chemistry that you can understand only if you know that ***alkyl groups are electron-donating***.

So, all alkyl groups are weak activators (methyl, ethyl, propyl, etc.)

Now we have seen all of the different categories of activators (strong, moderate, and weak). To review, this is what we saw:

Strong Activators	Lone pair next to ring
Moderate Activators	Lone pair next to ring, but tied up in resonance outside of ring as well
Weak Activators	Alkyl groups

Now, we will turn our attention to the different categories of *deactivators*. This time, we will start off with the *weak* deactivators and work our way toward *strong* deactivators (rather than starting with strong). There is a reason for using this order, and that reason will soon become clear.

Weak deactivators are the halogens. We have already seen that halogens are the one case where induction beats resonance (and therefore, we argued that the net effect of a halogen is to *withdraw* electron density from the ring). So, we saw that halogens are deactivators. But you should know that the competition between induction and resonance (in the case of the halogens) is a close competition, so halogens are only weakly deactivating.

We will summarize all of this information in one complete chart, but for now let's move on to moderate deactivators.

Moderate deactivators are groups that take electron density away from the ring by resonance. For example, consider the following group:

This group does *not* have a lone pair next to the ring (so it is *not* an activator). But it does have a pi bond next to the ring, and we see the following resonance structures:

When we look closely at these resonance structures, we can see that the group is *withdrawing* electron density from the ring:

Therefore, this group is a *moderate deactivator*. Numerous other similar groups can also withdraw electron density from the ring. Here is a list of many examples:

All of these examples are withdrawing electron density from the ring by resonance. They all have one feature in common: a pi bond to an electronegative atom. Take a close look at the last example. A cyano group is a pi bond (a triple bond) to an electronegative atom (nitrogen). So we see that a triple bond can also be included in this category.

And now for the last category: *strong* deactivators. There are three common functional groups in this category. We have already seen one of these groups: the nitro group. But there are two other groups as well:

Unfortunately, there is no one common reason why these three groups are all together in the same category. But these three groups are all somewhat different from each other, and so, we will have to explain each of them separately. We have already explained (using resonance) why the nitro group is so powerfully electron-withdrawing. We saw that earlier in this chapter. The nitro group is electron-withdrawing by resonance *and* induction.

So let's move on to the next group (the trichloromethyl group). To understand why this group is a strong deactivator, we need to focus on the collective inductive effects of all of the chlorine atoms:

The inductive effects of each chlorine atom add together to give one very powerful deactivating group. Be careful not to confuse this group with a halogen on a ring:

When a halogen is connected directly to the ring (above right), then there are lone pairs next to the ring, so there are resonance effects to consider. (We spent a lot of time talking about the competition between resonance and induction in the case of halogens.) The group we are talking about now (above left) does not have any resonance effects to consider because the lone pairs are *not* directly next to the ring. So, there is only an inductive effect to consider, and this inductive effect is very significant in this case (because there are three inductive effects adding together).

If we look at our third *strong deactivator*, we see a nitrogen atom with a positive charge next to the ring:

The nitrogen atom is so poor in electron density that it is practically sucking electron density out of the ring like a vacuum cleaner:

We have now explained each of the three strong deactivators.

Now we are ready to summarize everything we have seen into one chart:

Common Feature	*Some Examples*

ACTIVATORS

Strong   Lone pair next to ring

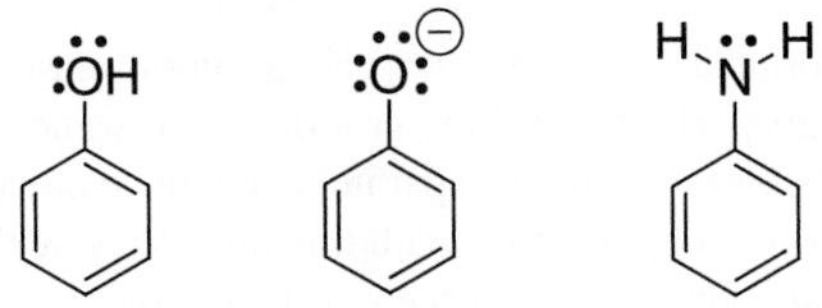

Moderate   Lone pair next to ring
But tied up in resonance
outside of ring as well

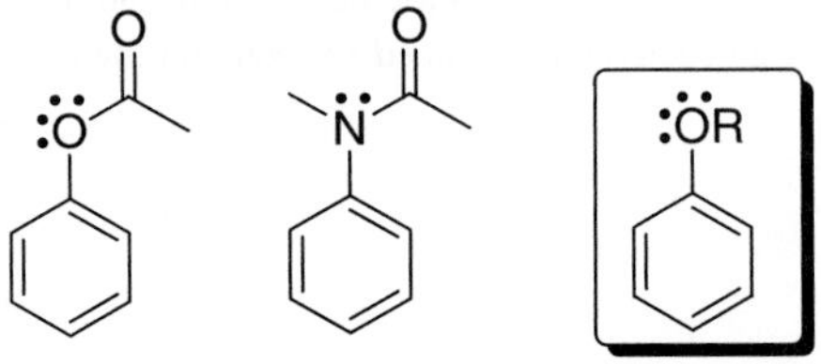

Does not fit
the pattern

Weak   Alkyl groups

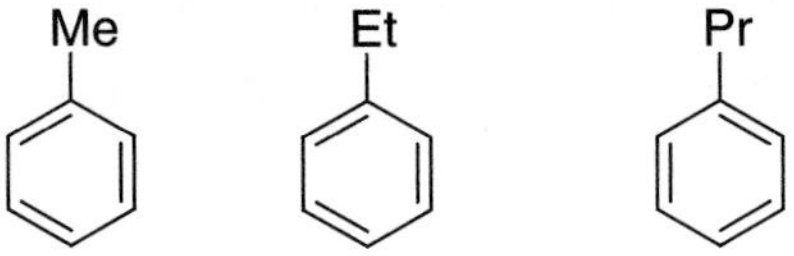

Me   Et   Pr

DEACTIVATORS

Weak   Halogens

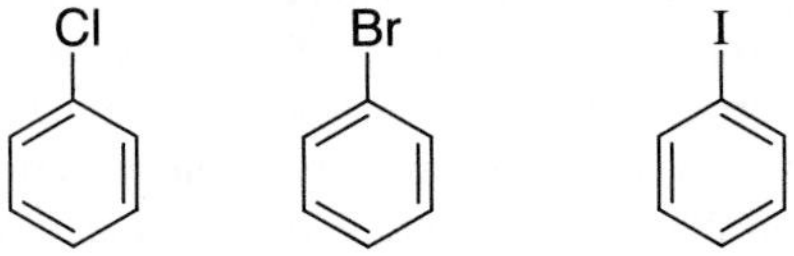

Cl   Br   I

Moderate   Pi bond to an Electronegative
atom (next to ring)

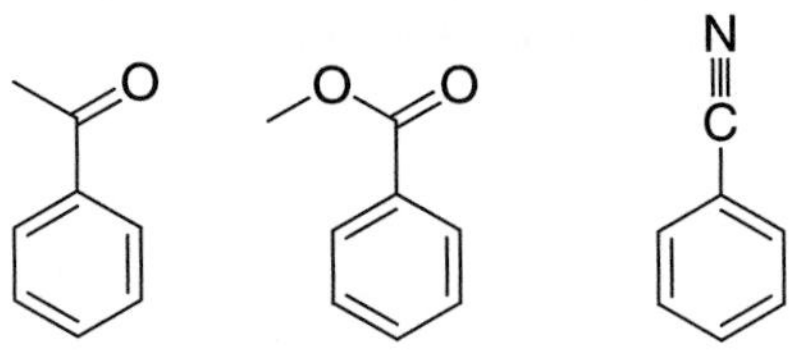

Strong   Very powerfully Electron-
withdrawing

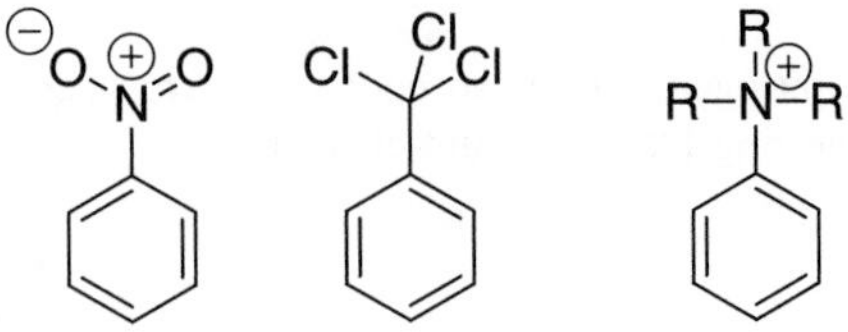

Take a close look at this chart and make sure that every category makes sense to you. As you look over the chart, you should be able to remember the arguments that we gave for each category. If you have trouble with this, you might want to review the last few pages of explanation.

And now we can understand why we looked at weak deactivators first (before strong deactivators). When we organize it like this, we can clearly organize the directing effects in our minds:

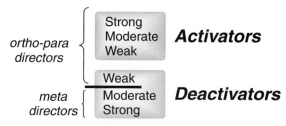

This chart shows that all activators are ortho-para directors and all deactivators are meta directors, with the exception of the weak deactivators (halogens).

EXERCISE 3.57.   Look closely at the following group:

OH
|
O=S=O

Try to predict what kind of group it is (a strong activator, a moderate activator, a weak activator, a weak deactivator, a moderate deactivator, or a strong deactivator).
  Use this information to predict the directing effects.

Answer: This group does not have a lone pair next to the ring, and it is not an alkyl group. Therefore, it is definitely not an activator. This group has a pi bond to an oxygen atom (next to the ring), and therefore, it is a moderate deactivator.
  Because all deactivators are meta-directors (except for weak deactivators—halogens), we predict the following directing effects:

OH
|
O=S=O

For each of the following groups, determine what kind of group it is (a strong activator, a moderate activator, a weak activator, a weak deactivator, a moderate deactivator, or a strong deactivator). Place your answer on the space provided. Try to do this **without** looking at the chart that we constructed. You won't have access to this chart on an exam. Try to remember and apply the explanations that we used.
  Then, use that information to predict the directing effects. Indicate the directing effects using arrows for pointing to the positions where you would expect to see an electrophilic aromatic substitution:

3.58.   Answer: _____   **3.59.**   Answer: _____

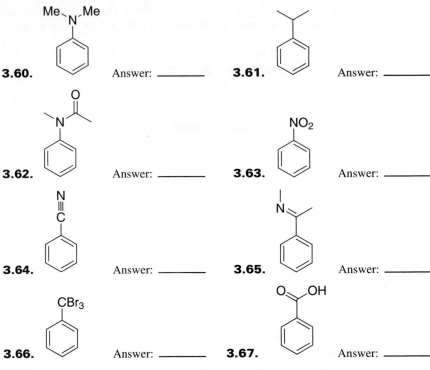

3.60. Answer: _____ **3.61.** Answer: _____

3.62. Answer: _____ **3.63.** Answer: _____

3.64. Answer: _____ **3.65.** Answer: _____

3.66. Answer: _____ **3.67.** Answer: _____

3.68. Can you explain why the following group is a strong activator:

(*Hint*: Think about what strong activators have in common.)

Now we can use the skills that we developed in this section to predict the products of a reaction. Let's see an example:

EXERCISE 3.69. Predict the product of the following reaction:

$$\xrightarrow[\text{H}_2\text{SO}_4]{\text{HNO}_3}$$

Answer: We look at the reagents to see what kind of reaction we expect. The reagents are nitric acid and sulfuric acid. These reagents give us NO_2^+, which is an excellent electrophile. So, we know that we will be putting a nitro group onto the ring. But the question is: where?

To answer this question, we must predict the directing effects of the group that is currently on the ring. We recognize that this group is a moderate deactivator, which means that it must be a meta director. So, we predict the following product:

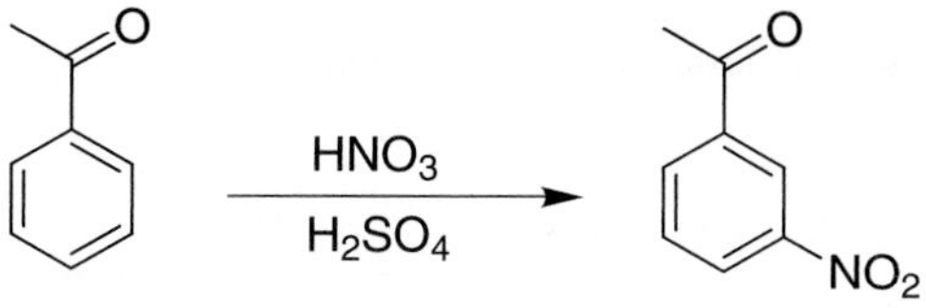

Predict the products of the following reactions:

3.70.

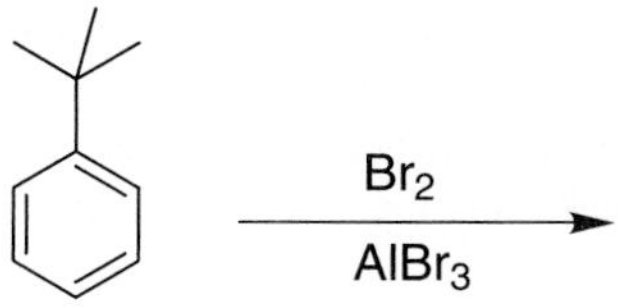

$$Br_2$$ / $$AlBr_3$$

3.71.

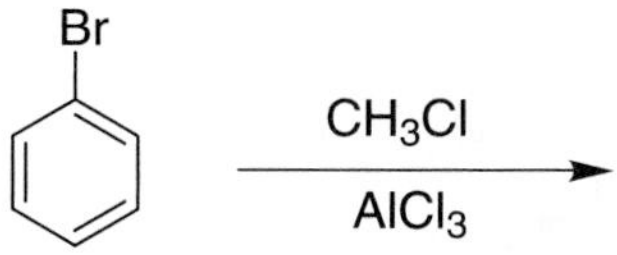

$$CH_3Cl$$ / $$AlCl_3$$

3.72.

$$Br_2$$

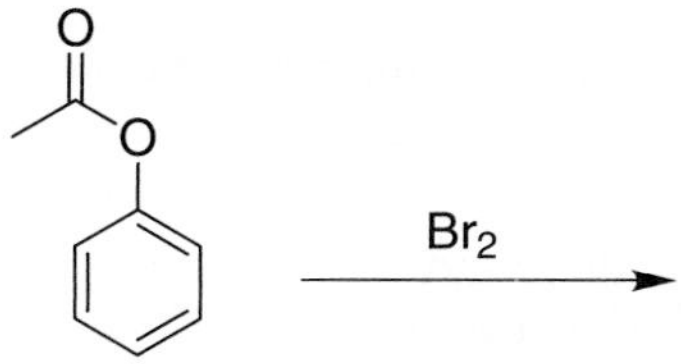

Notice that in this reaction, we do not need a Lewis acid. Can you explain why not?

3.73.

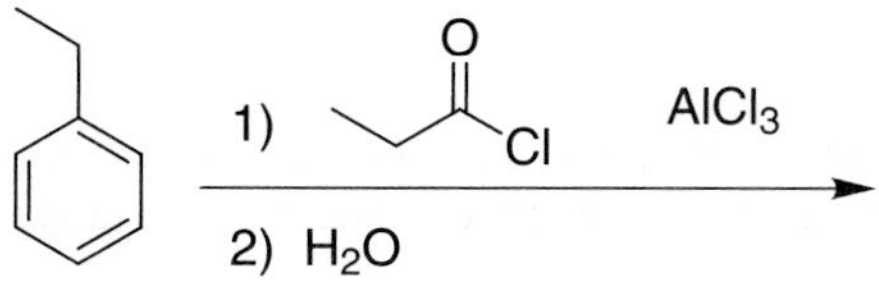

1) (acyl chloride) $$AlCl_3$$

2) $$H_2O$$

3.74.

$$HNO_3$$ / $$H_2SO_4$$

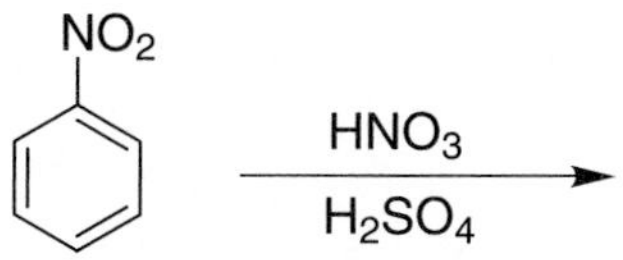

Now let's combine what we did in the previous section with the material in this section. Recall from the previous section that we learned how to predict the directing effects when you have more than one group on the ring. When the two groups are competing with each other, we saw that you can determine the directing effects by using the following two rules:

• *Ortho-para directors always beat meta-directors.*
• **Strong** *activators always beat* **weak** *activators.*

Now that we have learned how to categorize the various kinds of groups, let's do some real problems:

EXERCISE 3.75. Predict the product of the following reaction:

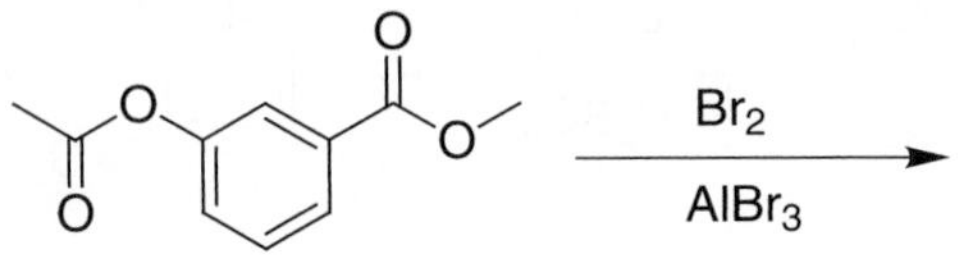

Answer: We look at the reagents to see what kind of reaction we expect. The reagents are bromine and aluminum tribromide. These reagents give us Br^+, which is an excellent electrophile. So, we know that we will be putting a Br onto the ring. But the question is: where?

To answer this question, we must predict the directing effects of both groups that are currently on the ring. The group on the left is a moderate activator (make sure that you know why), and therefore, it directs ortho-para to itself:

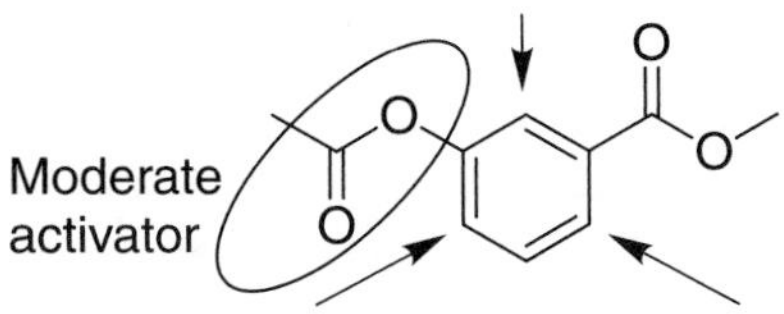

The group on the right is a moderate deactivator (make sure you know why), so it directs meta to itself:

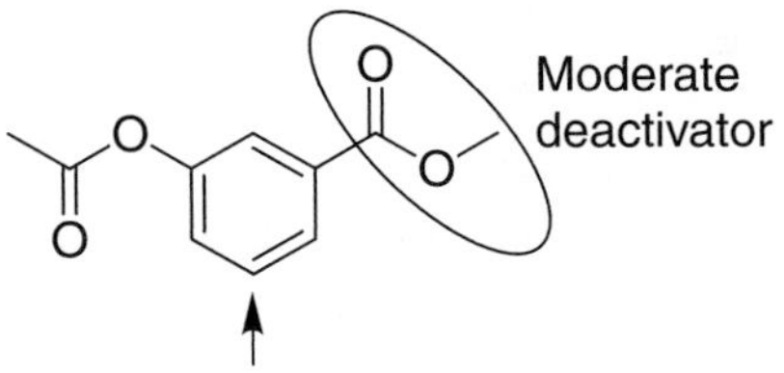

There is a competition between these two groups. Remember our first rule for determining which group wins: ortho-para directors beat meta directors. So, we expect the following products:

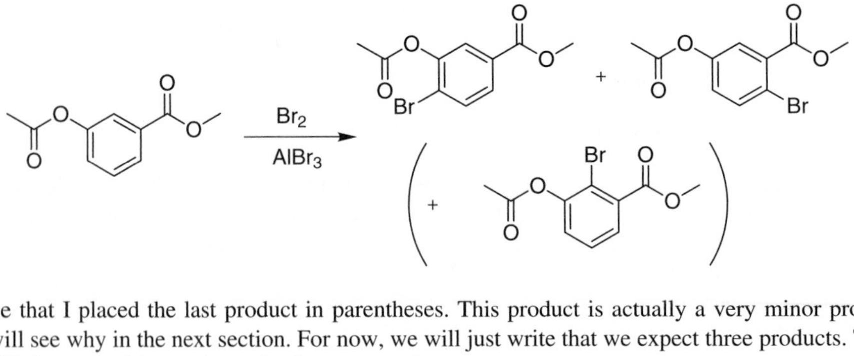

Notice that I placed the last product in parentheses. This product is actually a very minor product. We will see why in the next section. For now, we will just write that we expect three products. Then, we will fine-tune this prediction in the next section.

PROBLEMS.   Predict the products of the following reactions:

3.76.

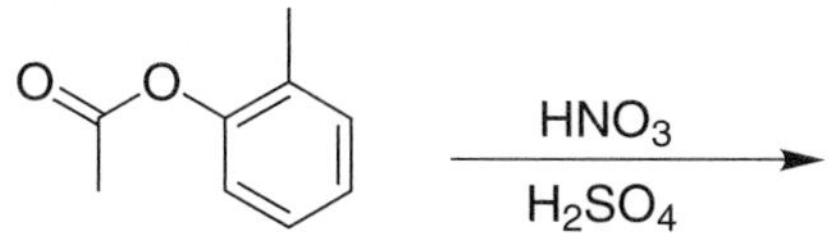

$\dfrac{HNO_3}{H_2SO_4}$

3.77.

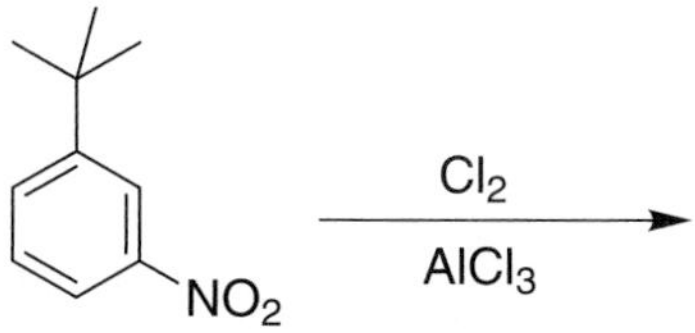

$\dfrac{Cl_2}{AlCl_3}$

3.78.

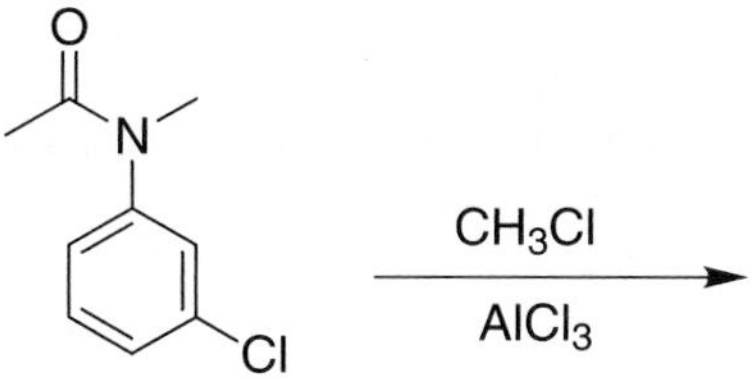

$\dfrac{CH_3Cl}{AlCl_3}$

3.79.

1) [acetyl chloride structure] $AlCl_3$

2) H_2O

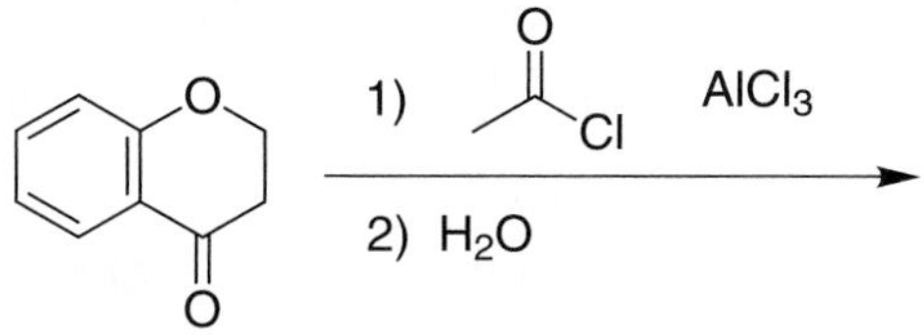

(*Hint*: Consider this aromatic ring as possessing two separate groups, and analyze each group separately.)

3.8   PREDICTING AND EXPLOITING STERIC EFFECTS

In the previous sections, we learned the skills that we need in order to predict the products of an elec-trophilic aromatic substitution. We saw many cases where there is *more* than one product. For exam-ple, if the ring is activated, then we expect ortho *and* para products. In this section, we will see that it is possible to predict which product will be the major product and which will be the minor product (or-tho vs. para). It is even possible *to control* the ratio of products (ortho vs. para). This is VERY impor-tant for synthesis problems, which will be the next (and final) section of this chapter.

Consider an electrophilic aromatic substitution with propyl benzene. The propyl group is a weak activator, and therefore, we expect the directing effects to be ortho-para:

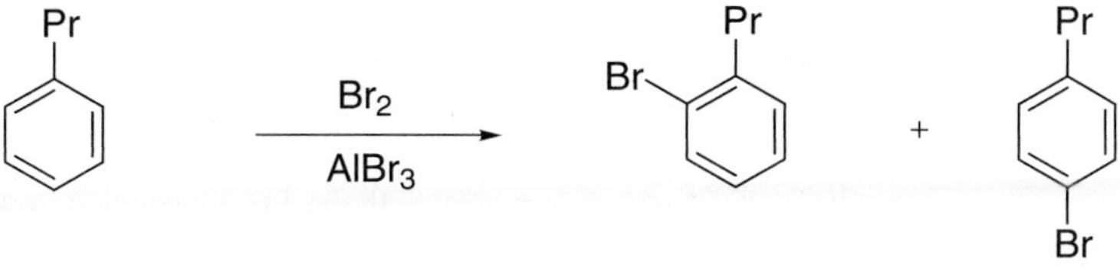

There are two products here. But let's try to figure out which one of these products is the major product? Ortho or para? At first, we might be tempted to say that the ortho product should be major. Let's see why. The propyl group is an ortho-para director, so there should be a total of three positions that can get attacked (two ortho positions and one para position):

Therefore, the chances of attacking an ortho position should be two-thirds (or 67%), while the chances of attacking the para position should be one-third (33%). From a purely statistical point of view, we should therefore expect our product distribution to be 67% ortho and 33% para. But there is one all-important factor that makes the product ratio different from what we might expect: *sterics*.

The propyl group is fairly large, and it partially "blocks" the ortho positions. You can still get ortho products, but you get less than 67%. In fact, the para product is the major product in this case:

This is usually the case (that para is the major product), except when the group on the ring is very, very small. So, when the group on the ring is a methyl group, it is possible to get slightly more ortho product than para product:

But this is only the case with a methyl group. With just about any other group, we should expect that the para product will be the major product. Keep that in mind because it is very important—para is usually the major product.

With that in mind, imagine that I asked you to propose an efficient synthesis for the following conversion:

This is simple to do. The *tert*-butyl group is so large that we expect the para product to be the major product. So we just use Br_2 and $AlBr_3$, and we get the product we want.

But suppose we wanted to make the ortho product:

How would we do this? When confronted with this problem, students often suggest using the same reaction as before, with the understanding that the ortho product will be a minor product (so *some* ortho product will definitely be formed). But you can't do that. Whenever you have a synthesis problem, you must choose reagents that give you the desired compound as the MAJOR product. If you propose a synthesis that would produce the desired product as a MINOR product, then your synthesis is not efficient. So we have a problem here. How do we run the reaction so that the ortho product will be the major product?

The answer is: we cannot do it in one step. There is no way to "turn off" steric effects. However, there is a way to do it in a few steps. At the start of this chapter, we learned about the sulfonation reaction (using fuming sulfuric acid to place an SO_3H group onto the ring). We saw that this group can be placed onto the ring, *and* it can be taken *off* the ring very easily. We said that this feature (reversibility) would be VERY important in synthesis problems. Now we are ready to see why.

If we do a sulfonation reaction first, we will expect the SO_3H group to go predominantly in the para position (the major product will be from para substitution):

Now think about what we have done. We have "blocked" the para position. Now if we brominate, the Br is forced to go in the ortho position (because the para position is already taken). So, the reaction places the Br where we want it to go:

Finally, we can do a desulfonation to take off the SO_3H group. To do so, remember that we need to use dilute sulfuric acid:

And now we have made our product. So, here is our entire synthesis:

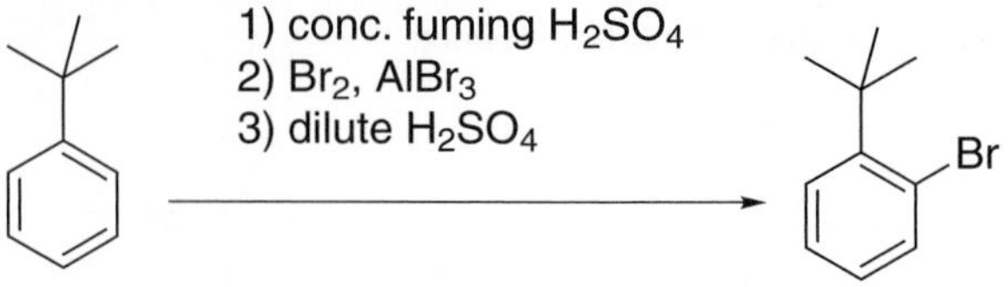

1) conc. fuming H_2SO_4
2) Br_2, $AlBr_3$
3) dilute H_2SO_4

Notice that it took us three steps (where the first step was used to block the para position and the third step was used to unblock the para position). Three steps might seem inefficient, BUT we did not need to rely on isolating minor products. At each step of the way, we were using the major product to move on to the next step.

If you think about what we have done, you should realize that this trick is really very clever. We recognized that we cannot just "turn off" the steric effects. So, instead, we developed a strategy that *uses* the steric effects. Notice that the SO_3H group is not in our final product at all. It was just used temporarily, as a "blocking group." This type of concept is very important in organic chemistry. As you move through the course, you will see a few other examples of blocking groups (in reactions that have nothing to do with electrophilic aromatic substitution). The basic strategy is applicable elsewhere. By temporarily blocking the position where the reaction would primarily occur (and then unblocking after you do the reaction you want), it is possible to form a product that would otherwise be the minor product.

Now let's make sure that we know how to use this:

EXERCISE 3.80. Propose an efficient synthesis for the following reaction:

Answer: We see that we need to place an acyl group in the ortho position. If we just do a Friedel-Crafts Acylation, we will expect the para product to be the major product (because of steric effects). So, we must use a sulfonation reaction to block the para position. Our answer is:

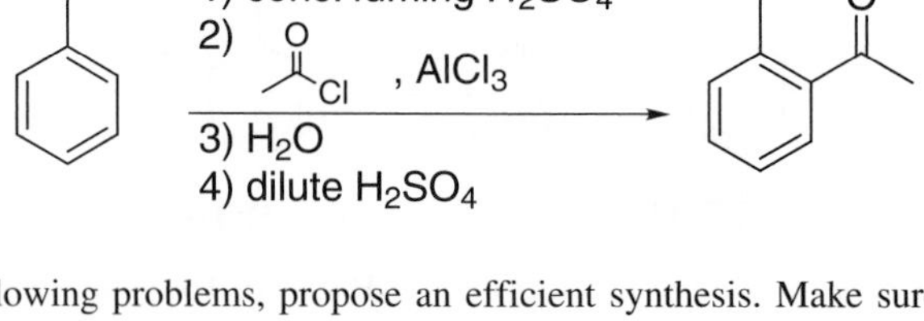

1) conc. fuming H_2SO_4
2) [acyl chloride], $AlCl_3$
3) H_2O
4) dilute H_2SO_4

For each of the following problems, propose an efficient synthesis. Make sure that sulfonation is necessary (I am purposefully giving you at least one problem that does not need sulfonation—to make sure that you understand *when* to use this blocking technique):

3.81.

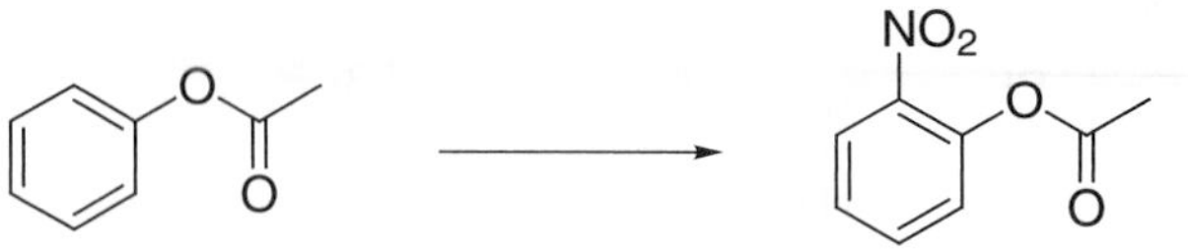

3.82.

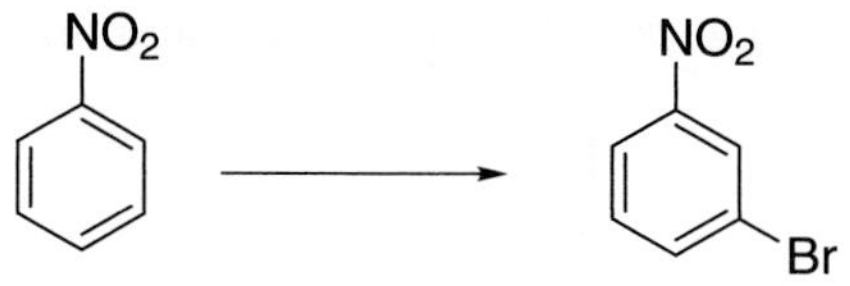

3.83.

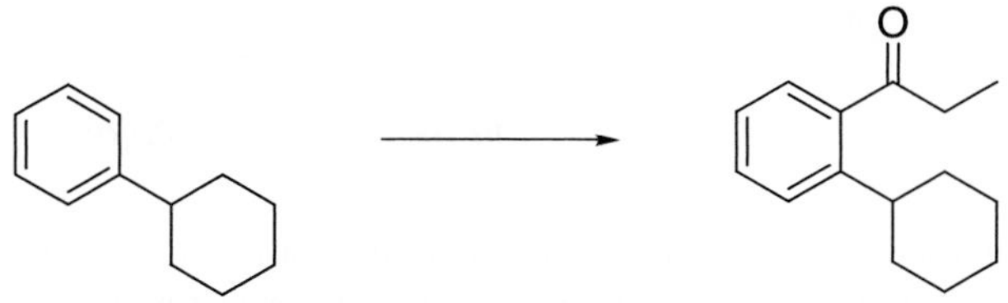

3.84.

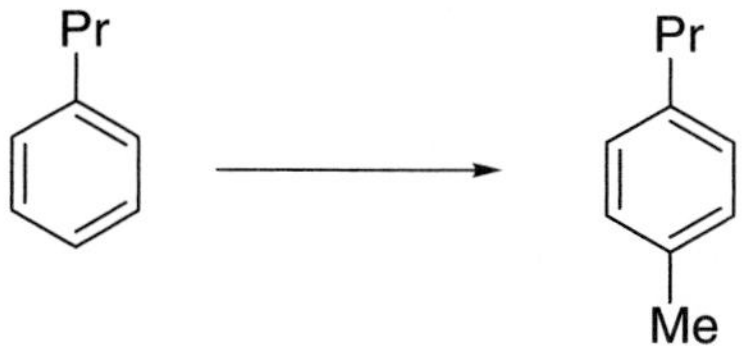

3.85.

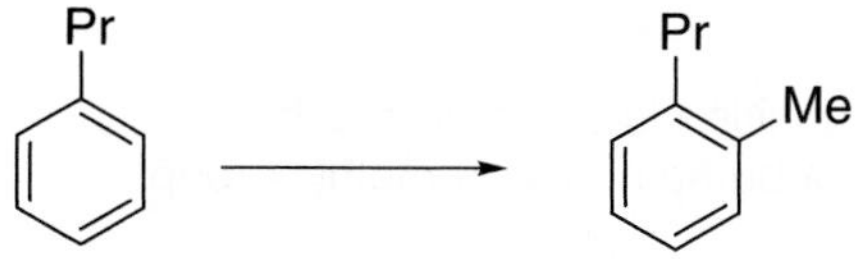

Before we move on to the final section of this chapter, you should be familiar with a few other steric effects. So far, we have seen the steric effects of ONE group on a ring. But what happens when we have two groups on a ring. For example, consider the directing effects of *meta*-xylene:

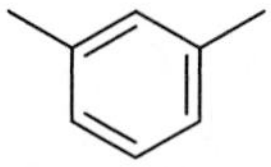

This compound has *two* methyl groups on the ring. Both methyl groups are directing to the same three positions:

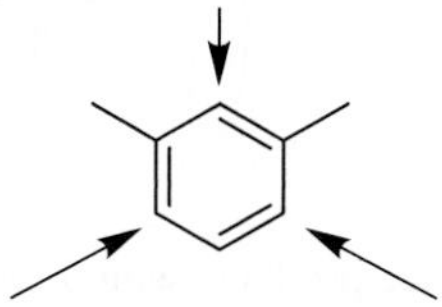

Two of these positions are essentially the same because of symmetry:

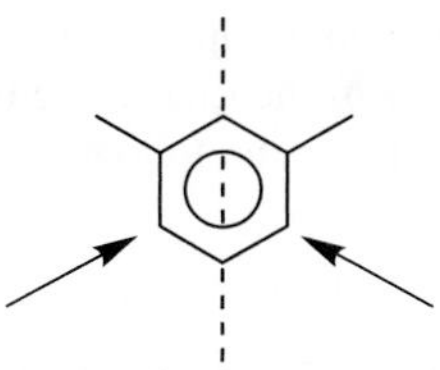

Attacking either of these two positions
would yield the same product.

So, if we brominate this compound, we will expect to get only two products (rather than three):

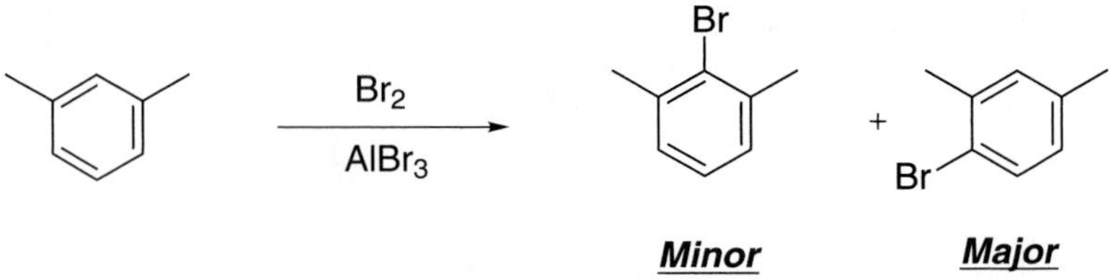
Br
Br2
AlBr3
+
Br
Minor
Major

Notice that we have indicated that one of the products is major. To understand why, we must consider the steric effects. The position in between the two methyl groups is more sterically hindered than the other positions. Therefore, we primarily get just one product.

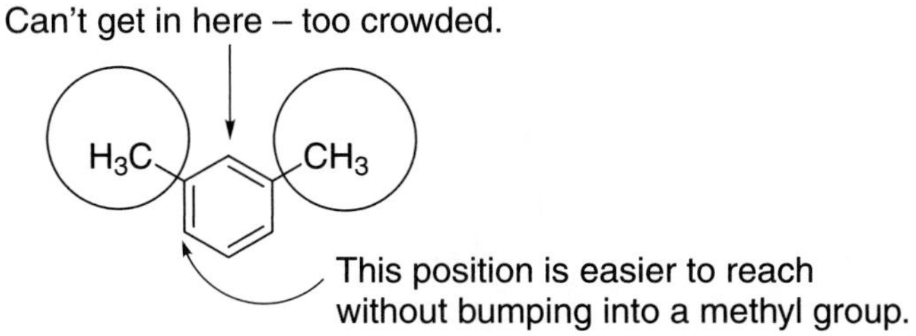
Can't get in here – too crowded.
H3C
CH3
This position is easier to reach
without bumping into a methyl group.

This type of argument can be used in a variety of similar situations. For example, you might remember that we saw the following reaction earlier in this chapter:

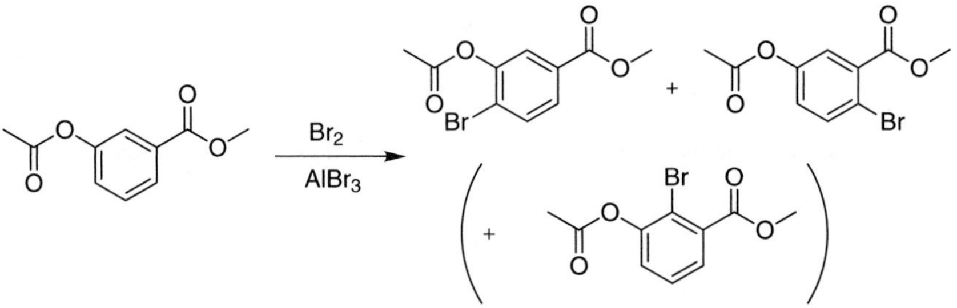
O
O
O
O
Br2
AlBr3
Br
Br
Br

At the time, we said that one of the three products would only be a minor product (the one shown above in parentheses). Now we can understand why it is a minor product. It is the same argument that we just saw—it is an argument of sterics. The starting compound has two groups that are meta to each other, so the spot in between the two groups is sterically hindered.

But suppose you have a disubstituted benzene ring where the two groups are *para* to each other. For example, consider the directing effects for the following compound:

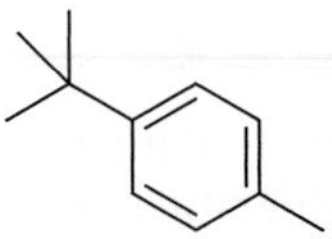

In this case, we have two groups that are *para* to each other. Here is a summary of the directing effects of each group:

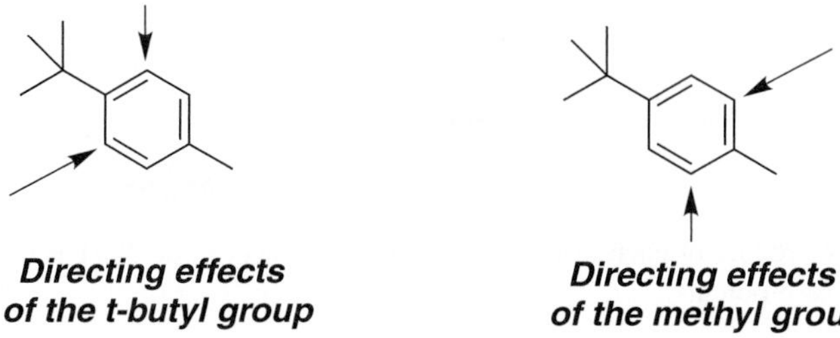

**Directing effects
of the t-butyl group**   **Directing effects
of the methyl group**

So, these two groups are directing to all four potential spots. Both groups are weak activators (alkyl groups). So, when we consider electronics, we don't really see any preference among the possible spots. However, when we consider sterics, we notice that the *tert*-butyl group is very large compared to the methyl group. As a result, we see the following results:

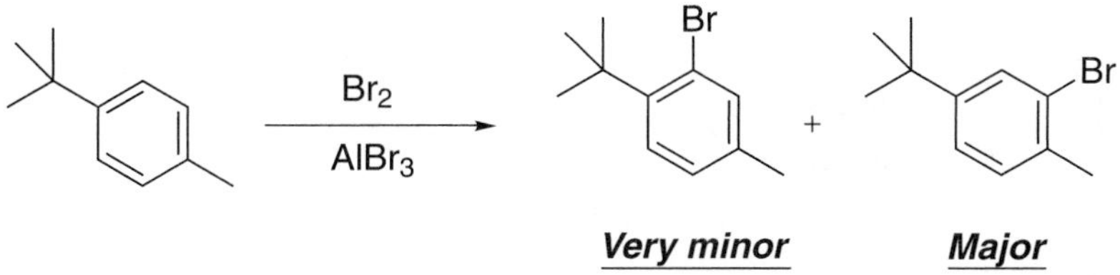

Very minor   **Major**   "

In fact, the *tert*-butyl group is so large that you will find some textbooks that do not even show the minor product above at all. It is so minor that it is almost not worth mentioning.

In this section, we have seen many examples where steric effects play a significant role in determining the product distribution (which are major and which are minor). Now let's see some problems that utilize these principles.

EXERCISE 3.86.   Predict the major product of the following reaction:

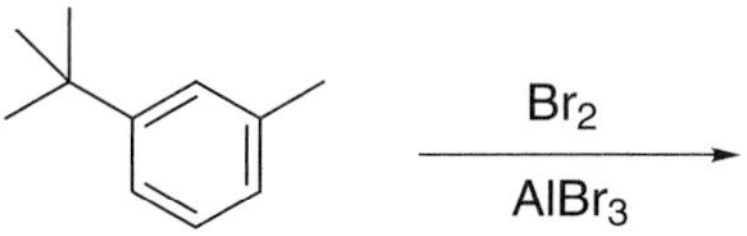

Answer: This example has two groups on the ring: a *tert*-butyl group, and a methyl group. Both are weak activators (ortho-para directors), and both groups are directing to the same positions:

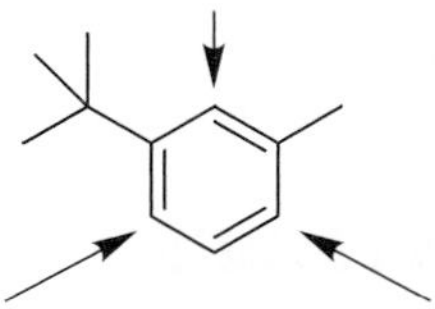

Of these three positions, the one in between the two groups is the most sterically hindered. We won't expect the reaction to take place at that spot very often. Also, the position next to the *tert*-butyl group is fairly hindered, so we won't expect the reaction to take place there either. Thus, we expect the reaction to take place most often at the position next to the methyl group:

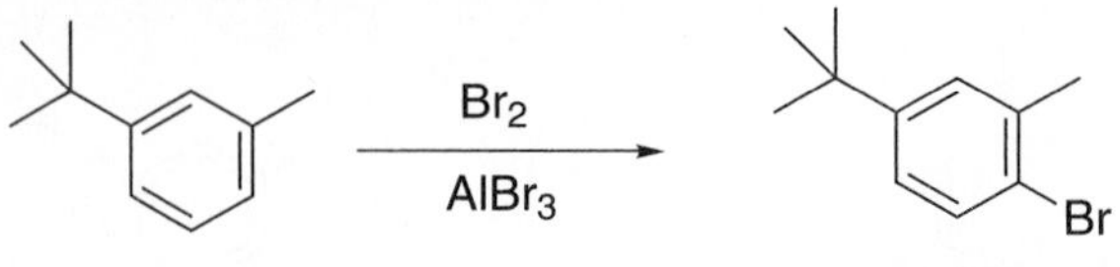

Major

Predict the MAJOR product of each of the following reactions (you do NOT need to show the minor products in these problems):

3.87.

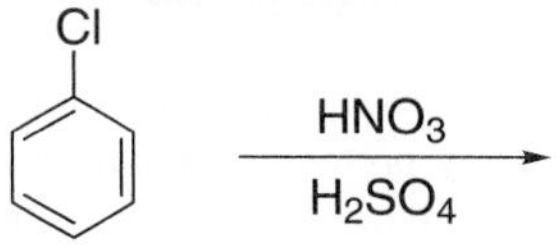

$$\xrightarrow[\text{H}_2\text{SO}_4]{\text{HNO}_3}$$

3.88.

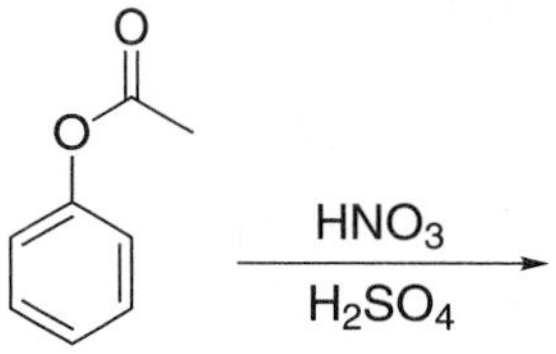

$$\xrightarrow[\text{H}_2\text{SO}_4]{\text{HNO}_3}$$

3.89.

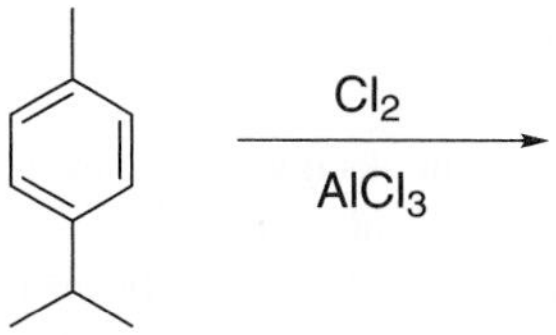

$$\xrightarrow[\text{AlCl}_3]{\text{Cl}_2}$$

3.90.

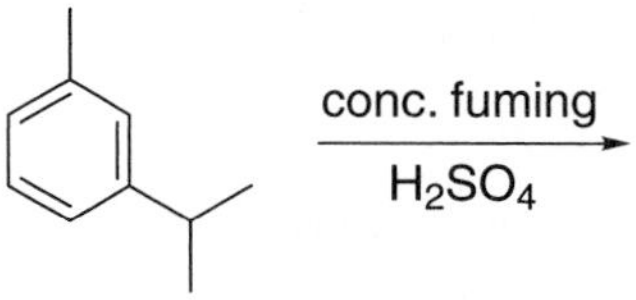

$$\xrightarrow[\text{H}_2\text{SO}_4]{\text{conc. fuming}}$$

3.91.

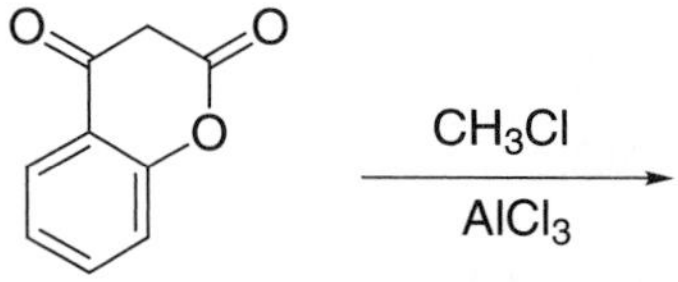

$$\xrightarrow[\text{AlCl}_3]{\text{CH}_3\text{Cl}}$$

3.92.

$$\xrightarrow[\text{AlBr}_3]{\text{Br}_2}$$

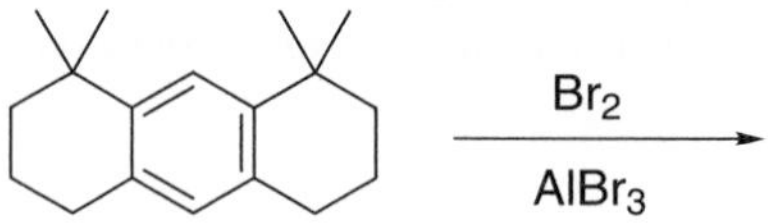

3.9 SYNTHESIS STRATEGIES

In this section, we will tackle some strategies for the toughest problems you can expect to see—synthesis problems. Let's begin with a quick review of the reactions we have seen earlier in this chapter. We have seen how to put many different groups onto a benzene ring:

Carefully look at the chart above and make sure that you know the reagents that you would use to do each of these transformations. If you are not familiar with the reagents, then you will be totally unable to do synthesis problems.

It would be nice if all synthesis problems were just one-step problems, like this one:

Usually, however, synthesis problems require a few steps, where you must place two or more groups onto a ring, like this:

When dealing with such problems, there are many considerations to keep in mind:

- Take a close look at the groups on the ring and make sure you know how to put each group on individually.

- Consider the order of events. In other words, which group do you put on first? After you put on the first group, the directing effects of that group will determine where the next group will go. This is an important consideration because it will affect the relative position of the two groups in the product. In the example above, the two groups are ortho to each other. So we must choose a strategy that places the two groups ortho to each other.

- Take steric effects into account (and determine when you need to use sulfonation as a blocking group).

There are certainly other considerations, but these will help you begin to master synthesis problems. The first consideration above is just a simple knowledge of the reagents necessary to put any group onto a ring. The last two considerations can be summarized like this: electronics and sterics (hopefully, this will make it easy for you to remember these considerations). Whenever you are solving any problem, you must always consider electronic effects and steric effects. As you move through this course, you will find the same theme in every chapter. You will find that you must always consider steric effects and electronic effects.

Let's try to use these considerations to solve the problem we just saw:

Let's begin by making sure we know how to put both of these groups on individually. There are two groups that we need to place onto the ring: a propyl group and a nitro group. The nitro group is easy—we just do a nitration (using sulfuric acid and nitric acid). The propyl group is a bit trickier because we *cannot* use a Friedel-Crafts Alkylation (remember carbocation rearrangements). Instead, we must use a Friedel-Crafts Acylation, followed by a reduction to pull off the C=O double bond. All-in-all then, we have a total of three steps that we will have to do: one step to put on the nitro group and two steps to put on the propyl group.

Now let's focus on electronic considerations. In this case, we can begin to appreciate the importance of "order of events." Imagine that we put the nitro group on first. The nitro group is a meta-director, so the next group will end up going meta to the nitro group. That doesn't work for us because we want the groups to be ortho to each other in the final product. So, we have decided that we cannot put on the nitro group first. Instead, let's try to put the propyl group on first. That should work because the propyl group is an ortho-para director. So the propyl group will direct the incoming nitro group into the correct position (ortho). BUT the propyl group will also direct to the para position. And this is where sterics comes into the picture.

When we look at the steric effects, we encounter a difficulty. The steric effects are not in our favor here. We should expect the following results:

Notice that the compound we want to make is the *minor* product. So, we need a way to get the ortho product as our major product. And we have seen exactly how to do that. We just use sulfonation to block the para position. So, our overall synthesis goes like this:

The answer that we just developed can be summarized like this:

1) (acetyl chloride structure), AlCl₃
2) H₂O
3) Zn [Hg] , HCl, heat
4) conc. fuming H₂SO₄
5) HNO₃ , H₂SO₄
6) dilute H₂SO₄

(benzene) → (product: ring with Pr and NO₂ ortho)

Remember that we arrived at our answer through careful analysis of electronic effects and steric effects. Let's try to use the same type of analysis now to solve a slightly different synthesis problem. Consider the following problem:

(benzene) — **?** → (ring with Pr and NO₂ meta)

In this case, we are dealing with the same two groups as in the previous problem: a propyl group and a nitro group. But in this problem, these two groups need to be *meta* to each other. If we start with the propyl group, we get the wrong directing effects. The propyl groups will direct the nitro group to be ortho and para. So we conclude that the nitro group must go first. The nitro group will direct toward the meta position, which should give us the product we want:

(benzene) — First put on the nitro group → (nitrobenzene) — The nitro group should direct the propyl group into the meta position → (ring with NO₂ and Pr meta)

BUT this doesn't work for another reason (a reason that we have not explained until right now). When we were learning about Friedel-Crafts reactions, you might remember me mentioning that there are a few important limitations to the Friedel-Crafts reaction. At the time, we only discussed a couple of limitations. But we are now ready to understand another important limitation of Friedel-Crafts reactions. It turns out that you cannot do a Friedel-Crafts reaction on a ring that is either moderately deactivated or strongly deactivated. You *can* do a Friedel-Crafts on a weakly deactivated ring (and certainly on an activated ring). But not on a significantly deactivated ring—the reaction just doesn't work. (You can do some other reactions with deactivated rings but not Friedel-Crafts reactions.) With that in mind, take a look at the synthesis we just proposed. It calls for us to use a Friedel-Crafts reaction on a strongly deactivated ring (nitrobenzene). That doesn't work.

So, it would seem that we have an unsolvable problem: if we put the propyl group on first, it will direct to the wrong spot; and if we put the nitro group on first, then we won't be able to put the propyl group on.

Based on everything we have learned, there *is* an answer to this problem. And this answer will help us truly appreciate why we must always carefully consider the order of events. Let's take a closer look.

In this problem, we need to put on two groups: the nitro group and the propyl group. The nitro group can be placed on the ring with one reaction. But we have seen that the propyl group requires two steps (in order to avoid car-bocation rearrangements, we must do a Friedel-Crafts Acylation, followed by a reduction):

The Two Steps for Putting on a Propyl Group

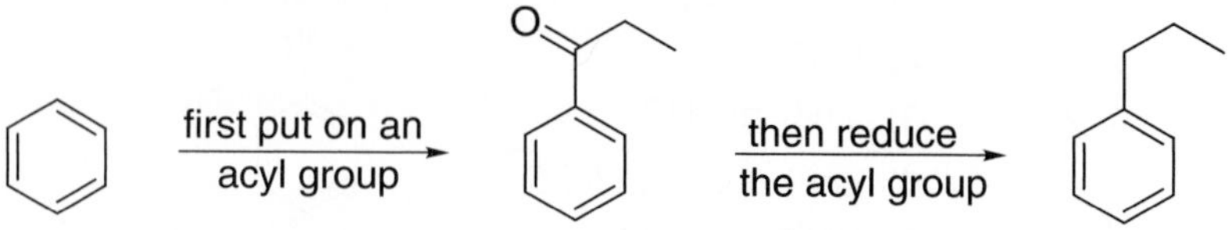

first put on an
acyl group

then reduce
the acyl group

You have probably assumed until now that these two steps (above) must be done one after the other. But that is not the case. It is possible to take a break and do a different reaction *in between* the two steps above. Consider the following order of events, where we place the nitration step in between the two steps for putting on the propyl group:

Step 1 of putting on the propyl group

Step 2 of putting on the propyl group

first put on an
acyl group

then do a nitration
before reducing
the acyl group

then reduce
the acyl group

Nitration goes *in between* the two steps for
putting on the propyl group

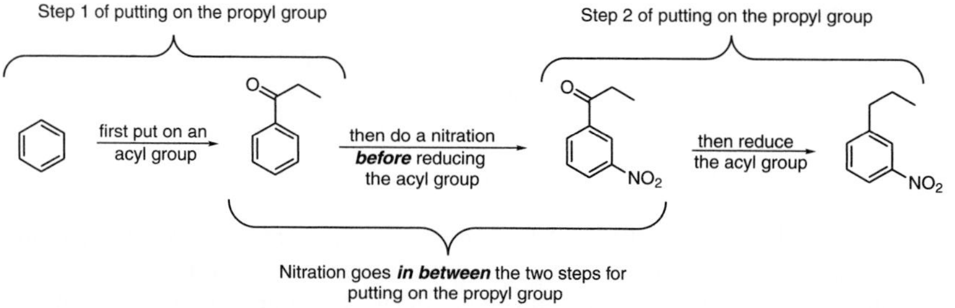

If we do it this way, we can take advantage of the meta-directing effects of the acyl group, and this places the nitro group into the meta position. So, the overall synthesis can be summarized as follows:

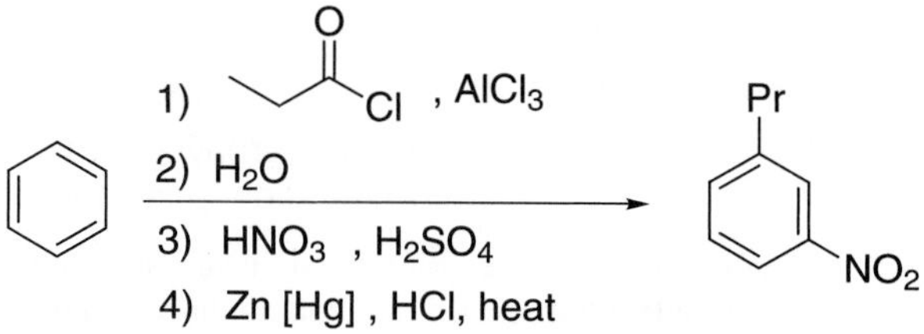

1) $\overset{O}{\underset{Cl}{\parallel}}$, $AlCl_3$

2) H_2O

3) HNO_3 , H_2SO_4

4) Zn [Hg] , HCl, heat

This synthesis teaches us the importance of "order of events." Whenever you are trying to solve a synthesis problem, you should always consider the order of events.

Before you try some problems yourself, let's do one more together:

EXERCISE 3.93. Propose an efficient synthesis for the following transformation:

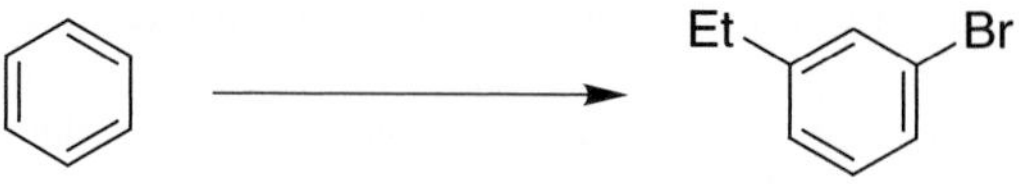

Answer: We must place two groups onto the ring: an ethyl group and a bromine. Let's first make sure that we know what reagents we would need to put each group onto the ring. To put the bromine onto the ring, we would use Br_2 and a Lewis acid. To put the ethyl group onto the ring, we would use a Friedel-Crafts Alkylation (or acylation, followed by a reduction). Whenever we put an ethyl group on a ring, we don't need to worry about carbocation rearrangements, so we can use a simple alkylation (rather than an acylation followed by a reduction).

But we immediately see a serious issue when we consider the directing effects of each group. The bromine is ortho-para directing, so we can't put the bromine on the ring first (if we did, we would not get the groups to be meta to each other). And the ethyl group is also ortho-para directing. So, whichever group we put on first, there would seem to be no way to get these two group to be meta to each other.

UNLESS we use an acylation (rather than an alkylation). If we do that, we will put an acyl group on the ring first. *And acyl groups are meta-directing.* That would allow us to put the bromine in the correct spot. So our strategy would go like this:

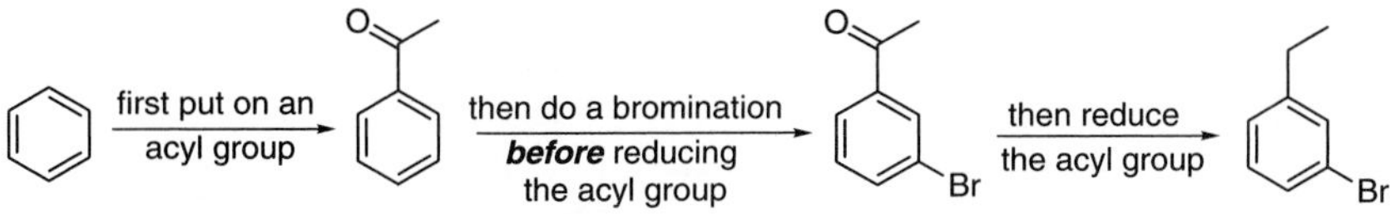

So, our synthesis goes like this:

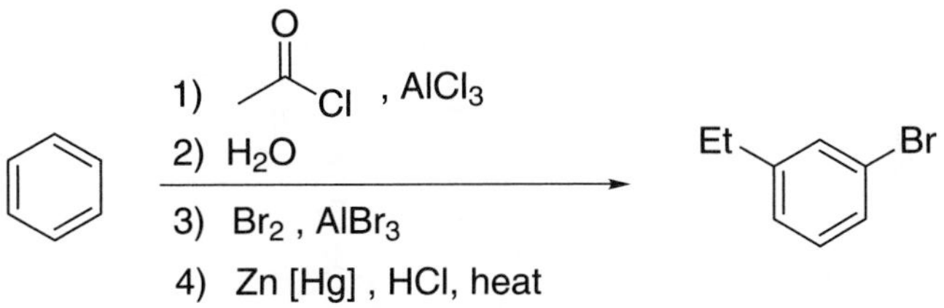

1) (acetyl chloride) , AlCl₃
2) H₂O
3) Br₂ , AlBr₃
4) Zn [Hg] , HCl, heat

For each of the following problems, propose an efficient synthesis. In each problem, you do **not** need to write down the products from each step of your synthesis. Simply write down a list of the reagents you would use and place that list on the arrow (just as we did in the previous examples when we summarized our solution). You might want to use a separate piece of paper to help you work through each of these problems.

3.94.

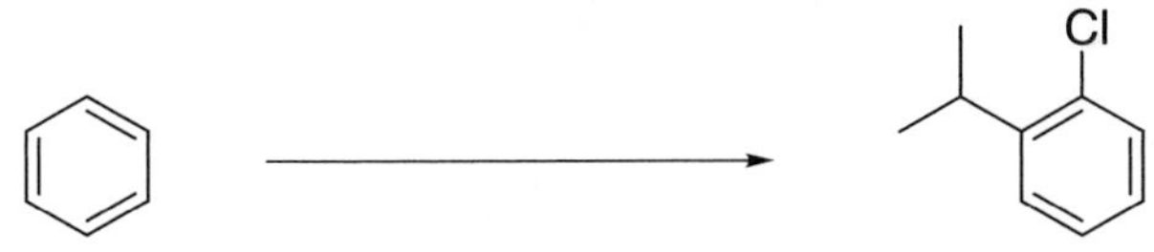

3.95.

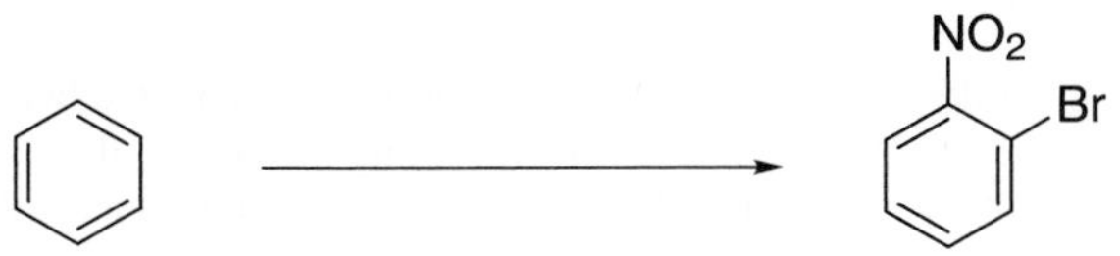

3.96.

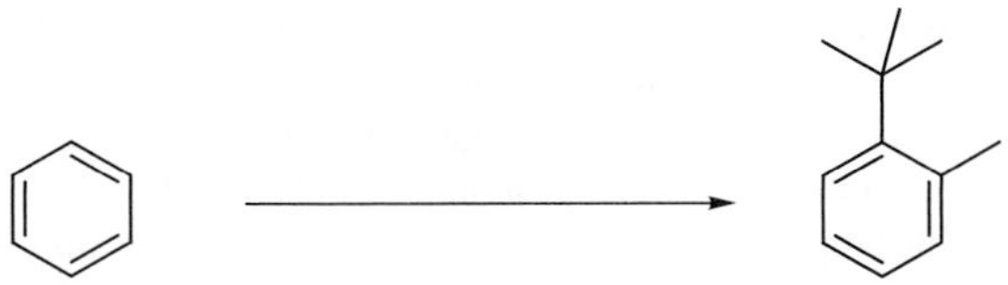

3.97.

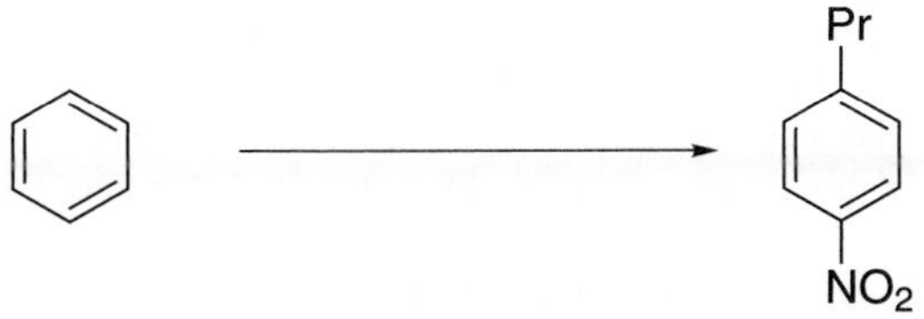

3.98.

3.99.

3.100.

3.101.

3.102.

3.103.

3.104.

Before we end this chapter, it is important that you realize what we have covered here and, more importantly, what we have ***not*** covered. We did not cover everything in your textbook chapter on electrophilic aromatic substitution. As you go through your lecture notes and your textbook chapter, you will find a few reactions that we did not cover here. You will need to go through your textbook and your notes carefully to make sure that you learn those reactions. You should find that we covered 80%, or even 90%, of what you read in your textbook.

The purpose of this chapter was not to cover everything but rather to serve as a foundation for your mastery of electrophilic aromatic substitution. If you went through this chapter, then you should feel comfortable with the steps involved in proposing mechanisms, predicting products, and proposing a synthesis. You should know how directing effects work and how to use them when solving synthesis problems. You should also know about steric effects and how to use them when solving problems.

With all of that as a foundation, you should now be ready to go through your textbook and lecture notes, and polish off the rest of the material that you must know for your exam. Through the foundation that we have developed in this chapter, you should find (hopefully) that the content in your textbook will seem easy.

Do the problems in your textbook. Do all of them. Good luck.

CHAPTER 4

NUCLEOPHILIC AROMATIC SUBSTITUTION

4.1 CRITERIA FOR NUCLEOPHILIC AROMATIC SUBSTITUTION

In the previous chapter, we learned all about *electrophilic* aromatic substitution reactions. At the time, we called it "electrophilic" aromatic substitution because the aromatic ring was attacking an electrophile. As an example, consider the following mechanism:

SIGMA COMPLEX

This mechanism shows that the aromatic ring functions as a nucleophile and attacks an electrophile (look at step 1 of the mechanism). We saw ways to make the electrophile more electrophilic, and we saw ways to make the nucleophile (the aromatic ring) more nucleophilic. But in all of those re-actions, the aromatic ring always functioned as the nucleophile. In this short chapter, we will look at the flipside: is it possible for an aromatic ring to function as an electrophile rather than a nucleo-phile? In other words, is it possible for the aromatic ring to be so electron-poor that it can get at-tacked by a nucleophile? The answer is: yes.

But to get this kind of reaction, we will need to meet three very specific, necessary criteria. Let's look closely at each one of these criteria:

1. The ring must have a very powerful electron-withdrawing group. The most common exam-ple is the nitro group:

We saw in the previous chapter that the nitro group is a strong deactivator toward electrophilic aromatic substitution because the nitro group very powerfully withdraws electron density from the ring (by resonance). This causes the electron density in the ring to be very poor:

In Chapter 3, we wanted the aromatic ring to function as a nucleophile. And we saw that the effect of a nitro group is to *deactivate* the ring. But now, in this chapter, we want the ring to act as an *electrophile*. So the effect of the nitro group is a very good thing. In fact, it is *necessary* to have a nitro group if you want the ring to function as an electrophile. The presence of the nitro group is the first criterion that you must have in order for the ring to function as an electrophile. Now, let's look at the second criterion:

2. There must be a leaving group that can leave.

To understand this, let's think back to what happened in the previous chapter, when the ring always functioned as a nucleophile. We saw that all the reactions from the previous chapter could be summarized like this: E^+ comes on the ring, and then H^+ comes off (or, in other words: attack, then deprotonate). But now, in this chapter, we want the ring to function as an electrophile. So we are trying to see if we can get a nucleophile (Nuc^-) to attack the ring. If we can make it happen, and a nucleophile (with a negative charge) actually does attack the ring, then something with a negative charge is going to have to come off of the ring. We can summarize it like this: Nuc^- comes on the ring, and X^- comes off.

There is one main difference between the mechanism here and the one we saw in Chapter 3. The difference is in the kind of charges we are dealing with. In the previous chapter, we dealt with something positively charged coming onto the ring to form a positively charged sigma complex, and then H^+ came off the ring to restore aromaticity. In those mechanisms, everything was positively charged. But now, we are dealing with negative charges. A nucleophile with a negative charge will attack the ring to form some kind of negatively charged intermediate. That intermediate must then expel something negatively charged. And that explains the second criterion for this reaction to occur: we need the ring to have some leaving group that can leave with a negative charge.

If there is no leaving group that can leave with a negative charge, then the ring will have no way of reforming aromaticity. And we cannot just kick off H^- because H^- is a terrible leaving group. NEVER kick off H$^-$. If you are a bit rusty on leaving groups, you might want to go back to first-semester material and quickly review which groups are good leaving groups.

3. And the final criterion is: the leaving group must be ortho or para to the electron-withdrawing group:

To understand why, we will need to take a closer look at the mechanism. In the upcoming section, we will explore the mechanism of this reaction so that we can understand this last criterion. For now, let's just make sure that we can identify when all three criteria have been met. Once again, the three criteria are:

1. There must be an electron-withdrawing group on the ring.
2. There must be a leaving group on the ring.
3. The leaving group must be ortho or para to the electron-withdrawing group.

Now let's get some practice looking for all three criteria:

EXERCISE 4.1. Predict whether the following conditions will produce a nucleo-philic aromatic substitution reaction.

Answer: In order to have a nucleophilic aromatic substitution, all three criteria must be met. We look for all three criteria.

We look at the ring, and we see that it does have a nitro group. Therefore, the first criterion has been met.

We then look for a leaving group. There is NO leaving group here. Methyl is NOT a leaving group. Why not? Because a carbon with a negative charge is a *terrible* leaving group. Never kick off C$^-$. So criterion 2 has not been met.

Therefore, we conclude that the conditions above will NOT produce a nucleophilic aromatic substitution reaction.

For each of the following problems, predict whether the conditions will produce a nucleophilic aromatic substitution reaction. If you determine that not all three criteria are met, then simply write "no reaction."

4.2.

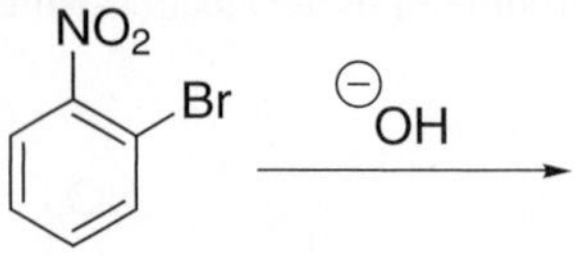

4.3.

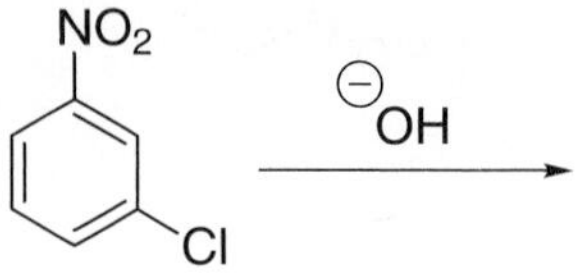

4.4.

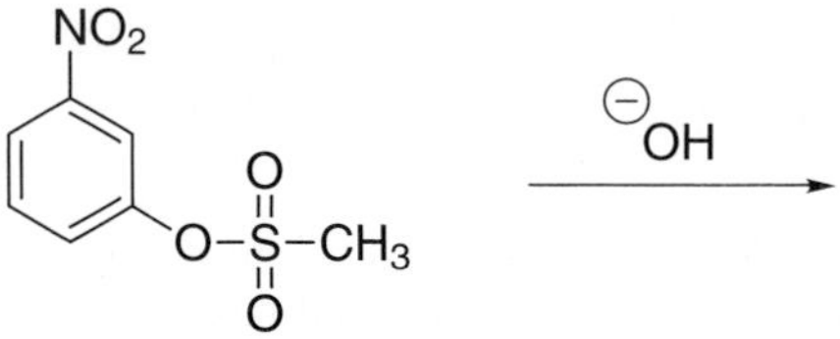

4.5.

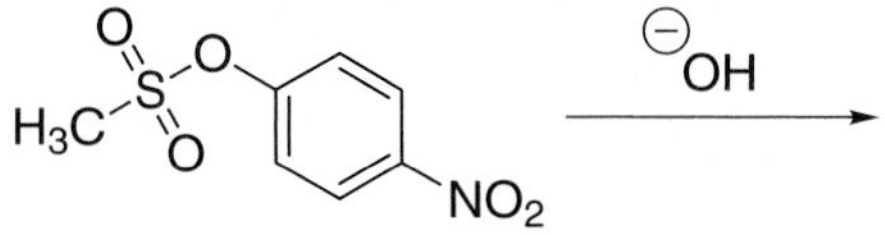

4.6.

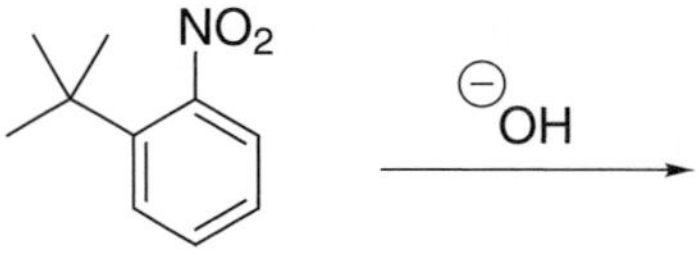

4.7.

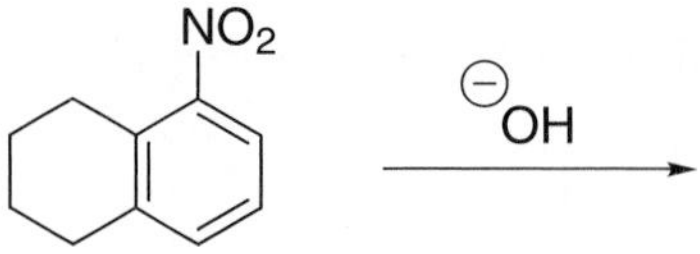

4.2 S$_N$AR MECHANISM

What is the mechanism of a nucleophilic aromatic substitution reaction? Let's explore the possibilities.

It cannot be an S$_N$2 mechanism because you cannot do an S$_N$2 reaction at an sp^2 hybridized carbon:

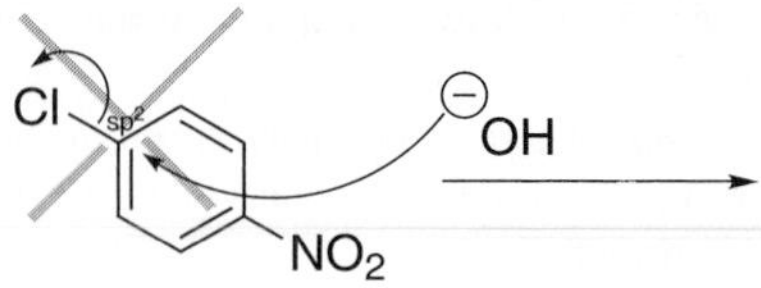

S$_N$2 reactions only work on sp^3 hybridized centers. So our reaction cannot be an S$_N$2 mechanism. What about an S$_N$1? That would require the loss of the leaving group *first* to form a carbocation:

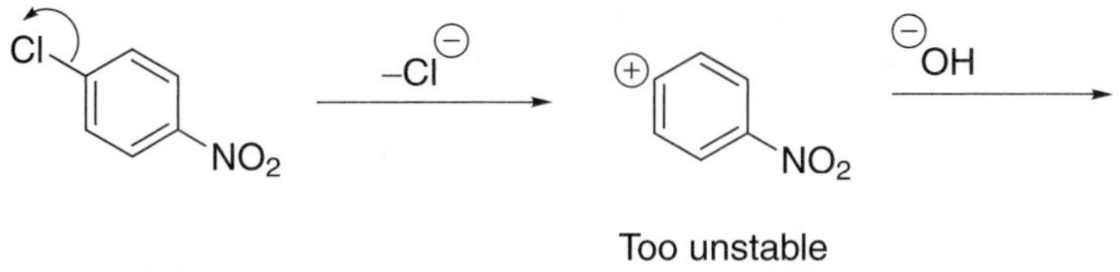

Too unstable

This kind of carbocation is not stabilized by resonance. Since it is unstable, it is therefore a very high-energy intermediate. So, we don't expect the leaving to leave if it means creating an unstable intermediate. The leaving group just won't leave. Therefore, we don't expect the mechanism to be an S$_N$1 mechanism either.

So, if it's not S$_N$2 and its not S$_N$1, then what is it? And the answer is: it's a new mechanism, called S$_N$Ar. In many textbooks, it is called an *addition-elimination* mechanism. And it goes like this:

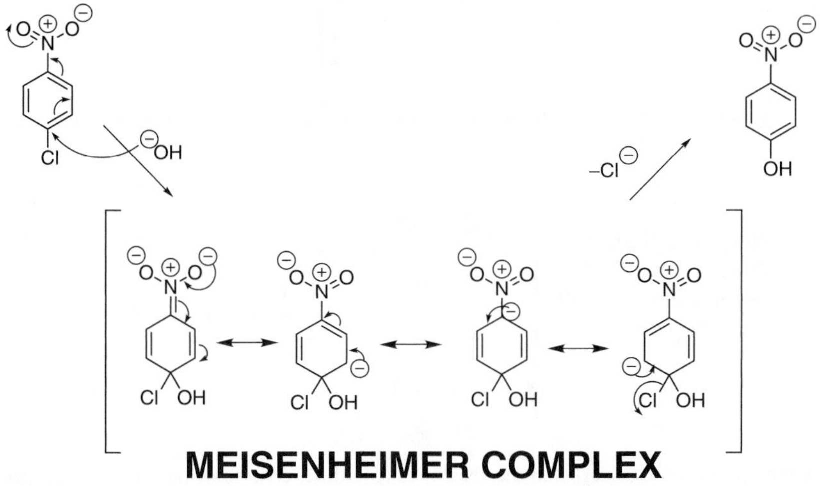

MEISENHEIMER COMPLEX

In step 1, the ring gets attacked by a nucleophile to form a resonance stabilized intermediate. This intermediate should remind us of the intermediate in an electrophilic aromatic substitution reaction (the sigma complex), but the main difference is that our intermediate here is *negatively* charged (a sigma complex is positively charged). So we can't call this a sigma complex. Instead, we give it a new name, and we call it a Meisenheimer complex. This Meisenheimer complex then loses a leaving group (chloride) to reform aromaticity.

Let's take a close look at the Meisenheimer complex, and let's focus our attention on one particular resonance structure:

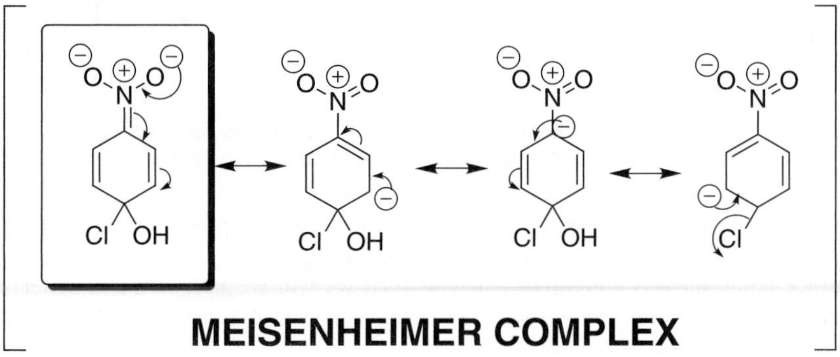

MEISENHEIMER COMPLEX

The resonance structure that has been circled is special because it places the negative charge on an *oxygen* atom. Since the negative charge is spread out over three carbon atoms *and an oxygen atom*, the negative charge is fairly stabilized by resonance. You should think of the reaction like this: A nucleophile attacks the ring, kicking the negative charge up into the reservoir:

Negative charge
goes up onto
the oxygen atom
in the nitro group

Reservoir

Then, the reservoir releases its load by pushing the electron density back down onto a leaving group:

Negative charge
comes down from
reservoir to kick off
a leaving group

And now we are ready to understand the reason for the third criterion (that the leaving group must be ortho or para to the electron-withdrawing group). Now we can understand that the reservoir is available only if the nucleophile attacks at the ortho or para and positions. If we attack at the meta position, there is no way to place the negative charge up onto the reservoir:

Meta

In this intermediate, the negative charge is spread out over three
carbon atoms, but *not* on the other oxygen atom in the nitro group

Therefore, the intermediate is not stabilized. So the reaction doesn't happen. And that is why the leaving group must be ortho or para to the electron-withdrawing group.

As a side point, there is a group of reactions (that we will see later in the semester) that are VERY similar to the reaction we just saw. For example, consider the following mechanism:

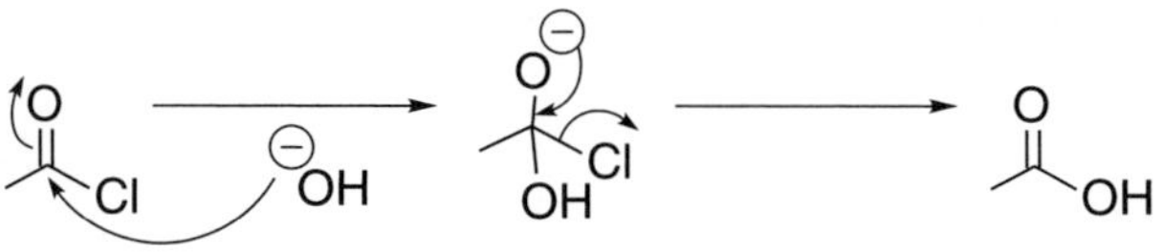

This reaction follows the same sequence of basic moves as an S$_N$Ar mechanism. The sequence is: attack, then loss of a leaving group. Also notice the reservoir concept here. The nucleophile pushes its negative charge up onto the oxygen, which functions as a momentary reservoir for the charge:

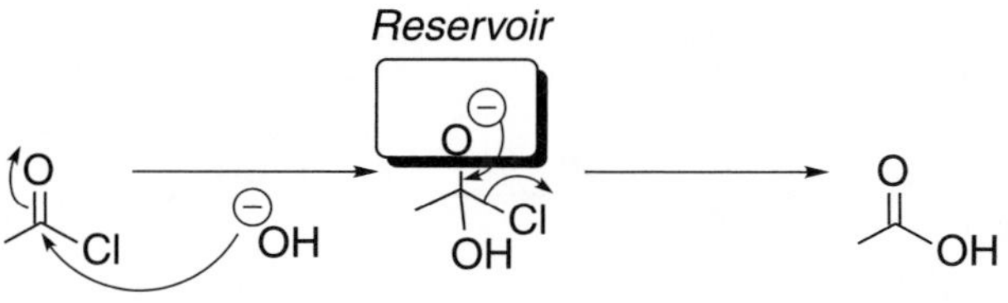

Reservoir

Then the reservoir releases the charge by pushing it down onto a leaving group.

EXERCISE 4.8.   Draw the mechanism for the following reaction:

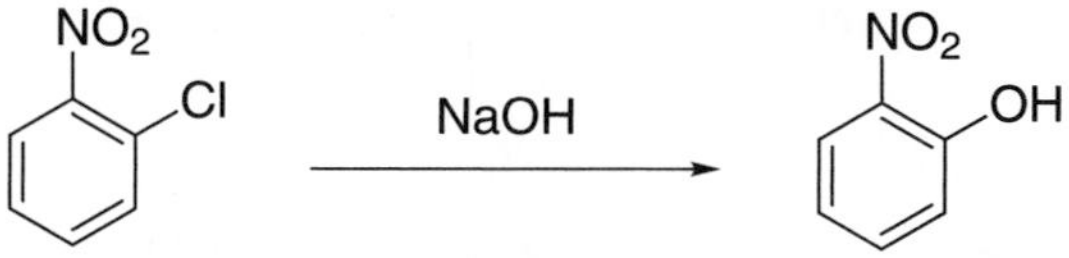

Answer: We have all three criteria for an S$_N$Ar mechanism: we have (1) an electron-withdrawing group (NO$_2$) and (2) a leaving group, and (3) they are ortho to each other.

In a nucleophilic aromatic substitution reaction, the nucleophile (hydroxide) attacks at the position where the leaving group is pushing the electrons up onto the reservoir:

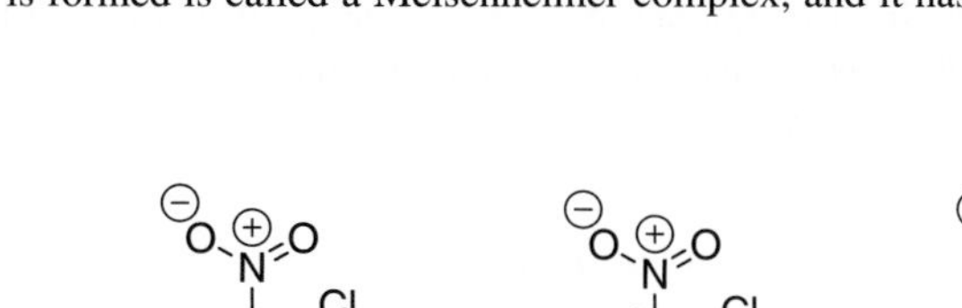

This intermediate that is formed is called a Meisenheimer complex, and it has resonance structures that we should draw:

Finally, the leaving group gets kicked off to give the product. So, the entire mechanism looks like this:

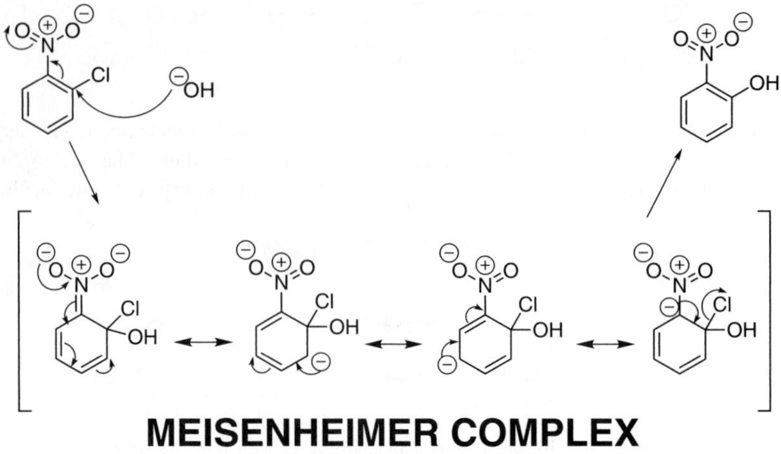
MEISENHEIMER COMPLEX

Propose a mechanism for each of the following transformations:

4.9. 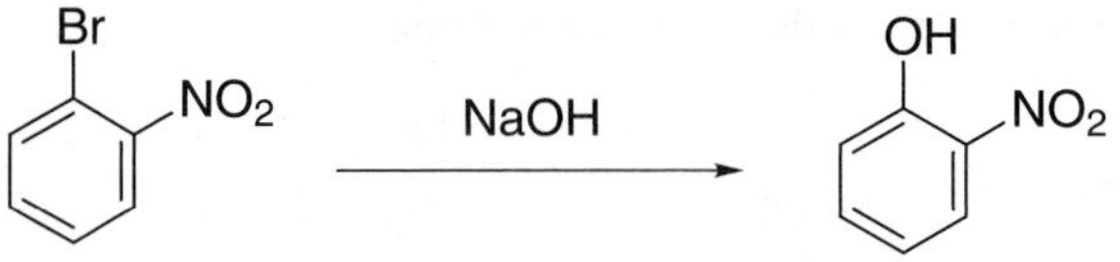

4.10. 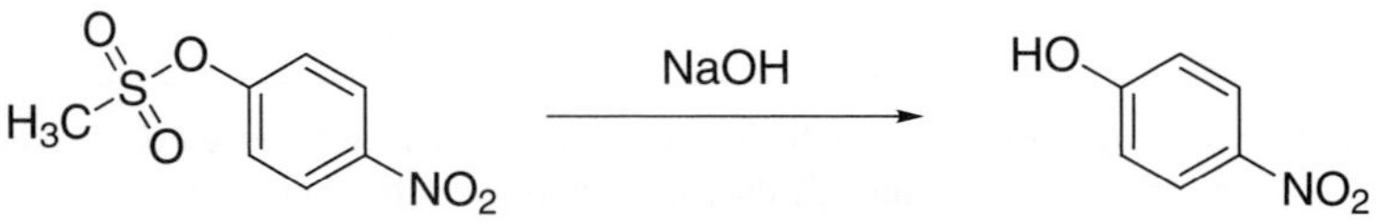

4.11. 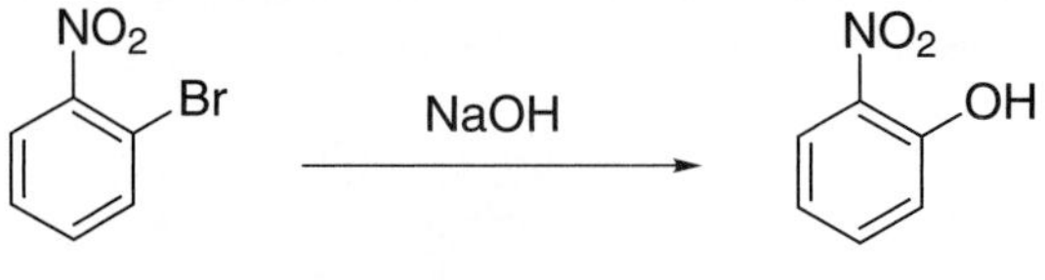

4.3 ELIMINATION-ADDITION

In the previous section, we discussed the three criteria that you need in order to get an S_NAr mechanism. The obvious question is: can you ever get a reaction if you do not have all three criteria? For example, what if there is no electron-withdrawing group?

If we take chlorobenzene and we mix it with hydroxide, we do not get a reaction:

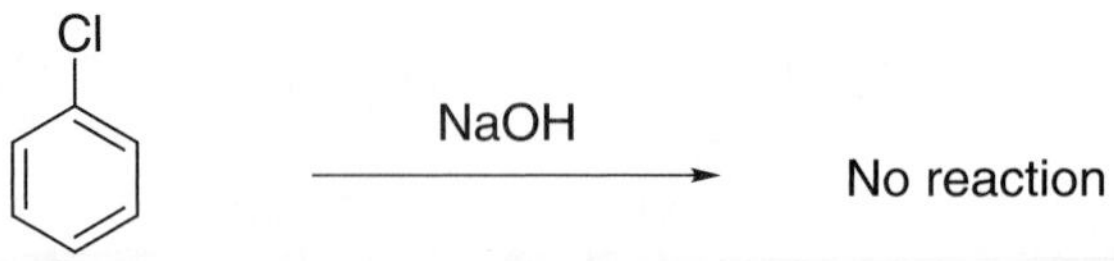
Cl
NaOH
No reaction

The hydroxide does not attack to kick off the leaving group because there is no "reservoir" to hold the electron density for a moment. In fact, if we try heating the reaction up a little bit, we still don't get a reaction.

But then we keep heating it, and when we get to 350° C, we do get a reaction:

This reaction is commercially important because it is an excellent way of making phenol. The reaction above is called the Dow Process.

We can use this same process to make *aniline* (that is the common name for aminobenzene):

Aniline

We don't even need high temperatures to make aniline. We just use NH_2^- in liquid ammonia. So we have a serious question: if there is no "reservoir" to temporarily hold the electron density, then how does this reaction work? What is the mechanism?

To understand the mechanism, chemists have used an important technique called isotopic labeling. All elements have isotopes (for example, deuterium is an isotope of hydrogen because deuterium has an extra neutron in the nucleus). Carbon also has some important isotopes. ^{13}C is an important isotope because we can easily determine the position of a ^{13}C in a compound using NMR spectroscopy. So if we enrich a specific spot with ^{13}C, then we can follow where that carbon atom goes during the reaction. For example, let's say we take chlorobenzene, and we enrich one particular site with ^{13}C:

The site with the asterisk is the place where we put the ^{13}C. When we say that we enriched that spot with ^{13}C, we mean that most of the molecules in the flask have a ^{13}C atom in that position.

Now let's see what happens to that isotopic label as the reaction proceeds. After running the reaction, here are the results that we see:

This seems quite strange: how does the isotopic label "move" its position? We will not be able to explain this with a simple nucleophilic aromatic substitution. Even if we could somehow ignore the issue of not having a reservoir where we can place the electron density during the reaction, we would still not be able to explain the isotopic labeling results.

So here is a proposal that explains the isotopic labeling experiments. Imagine that in step 1, the hydroxide ion acts as a base first, instead of acting as a nucleophile, and we get an elimination reaction:

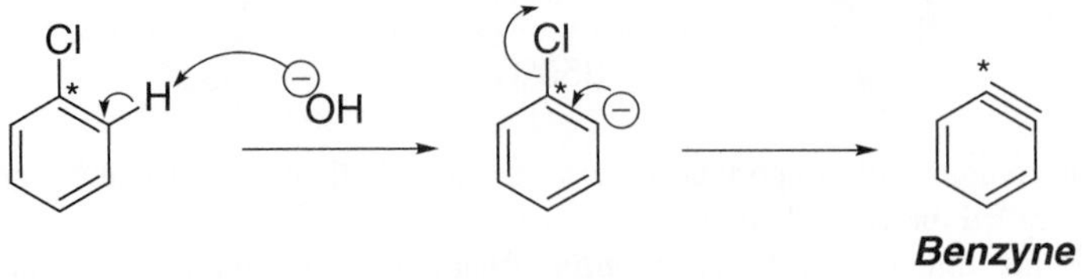

Benzyne

This would form a very strange-looking (and very reactive) intermediate, which we will call benzyne. Next, another hydroxide ion swoops in, this time acting as a nucleophile to attack benzyne. But there are two places it can attack:

Hydroxide can attack like this

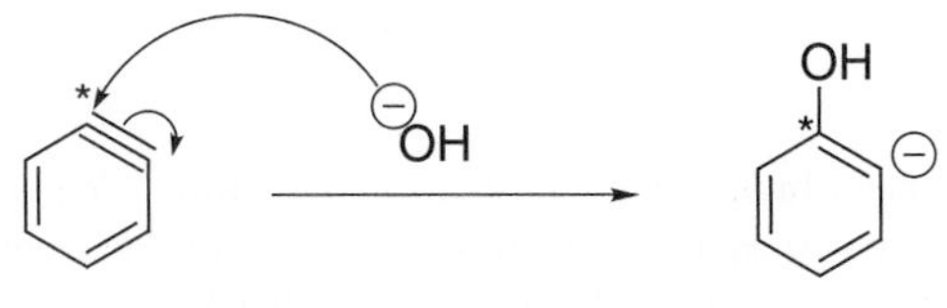

or like this

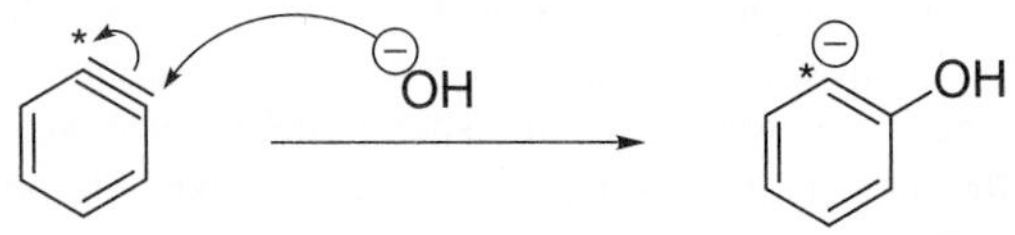

And there is no reason to prefer one site over the other, so we must assume that these two pathways occur with equal probability. That would give a 50–50 mixture of the following two anions:

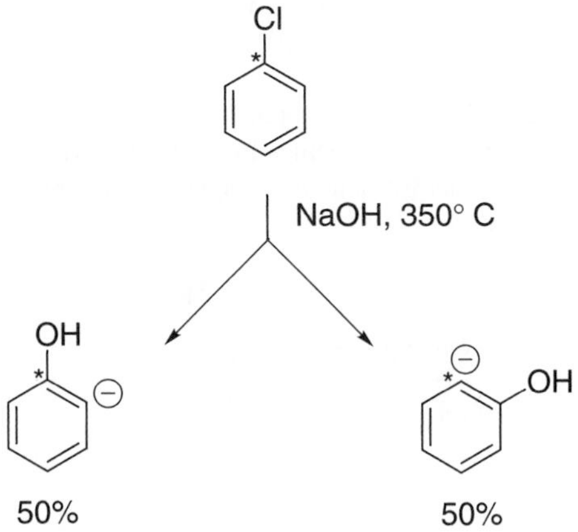

NaOH, 350° C

50%   50%

In the last step, we can provide these anions with a proton from water (water was formed in the first step when the hydroxide ion acted as a base to pull off a proton):

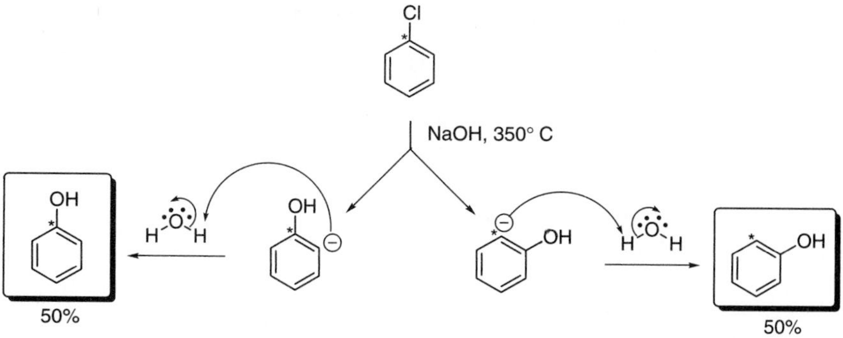

50%   50%

This proposed mechanism is essentially an elimination followed by an addition. So, it makes sense that we call this mechanism an **elimination-addition** reaction (as compared with the S$_N$Ar mechanism, which was called *addition-elimination*). This mechanism certainly seems a bit off the wall when you think about. Benzyne? It looks like a terrible intermediate. But chemists have been able to show (with other experiments) that benzyne is in fact the intermediate of this reaction. Your textbook or instructor will most likely provide the additional evidence for the short-lived existence of benzyne (we use a trapping technique involving a Diels-Alder reaction). If you are curious about the evidence, you can look in your textbook. For now, let's make sure that we can predict products of this reaction.

EXERCISE 4.12.   Predict the products for the following reaction:

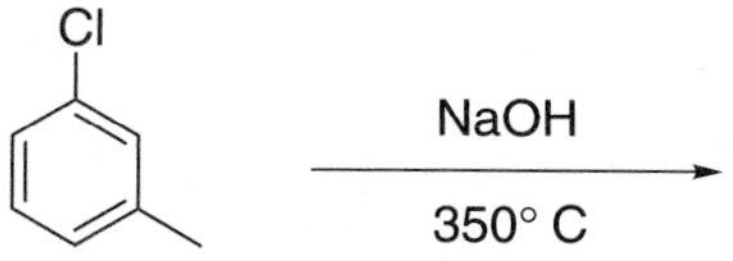

Answer: The ring does *not* have an electron-withdrawing group, so we are not dealing with an S$_N$Ar mechanism. Rather, we must be dealing with an elimination-addition mechanism.
  The first step would be to do the elimination, which could happen to either side of the chlorine atom:

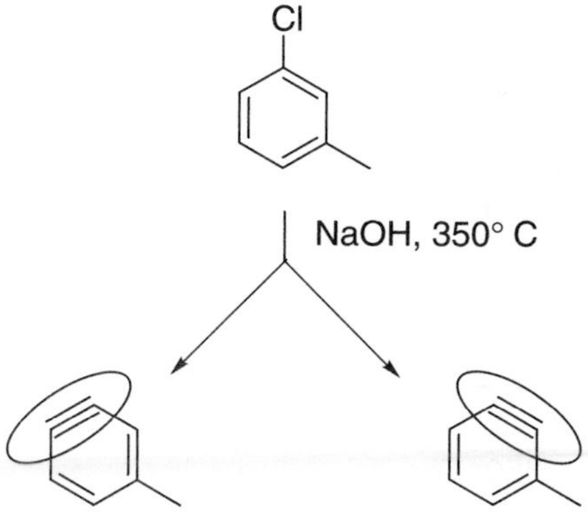

Then, we can add across either triple bond above, giving us the following products:

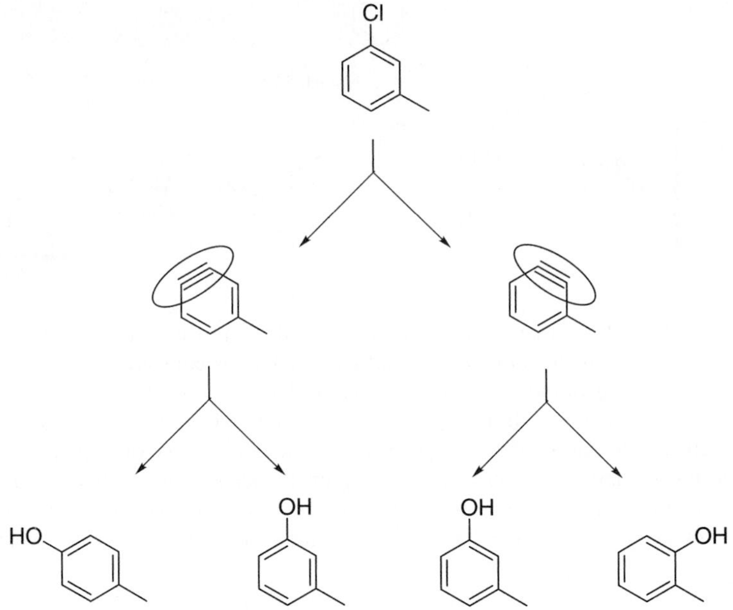

If you look closely at these products, you will see that the middle two are the same. So, we expect the following three products from this reaction:

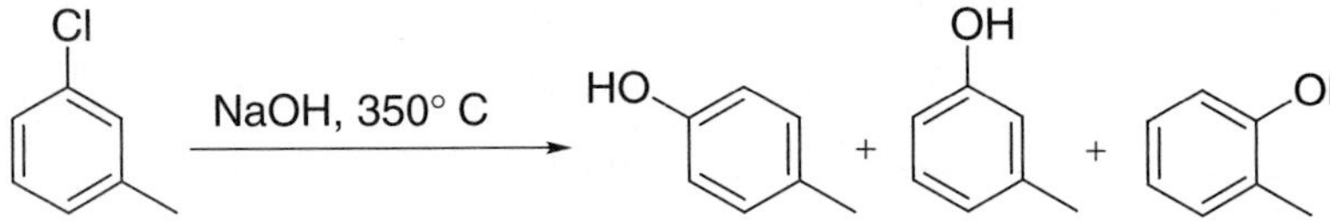

Predict the products for each of the following reactions. Just to keep you on your toes, we will throw in some problems that go through an addition-elimination mechanism (S_NAr), rather than an elimination-addition. In each case, you will have to decide which mechanism is responsible for the reaction (based on whether or not you have all three criteria for an S_NAr mechanism). Your products will be based on that decision.

4.13.

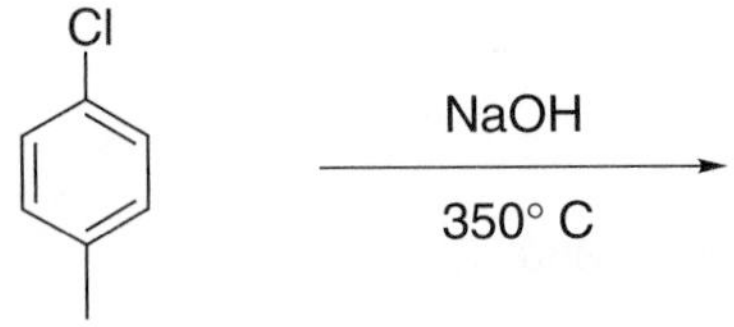

4.14.

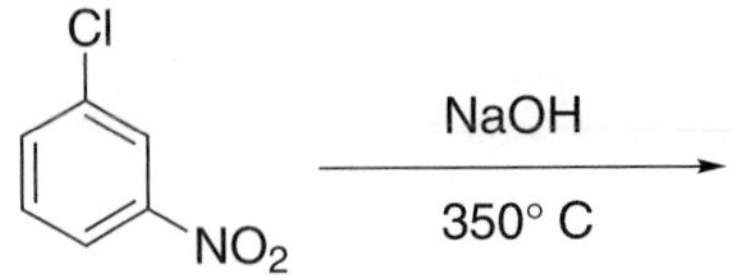

4.15.

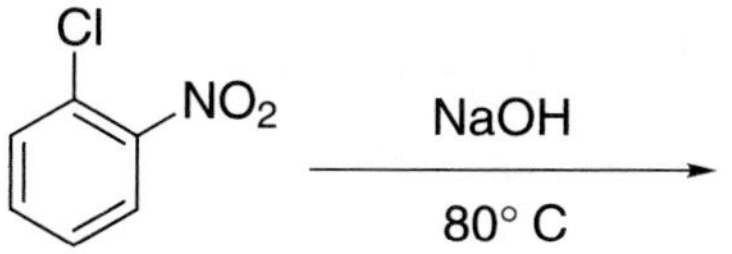

4.16.

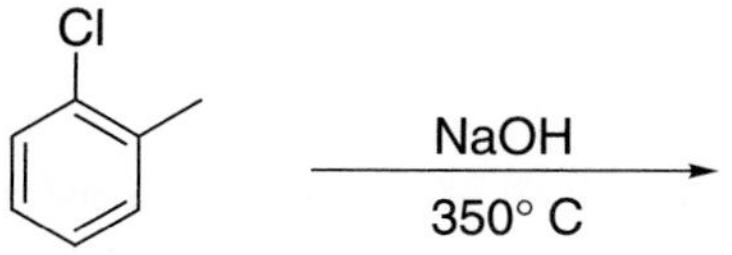

4.17.

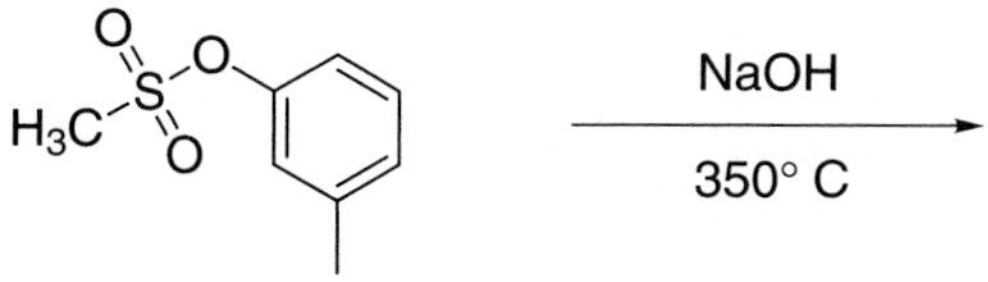

4.18.

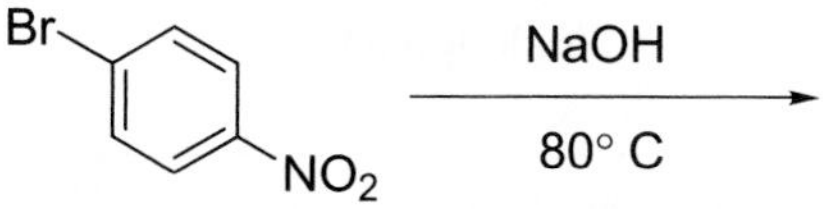

   In this section, we have seen how to place an OH or an NH_2 onto an aromatic ring. This is important because we did not see how to do either of these two things in the previous chapter. Here is a summary of how to put an OH or an NH_2 onto a ring. In each case, it is a two-step process that starts with putting a Cl on the ring:

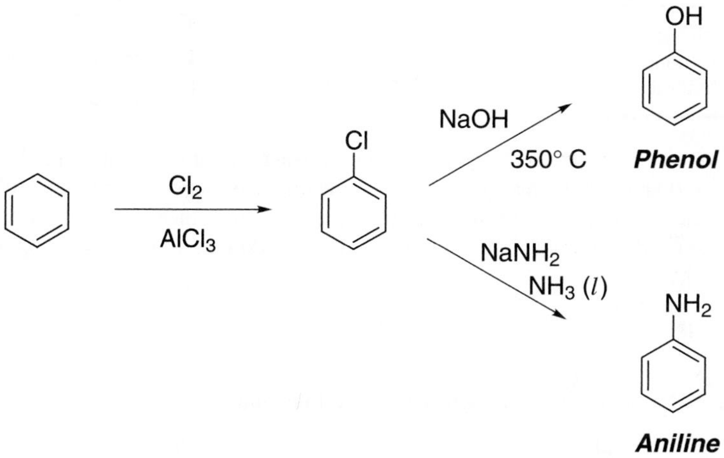

We begin with chlorination of benzene (which is just an electrophilic aromatic substitution reaction), followed by an elimination-addition reaction. When we do the elimination-addition step, we must carefully choose the reagent to use. If we use NaOH, we will form phenol, and if we use $NaNH_2$, we will form aniline (shown in the scheme above).

   The synthesis of aniline, shown above, will show up again later in this course. When we learn about the chemistry of amines (in Chapter 8), we will see many reactions that use aniline as a starting material. When we get to that part of the course, it will be very helpful (for solving synthesis problems) if you remember how to make aniline from benzene. So make sure to remember this reaction. You will definitely see it again later.

4.4 MECHANISM STRATEGIES

So far, we have seen three different mechanisms involving aromatic rings:

1. Electrophilic aromatic substitution
2. S_NAr (sometimes called addition-elimination)
3. Elimination-addition.

When you are given a problem, you must be able to look at all of the information and determine which of the three mechanisms is operating. This is not difficult to do. Here is a simple chart that shows the thought processes involved:

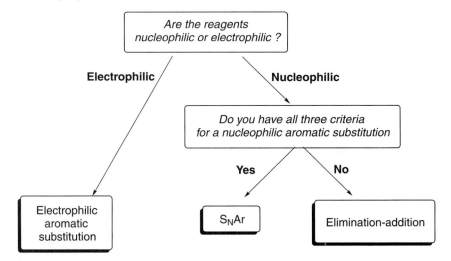

You first look at the reagents that you are using to react with the aromatic ring. If the reagents are electrophilic (like all of the reagents we saw in the previous chapter), then you will get an electrophilic aromatic substitution. But if the reagents are nucleophilic, then you have to decide between an S_NAr reaction and an elimination-addition reaction. To do that, look for the three criteria necessary for an S_NAr reaction.

Let's see an example:

EXERCISE 4.19. Propose a mechanism for the following reaction:

Cl→(ring with NO₂) + NaOH, 350° C → OH (ring with NO₂)

Answer: We begin by looking at the reagents. Hydroxide is a nucleophile, so we do NOT get an electrophilic aromatic substitution. Now, we must decide between an S_NAr mechanism and an elimination-addition mechanism. So, we look for the three criteria that we need for an S_NAr reaction. (1) We do have an electron-withdrawing group, and (2) we do have a leaving group, BUT (3) the electron-withdrawing group and the leaving group are NOT ortho or para to each other. That means that we cannot get an S_NAr mechanism. (When the electron-withdrawing group and the leaving group are meta to each other, we don't have the "reservoir" to use.) Therefore, the mechanism must be an elimination-addition:

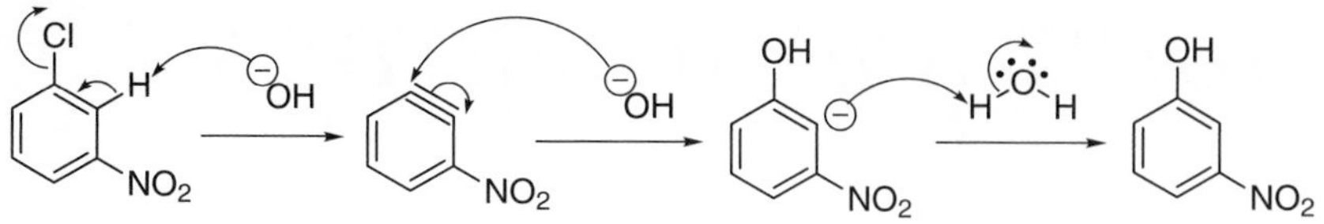

In this particular example, it is true that we would expect a total of three products from an elimination-addition mechanism:

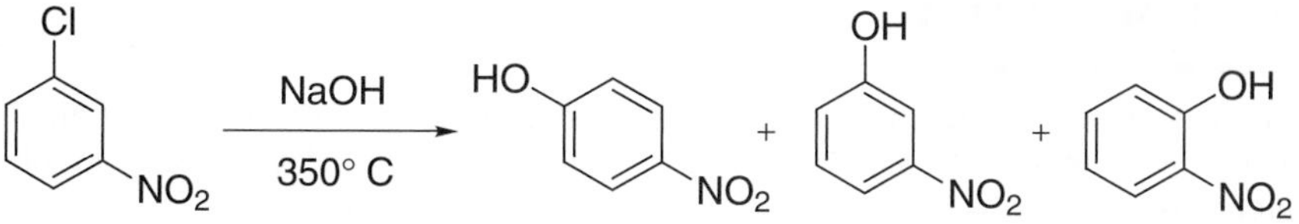

But keep in mind that mechanism problems do not always show you all of the products. The problem will typically show you just one product, and you will need to show the mechanism for forming that product (and only that product). In some cases, it might even be a minor product. But the problem is not making any claims that the product is major or minor. A mechanism problem is simply asking you to justify "how" the product was formed, regardless of how much of it was actually formed in the reaction.

Propose a mechanism for each of the following reactions:

4.20.

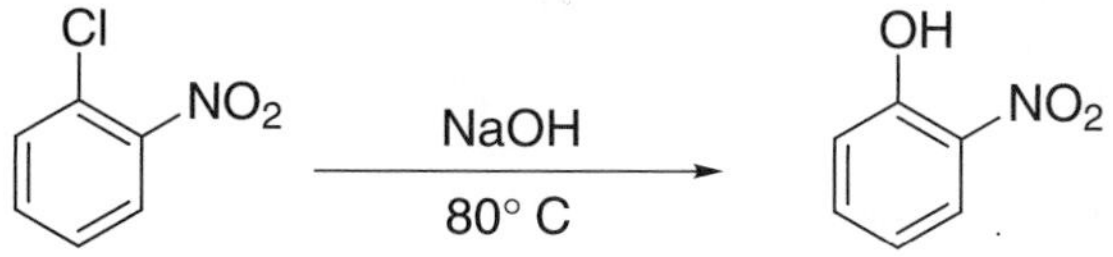

4.21.

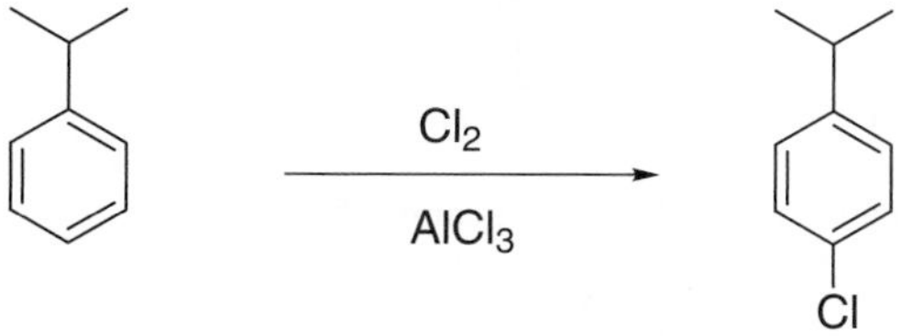

4.22.

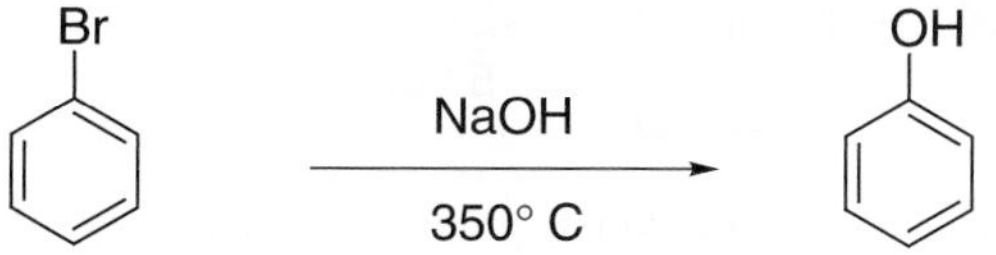

4.23.

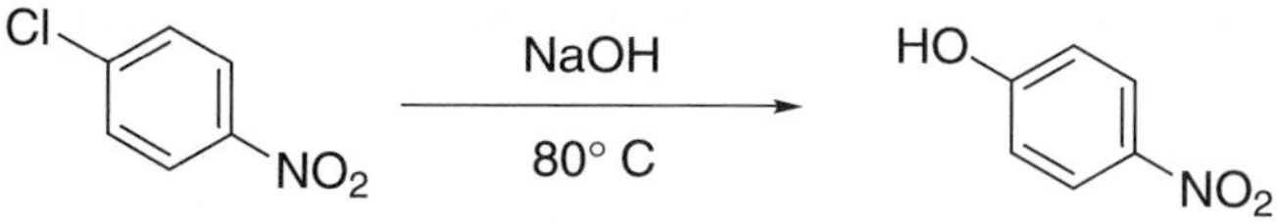

4.24.

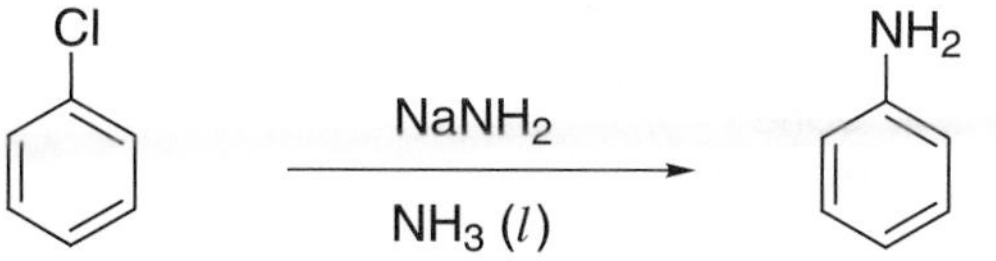

CHAPTER 5

KETONES AND ALDEHYDES

5.1 PREPARATION OF KETONES AND ALDEHYDES

Before we can learn about reactions of ketones and aldehydes, we must first make sure that we know how *to make* ketones and aldehydes. That information will be vital when we try to solve synthesis problems.

We can make ketones and aldehydes in many ways as shown in a number of textbooks. In this book, we will only see a few of these methods. These few reactions should be sufficient to help you solve synthesis problems in which a ketone or aldehyde needs to be made.

When solving synthesis problems, the most useful type of transformation you need to know is how to form a C=O double bond from an alcohol.

Primary alcohols can be oxidized to form aldehydes:

Secondary alcohols can be oxidized to form ketones:

Tertiary alcohols cannot be oxidized because the carbon atom cannot form five bonds:

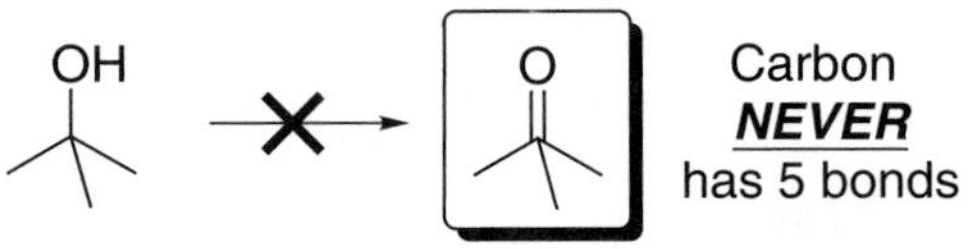

So, we need to be familiar with the reagents that will oxidize primary and secondary alcohols (to form aldehydes or ketones, respectively). Let's start with secondary alcohols.

To convert a secondary alcohol into a ketone, we can use chromic acid, or we can use the Jones reagent:

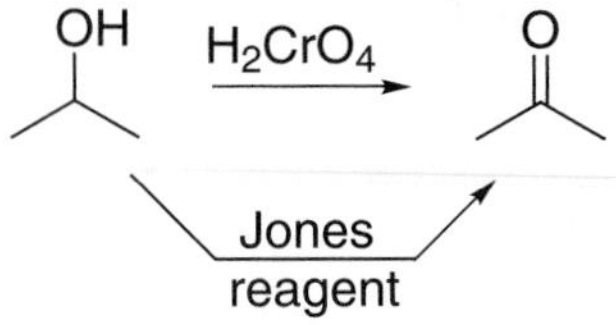

Some instructors will allow you to simply write out the word "Jones" on a synthesis problem (as in the above), whereas other instructors will require that you actually show what the Jones reagent is, such as:

You should look through your lecture notes to determine whether you need to memorize the actual reagents for a Jones oxidation.

Whether you use chromic acid or the Jones reagent, you are essentially doing an oxidation that involves chromium (the alcohol is being oxidized and the chromium is being reduced). The mechanisms for chromium oxidations are sometimes covered and sometimes not covered, depending on the course you are taking. You should look through your lecture notes to see if your instructor gave you the mechanisms for these oxidation reactions (or if they are in your textbook). Whatever the case, you should definitely have these reagents at your fingertips because you will encounter many synthesis problems in which you will need to convert an alcohol into a ketone or aldehyde.

Chromium oxidations work well for secondary alcohols, but we run into a problem when we try to do a chromium oxidation on a primary alcohol. The initial product formed is an aldehyde:

But under these strong oxidizing conditions, the aldehyde does not survive. The aldehyde is further oxidized to give a carboxylic acid:

So clearly, we need a way to oxidize a primary alcohol into an aldehyde, under conditions that will *not* further oxidize the aldehyde. To do this, we use the following conditions:

The reagent that we used is called Pyridinium Chlorochromate (or PCC). This reagent provides milder oxidizing conditions, and therefore, the reaction stops at the aldehyde. PCC will oxidize a primary alcohol to give an aldehyde:

There is one other common way to form a C=O double bond (other than oxidation of an alcohol). You might remember the following reaction from last semester:

This reaction is called an ozonolysis. It essentially takes every C=C double bond in the compound and breaks it apart into two C=O double bonds:

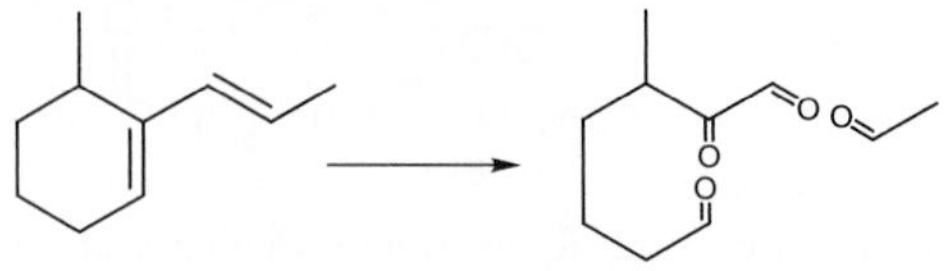

If you focus on the reagents for an ozonolysis, you will find that there are many variations for the second step:

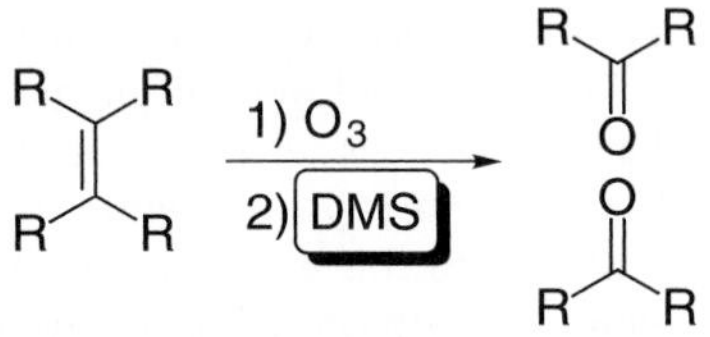

Many other reagents can be used for step 2 (other than DMS), all of which will give the exact same products. You should look in your lecture notes to see what reagents your instructor (or textbook) used for step 2 of an ozonolysis. Some reagents for step 2 would give you different products than the ones shown above. These reagents are typically not covered in the first year of organic chemistry, so we will not cover them here. We will only focus on a simple ozonolysis because the products will always be ketones or aldehydes.

We have only seen a few ways to make a C=O double bond. First, we saw reagents for oxidizing a secondary alcohol (chromic acid or Jones). Then, we saw reagents for oxidizing a primary alcohol (PCC). And finally, we saw reagents for doing an ozonolysis. Let's just make sure that you are comfortable with these reagents.

EXERCISE 5.1. Predict the major product of the following reaction:

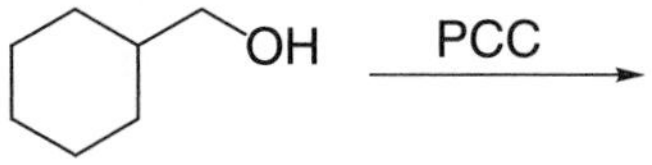

Answer: PCC is used to convert a primary alcohol into an aldehyde. So our product is:

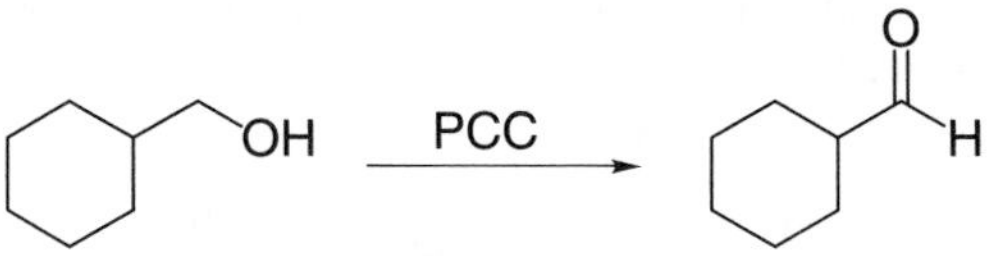

For each of the following problems, predict the major product of the reaction:

5.2.

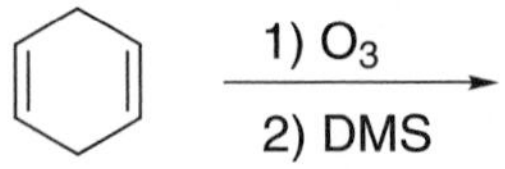

5.3.

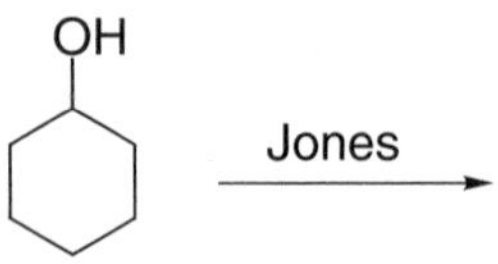

5.4. $\xrightarrow{\text{H}_2\text{CrO}_4}$

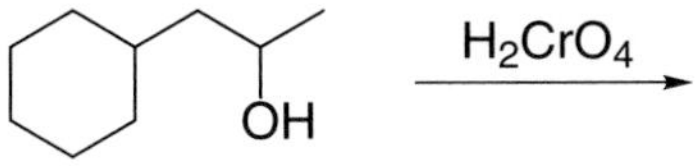

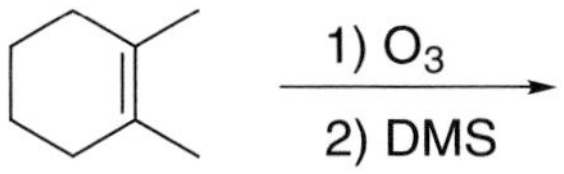

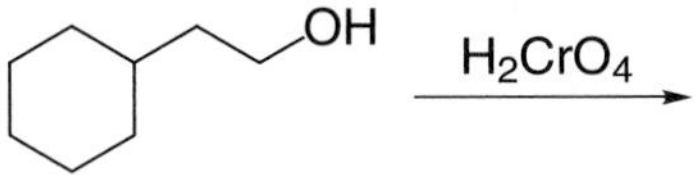

5.5. $\xrightarrow[\text{2) DMS}]{\text{1) O}_3}$

5.6. $\xrightarrow{\text{H}_2\text{CrO}_4}$

5.7. $\xrightarrow{\text{PCC}}$

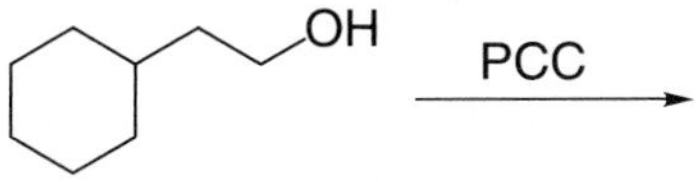

It is not enough simply to "recognize" the reagents when you see them (as we did in the previous problems). But you actually need to know the reagents well enough to write them down when they are not in front of you, as in the following exercise.

EXERCISE 5.8.   Identify the reagents you would use to do the following transformation:

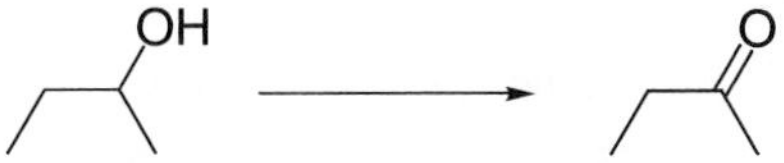

Answer: In this reaction, we are converting a secondary alcohol into a ketone, so we don't need to use PCC. We would only need PCC if we were trying to convert a *primary* alcohol into an aldehyde. But in this case, PCC is unnecessary. Instead, we would use chromic acid or the Jones reagent:

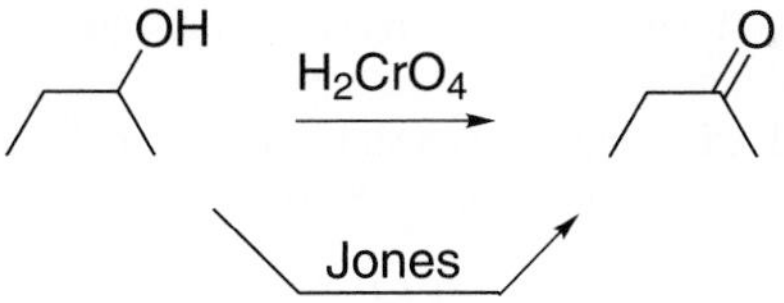

For each of the following transformations, identify the reagents you would use. Try not to look back at the previous problems while you are solving these problems.

5.9.

5.10.

5.11.

5.12.

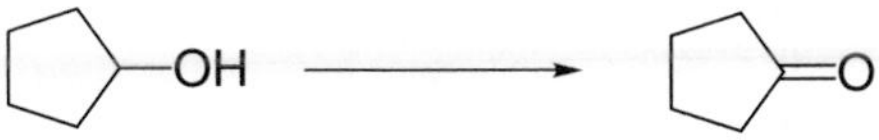

5.2   STABILITY AND REACTIVITY OF THE CARBONYL

Ketones and aldehydes are very similar to each other in structure:

Ketone   *Aldehyde*

Therefore, they are also very similar to each other in terms of reactivity. Most of the reactions that we see in this chapter will work just as well with ketones or aldehydes. So, it makes sense to learn about ketones and aldehydes in the same breath.

But before we can get started, we need to know some basics about C=O double bonds. Let's start with some terms that we will use throughout the entire chapter. Instead of constantly using the expression C=O double bond, it will be faster if we have a special term for this bond. The term we use is ***carbonyl***. This term is NOT used for nomenclature. You will never see the term *carbonyl* appearing in the IUPAC name of a compound. Rather, it is just a term that we use when we are talking about mechanisms, so that we can quickly refer to the C=O double bond without having to say "C=O double bond" all of the time.

Don't confuse the term *carbonyl* with the term *acyl*. The term *acyl* is used to refer to a carbonyl *together with* one alkyl group:

Carbonyl   *Acyl*

We will use the term *acyl* in the next chapter; in this chapter, we will focus on the carbonyl group.

If we want to know how a carbonyl group will react, we must look at its electronics. Whenever you want to know the electronics of a group (i.e., you want to know the locations of $\delta+$ and $\delta-$), you must always look at two factors: induction and resonance. If we start with induction, we notice that oxygen is more electronegative than carbon, and therefore, the oxygen will pull on the electron density:

This makes the carbon $\delta+$ and the oxygen $\delta-$.

Next, we look at resonance:

And we see, once again, that the carbon is $\delta+$ and the oxygen is $\delta-$, this time because of resonance. Thus, both induction and resonance paint the same picture:

This means that the carbon atom is very electrophilic and the oxygen atom is very nucleophilic. Although many reactions involve the oxygen atom functioning as a nucleophile, we will not see any of those reactions in this course. Accordingly, we will focus all of our attention in this chapter on the carbon atom. We will see *how* the carbon atom functions as an electrophile, *when* it functions as an electrophile, and *what happens after* it functions as an electrophile.

The geometry of a carbonyl group also facilitates the functioning of the carbon atom as an electrophile. We saw in the first semester of organic chemistry that sp^2 hybridized carbon atoms have a trigonal planar geometry:

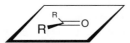

This makes it easy for a nucleophile to attack the carbonyl because there is no steric hindrance that would block the incoming nucleophile:

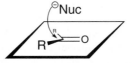

In this chapter, we will see many different kinds of nucleophiles that can attack a carbonyl. In fact, this entire chapter will be organized based on the kinds of nucleophiles that can attack. We will start with hydrogen nucleophiles. Then we will look at oxygen nucleophiles, followed by sulfur nucleophiles, nitrogen nucleophiles, and, finally, carbon nucleophiles. This approach—dividing the chapter based on the kinds of nucleophiles—might be somewhat different from that of your textbook. But the order of coverage should be very similar to your textbook. And hopefully, the order that we use here will help you to appreciate the similarity between the reactions.

We need to mention one more feature of carbonyl groups before we can get started. Carbonyl groups are thermodynamically very stable. In other words, when we form a carbonyl, we are going downhill in energy. When we destroy a carbonyl (turning it into a C—O single bond), we are going uphill in energy. As a result, the formation of a carbonyl is often the driving force for a reaction. We will use that argument many times in this chapter, so make sure you are prepared for it. The mechanisms in this chapter will be explained in terms of the stability of carbonyl groups.

Now let's quickly summarize the important characteristics that we have seen so far. The carbon atom (of a carbonyl) is electrophilic, and it is ready to be attacked by a nucleophile (and there are MANY different kinds of nucleophiles that can attack it). We have also seen that a carbonyl group is very stable. So, if it does get attacked by a nucleophile:

then the intermediate formed will be high in energy, and it will want to re-form the carbonyl if at all possible.

These two principles will drive all of the chemistry we are about to see: (1) carbonyl groups want to be attacked because they are very electrophilic, but (2) when they do get attacked, they want to do whatever it takes to become a carbonyl again.

5.3 H-NUCLEOPHILES

We will now look at the various nucleophiles that can attack ketones and aldehydes. We will divide all nucleophiles into categories, and in this section, we will focus on hydrogen nucleophiles. I call them "hydrogen" nucleophiles, because they possess a negatively charged hydrogen atom (which we call a "hydride" ion) that can attack a ketone or aldehyde. The simplest way to get a hydride ion is from sodium hydride (NaH). This compound is ionic, so it is composed of Na^+ and H^- ions (very much the way NaCl is composed of Na^+ and Cl^- ions). Thus, NaH is a good source of hydride ions.

However, you will not see any reactions where we use NaH as a source of hydride *nucleophiles*. As it turns out, NaH is an excellent base, but it is not a very good nucleophile. This is an excellent example of how basicity and nucleophilicity do NOT completely parallel each other. The reason for this goes back to something we learned during the first semester of organic chemistry. Try to remember back to the difference between basicity and nucleophilicity, and let's review it quickly.

The strength of a base is determined by the *stability* of the negative charge. A strong base is when you have an unstable negative charge; a weak base is when you have a stabilized negative charge (since it is stable, it is not as eager to run around looking for a proton to grab). But nucleophilicity is NOT based on stability. Nucleophilicity is based on *polarizability*. When a negative charge is contained in an orbital *of a large atom* (such as sulphur or iodine), then the electron density in that orbital can move around freely within a fairly *large* volume of space. This "moving around" is called polarizability. Larger atoms (which are polarizable) are strong nucleophiles, whereas smaller atoms (which are not polarizable) are not good nucleophiles.

With that in mind, we can understand why H^- is a strong base but not such a strong nucleophile. It is a strong base because the negative charge is on a hydrogen atom, and hydrogen does not stabilize the charge well (hydrogen is not a very electronegative atom). We just said that basicity is dependent on stability. Since the charge is *not* stabilized, it will be a strong base. But when we consider the nucleophilicity of H^-, we must look at the polarizability of the hydrogen atom. Hydrogen is the smallest atom, and therefore, it is not very polarizable at all. Therefore, H^- is not such a good nucleophile.

Now we can understand why we don't use NaH as a source for a hydrogen nucleophile. It is true that it is an excellent base, and you will see NaH used several times this semester. But it will always be used as a strong *base* and never as a *nucleophile*. So how do we form a hydrogen nucleophile?

H^- itself is not a good nucleophile, but certain reagents can serve as a "delivery agent" of H^-. For example, consider the following compound, called sodium borohydride ($NaBH_4$):

$$Na^{\oplus} \qquad \begin{array}{c} H \\ | \\ H-\overset{\ominus}{B}-H \\ | \\ H \end{array}$$

If we look at a periodic table, we see that boron is in the third column. Therefore, it has three valence electrons. That means that it can comfortably form three bonds. But in sodium borohydride (above), the central boron atom has *four* bonds. That means that it has one extra electron. One way to think of it is as follows:

$$H^{\ominus} \overset{\frown}{} \begin{array}{c} H \\ | \\ B-H \\ | \\ H \end{array} \quad \longrightarrow \quad \begin{array}{c} H \\ | \\ H-\overset{\ominus}{B}-H \\ | \\ H \end{array}$$

When we think of it this way, we see that sodium borohydride is really just a combination of BH_3 and H^- (you can ignore the sodium ion, Na^+, because it is just the counterstabilizing cation). So, this reagent can serve as a delivery agent of H^-, as in:

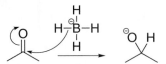

Notice that H^- never really exists by itself in this reaction; rather, H^- is **delivered** from one place to another. That is a good thing because H^- by itself would not be a good nucleophile (as we saw earlier). But sodium borohydride can serve as a source of a hydrogen *nucleophile* because the central boron atom is somewhat polarizable. This polarizability of the boron atom allows the entire compound to serve as a nucleophile, and **deliver** a hydride ion to attack the ketone. Now, it is true that boron is not so large, and therefore, it is not very polarizable. As a result, $NaBH_4$ is a somewhat tame nucleophile. It is not VERY nucleophilic. We will soon see that $NaBH_4$ is selective about what it reacts with. It will not react with all carbonyl groups (for example, it will not react with an ester). But it will react with ketones **and** with aldehydes (and that is all we care about right now in this chapter).

Another common reagent is very similar to sodium borohydride, but it is much more reactive. This reagent is called lithium aluminum hydride ($LiAlH_4$, or LAH for short):

$$Li^{\oplus} \quad H{-}\overset{\ominus}{\underset{\displaystyle H}{\overset{\displaystyle H}{Al}}}{-}H$$

This reagent is very similar to $NaBH_4$ because aluminum is also in the third column of the periodic table (directly beneath boron). So, it also has three valence electrons. And in the compound above, the aluminum atom has four electrons, which is why it has a negative charge. As is true of $NaBH_4$, LAH is also a source of nucleophilic H^-. But compare these two reagents to each other—aluminum is larger than boron. That means that it is more polarizable, and therefore, LAH is a much better nucleophile than $NaBH_4$. LAH will react with almost any carbonyl (not just ketones and aldehydes).

It will soon become very important that LAH is more reactive than $NaBH_4$. But for now, we are talking about the nucleophilic attack of ketones and aldehydes, and both $NaBH_4$ and LAH will react with ketones and aldehydes.

There are other sources of H nucleophiles in addition to $NaBH_4$ and LAH, but these two are the most common reagents. You should look through your textbook and lecture notes to see if you are responsible for being familiar with any other hydrogen nucleophiles.

Now let's take a close look at the mechanism for a hydrogen nucleophile attacking a carbonyl. As we have seen, the reagent (either $NaBH_4$ or LAH) can deliver a hydride ion to the carbonyl, as in the following:

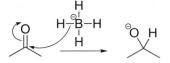

In the beginning of this chapter, we said two important things about a carbonyl:

- It is easily attacked by nucleophiles (and we have now just seen that happen).
- After it is attacked, it will try to re-form the carbonyl, if possible. Now we need to understand what we mean when we say: "if possible."

When we re-form the carbonyl, we have a problem that the central carbon atom cannot have five bonds:

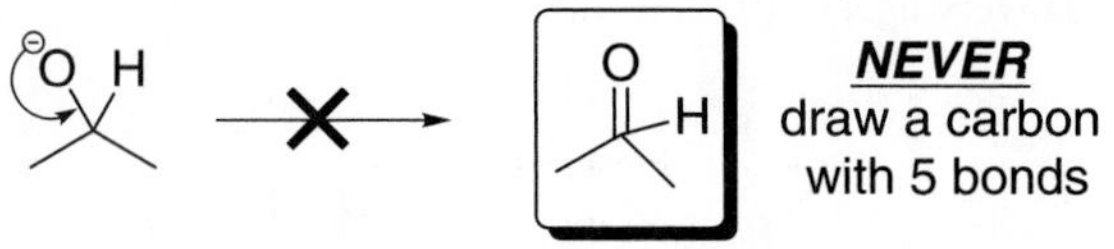

That would be impossible because carbon only has four valence orbitals to use. So, if we want to re-form the carbonyl, we must kick off a leaving group, as in:

So we just need to know what groups will leave and what groups will not leave. Fortunately, one simple rule will tell you the answer most of the time—NEVER kick off H$^-$ or C$^-$. There are a few exceptions to this rule, which we will see later, but unless you recognize that you are dealing with one of the rare exceptions, do NOT kick off H$^-$ or C$^-$. For example, never do this:

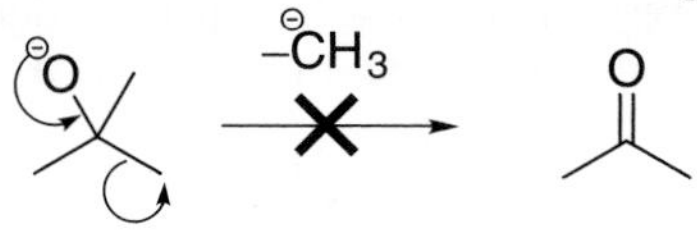

or this:

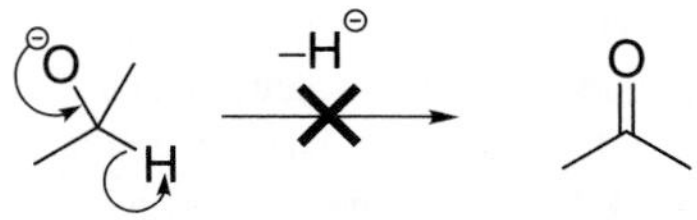

We have just learned a simple general rule. Now let's try to apply this rule to our discussion of a hydrogen nucleophile attacking a ketone. Let's use this general rule to show us what the products will be.

We saw that the first step of the mechanism was for the hydrogen nucleophile to attack:

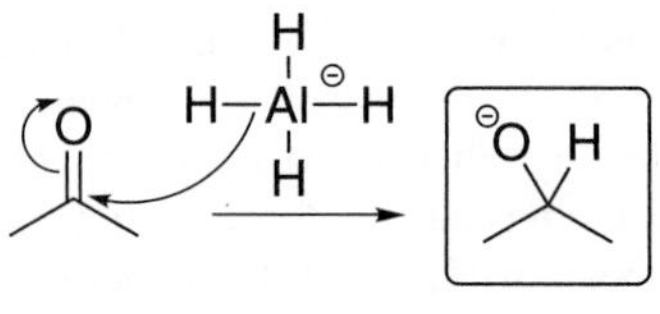

Now, what do we do next? In order to re-form the carbonyl, we must kick off a leaving group. But there are no leaving groups here. We cannot re-form the carbonyl to kick off a C$^-$:

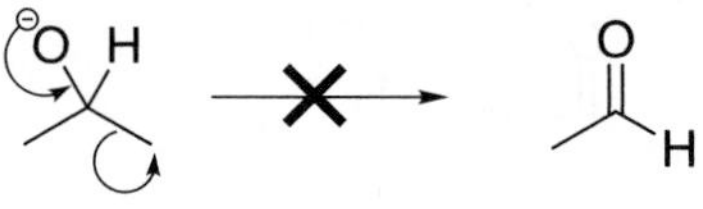

And we cannot re-form the carbonyl to kick off an H$^-$:

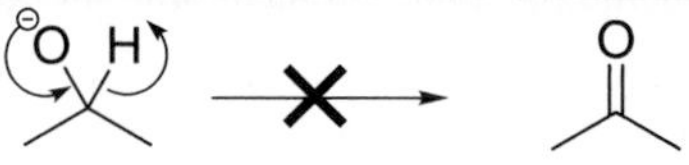

And we cannot re-form the carbonyl to kick off a C^-:

So that means that we are stuck. Once a hydrogen nucleophile delivers H^- to the carbonyl, then it will not be possible to re-form the carbonyl. The reaction is therefore over, and it just waits for us to add a source of protons to quench the reaction (to give the negative charge a proton to grab). For this reaction, we use water as the source of protons:

So we see that the product of this reaction is an alcohol:

Whenever you are using this transformation in a synthesis problem, you must show that the proton source is added AFTER the reaction has taken place:

In other words, it is important to show that LAH and water are *two separate steps*. Do *not* show it like this:

This would mean that LAH and H_2O are added at the same time, and that would *not* work. LAH would react very violently with water to form H_2 gas (because you would have H^+ and H^- reacting with each other).

As it turns out, $NaBH_4$ is a milder source of hydride, and therefore, $NaBH_4$ can be added together at the same time as the proton source:

The common proton sources are MeOH or water (sometimes you might see EtOH). Notice that we didn't show it as two separate steps. When you are dealing with LAH, you must show two steps (one step for LAH and another step for the proton source); but when you are dealing with $NaBH_4$, you can show the proton source in the same step as $NaBH_4$.

LAH and NaBH$_4$ are very useful reagents. They allow us to *reduce* a ketone or aldehyde, which is important when you realize that we have already learned how to *oxidize* an alcohol to get a ketone:

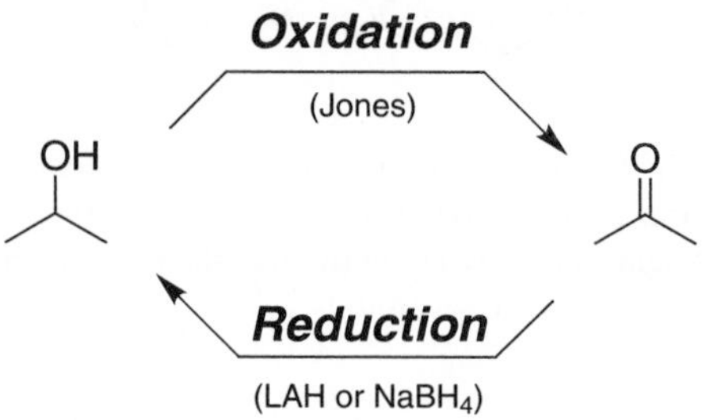

Oxidation

(Jones)

OH

Reduction

(LAH or NaBH$_4$)

These two transformations will be *tremendously* helpful when you are trying to solve synthesis problems later on. You would be surprised just how many synthesis problems involve the interconversion between alcohols and ketones. You need to have these two transformations at your fingertips.

EXERCISE 5.13. Predict the major product of the following reaction:

NaBH$_4$

MeOH

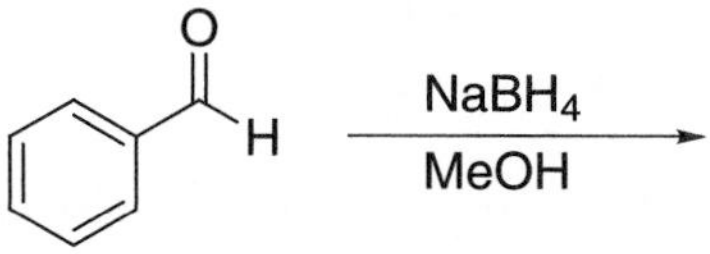

Answer: The starting compound is an aldehyde, and it is reacting with sodium borohydride. This hydrogen nucleophile will *deliver* H$^-$ to the aldehyde, and the carbonyl will not be able to re-form because there is no leaving group. So it just has to wait until we add water to give it a proton. We get an alcohol as our product:

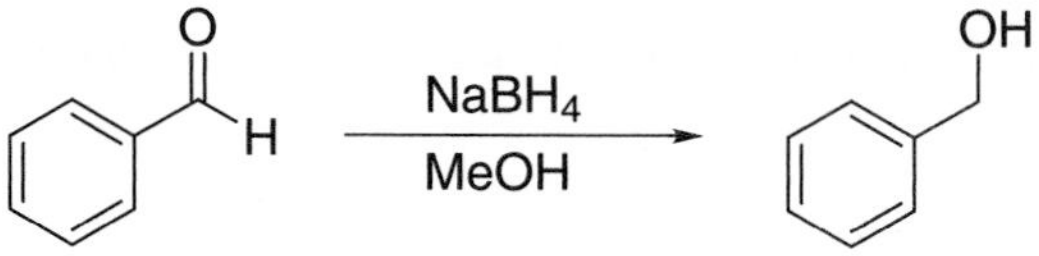

NaBH$_4$

MeOH

OH

Predict the major product for each of the following reactions:

5.14.

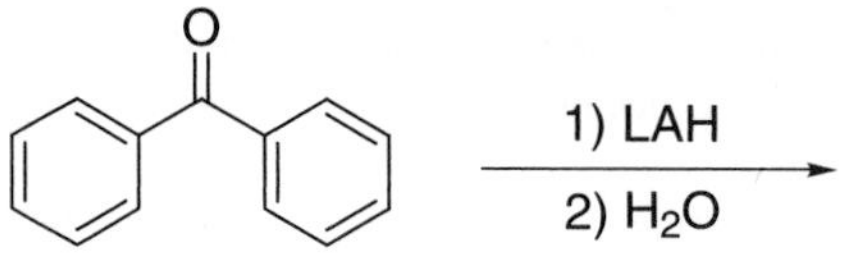

1) LAH

2) H$_2$O

5.15.

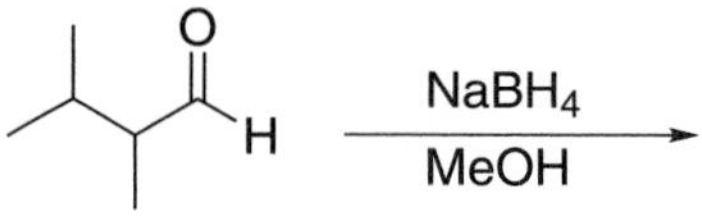

NaBH$_4$

MeOH

5.16.

OH

PCC

H

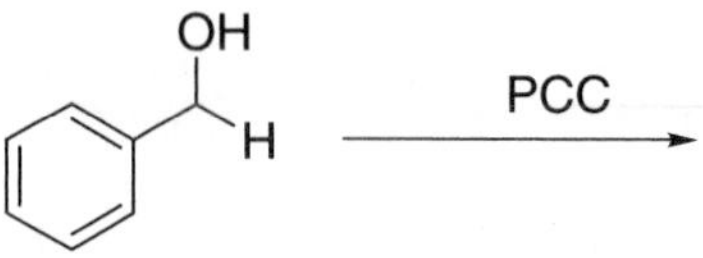

5.17.   NaBH$_4$ / MeOH

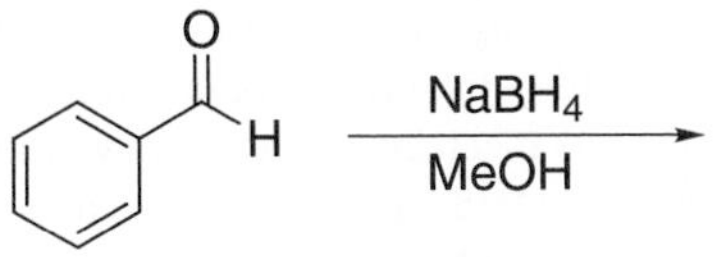

5.18.   1) O$_3$
2) DMS
3) LAH
4) H$_2$O

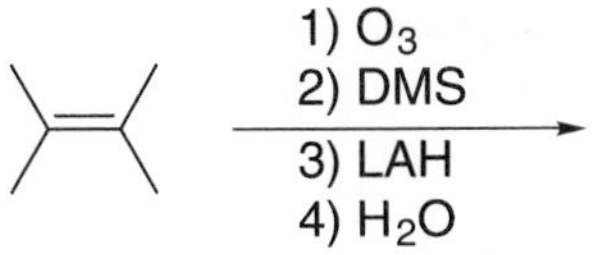

EXERCISE 5.19.   Draw the mechanism of the following reaction:

1) LAH
2) H$_2$O

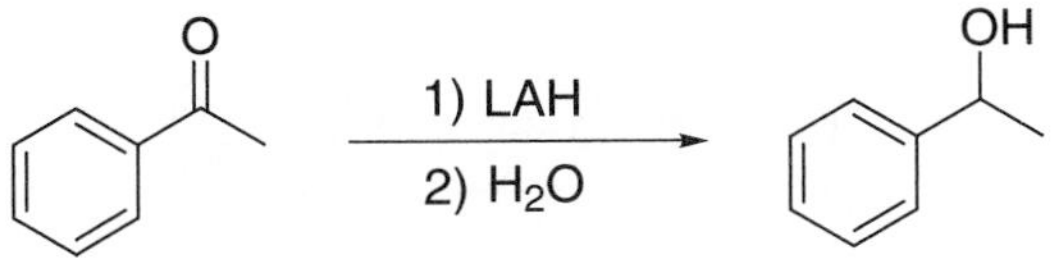

Answer: First, LAH delivers a hydride to the ketone, and the carbonyl is not able to re-form, so the intermediate waits until we give it a proton from water.

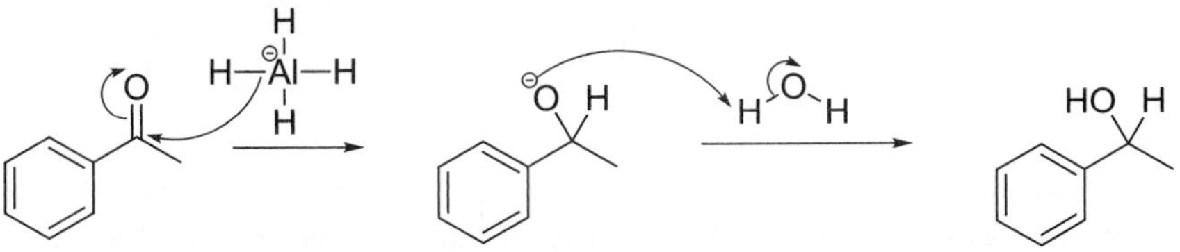

Propose a mechanism for each of the following reactions. The following problems will probably seem too easy, but just do them anyway. These basic arrows need to become *routine* for you because as the problems become more complex in the next section, you will want to have these basic skills down cold:

5.20.   1) LAH
2) H$_2$O   CH$_3$OH

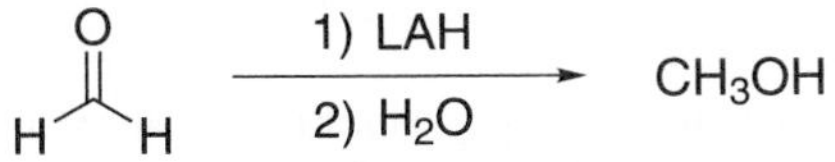

5.21.   NaBH$_4$ / MeOH

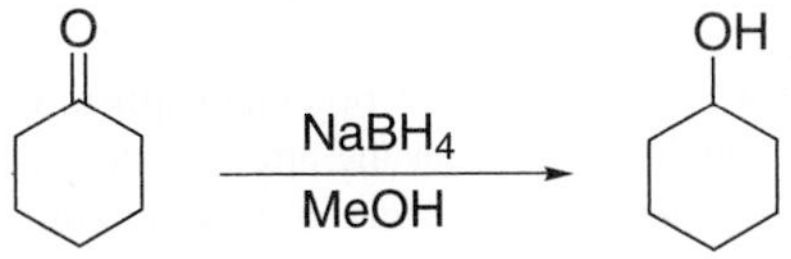

5.22.   1) LAH
2) H$_2$O

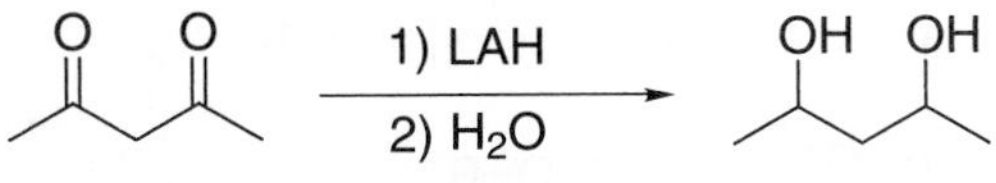

5.4   O-NUCLEOPHILES

In this section, we will focus on oxygen nucleophiles, specifically, the attack of an alcohol (ROH) on a ketone or aldehyde.

Be warned: the mechanism we are about to see is VERY long. It is probably one of the longest mechanisms that you will see in this course, but it is incredibly important because it lays the foundation for so many other mechanisms. If you can master this mechanism, then you will be in good shape to move on. Actually, you have no other option: you MUST master this mechanism. So be prepared to read through the next several pages slowly, and then be prepared to reread those pages as many times as necessary until you know this mechanism intimately.

Alcohols are nucleophilic because the oxygen has lone pairs that can attack an electrophile:

$$R-\ddot{O}-H \qquad E^{\oplus}$$

When an alcohol attacks a carbonyl, we go up to an intermediate that reminds us of the intermediate that was formed in the previous section:

Notice how similar this is to the hydride attack we saw earlier:

But there is one major difference here. When we saw the attack of a hydrogen nucleophile in the previous section, we argued that the carbonyl could not re-form after the attack because there was no leaving group. But here in this section (with the attack of an alcohol), there is a leaving group. So, it *is* possible for the carbonyl to re-form:

The nucleophile that attacked (ROH) can function as the leaving group. But of course, that gets us right back to where we started. As soon as a molecule of alcohol attacks the carbonyl, it just gets spit out immediately, and there is no net reaction.

So let's explore other possible avenues, so that we can see if a reaction can happen. First of all, we should realize that the attack of an alcohol is much slower than the attack of a hydrogen nucleophile because alcohols do not have a negative charge. They are neutral, and therefore, they are not incredibly strong nucleophiles. So, if we want to speed up this reaction, we would want to make the nucleophile more nucleophilic (for example, using RO⁻ instead of ROH):

Theoretically, this would speed up the reaction, but under these conditions, we would have the same problem that we just had a moment ago. We cannot prevent the carbonyl from re-forming. The initial intermediate will just spit out the nucleophile, and we would get right back to where we started:

So, we will take a slightly different approach. Rather than making the nucleophile more nucleophilic, we will focus on making the electrophile more electrophilic. So let's focus on the electrophile of our reaction:

How do we make a carbonyl group even more electrophilic? The answer is, we protonate it:

This is VERY IMPORTANT because we will see this again and again and again. When you protonate a ketone, you make the entire compound positively charged. This makes the carbonyl group even more electrophilic.

So, we protonate our ketone, and *then* we attack using the alcohol:

This gives us an intermediate that has a tetrahedral geometry. (The starting ketone was sp^2 hybridized, and therefore trigonal planar; but this intermediate is now sp^3 hybridized, and therefore tetrahedral.) So, we will refer to this intermediate as a "tetrahedral intermediate."

But doesn't this tetrahedral intermediate give us the same problem? Doesn't it kick off a leaving group to re-form the protonated ketone?

Yes, this *can* happen. In fact, it *does* happen—and most of the time. That is why we need to use equilibrium arrows:

So it is true that an equilibrium exists between the forward and reverse processes. But every now and then, something else can happen to the tetrahedral intermediate. A different carbonyl can be re-formed:

What about kicking off this leaving group ?

In other words, we are looking to kick off OH⁻ as our leaving group, which should theoretically work because we said before that you can kick off anything except for H⁻ and C⁻. So, OH⁻ could theoretically leave, but we cannot kick off OH⁻ *in acidic conditions*. Rather, we will have to protonate it first in order to turn it into a better leaving group. (This is a BIG DEAL—make sure that this rule becomes part of the way you think. NEVER kick off a negatively charged oxygen into acidic conditions; always protonate it first.) So we do the following proton transfers:

Notice that we first took a proton off to form an intermediate that has no charge, and only then we put another one on. We did this to avoid having two positive charges on our intermediate. (That is another subtle rule that should become part of the way you think from now on.) And don't just try to move over that one proton all in one step (intramolecularly), as follows:

That doesn't happen because the ends of the molecule are just not close enough together in space to transfer a proton intramolecularly. So, you must first take a proton off, and only then do you put another proton back on (and it is probably not going to be even the same exact proton that you took off).

We therefore do two separate proton transfers, and that gets us to this intermediate:

Now we are ready to kick off our leaving group (which is now water) to re-form a carbonyl, as in:

This new intermediate now *does* have a carbonyl, *but* there is no easy way to get rid of the charge. You can't just lose R⁺ the way you can lose a proton:

But there is another way to get rid of the charge. This intermediate can get attacked by *another* molecule of alcohol, just as we attacked the protonated ketone at the beginning of our mechanism:

And finally, we can lose a proton to get our product:

So the overall transformation is:

To make sure that we understand some of the key features of this mechanism, let's take a close look at the whole thing all at once. It is very long:

Let's point out a few important features. First, let's take a close look at the proton transfers. The third step of the mechanism above shows two proton transfers in one step (without showing the curved arrows for those two steps; we showed those two steps earlier in our discussion). Some textbooks and instructors will do it this way. Others will insist that you actually draw both proton transfer steps, showing the curved arrows. If your instructor and textbook do it the way I have done it above, then you can take this short-cut as well. But be careful: some instructors want you to show both steps because it allows you to draw one extra intermediate that is actually quite important later on in biochemistry:

This intermediate has a special name that we will soon see.

Now let's focus our attention on *all* of the proton transfers in the entire mechanism, and we should notice that the acid is a *catalyst* here (throughout the mechanism, we put on two protons, and we took off two protons). So, in the end, we did not consume the acid (it is a catalyst), and we show that by placing brackets around the H^+:

When we focus on the proton transfers, we realize that *most* of the mechanism is just proton transfers. There are only three major steps (other than proton transfers): ROH attacks, water leaves, and then ROH attacks again. All of the proton transfers are simply used to facilitate these three steps. (We use proton transfers to make the carbonyl more electrophilic, to produce water as a leaving group instead of hydroxide, and to avoid multiple charges.) It is important that you see the reaction in this way. It will greatly simplify the whole mechanism in your mind.

The drawing below is NOT a mechanism; the arrows in this drawing are only being used to help you see all three critical steps in your mind at once:

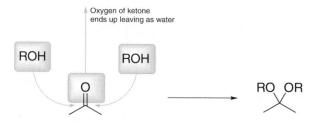

The product of this reaction is called a *ketal*. When we form a ketal from a ketone, there is one intermediate that gets a special name because it is the only intermediate that does not have a charge. It is called a *hemiketal*, and you can think of it as "halfway" toward making a ketal:

Hemiketal   Ketal

We give it a special name because it is theoretically possible to isolate it and store it in a bottle (although in most cases, this is very difficult to do) and because this type of intermediate will be important if and when you learn biochemistry.

Let's present one more bit of terminology before we move on. We have just seen that you get a *ketal* when you start with a *ketone* (via the hemiketal). But when you start with an *aldehyde*, then we call the product an *acetal* (and it is formed via a hemiacetal):

Hemiacetal   Acetal

Now we have seen the reaction and the terminology that describes it, but one important issue remains to be discussed. Notice that a ketal does not have a carbonyl in it. This means that the equilibrium will lean toward the starting materials rather than the products:

In other words, if we try to do this reaction in a lab, we will get very little (if any) product. So the question is: how can we force the reaction to form the ketal? And we have a clever trick for doing this.

It is possible to remove water from the reaction as the reaction proceeds. If we remove water as it is formed, we will essentially stop the reverse path at a particular step (shown below in the mechanism where there is an X through an arrow). It is like putting up a brick wall that prevents the reverse reaction from occurring:

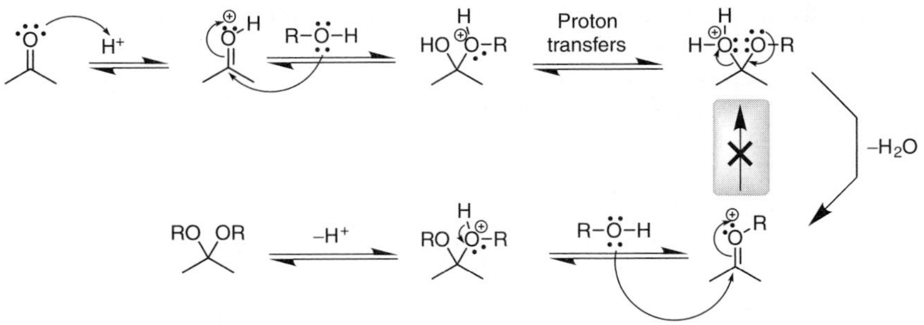

By removing water as it is being formed, then we force the reaction to a certain point. And then by depriving the system of water, the reaction cannot be reversed to re-form the starting ketone. In order to be reversed, the system would need water, and we have removed all of the water. This very clever trick allows us to force the equilibrium to favor the products even though they are less stable than the reactants (and the equilibrium does not naturally favor the products). Now look at the part of the mechanism that is after this critical step:

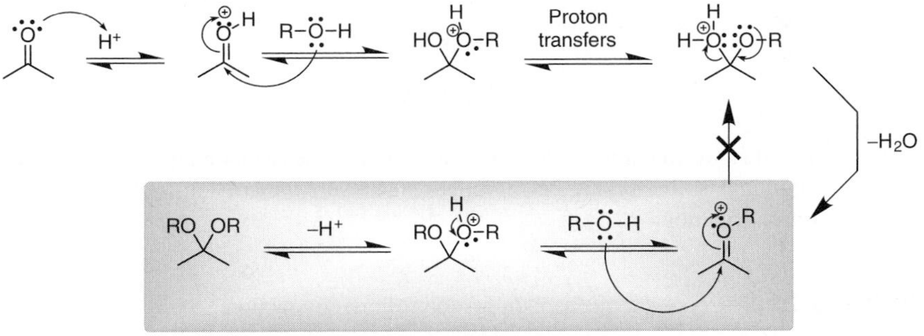

We see that there are three structures in equilibrium with each other. Two of them are positively charged, and one of them (the product) is neutral. This equilibrium now favors formation of the product.

In your textbook and in your lectures, you will probably learn how chemists remove water from the reaction as it proceeds. It is called azeotropic distillation, and a special piece of glassware is used (called a Dean-Stark trap). I will not go into the details of azeotropic distillation here, but I wanted to just briefly mention it because you should know how to indicate the removal of water. There are two ways to show it:

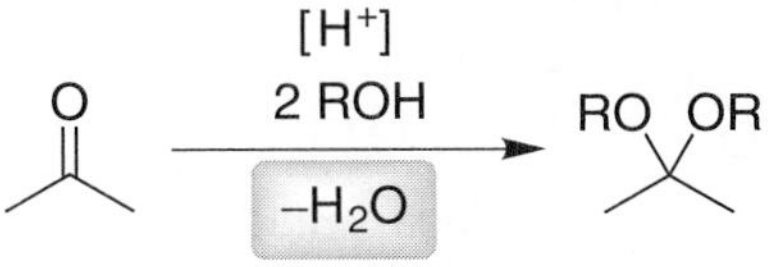

or like this:

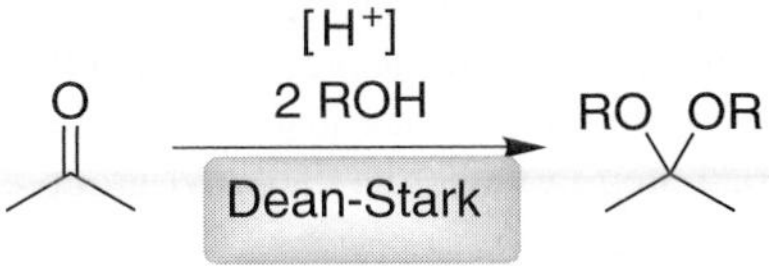

By just writing the words "Dean-Stark," you are indicating that you understand that it is necessary to remove water in order to form the ketal.

Now we can also appreciate how you would reverse this reaction. Suppose you have a ketal, and you want to convert it back into a ketone. You would just add water with a pinch of some acid, and presto, it goes back to the ketone in a flash:

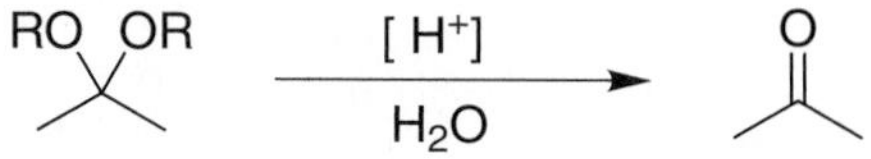

This forces the equilibrium to push everything back toward the ketone. So, now we know how to turn a ketone into a ketal, and we know how to turn the ketal back into a ketone:

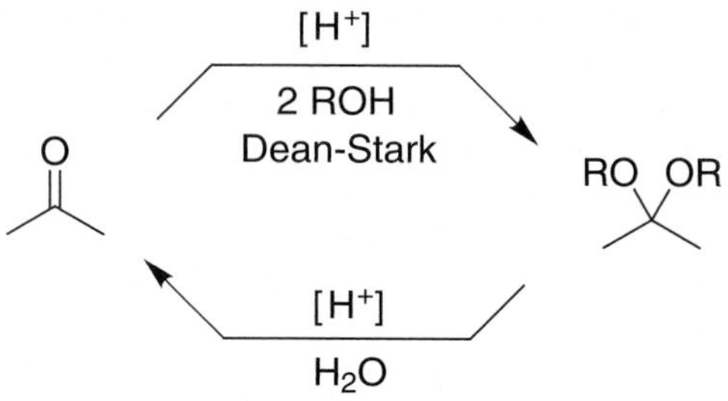

It is very important that we can control the conditions to push the reaction either way we want. We will soon see why this is so important, but first, let's make sure we are comfortable with the basic mechanism of ketal formation:

EXERCISE 5.23. Propose a mechanism for the following reaction.

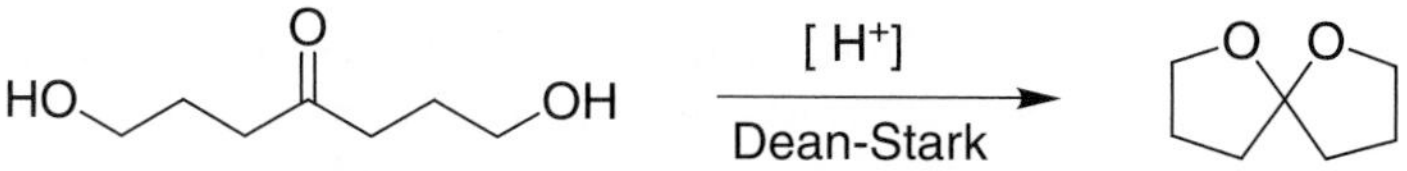

Answer: Notice that we are starting with a ketone, and we are ending up with a ketal. It is a bit tricky to see because it is all happening intramolecularly. In other words, the two alcoholic OH groups are present in the same molecule as the ketone (the OH groups are *tethered* to the ketone):

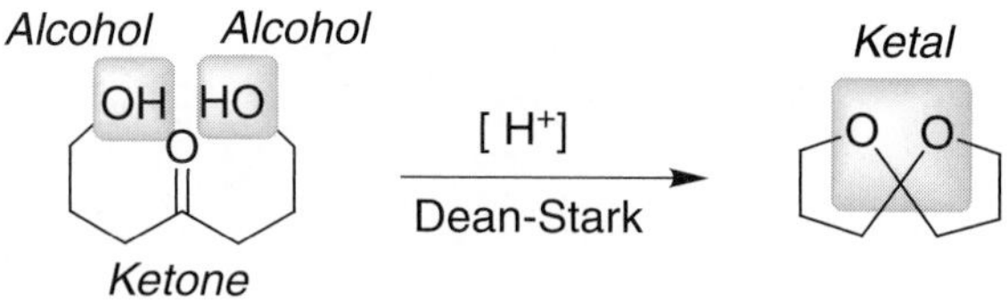

So the mechanism follows the same exact steps as the mechanism we have already seen. Namely, there are three critical steps (OH attacks, water leaves, and the other OH attacks) surrounded by a bunch of proton transfers. The proton transfers are just there to facilitate these three steps. We use a proton transfer in the very first step to make the carbonyl group even more electrophilic. Then, we use proton transfers to form water (so that it can leave). And finally, we use proton transfers to facilitate the second attack of OH.

Perhaps you should try to draw the mechanism for this reaction on a separate piece of paper. Then, when you are done, you can compare your work to this answer:

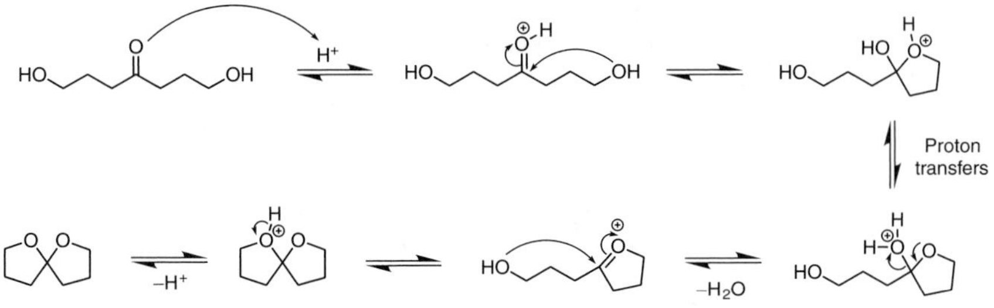

Earlier we said that this type of mechanism is so incredibly important because so many more reactions will build on the concepts that we developed in this mechanism. To get practice, you should work through the following problems slowly and methodically. Make sure that you go through these problems with the goal of becoming the master of this mechanism. I cannot stress this enough.

For each of the following problems, propose a plausible mechanism:

5.24.

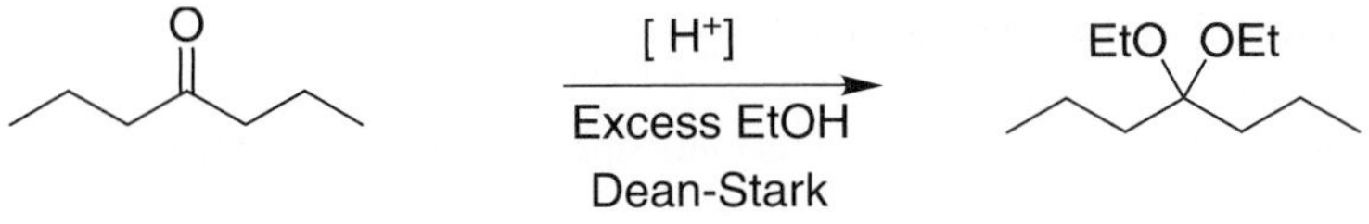

5.25.

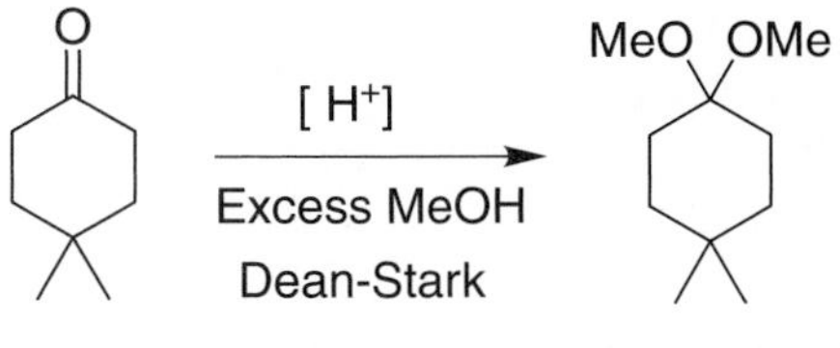

5.26.

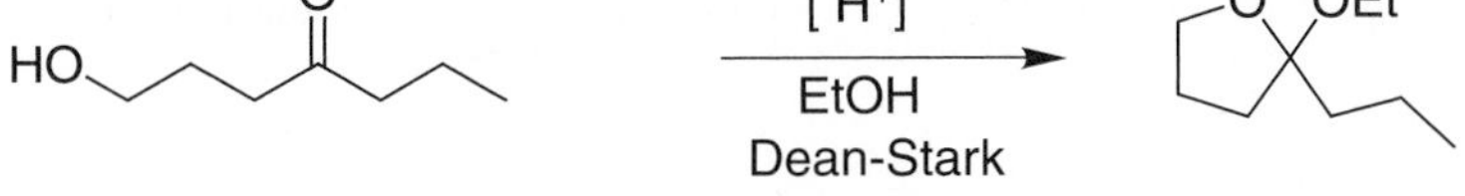

There is one sure way to know whether or not you have mastered a mechanism forwards and backwards: you should try to actually draw the mechanism backwards. That's right, backwards. Consider the following problem:

PROBLEM 5.27.   Draw the mechanism of the following reaction:

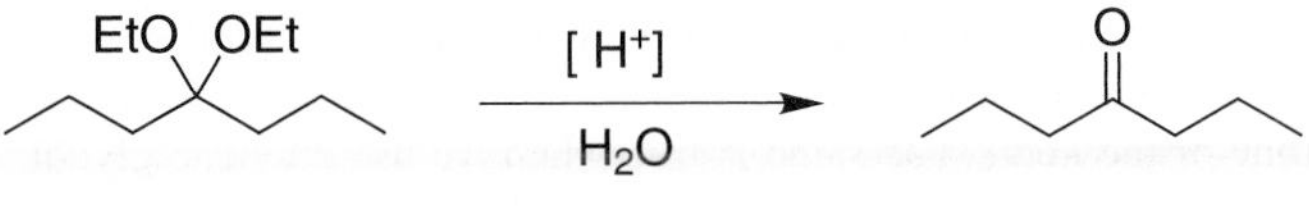

To help yourself, you can start at the end (with the ketone), and then draw the intermediate you would obtain if you were converting the ketone into a ketal (remember that the first step was to protonate the ketone), as in:

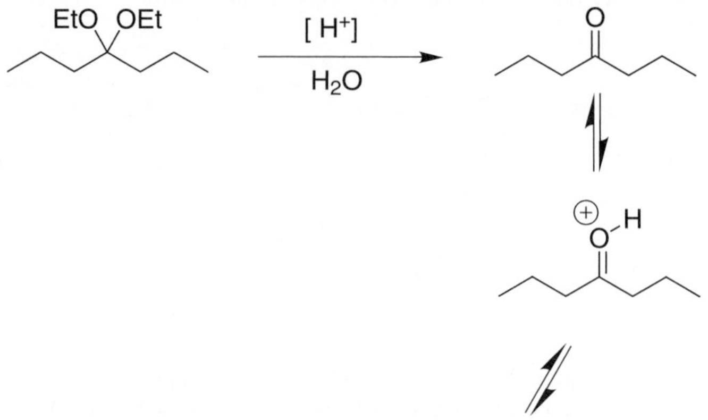

Slowly work your way backwards, until you arrive at the ketal. But *don't* draw in any arrows yet. Just draw the intermediates, working backwards from the ketone to the ketal. Then, once you have all of the intermediates drawn, come back and try to fill in arrows, STARTING with the ketal. Use a separate sheet of paper to draw your mechanism. When you are finished, you can compare your answer to the answer in the back of the book.

In this section, we have seen the reaction that takes place between a ketone and *two* molecules of ROH:

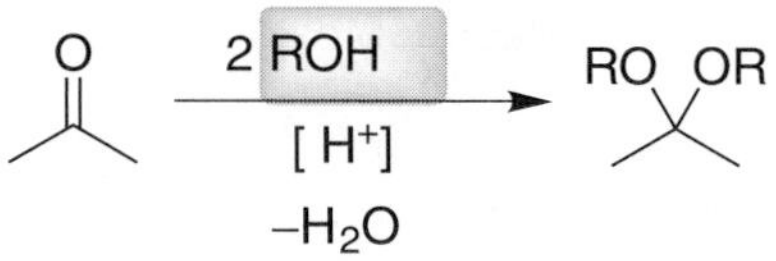

We saw the mechanism, step-by-step, and we saw that the ketone gets attacked twice. This same reaction can be done when both alcoholic OH groups are in the same molecule. This produces a *cyclic* ketal :

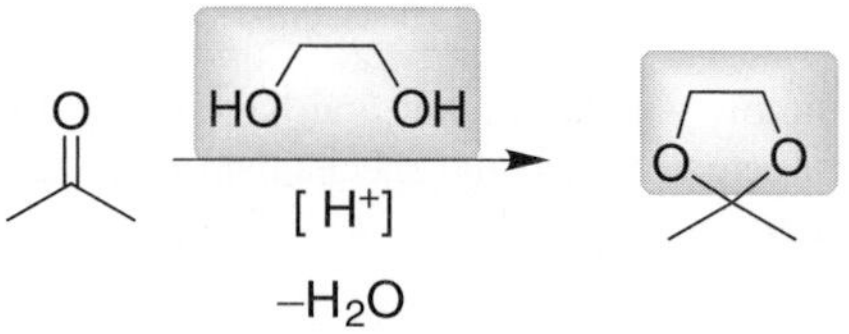

This type of reaction is used quite a lot in this course, so you will need to be familiar with it. The diol in the reaction above is called ethylene glycol.

Now let's examine one reason why this reaction with ethylene glycol is so important.

We have already seen that we can manipulate the conditions of the reaction to control whether we form the ketone or the ketal. The same is true when we use ethylene glycol to form a cyclic ketal:

This is important because it allows us to *protect* a ketone from an undesired reaction. Let's see a specific example of this. (It will take us two pages to build up this argument with a concrete example, so please be patient as you read through this material.)

Suppose you have the following compound:

When this compound reacts with LAH, both carbonyl groups get reduced:

1) LAH
2) H_2O

LAH attacks the ketone *and* the ester. It may be difficult to see why the ester was converted into an alcohol—we will focus on that in the next chapter. But if you are curious to test your abilities, you have actually learned everything you need in order to figure out how an ester turns into an alcohol in the presence of LAH. (Remember that you should always re-form a carbonyl if you can, but never kick off H^- or C^-.)

So, we see that LAH will reduce both carbonyl groups in the compound above. If instead, we mix the starting compound with $NaBH_4$, we see that only the ketone gets reduced:

$NaBH_4$
MeOH

The ester is *not* reduced because $NaBH_4$ is a milder source of hydride (as we have explained earlier). In the next chapter we will see that $NaBH_4$ will not react with esters (only with ketones and aldehydes) because the carbonyl group of an ester is less reactive than the carbonyl group of a ketone.

But suppose you want to do the following transformation:

?

Essentially, you want to reduce the ester *but not* the ketone. That would seem impossible because esters are less reactive than ketones. Any reagent that reduces an ester should also reduce a ketone.

But there is a way to do it. Suppose we *protect* the ketone by turning it into a ketal:

Only the ketone turns into a ketal. The carbonyl of the ester does ***not*** turn into a ketal (because esters are less reactive than ketones). So we are using the reactivity of the ketone to our advantage, by selectively *blocking* the ketone. Now, we can react this compound with LAH, and the ketal will not be affected (ketals do not react with bases or nucleophiles):

Notice that we use water above (as we have done every other time that we used LAH). In the presence of water, the ketal is removed (and we should really add a bit of acid to catalyze this process):

In the end, we have a three-step process for reducing the ester ***without affecting*** the ketone:

This trick (protecting the ketone by turning it into a ketal, then running the desired reaction, and finally pulling off the protecting group) is actually quite similar to something we saw when we learned about electrophilic aromatic substitution. Recall that the para position was usually favored over the ortho position (because of sterics). So we had a problem if we wanted to put a group into the ortho position:

To do this, we used a clever strategy. Our strategy was to *block* the para position with a sulfonation reaction. Then we would run the reaction we wanted to run, placing the desired group into the ortho position. And finally, we would *unblock* the para position:

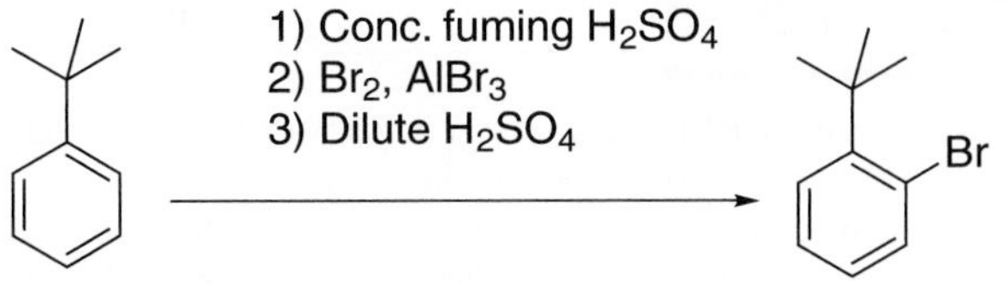

This strategy is the same one we are using now in this chapter. It is also a three-step strategy: (1) we *block* the ketone from reacting by converting it into a ketal, then (2) we run the reaction we want to run, and finally (3) we unblock to turn the ketal back into a ketone.

We will talk more about this strategy in the next chapter. For now, let's just focus on knowing the reactions well enough to predict products:

EXERCISE 5.28. Predict the major product of the following reaction:

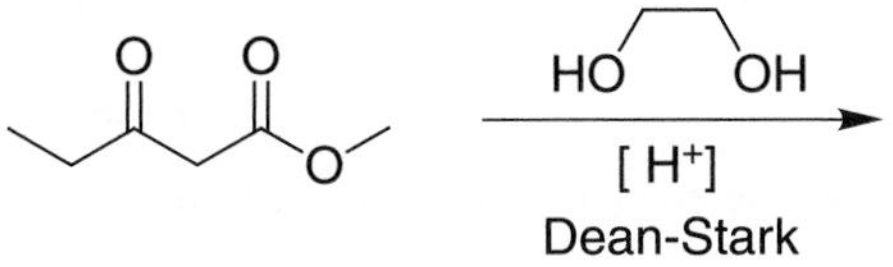

Answer: In this reaction, we have ethylene glycol, so we will be forming a cyclic ketal. Our starting compound has two carbonyl groups. One is a ketone, and the other is an ester. We have seen that we only form ketals from ketones. We do *not* form ketals of esters. So, our major product is:

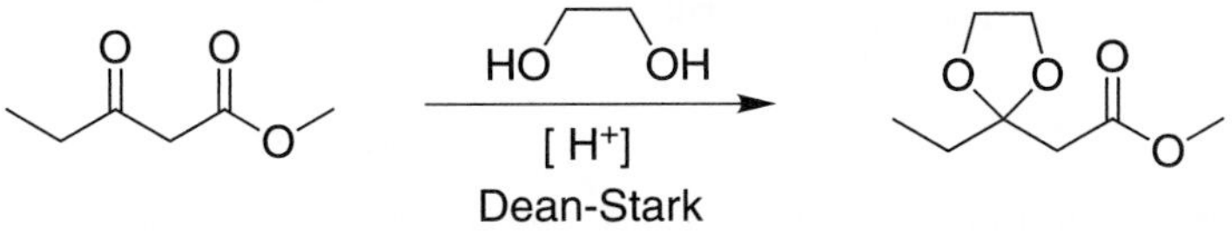

For each of the following problems, predict the major product of the reaction:

5.29.

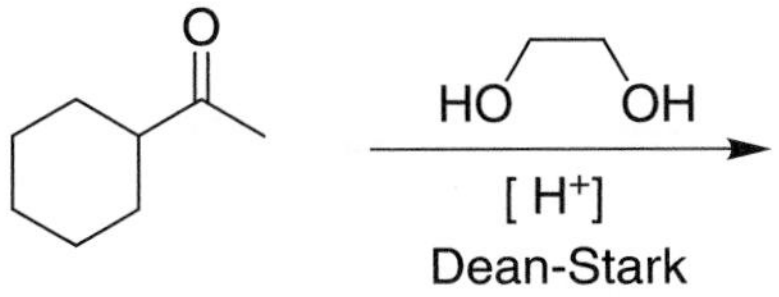

5.30.

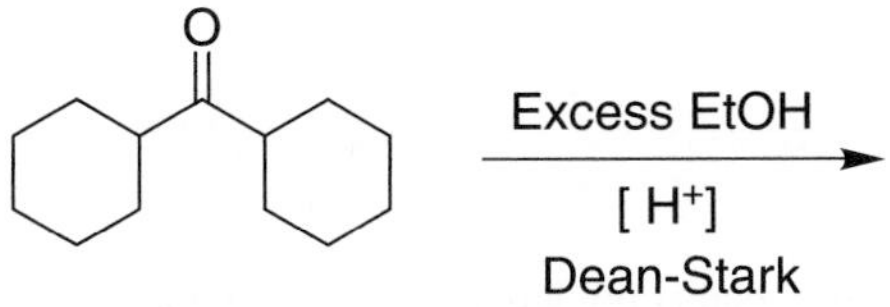

5.31.

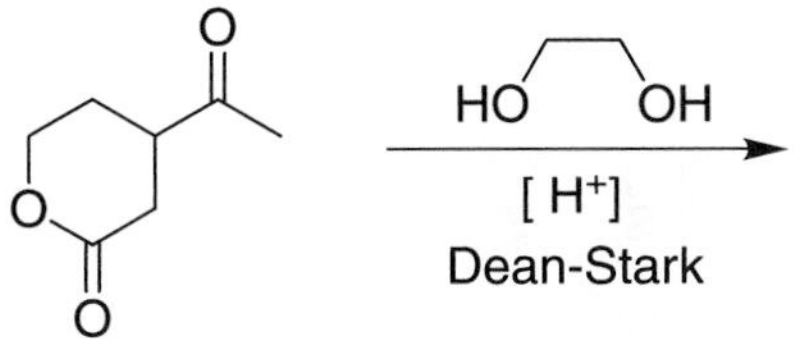

5.32.

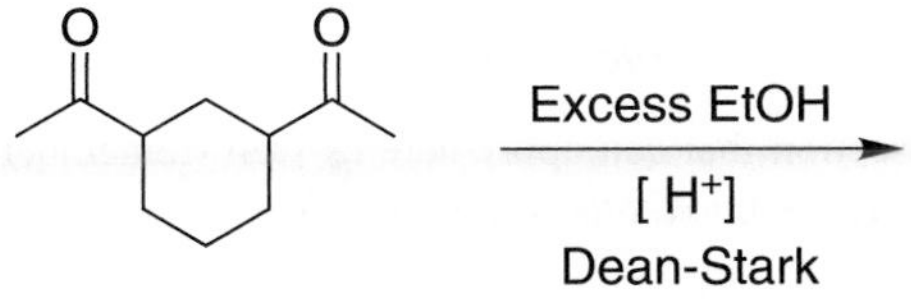

5.5 S-NUCLEOPHILES

Some instructors might skip sulfur nucleophiles, so you might want to look through your notes and textbook to find out if you are responsible for the content in this section.

Sulfur is directly below oxygen on the periodic table (in the sixth column). Therefore, the chemistry of sulfur-containing compounds is very similar to the chemistry of oxygen-containing compounds. We will see the similarities, and more importantly, we must look at the subtle differences.

Let's start with the similarities. We have seen that you can use ethylene glycol to turn a ketone into a ketal:

Ketal

In the same way, you can also use ethylene *thio*glycol to form a *thio*ketal (*thio* means sulfur instead of oxygen):

Thioketal

The only difference is that we use BF_3 instead of H^+ to make the carbonyl more electrophilic. Compare the first step of each mechanism:

Other than this small difference, making a *thio*ketal is not any different than making a ketal. After all, they are very similar in structure:

Ketal   *Thioketal*

We saw some terminology in the previous section that can apply here as well (remember the difference between ketals and acetals). We can make thio*ketals* or thio*acetals*:

*Thio**ketal***   *Thio**acetal***

To make a thio***ketal***, you would just start with a *ketone*:

And to make a thio***acetal***, you would just start with an *aldehyde*:

In the previous section, we saw that we could use ketal formation as a way to protect ketones, but we don't really use thioketals for protecting ketones. We use thioketals and thioacetals for two other important reactions, which we will explore now.

Both of these reactions are very powerful synthetic tools, and you will want to learn them both very well. They will prove very useful to have in your back pocket when you solve synthesis problems. Let's start with the first reaction:

We just saw that thioacetals have an extra proton that thioketals do not have:

*Thio**acetal***

This extra proton is mildly acidic. It can be pulled off with a very strong base (we usually use butyl-lithium, which can be thought of as Bu^- and Li^+):

To see why this proton is acidic, we must look at the conjugate base. The neighboring large sulfur atoms withdraw electron density and stabilize the conjugate base:

This anion can now function as a nucleophile to go and attack something, as in the following:

And then after the attack, we can rip off the cyclic thioketal, using the following reagents:

HgCl$_2$ (shown above) is just a catalyst for speeding up this reaction (this catalyst is *not* necessary when deprotecting ketals or acetals; it is only necessary here, when deprotecting a *thio*ketal or a *thio*acetal).

To summarize, we have developed a three-step synthesis strategy here:

1. We convert an aldehyde into a thioacetal.
2. We then pull off a proton (with BuLi) to form an anion, which then attacks an alkyl halide (this hooks on the new alkyl group).
3. And finally, we convert back to the ketone.

These three steps give us the ability to do the following synthetic transformation:

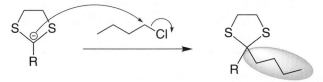

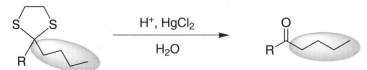

Notice that we can convert an aldehyde into a ketone in other ways. This might be useful to you in synthesis problems.

You might wonder why we have to use the thioketal step at all. In other words, why can't we skip the thioketal step and deprotonate the aldehyde directly?

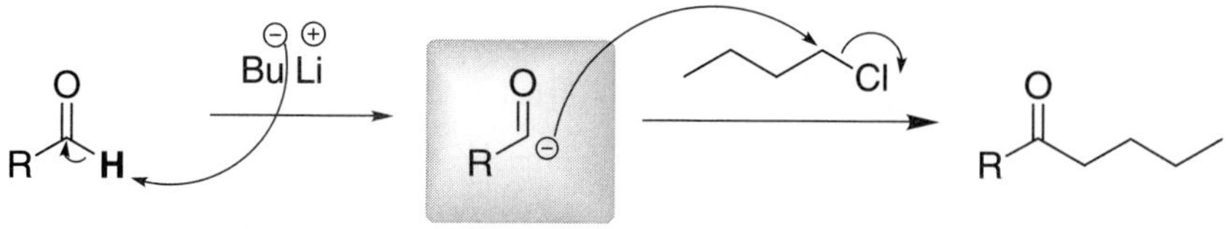

Too unstable
Too high in energy
Does not form

This doesn't work because an acyl anion is not a stabilized negative charge. Being too high in energy, it cannot form. Therefore, it is *very* difficult to deprotonate an aldehyde. If we want to pull off that proton and replace it with an alkyl group, then we will need to use the strategy that we learned in this section. To convert an aldehyde into a ketone, you can use the three-step method that we just saw.

In fact, you can construct an entire ketone by putting *both* sides on, one after another, as in the following:

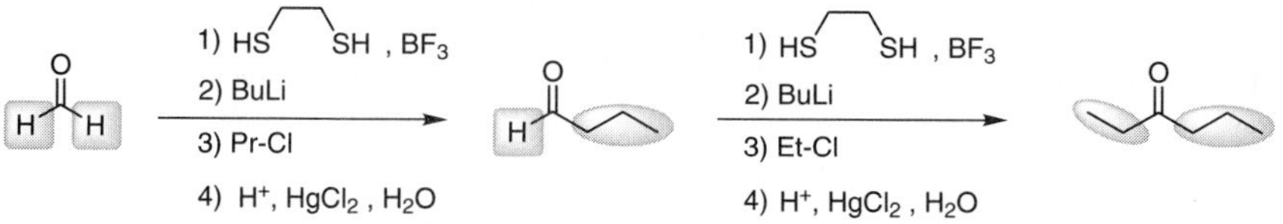

Starting with formaldehyde, you could put one alkyl group on; forming an aldehyde; and then you could put another group on, forming a ketone. In the preceding example, we put a propyl group on one side and an ethyl group on the other side.

It turns out that you don't actually have to start with formaldehyde and then turn it into the thioacetal because there is a thioacetal that is commercially available:

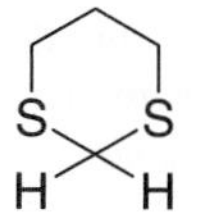

This compound
can be purchased

The sulfur atoms are bridged to each other by three carbon atoms rather than two. But that's OK. It works just as well. This compound is called 1,3-dithiane, and if you use this as your starting material, you can make any ketone you want.

Well, let's be careful about that, because you can't actually make *any* ketone. There are some limitations on this method. Focus back on the step where the anion attacked an alkyl halide:

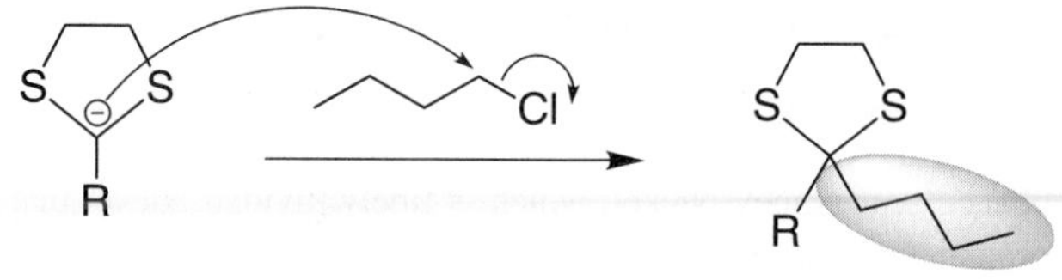

This was an S_N2 reaction; therefore, it will work best with primary alkyl halides. Secondary alkyl halides are slower to react, and tertiary alkyl halides won't react at all. So you could make a compound like this:

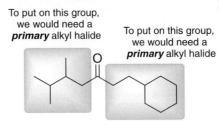

But you could **not** make a compound like this:

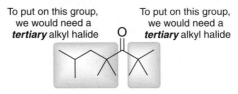

So far, we have seen one very important use for thioacetals. We can use them to make ketones (either from an aldehyde, or just from dithiane). Similarly, you could use this technique to make an aldehyde:

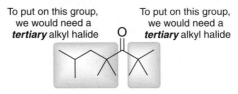

It is even easier to make an aldehyde than a ketone because you don't have to hook a group onto both sides of a carbonyl. You only have to hook a group onto one side of the carbonyl. But the same restrictions apply. You should use a **primary** alkyl halide to make an aldehyde by this method.

We will need to do some problems to make sure that you can get accustomed to using thioacetals for **making ketones**. But first, we need to explore one more reaction. This reaction is the opposite of what we have seen so far. Rather than making ketones, we will now see how thioketals can be used to **destroy ketones**. It goes as follows:

First you take a ketone and convert it into a thioketal:

Then you blast away the sulfur atoms, replacing them with hydrogen atoms:

This step is accomplished with Raney Nickel, which is finely divided Nickel that has hydrogen added to it. The mechanism for this reaction (blasting away the sulfur atoms by using Raney Nickel) and the scope of this course. But this is a VERY useful synthetic transformation, and so it is worth remembering, even if you don't know the mechanism. It allows you to completely reduce a ke-tone down to an alkane:

We have already seen one way to do this kind of transformation. It was called the Clemmensen re-duction, and we learned about this reaction in the chapter covering electrophilic aromatic substitution (Chapter 3). We will also see one more way to perform this transformation in the upcoming section.

Why do we need three different ways to do the same thing? Because each of these methods involves a different set of conditions. The Clemmensen reduction takes place under *acidic* conditions. The method we learned just now (desulfurization with Raney Nickel) is under *neutral* conditions, and the method in the upcoming section involves *basic* conditions. As we move through the course, at times it won't be good to subject an entire compound to acidic conditions, and at other times it won't be good to sub-ject an entire compound to basic conditions. For example, let's say your compound has a ketal some-where in it, and you want the compound to remain a ketal (the reaction you are doing is meant to in-volve another region within the compound). Acidic conditions would inadvertently convert the ketal into a ketone. So, you would not want to subject your compound to those acidic conditions.

When in doubt whether it is bad to use acidic conditions or basic conditions, you can always just use a desulfurization with Raney Nickel, which employs neutral conditions.

The following problems are designed to accustom you to using both strategies that we have seen in this section:

1. Using thioacetals to *make* ketones.

2. Using thioacetals to *destroy* ketones.

Let's start with some simple predict-the-product problems:

EXERCISE 5.33. Predict the major product of the following reaction:

Answer: The starting material is 1,3-dithiane. In step 1, we pull one proton off dithiane, and then in step 2, we alkylate:

Steps 3 and 4 are very similar. We pull off the other proton, and then alkylate again (this time, with a different alkyl group):

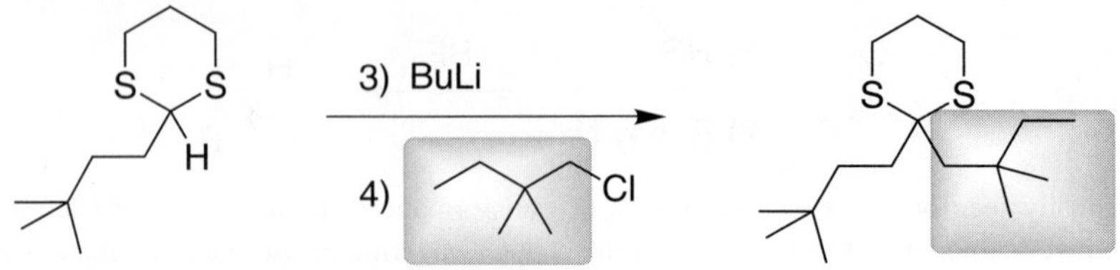

3) BuLi

4) [structure with Cl]

And finally, we pull off the thioketal, to produce a ketone:

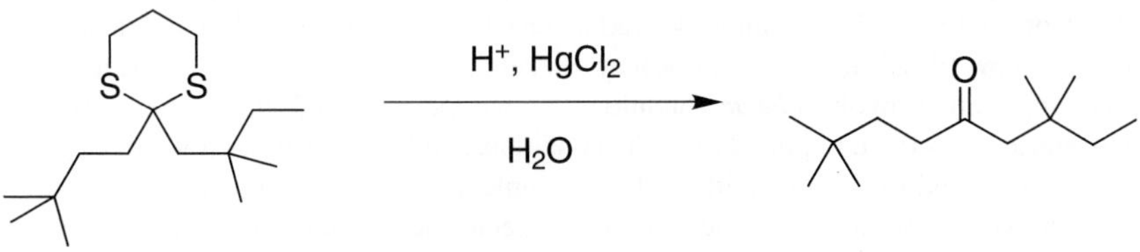

H⁺, HgCl₂

H₂O

Predict the major product for each of the following reactions:

5.34.

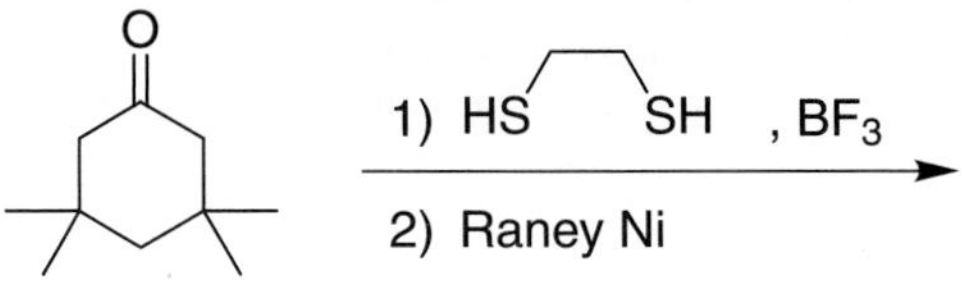

1) HS⌒SH , BF₃

2) Raney Ni

5.35.

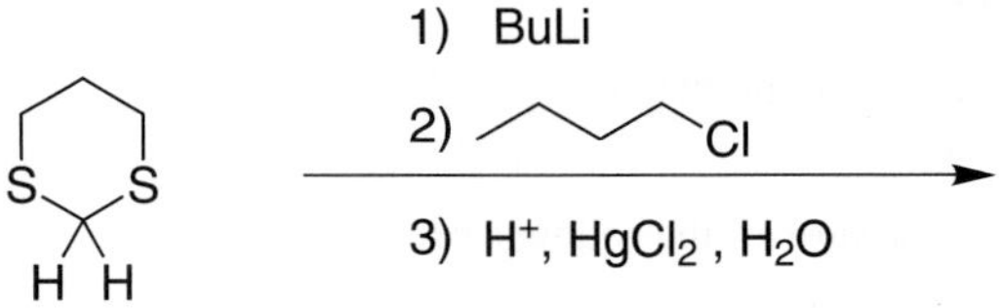

1) BuLi

2) ⌒⌒Cl

3) H⁺, HgCl₂ , H₂O

5.36.

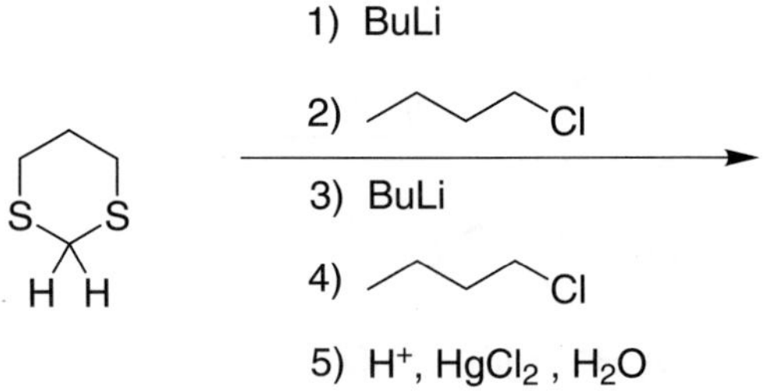

1) BuLi

2) ⌒⌒⌒Cl

3) BuLi

4) ⌒⌒Cl

5) H⁺, HgCl₂ , H₂O

5.37.

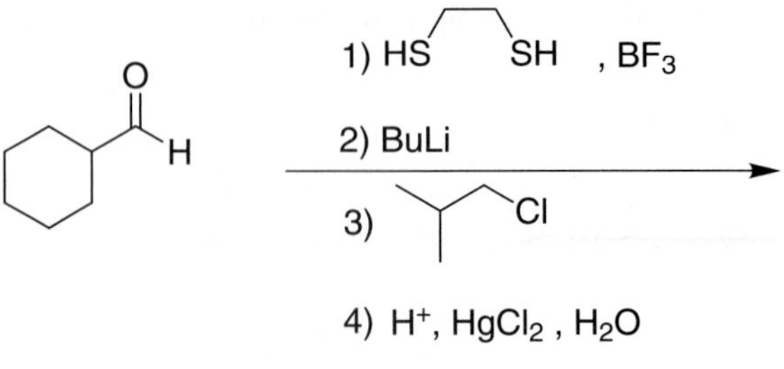

1) HS⌒SH , BF₃

2) BuLi

3) [structure with Cl]

4) H⁺, HgCl₂ , H₂O

5.38.

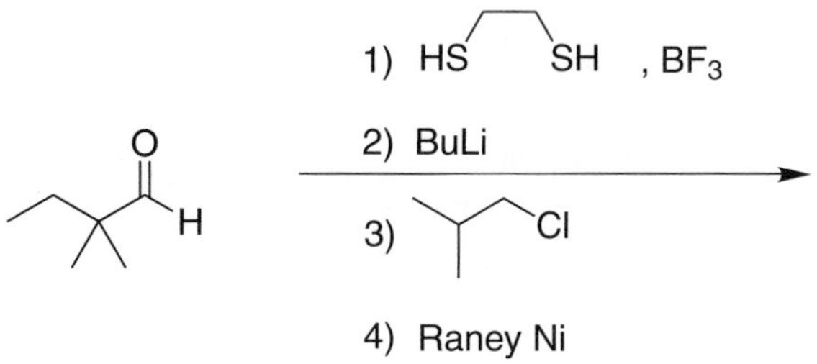

1) HS⌒SH , BF₃

2) BuLi

3) ⅄Cl

4) Raney Ni

Now let's move on to some synthesis problems. We will start off with a tough problem:

EXERCISE 5.39. What reagents would you use to do the following transformation?

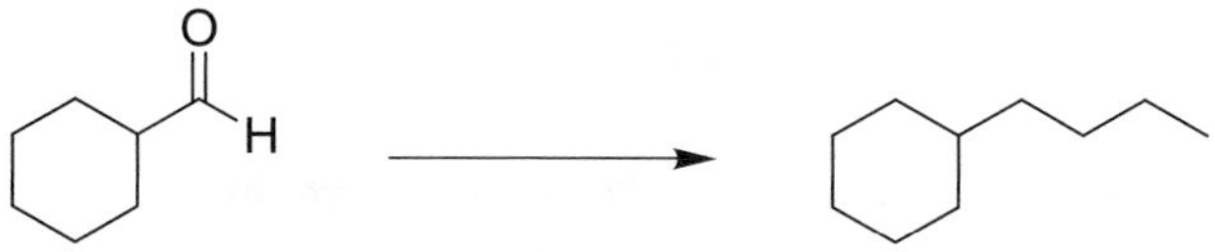

Answer: You should always begin a synthesis problem by looking at exactly how the starting material has changed. In this problem, the starting material has a carbonyl group, and the product has no carbonyl group at all (not even an OH group). So, we know that we will need to reduce the carbonyl down to an alkane.

One other major difference exists between the starting material and the product: the product has three extra carbon atoms.

So, our synthesis needs to accomplish two things: blast away the carbonyl and hook on a propyl group. Both of these transformations can be accomplished using the reactions we have seen.

If we blast away the carbonyl first, then we reach a dead-end:

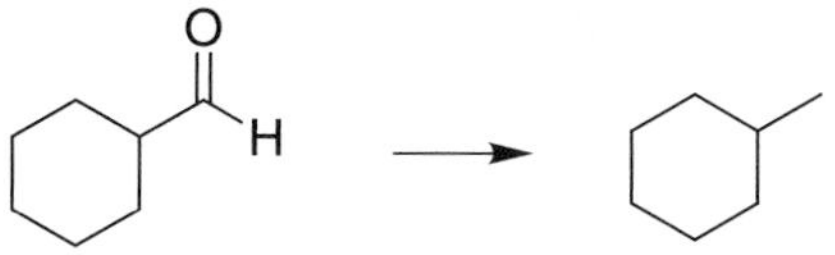

Now there is no way
to hook on
the propyl group

So instead, we hook on the propyl group first. To accomplish this, we turn the aldehyde into a thioacetal:

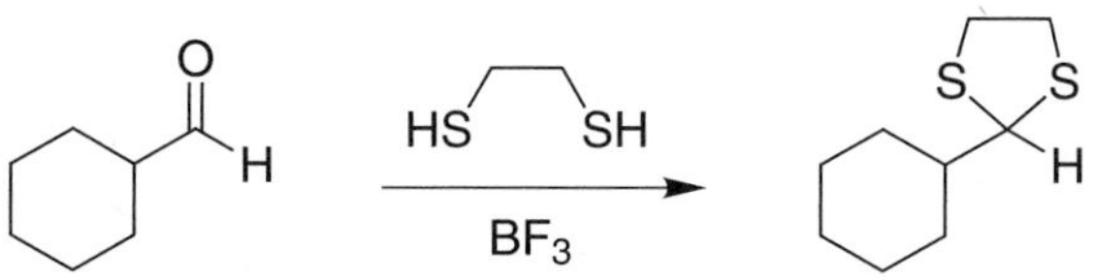

HS⌒SH

BF₃

and then we must deprotonate and react with propyl chloride (this requires two steps):

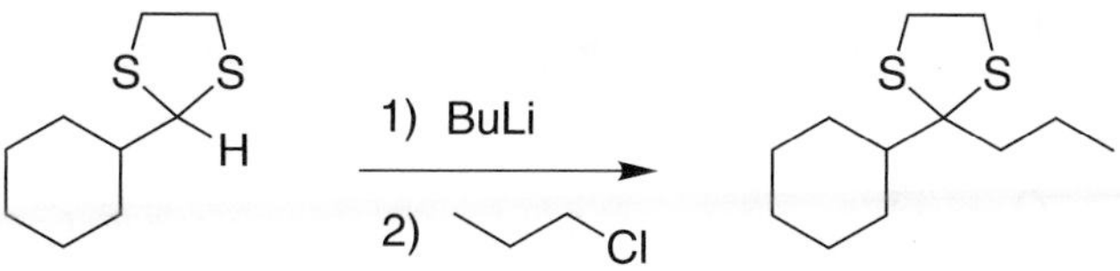

1) BuLi

2) ⌒⌒Cl

Finally, we pull off the thioketal, to get our product:

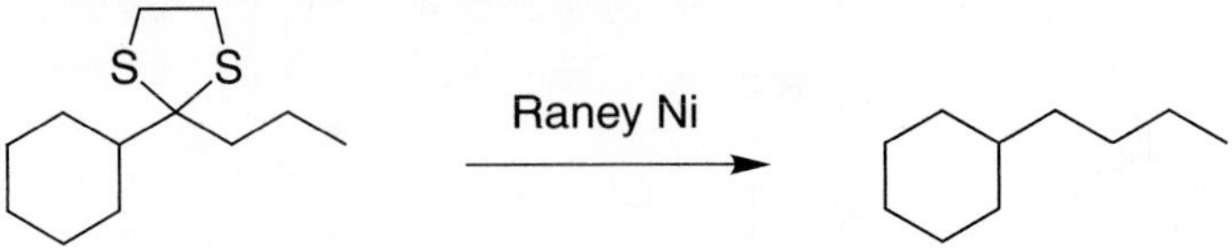

Raney Ni

So our proposed synthesis is:

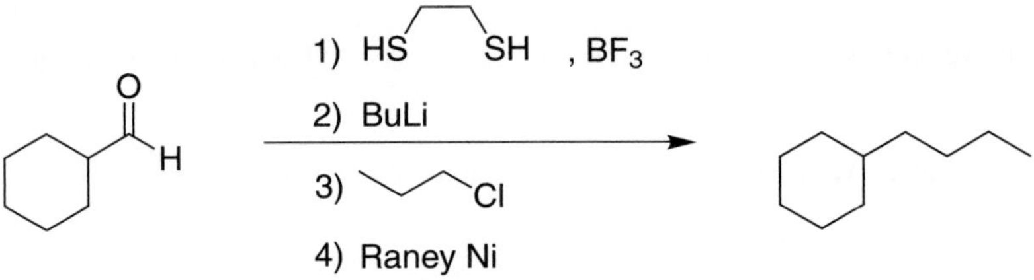

1) HS   SH , BF$_3$

2) BuLi

3)   Cl

4) Raney Ni

Propose a plausible synthesis for each of the following transformations:

5.40.

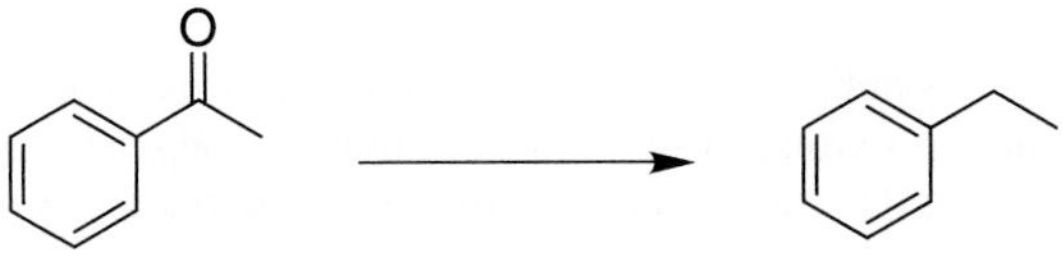

5.41.

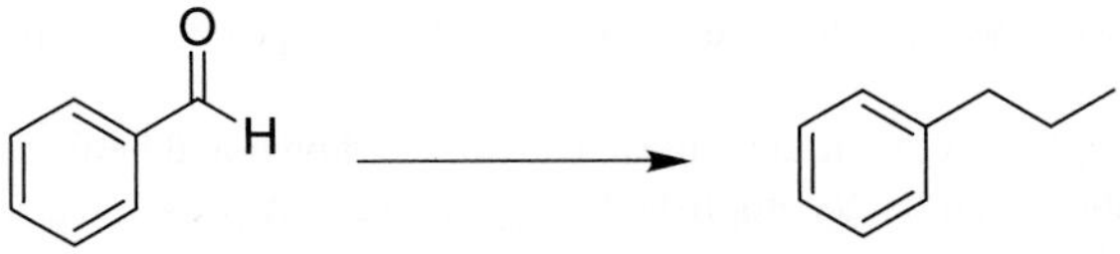

5.42.

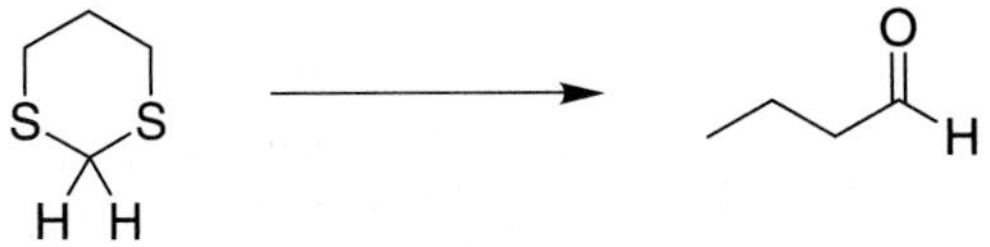

5.43.

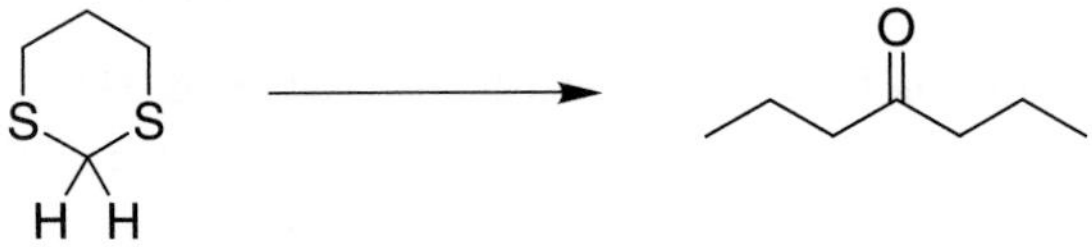

5.44.

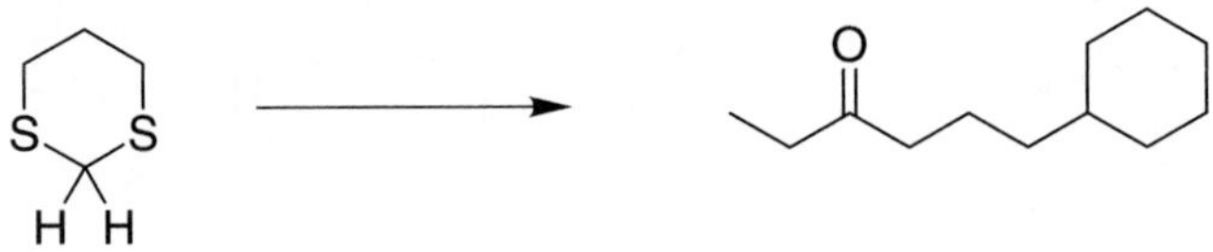

5.45.

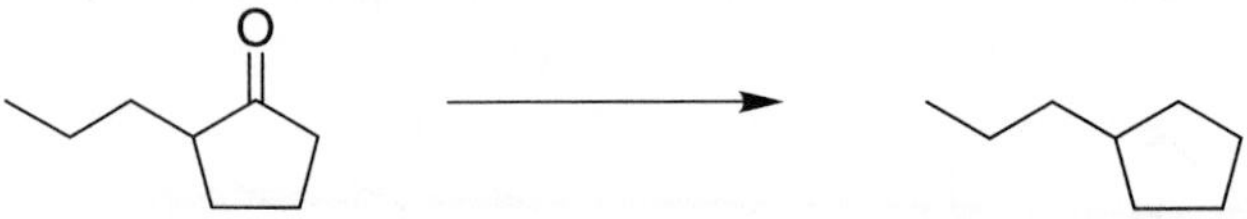

5.46.

5.6 N-NUCLEOPHILES

When we first learned about ketal formation, we said that the mechanism would serve as a foundation for other reactions in this chapter. Now, we will see how mechanisms can help us understand the similarities between reactions.

Compare the products of the following reactions:

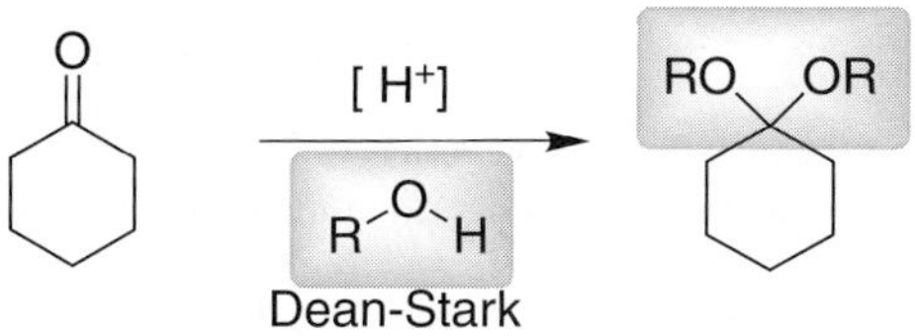

The products look very different from each other. When you use ROH as the nucleophile, you get a ketal; but when you use a primary amine as the nucelophile, you get a very different product.

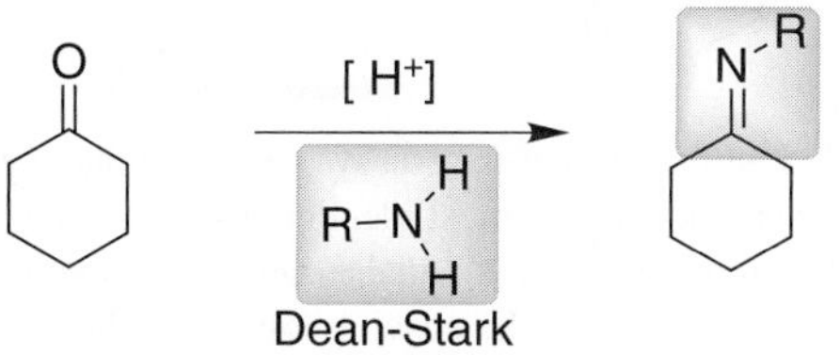

If you use a *secondary* amine as the nucleophile, then you get yet another product:

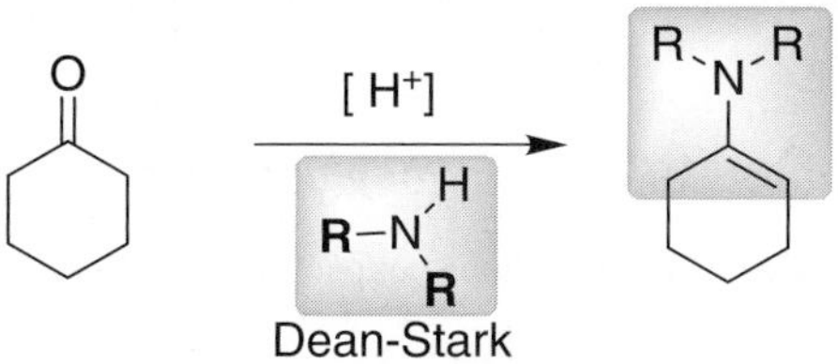

The products of these reactions look very different, but when we analyze the mechanisms, we will see that they are all *exactly* the same up until the very end of the mechanism. It is the last step of each mechanism that makes them different from each other. Let's take a closer look. We'll start with primary amines.

When a primary amine attacks a ketone, the mechanism starts off exactly the same as the mechanism of ketal formation. Shown below is the first two-thirds of the mechanism showing what happens when ROH attacks a ketone. And directly below it, you will see most of the mechanism showing what happens when RNH$_2$ attacks a ketone. Compare both mechanisms, step-by-step:

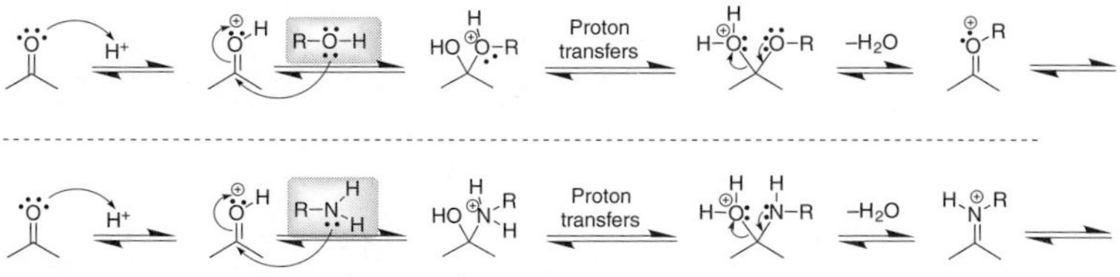

So far, both mechanisms are exactly the same!!! In both mechanisms, the first step is a proton transfer (to make the carbonyl more electrophilic). Then, we attack with the nucleophile, followed by more proton transfers and then loss of water. But here is where our mechanisms depart from each other. Let's try to understand why.

In the first mechanism (ketal formation), we had to attack with another molecule of ROH because there was no other way to get rid of the positive charge:

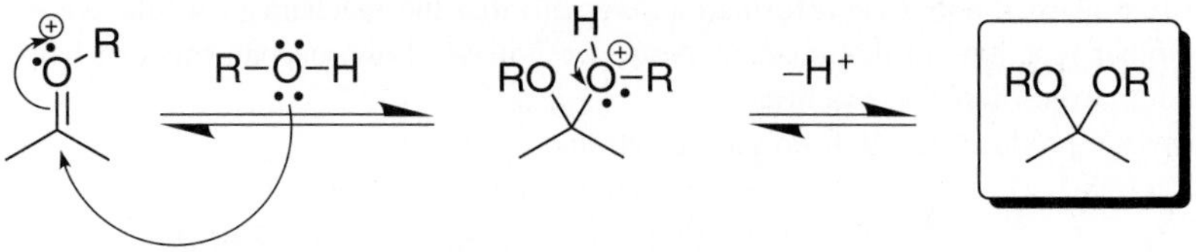

But when we attack with a primary amine, there is an easier way to get rid of the positive charge. Rather than attack with another molecule of amine, we can get rid of the charge by simply pulling off a proton:

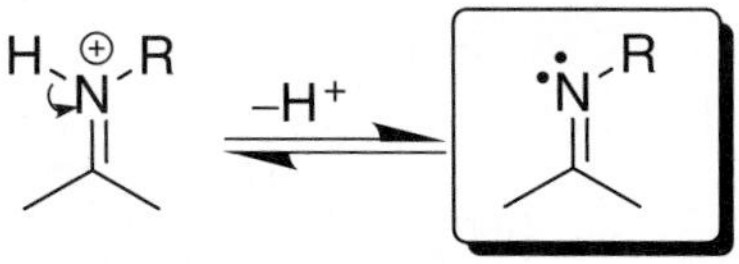

And this is our product. It is called an *imine* (that is what we call compounds that have a C=N double bond). So the mechanism of this reaction is almost identical to the mechanism of ketal formation, except for the very end. And the difference at the end makes sense when you really think about it.

When we do this reaction, we just need to take special notice of whether or not the starting ketone is symmetrical:

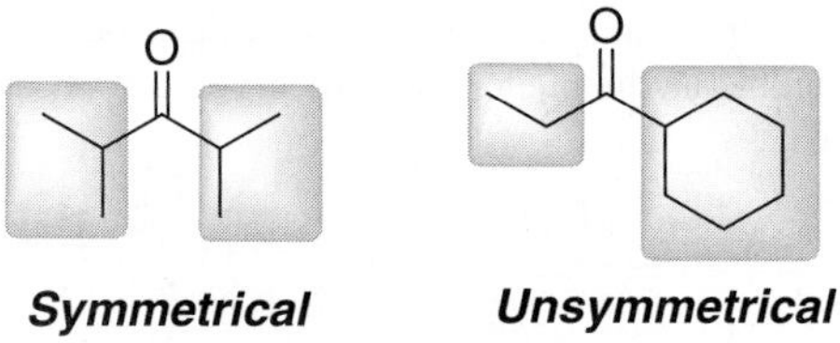

Symmetrical   **Unsymmetrical**

If the starting ketone is *un*symmetrical, then we should expect to form two diastereomeric imines:

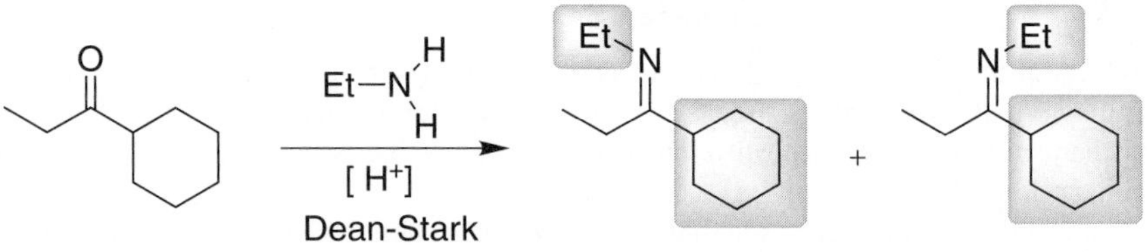

So far, we have seen what happens when a *primary* amine attacks a ketone. Now, let's see what happens when we attack with a *secondary* amine. Let's compare it to the mechanisms that we have seen so far:

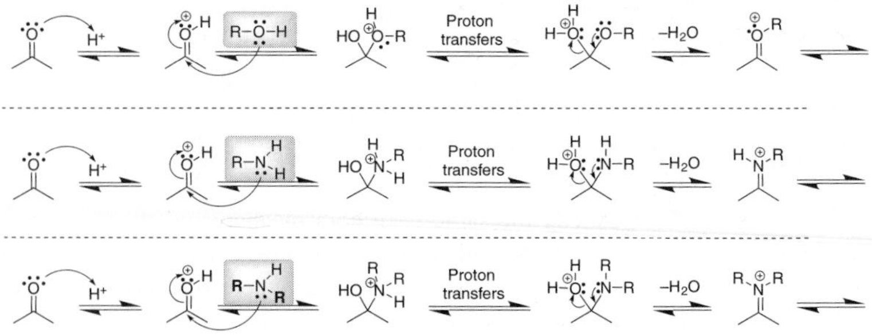

None of these mechanisms is complete: all three of them are missing the very last steps. But compare the steps that are shown and notice that, once again, these mechanisms are identical up until the very end of each mechanism. And it is right at the end where we see differences in the final products. In the first mechanism (ROH as the nucleophile), we saw that another molecule of ROH attacks. In the second mechanism, we saw that we were just able to lose a proton to form an imine. But in the third reaction, we cannot just lose a proton the way we did in the second mechanism. We might therefore be tempted to do what we did in the first mechanism (ketal formation). We might be tempted to say that another molecule of amine should come in and attack. But there is something else that happens instead:

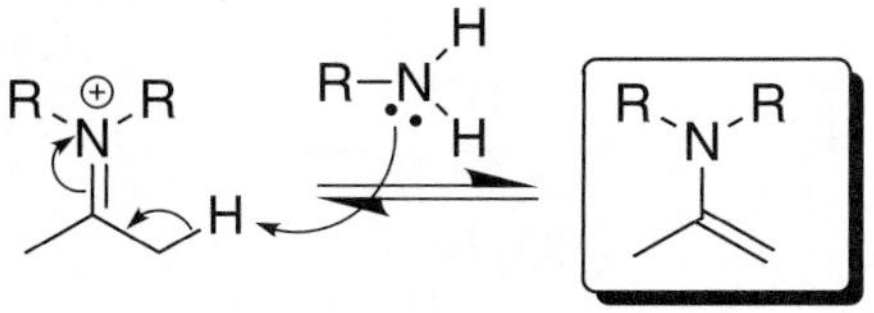

We use another molecule of the secondary amine to function as a base, rather than a nucleophile, and we do in fact find a proton to pull off. This gives us a product called an *enamine* ("en" because there is a double bond and "amine" because there is an NH_2 group).

Once again, we need to observe whether the ketone is unsymmetrical. If it is, then there will be two ways to form the double bond in the last step of the mechanism. This will give us two different enamine products. Here is an example:

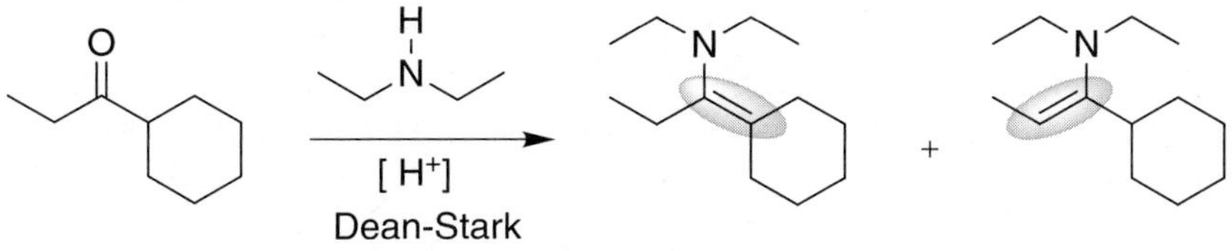

In a situation where we start with an unsymmetrical ketone, the major product will be the enamine with the *less* substituted double bond:

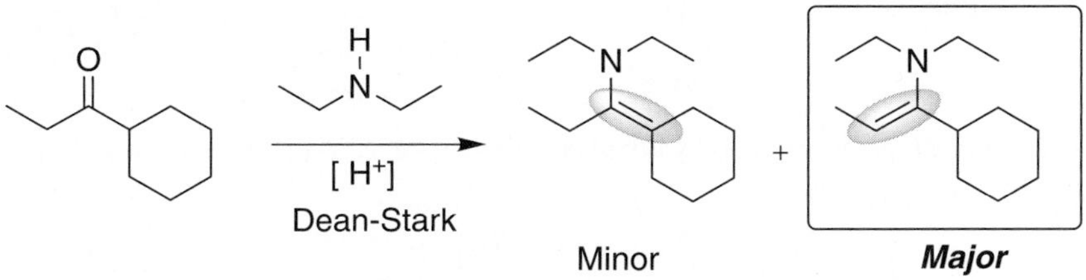

So far in this section, we have seen two new reactions (with primary amines, and with secondary amines), and we have seen the similarities in the mechanisms. Now, we will revisit the first reaction (the reaction between a ketone and a *primary* amine):

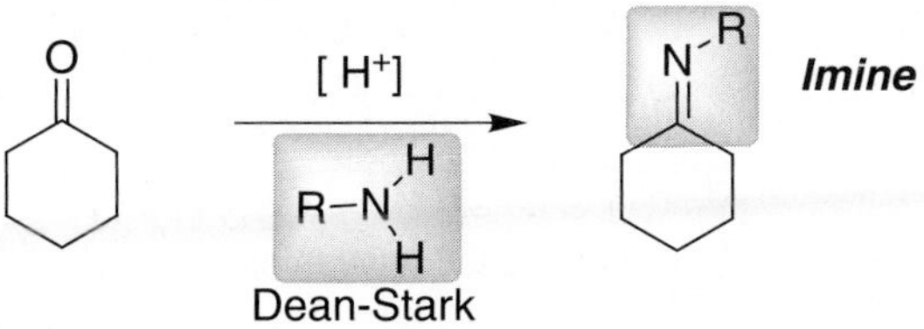

We normally think of the R (in RNH$_2$) as referring to an alkyl group (that is usually what R means). But we can also think of R as being something *other than an alkyl group*. For example, let's say we define R as being OH. In other words, we are starting with the following amine:

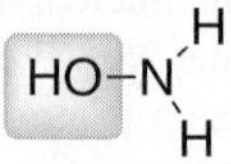

$$ HO-N\begin{matrix} H \\ \\ H \end{matrix} $$

This compound is called hydroxylamine, and the product that it forms (when it reacts with a ketone) is not surprising at all:

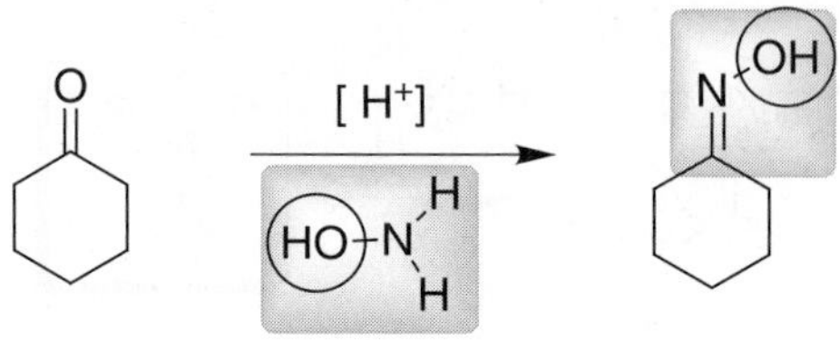

It is the exact same reaction as if it were a primary amine reacting with the ketone. But instead of getting an imine, we get something that we call an oxime:

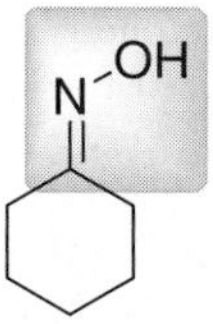

Oxime

Remember always to look if the starting ketone is unsymmetrical. If it is, we should expect to form two diastereomeric oximes:

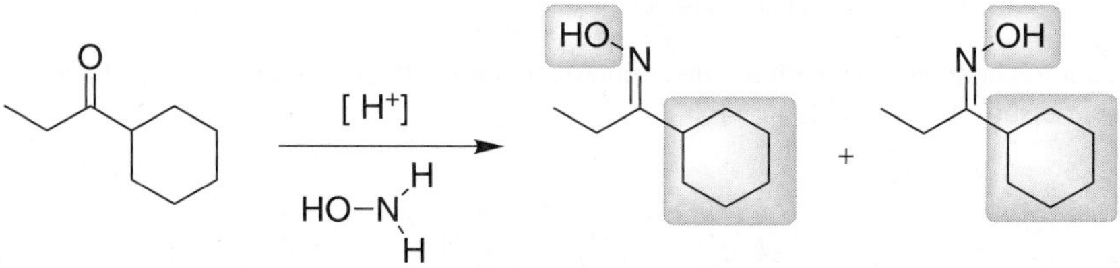

When you see this type of reaction, hydroxylamine might be shown in several ways. Here are the different ways you might see it:

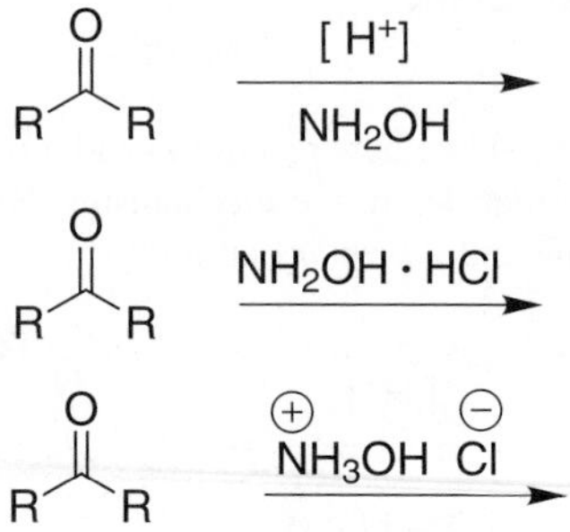

$$ \underset{NH_2OH}{\overset{[\,H^+]}{\longrightarrow}} $$

$$ \xrightarrow{NH_2OH \cdot HCl} $$

$$ \xrightarrow{\overset{\oplus}{NH_3OH}\ \overset{\ominus}{Cl}} $$

All of these are just different ways of showing the same thing.

Now that we have seen this special N-nucleophile (RNH$_2$ where R = OH), let's take a close look at one more special N-nucleophile. Let's look at a case where R = NH$_2$. In other words, we are looking at the following nucleophile:

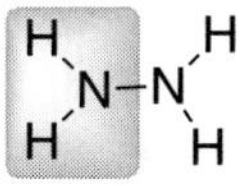

This compound is called hydrazine, and the product that it forms (when it reacts with a ketone) is not surprising at all:

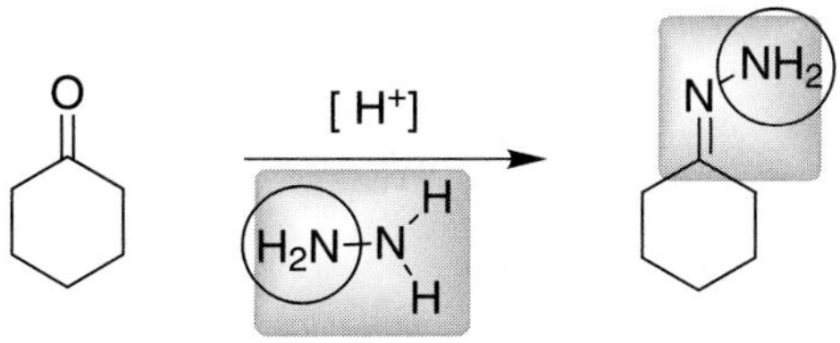

It is the exact same reaction as if it were a primary amine reacting with the ketone. But instead of getting an imine or an oxime, we get something that we call a hydrazone:

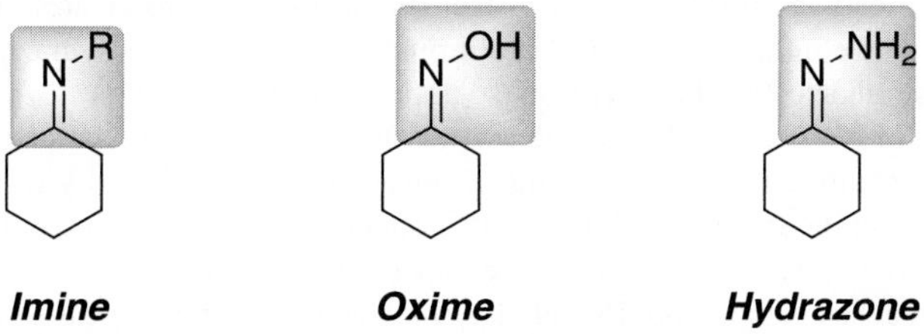

Imine   **Oxime**   **Hydrazone**

As is true of all the other reactions we have seen in this section, we need to take special notice of whether or not the starting ketone is symmetrical. If the starting ketone is unsymmetrical, then we should expect to form two diastereomeric hydrazones:

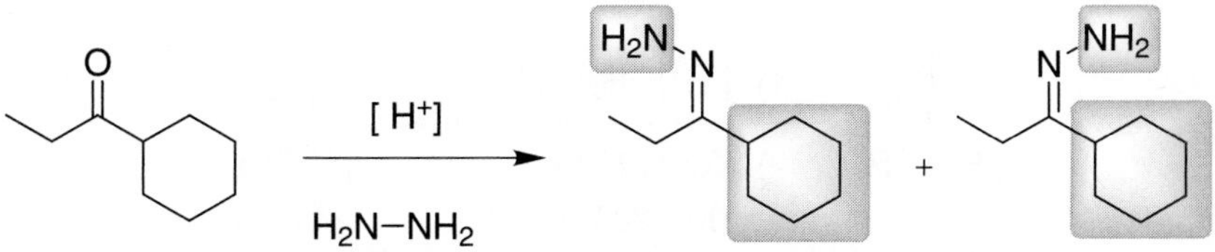

Hydrazones are useful for many reasons. In the past, chemists formed hydrazones as a way of identifying ketones, but, with the advent of NMR techniques, no one uses hydrazones that way anymore. But there is still one practical use in modern-day organic chemistry. A hydrazone can be reduced to an alkane under basic conditions:

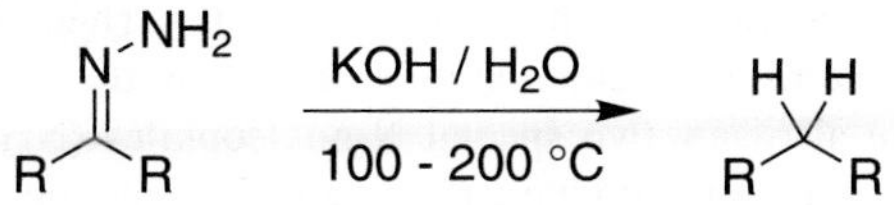

The mechanism goes like this:

The formation of a carbanion (highlighted in the second-to-last step) creates an uphill battle (in terms of energy), so we might expect that equilibrium should favor the hydrazone (starting material) rather than the alkane (product). However, notice that the formation of the carbanion is accompanied by loss of N_2 gas (also highlighted in the mechanism above). This explains why the reaction goes to completion. The small amount of nitrogen gas (produced by the equilibrium) will bubble out of the solution and escape into the atmosphere. That forces the equilibrium to produce a little bit more nitrogen gas, which also then escapes into the atmosphere. The process continues until the reaction reaches completion. Essentially, we are removing a reagent as it is being formed, and that is what pushes the equilibrium over the high-energy barrier created by the instability of the carbanion. If you think about it, this concept is not so different from those in the previous sections where we removed water from a reaction as it was being formed (as a way of pushing the equilibrium toward formation of the ketal).

This now gives us a new, two-step way of reducing a ketone to an alkane:

We have already seen two other ways to do this kind of transformation (the Clemmensen reduction and desulfurization with Raney Nickel). This is now our third way to reduce a ketone to an alkane, and it is called a Wolff-Kishner reduction.

In this section, we have only seen a few reactions. Here is a short summary. We first saw how a ketone can react with a *primary amine* to form an *imine* (and we saw that the mechanism was very similar to ketal formation, except for the very end). Then, we saw how a ketone can react with a *secondary amine* to form an *enamine* (once again, the mechanism was very similar up until the very end). We also saw two special N-nucleophiles (NH_2OH and NH_2NH_2), both of which gave us products that we would have expected. The reaction with NH_2NH_2 was of special interest because it gave us a new way to reduce ketones to alkanes.

Now let's solve some problems to make sure you are familiar with the reagents and the mechanisms for the reactions that we have seen in this section. Let's start with mechanisms:

EXERCISE 5.47. Propose a plausible mechanism for the following reaction:

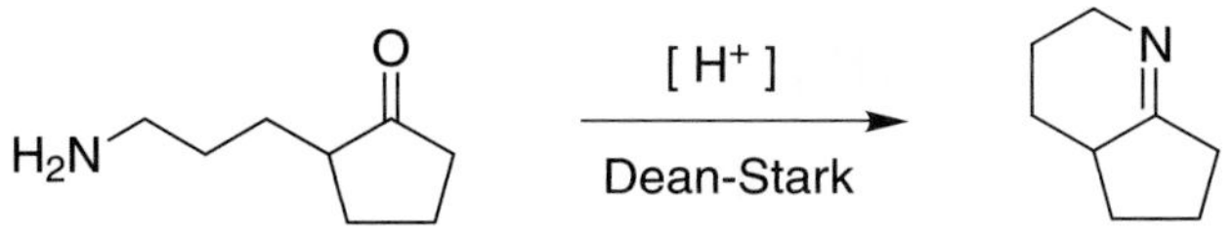

Answer: We begin by looking at the starting material. We see that it has two functional groups: it is a primary amine, ***and*** it is a ketone. Therefore, it could theoretically attack itself. (When two regions of the same molecule react with each other, we call it an ***intramolecular*** reaction.) Then we look at the reagents (acid and Dean-Stark conditions), and we notice that we are missing a nucleophile. This further supports the idea that we are doing an intramolecular reaction here. The starting material can function as both the nucleophile and the electrophile. Finally, we look at the product, and we see that it is an imine, which is the type of product that results from the reaction between a primary amine and a ketone. With all of this evidence, we conclude that it is an intramolecular mechanism.

The mechanism will therefore have exactly the same steps as any other mechanism involving a primary amine attacking a ketone (protonate, attack, proton transfers, lose water, and then deprotonate):

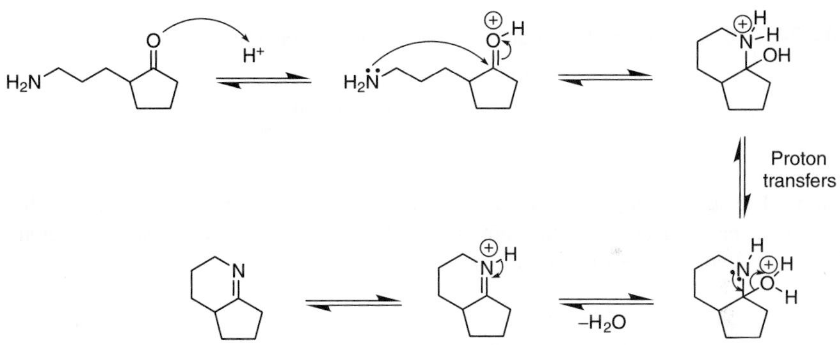

Propose a plausible mechanism for each of the following reactions. You will need a separate piece of paper to record your answers.

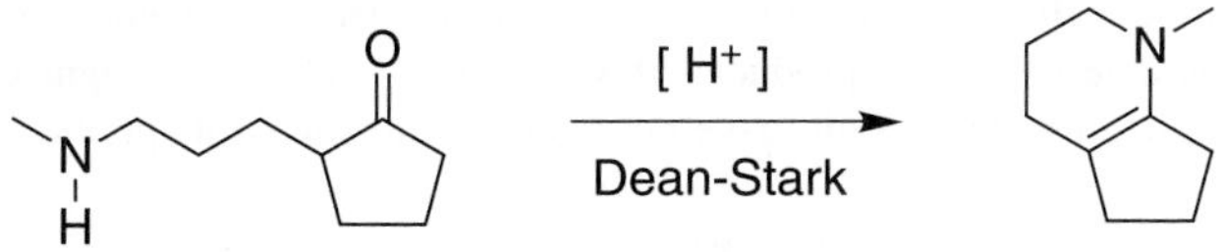

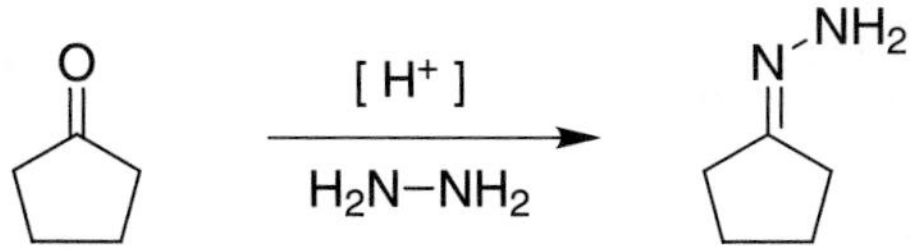

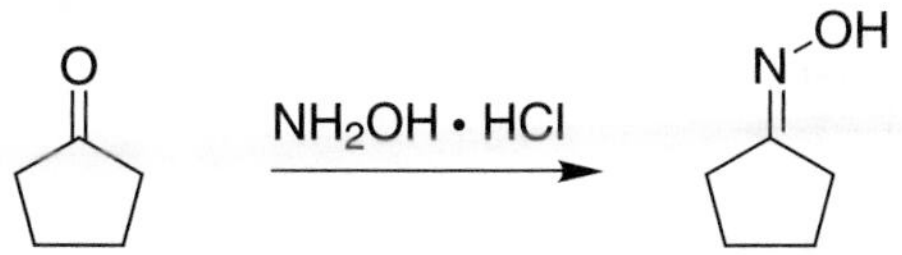

5.51.

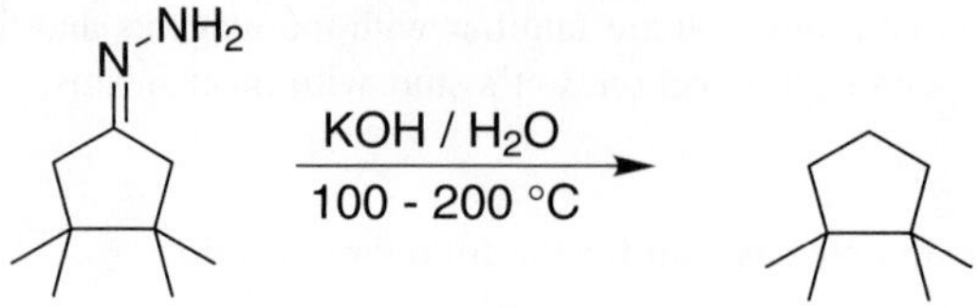

$$\xrightarrow[\text{100 - 200 °C}]{\text{KOH / H}_2\text{O}}$$

5.52.

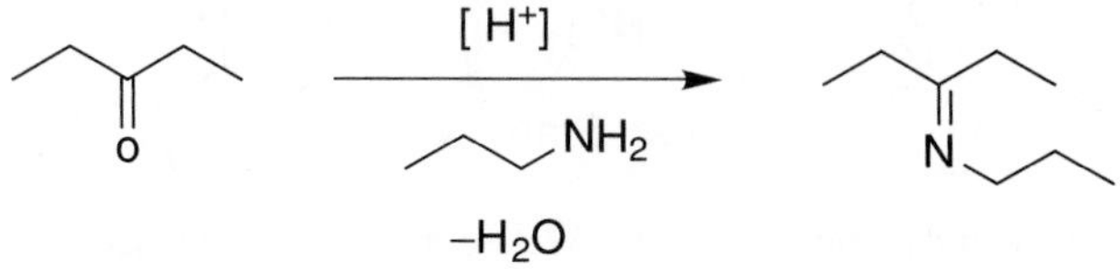

$$\xrightarrow[\substack{\text{NH}_2 \\ -\text{H}_2\text{O}}]{[\text{H}^+]}$$

5.53.

$$\xrightarrow[\substack{[\text{H}^+] \\ \text{Dean-Stark}}]{\substack{\text{H} \\ \text{N}}}$$

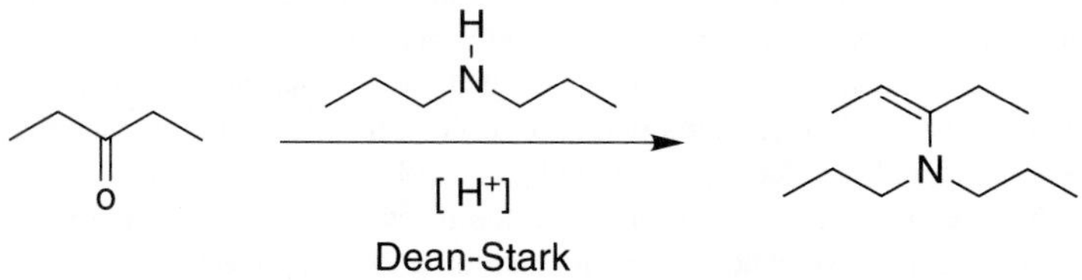

Now let's get some practice with predicting products.

EXERCISE 5.54. Predict the products of the following reaction:

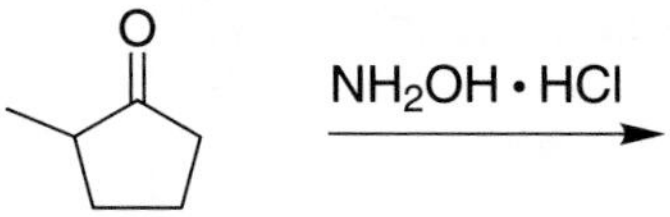

$$\xrightarrow{\text{NH}_2\text{OH} \cdot \text{HCl}}$$

Answer: The starting material is a ketone, and the reagent is hydroxyl amine. As we have seen in this section, the product of this reaction should be an oxime. Since the starting ketone is un-symmetrical, we would expect two oxime products:

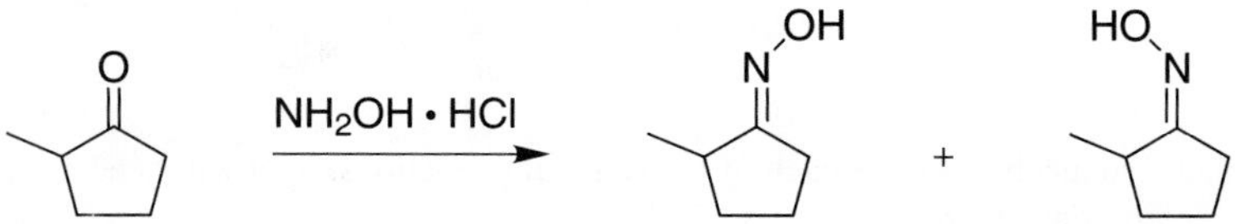

$$\xrightarrow{\text{NH}_2\text{OH} \cdot \text{HCl}} \quad + \quad$$

We saw several reactions in this chapter. You must be able to recognize the reagents for these re-actions, so that you will be able to predict products. If you were not able to recognize that the reagent here is hydroxyl amine, then you would have not been able to solve this problem.

Predict the products for each of the following transformations:

5.55.

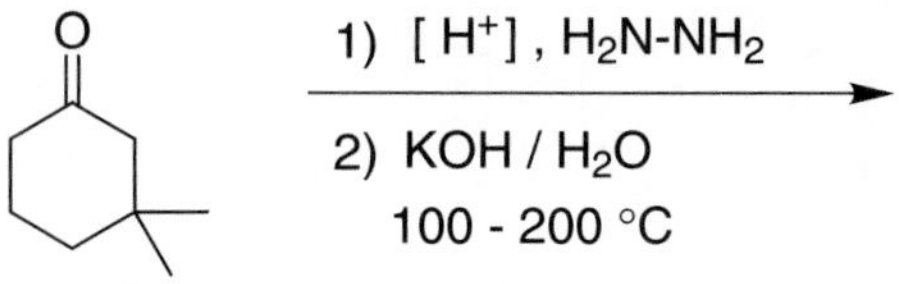

$$\xrightarrow[\substack{\text{2) KOH / H}_2\text{O} \\ \text{100 - 200 °C}}]{\text{1) } [\text{H}^+] \text{, H}_2\text{N-NH}_2}$$

5.56.

$$\xrightarrow{\text{NH}_2\text{OH} \cdot \text{HCl}}$$

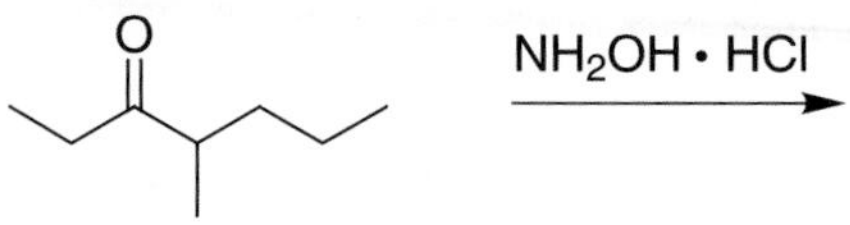

5.57.

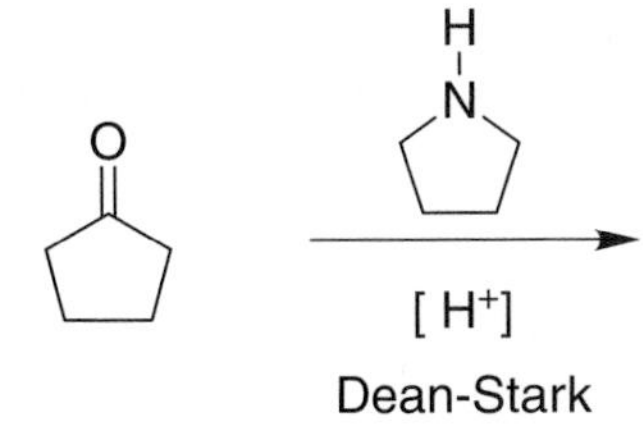

[H⁺]

Dean-Stark

5.58.

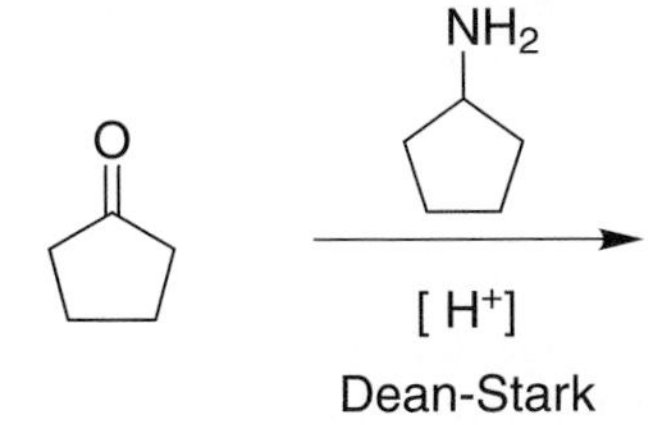

[H⁺]

Dean-Stark

5.59.

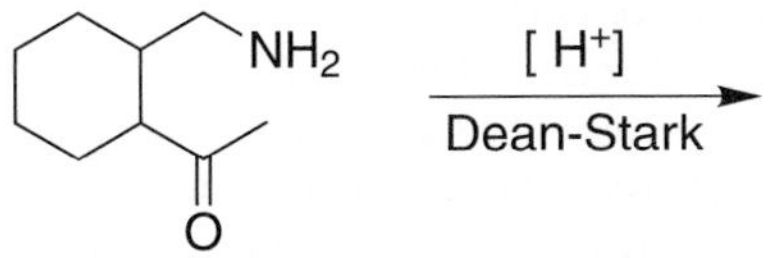

[H⁺]

Dean-Stark

5.60.

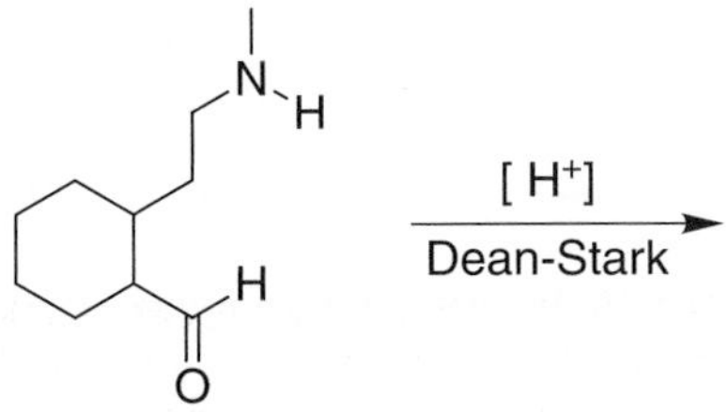

[H⁺]

Dean-Stark

5.7 C-NUCLEOPHILES

In this chapter, we have seen that many different kinds of nucleophiles can attack ketones and aldehydes. We started with hydrogen nucleophiles, and then we moved on to oxygen nucleophiles and
sulfur nucleophiles. In the previous section, we covered nitrogen nucleophiles. In this section we
will discuss three types of carbon nucleophiles.

Our first carbon nucleophile is the Grignard reagent. You may have been exposed to this reagent
in the first semester, but if you weren't, here is a quick overview:

Alkyl halides will react with magnesium in the following way:

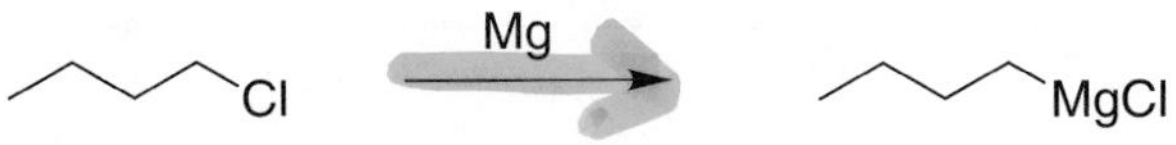

Essentially, an atom of magnesium positions itself in between the C-Cl bond (this reaction works
with other halides as well, such as Br or I). This magnesium atom has drastic effects on the electronics of the carbon atom to which it is attached. To see the effects, consider the electronics of the
alkyl halide (before Mg entered the picture):

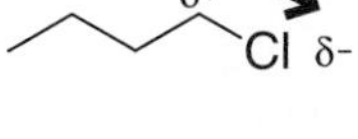

The carbon atom (connected to the halogen) is poor in electron density, or $\delta+$, because of the inductive effects of the halogen. After the magnesium is placed between C and Cl, however, the story changes very drastically:

Carbon is much more electronegative than magnesium. Therefore, the inductive effect places a lot of electron density on the carbon atom, making it very $\delta-$. This bond is polar covalent, but for purposes of simplicity, we will treat it like an ionic bond:

Because carbon is not very good at stabilizing a negative charge, this reagent (called a Grignard reagent) is highly reactive. It is a very strong nucleophile and a very strong base. Now let's see what happens when a Grignard reagent attacks a ketone or aldehyde.

In the previous section, we always started each mechanism by protonating the ketone (turning it into a better electrophile). That is not necessary here because the Grignard reagent is such a strong nucleophile that it has no problem attacking a carbonyl group. We could **not** use acid catalysis here (even if we wanted to) because protons destroy Grignard reagents. For example, consider what happens when a Grignard is exposed to water:

The Grignard reagent acts as a base and grabs a proton from water to form a more stable hydroxide ion. The negative charge is MUCH happier on an electronegative atom (oxygen), and as a result, the reaction essentially goes to completion. This means that you can never use a Grignard reagent to attack a compound that has protons connected to electronegative atoms. For example, you could never do the following reaction:

Because this would happen instead:

In general, proton transfers are faster than nucleophilic attack. And when the Grignard reagent grabs a proton, it irreversibly destroys the Grignard reagent. Similarly, you could never prepare the following kinds of Grignard reagents:

These reagents could not be formed because each of these reagents could react with itself to get rid of the negative charge on the carbon atom, for example:

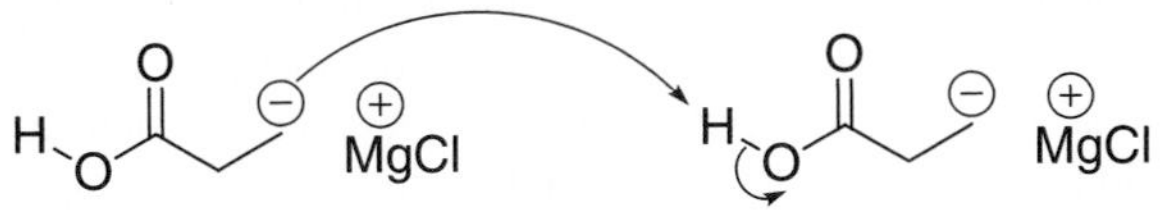

All of that was a quick review of Grignard reagents. Now let's see how Grignard reagents can attack a ketone or aldehyde. In the first step, the Grignard attacks the carbon of the carbonyl group:

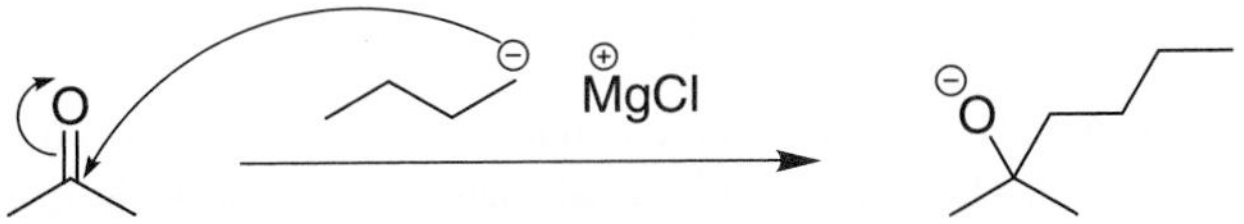

This intermediate then wants to re-form the carbonyl, if it can, but let's see if it can. Remember our rules from the beginning of this chapter: re-form the carbonyl if you can, but never kick off H^- or C^-. This intermediate is NOT able to re-form the carbonyl because there are no leaving groups to kick off. This is true whether we attack a ketone or an aldehyde:

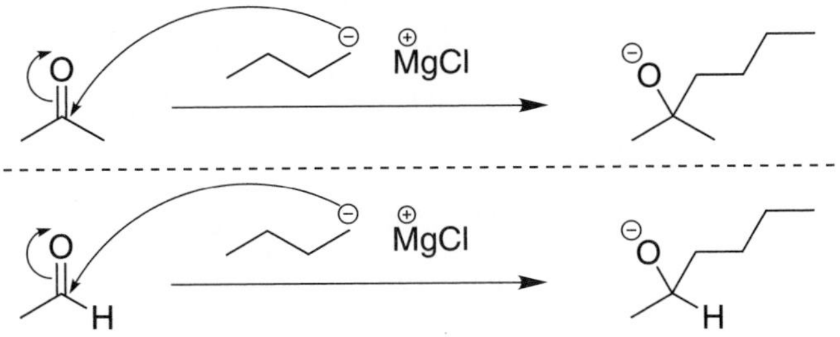

So, in either case, the reaction is over, and we must now give the intermediate a proton to get to our final product, which is an alcohol:

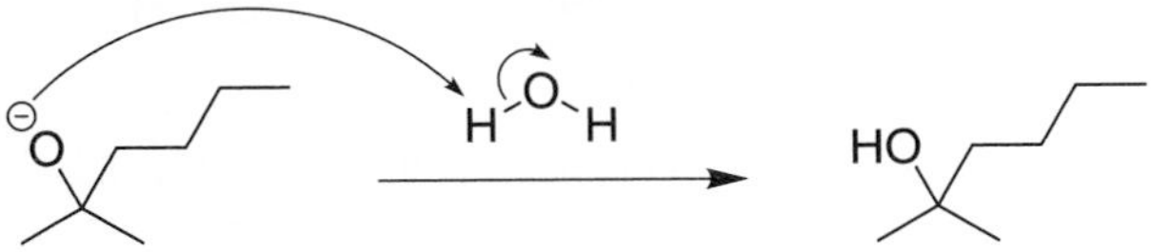

This reaction is not so different from the reactions we saw earlier in this chapter when we looked at hydrogen nucleophiles ($NaBH_4$ and LAH). We saw the exact same scenario there: the nucleophile attacked, and then the carbonyl was NOT able to re-form because there was no leaving group. Compare one of those reactions to this reaction:

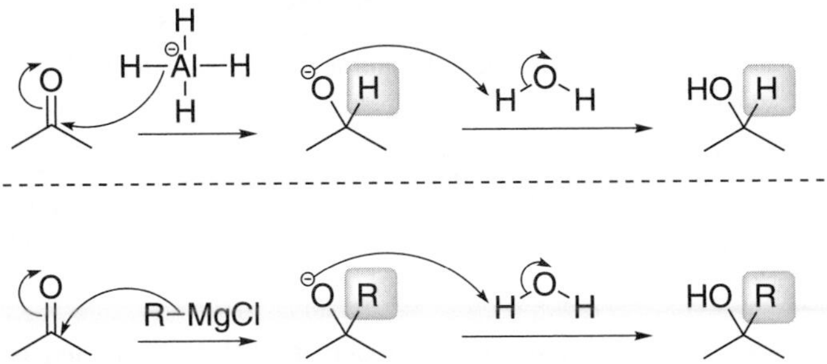

Notice that the mechanisms are identical. And it is worth a minute of time to think about why these reactions are so similar (while the other reactions in this chapter were different from these two reactions). What is special about these two reactions that makes them so similar? Remember our golden rule: never kick off H⁻ or C⁻. So, if we attack a ketone (or aldehyde) with either H⁻ or C⁻, then the carbonyl will be unable to re-form. And that is what these two reactions have in common.

When you write down the reagents of a Grignard reaction (in a synthesis problem), make sure you show the proton source *as a separate step*:

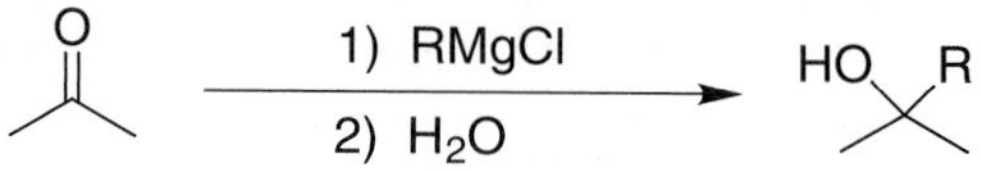

We saw this important subtlety when we learned about LAH, where we also had to show the proton source as a separate step. The same subtlety exists here because (as we have very recently seen) a Grignard reagent will not survive *in the presence* of a proton source. The proton source must come AFTER the reaction is complete (after the Grignard reagent has been consumed by the reaction).

In order to add this reaction to your toolbox of synthetic transformations, let's compare it one more time to the reaction with LAH. But this time, let's focus on comparing the products rather than comparing the mechanisms:

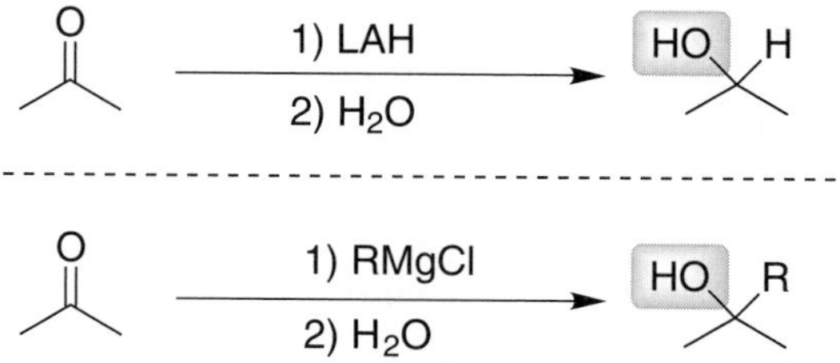

Notice that in both reactions we are reducing the ketone to an alcohol. But in the case of a Grignard reaction, the reduction is accompanied by the introduction of an alkyl group:

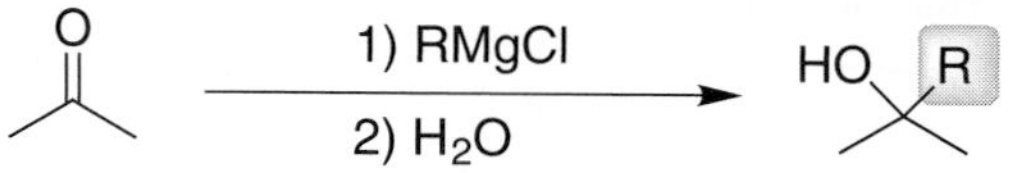

This will be helpful as we explore synthesis problems at the end of this chapter.

EXERCISE 5.61.   Predict the products of the following reaction:

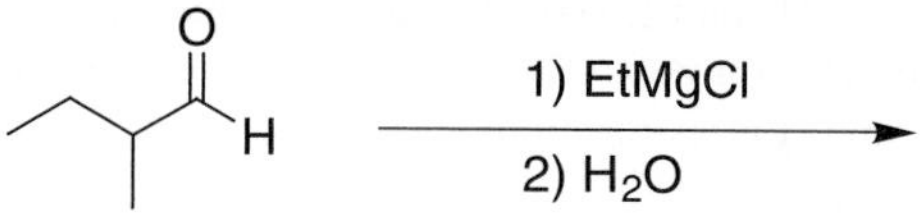

Answer: The starting material is an aldehyde, and it is going to react with a Grignard reagent. First, the Grignard attacks:

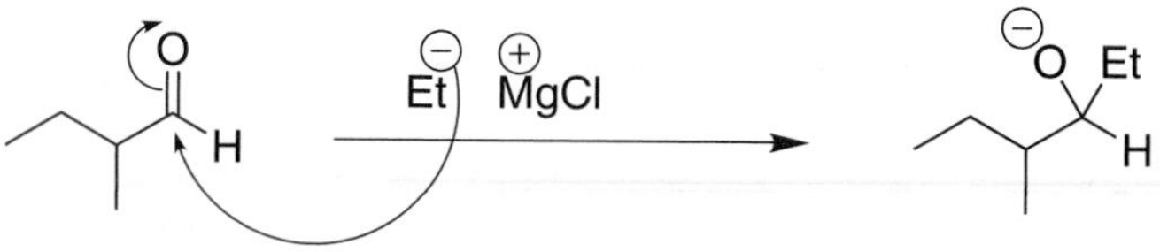

We cannot re-form the carbonyl because we cannot kick off H⁻ or C⁻, so there is nothing to kick off. All that we can do is take a proton from water to get our product:

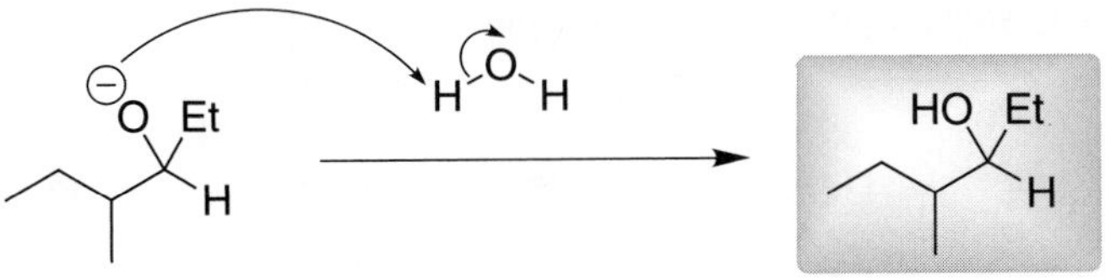

Because you don't have to show hydrogen atoms, you can redraw your product like this:

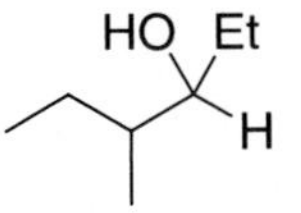

HO Et *is the same as* OH

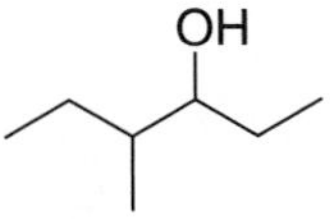

Predict the products for each of the following reactions:

5.62.

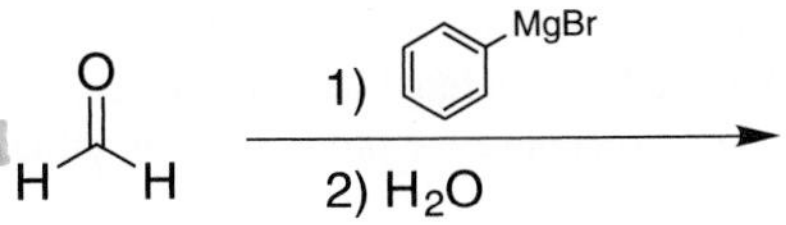

1) [C₆H₅]MgBr

2) H₂O

5.63.

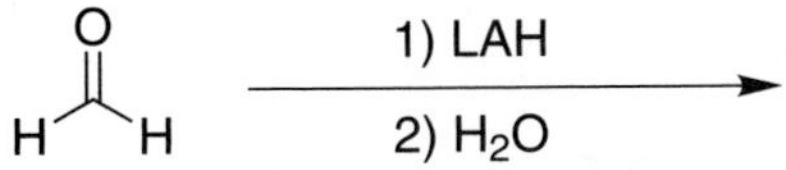

1) LAH

2) H₂O

5.64.

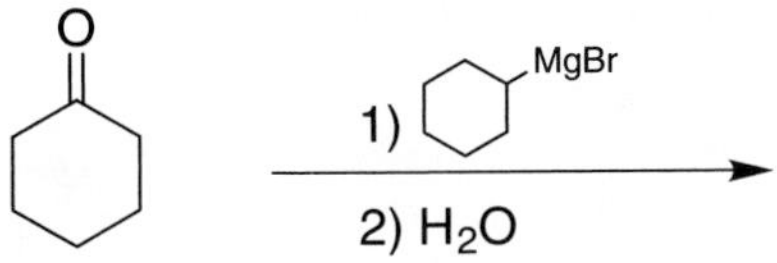

1) [cyclohexyl]MgBr

2) H₂O

5.65.

1) [C₆H₅]MgBr

2) H₂O

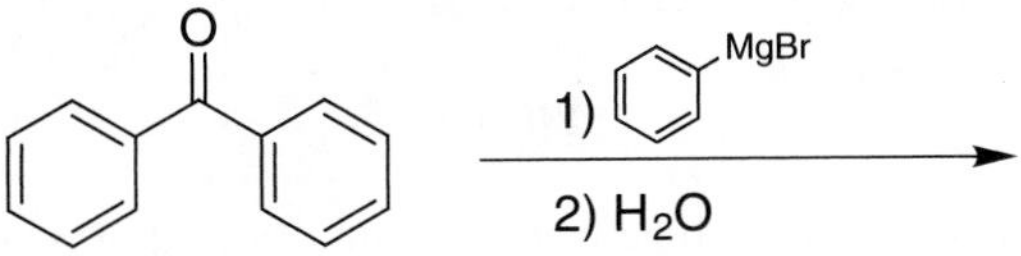

We must look at two more carbon nucleophiles now, both of which are different from the Grignard reagent. You will have to add these two new reactions to your toolbox. Both reactions involve "ylides" (pronounced "il – ids"). Let's take a close look at the first ylide:

We start with a compound called triphenylphosphine:

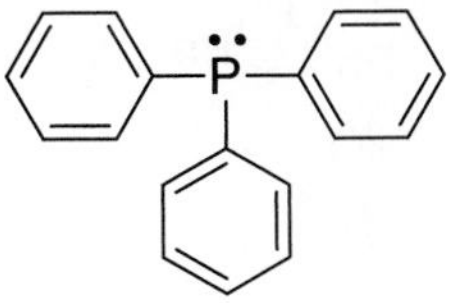

which can be drawn more quickly, like this Ph—P̈—Ph
                                                 |
                                                 Ph

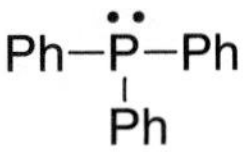

and we use this reagent to attack an alkyl halide (in an S$_N$2 reaction):

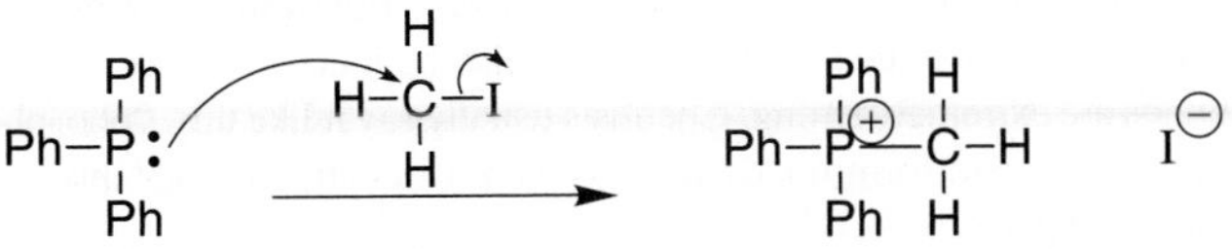

Ph
|
Ph—P: H—C—I
| |
Ph   H

Ph H
| |
Ph—P⁺—C—H   I⁻
| |
Ph H

Then, we use a very strong base to pull off a proton:

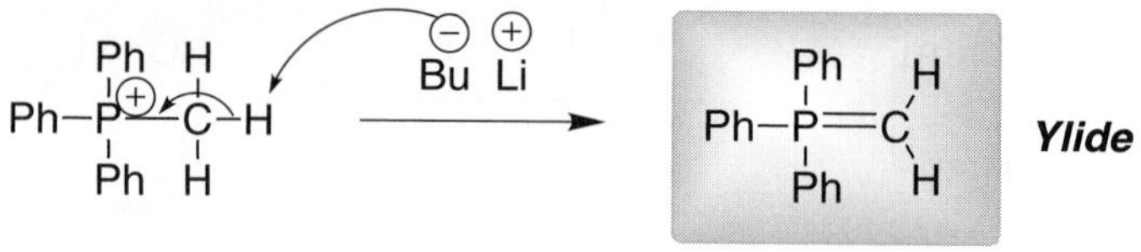

This compound is the important reagent that we call an ylide. It has two important resonance structures:

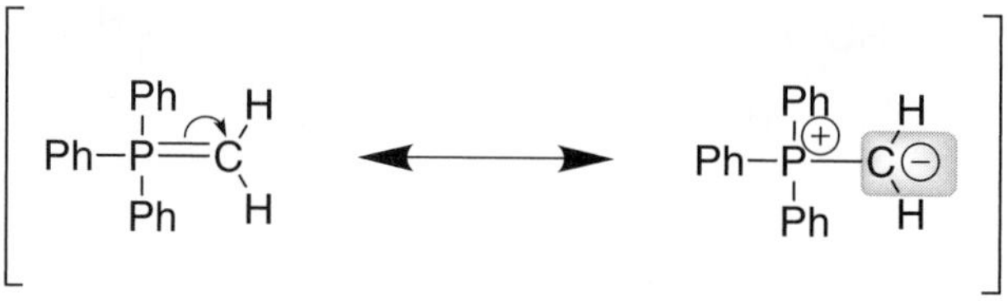

If we look closely, we will see that this compound has a region of high-electron density on a carbon atom. This means that we have now made a new carbon nucleophile, and this type of carbon nucleophile is called an ylide. Let's quickly review the procedure for preparing this ylide:

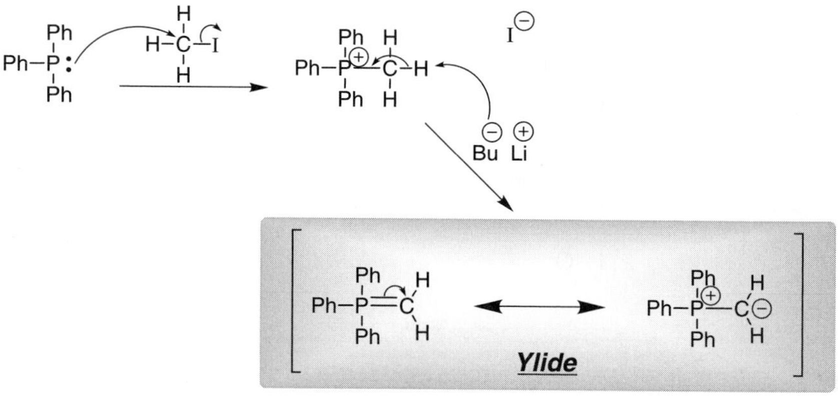

In a few minutes, we will see another type of ylide (one that uses sulfur instead of phosphorus). Our ylide here (based on phosphorus) has a special name—a Wittig reagent (pronounced "Vittig"). When we use a Wittig reagent to attack a ketone or aldehyde, we get a reaction that we call a Wittig reaction. So let's take a close look at the mechanism of a Wittig reaction:

The Wittig reagent attacks the carbonyl group in the same way that any nucleophile would attack a carbonyl:

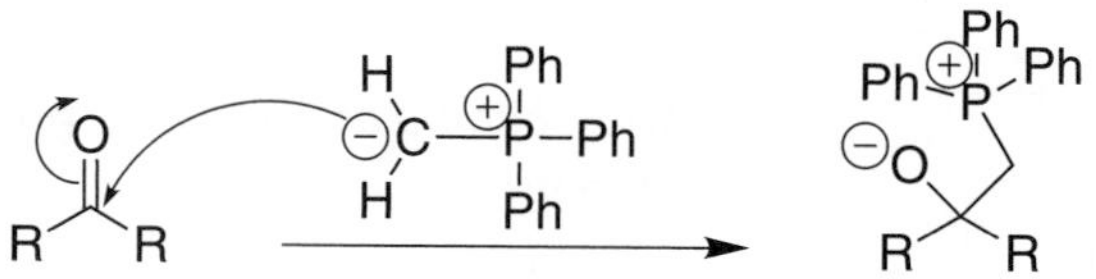

We have talked about how C=O double bonds are thermodynamically very stable, and therefore, the formation of a carbonyl is a driving force in reactions. In this reaction, we cannot re-form the carbonyl because we are not allowed to kick off H$^-$ or C$^-$. But something else can happen.

Another type of bond, which is also very stable, can also be a driving force in reactions. P-O single bonds and double bonds are extremely strong. Chemists usually say it like this: "phosphorus is oxophilic," meaning that phosphorus likes to form bonds with oxygen, if it can. And our intermediate is now perfectly set up for that to happen:

We can continue and form a P=O double bond now like this:

An alkene

And that gives us our product. Notice that it is an alkene.

This reaction is incredibly useful when you are solving synthesis problems. We already saw how to convert an *alkene* into a *ketone*, using an ozonolysis reaction. Now, with a Wittig reaction, we can go either way:

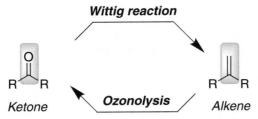

Wittig reaction

Ketone

Ozonolysis

Alkene

You should always take special notice whenever you learn how to interconvert two functional groups (going in either direction), like above. We have seen several cases like this in this book so far.

EXERCISE 5.66.   Predict the product of the following reaction.

Answer: We recognize the reagent to be a Wittig reagent. The tell-tale sign is the C=P double bond, but this reagent is slightly different from the one we saw before. Compare this reagent to the one we saw on the previous page. This reagent has one extra carbon. The way to form a reagent like this one is to use Et-I instead of Me-I when you are making the Wittig reagent.

The extra carbon atom comes along for the ride, and the final product looks like this:

You should try to draw out the mechanism of this reaction to make sure that you can "watch" the extra carbon atom coming along for the ride.

Predict the products of each of the following reactions:

5.67.

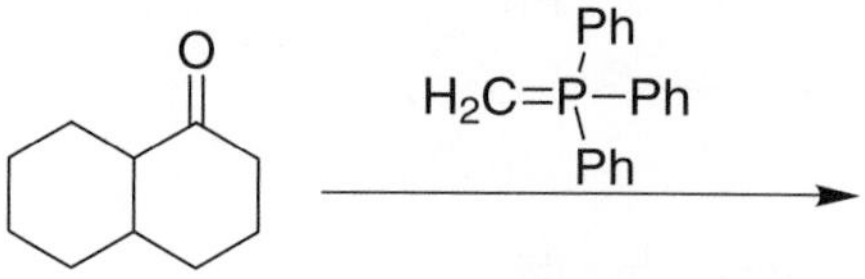

$$H_2C=P{-}Ph \quad (\text{with } Ph, Ph)$$

5.68.

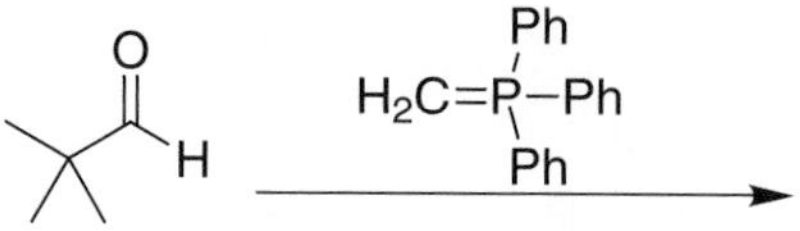

$$H_2C=P{-}Ph \quad (\text{with } Ph, Ph)$$

5.69.

$$H_2C=P{-}Ph \quad (\text{with } Ph, Ph)$$

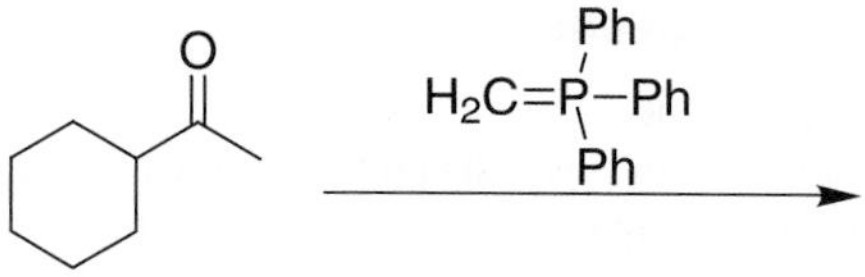

We now need to see one more ylide. This time, it is a *sulfur* ylide rather than a *phosphorus* ylide. Some instructors might skip sulfur ylides, so you might want to look through your notes and text-book to find out if you are responsible for the following reaction.

The mechanism for forming a *sulfur* ylide is very similar to the mechanism for forming a *phosphorus* ylide. Let's compare them:

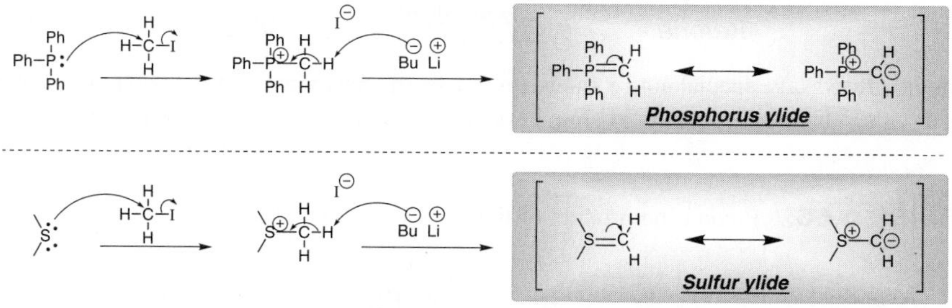

Phosphorus ylide

Sulfur ylide

To form a phosphorus ylide, we start with dimethyl sulfide (DMS). From that point on, everything is the same: we attack an alkyl halide, and then we deprotonate with a very strong base to form the ylide.

When a sulfur ylide attacks a ketone (or aldehyde), however, we get a very different product. We *don't* get an alkene as our product (as we did in the Wittig reaction). Instead, we get an epoxide:

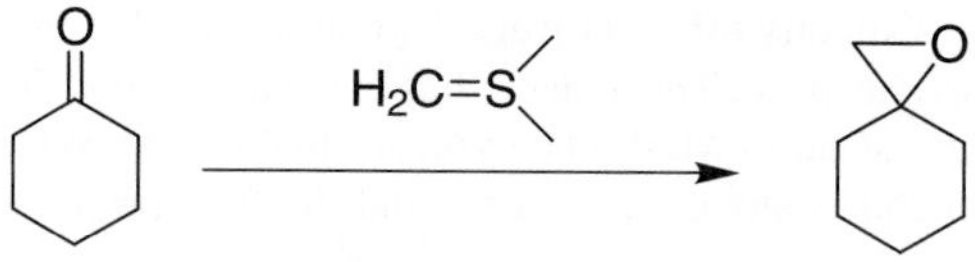

$$H_2C=S{\Big\langle}$$

Let's see how we get this strange product. The sulfur ylide attacks the carbonyl, just as any other nucleophile attacks a carbonyl:

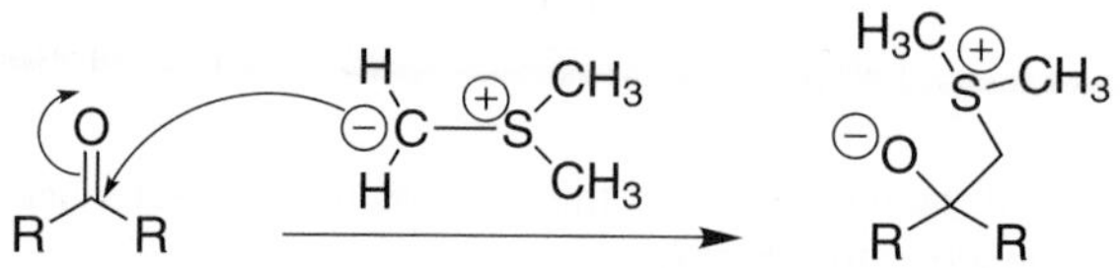

But here is where the reaction is different from a Wittig reaction because sulfur and oxygen do not form a bond the way phosphorus and oxygen did. Instead, the oxygen attacks in an intramolecular S_N2 reaction, kicking off DMS as a leaving group:

This reaction is very useful because it allows us to make epoxides from ketones. You should remember from the first semester that you learned a couple of ways to convert an *alkene* into an epoxide. But now we see that we can make an epoxide from a ketone as well:

We have now seen three carbon nucleophiles in this section. We started with Grignard reagents, and then we moved on to ylides (phosphorus ylides and sulfur ylides). We have seen that phosphorus ylides and sulfur ylides produce very different products:

The products are very different from each other, but the mechanisms are identical up until the very last step. If you don't see that, then go back to the mechanisms of each reaction and compare them. You will see for yourself that only the last step is different.

We have now seen the same common idea many times in this chapter. We have seen that you can often have two reactions that produce very different products, but when you look at the mechanisms carefully, you see that they are extremely similar.

EXERCISE 5.70.   Predict the major product of the following reaction:

Answer: This reagent is just a sulfur ylide. The tell-tale sign is the C=S double bond (where the sulfur has two other groups attached to it). Sulfur ylides are used to convert ketones into epoxides. So, our product is:

Predict the products of each of the following reactions:

5.71.

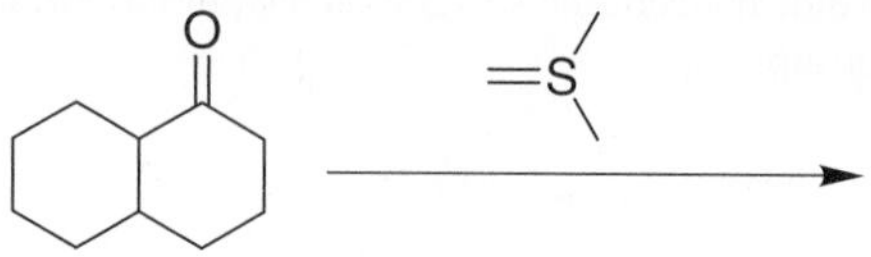

5.72.

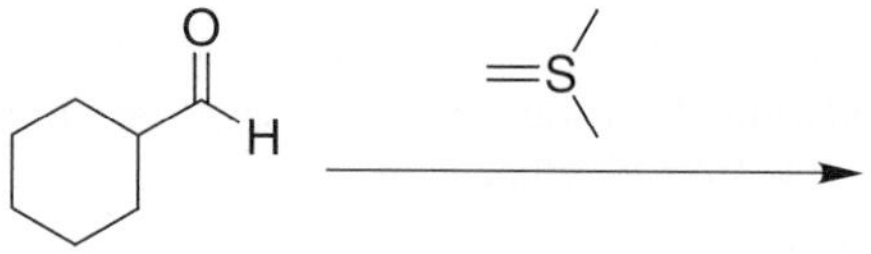

5.73.

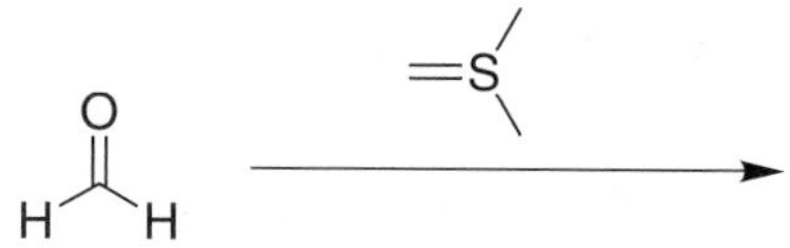

5.8 SOME IMPORTANT EXCEPTIONS TO THE RULE

In the beginning of the chapter, we saw a golden rule that helped us understand most of the chemistry that we saw. That rule was: always re-form the carbonyl if you can, but never kick off H^- or C^-. Yet there are a few, rare exceptions to this rule. In this section, we will look at two of these exceptions.

In the Cannizzaro reaction, it seems that we are kicking off H^- to re-form a carbonyl. We will not cover the Cannizzaro reaction in great detail because this reaction has very little synthetic utility. You will not likely use this reaction more than once, if at all, so we will just mention it in passing. Look up that reaction in your textbook and in your lecture notes. If you don't need to know that reaction, then you can ignore it. But if you are responsible for knowing that reaction, then you should look carefully at the mechanism in your textbook. If you focus on the step where H^- gets kicked off, you will see that H^- is not kicked off completely. It is simply transferred from one place to another, and in that sense, we can understand it a bit better. It is true that H^- is too unstable to ever leave as a leaving group. For that reason we never kick off H^- into solution. In the Cannizzaro reaction, however, it never actually leaves as a leaving group.

We will explore one other exception in much more detail. In one reaction it seems that we are re-forming a carbonyl to kick off C^-. This reaction, called the Baeyer-Villiger reaction, is extremely useful. If you know how to use it properly, you will find that you might use it many times to solve synthesis problems in this course. So, we should spend a bit of time covering that reaction right now.

The Baeyer-Villiger reaction uses a ***per***-acid as the reagent:

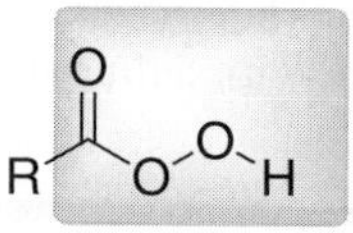

A per-acid has one more oxygen atom
than a regular carboxylic acid

R can be anything: it can be a methyl group or a much larger group. So, there are many common per-acids, the most common of which is MCPBA (*meta*-chloro **per**benzoic **a**cid). Thus, when you see the letters MCPBA, you should recognize that we are talking about a per-acid:

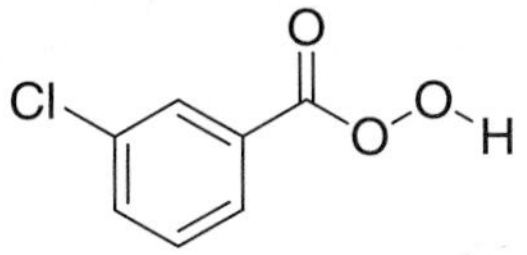

This reagent (or any other per-acid) is used to insert an oxygen atom next to the carbonyl group of a ketone, producing an ester:

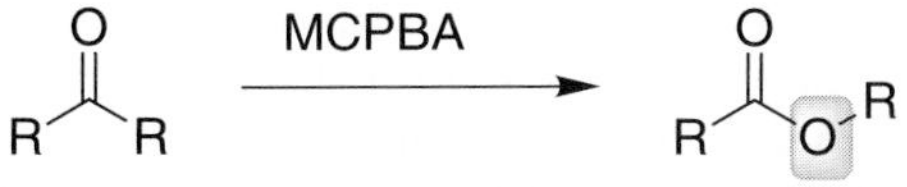

You can use the same reaction to convert an aldehyde into a carboxylic acid:

Once again, the outcome of this reaction is to "insert" an oxygen atom next to the carbonyl group. This is a very useful synthetic trick, so let's see the mechanism of how it works.

In the first step of the mechanism, the per-acid attacks the carbonyl, just like any other nucleophile:

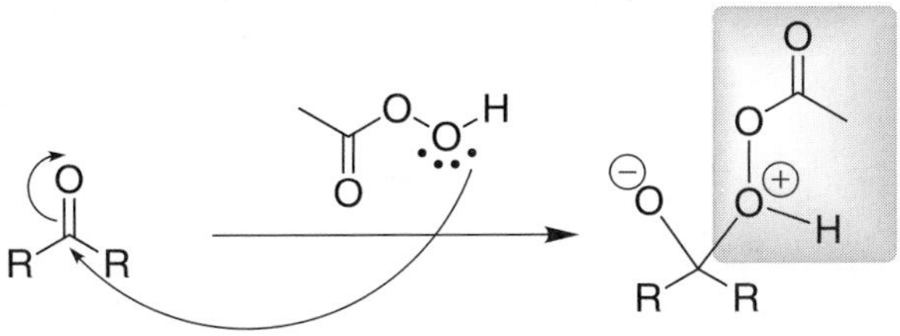

When we look at this intermediate, we can conclude that the only way to re-form the carbonyl will be to kick off the nucleophile that just attacked (shown above in gray). It is true that this probably happens, but we don't see any product when this happens. We therefore apply our golden rule to see if anything *else* can happen. In other words, we look to see if we can kick off any other leaving group. Our golden rule tells us to never kick off H⁻ or C⁻, and we don't see any other groups to kick off. BUT this is the exception to the golden rule. When you attack a carbonyl with a per-acid, something else can actually happen. It is unique to this situation, and you will not see this in any other mechanism (so don't worry about trying to apply this next step to any other mechanism). We get a rearrangement like the following:

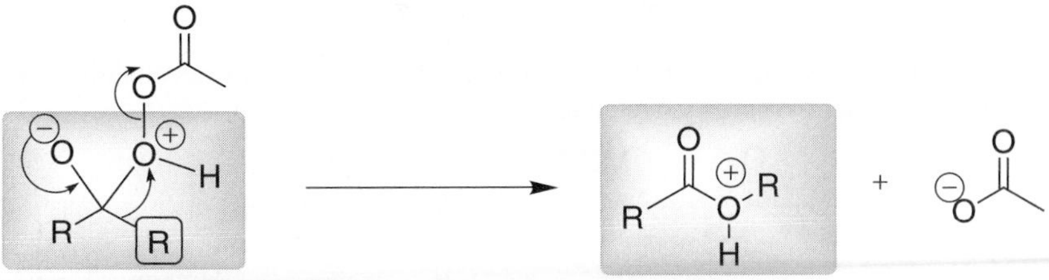

This R migrates

Look carefully at the R groups. Notice that the carbonyl group is re-forming to kick off one of the R groups, which migrates to the nearby oxygen. In other words, it looks like we are kicking off C^-. But the truth is that we are not *really* kicking off C^- into solution. C^- is too unstable to be kicked off into solution. Rather, it is just *migrating* over from one place to another (it is migrating over to attack the positively charged oxygen). It never really becomes C^- for any period of time. And that explains how we can have an exception here.

This mechanism is truly bizarre, and you should not worry if you feel that you would not be able to predict when this could happen in other situations. This is probably the first time you are ever seeing a rearrangement that does not involve a carbocation. So, this is truly different (although it is similar—we have an alkyl group migrating over to a positively charged center, which is essentially what happens in a carbocation rearrangement). You will not need to apply this mechanism to any other situations. (We will see one other type of rearrangement like this when we learn about amines, and I will point it out when we get there.) For now, however, don't focus too much on this mechanism because it might just make you upset. Instead, let's focus on how to use this reaction when you are solving synthesis problems because it will be a very useful reaction for you to have in your back pocket.

In order to use this reaction properly, you will need to know how to predict which side the oxygen goes on. For example, consider the following ketone:

This ketone is unsymmetrical, so we must decide where to put in the oxygen if we do a Baeyer-Villiger reaction. Which of the following two products do we get?

To answer this question, we need to know which R group is more likely to migrate. If you look back at the mechanism above, you will see that the migrating R group is the one that ends up next to the oxygen in the product. So, we just have to decide which R group can migrate faster.

There is an order to how fast R groups can migrate in this reaction, and we call it "migratory aptitude":

$$H \; > \; Ph \; > \; 3° \; > \; 2° \; > \; 1°$$

This means that H migrates the fastest. That explains how we can use this reaction to convert aldehydes into carboxylic acids (as we saw a couple of pages ago):

H migrates faster than any other group, so it doesn't even matter what the other group is.

If there is no H in the compound (in other words, if you are starting with a ketone instead of an aldehyde), then you should look for a phenyl group. Phenyl groups are the second fastest at migration. This answers the question that we asked above regarding an unsymmetrical ketone. The answer is:

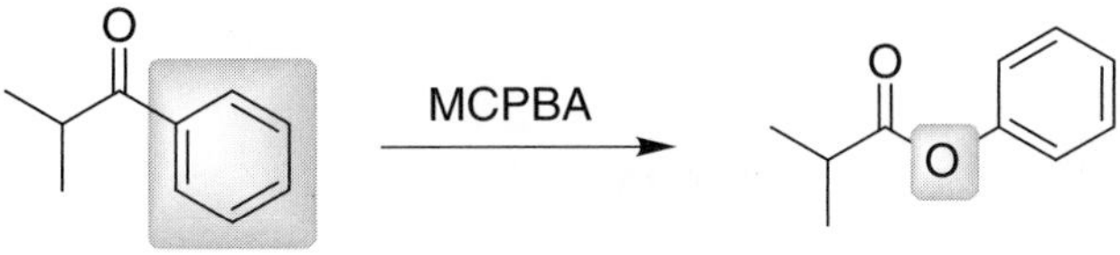

So, we placed the oxygen next to the phenyl group.

If you don't have a hydrogen or a phenyl group attached to the carbonyl, then you look for the most substituted alkyl group. For example:

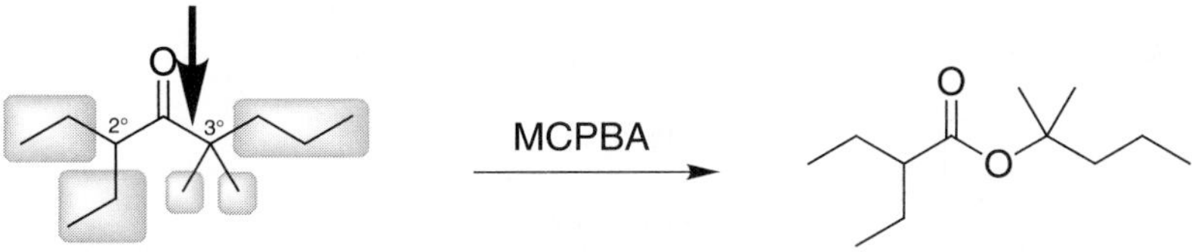

Notice that we put the oxygen on the side that is more substituted.

In order to use this reaction in synthesis problems, you must make sure that you can predict where to place the oxygen atom for any specific example. Let's do some problems to make sure you got it:

EXERCISE 5.74.   Predict the product of the following reaction:

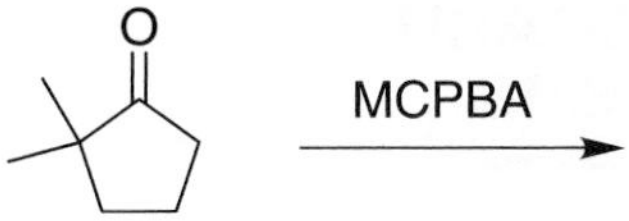

Answer: We see that because we have a per-acid reacting with a ketone, we expect a Baeyer-Villiger reaction to occur. We look closely at our starting ketone, and we see that it is unsymmetrical; we must therefore predict where the oxygen will insert itself. We look at both sides, and we see that the left side is tertiary, while the right side is primary. The tertiary R group will migrate faster, and that is where the oxygen will go.

This specific problem is an interesting example, because the insertion of an oxygen atom causes a ring expansion:

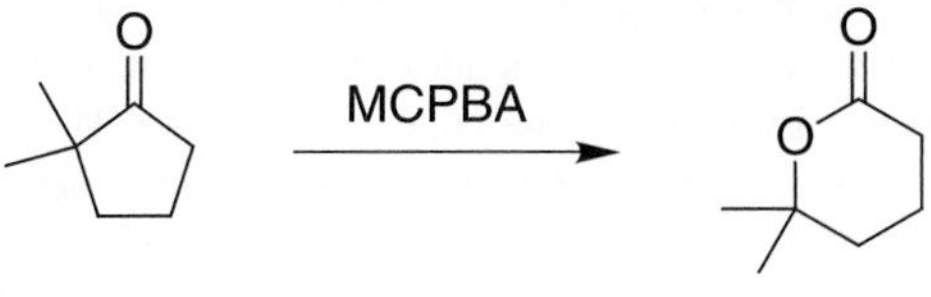

The product is not just an ester but a *cyclic* ester. Cyclic esters have a fancy name—*lactones*. Whenever you do a Baeyer-Villiger on a cyclic ketone, your product will be a lactone.

Predict the major product for each of the following reactions:

5.75.

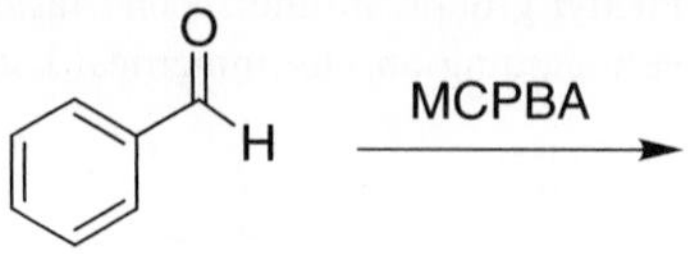

MCPBA

5.76.

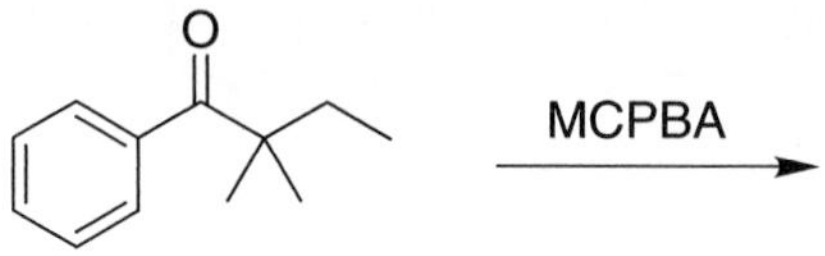

MCPBA

5.77.

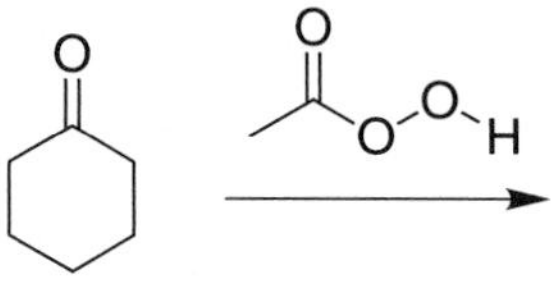

5.78.

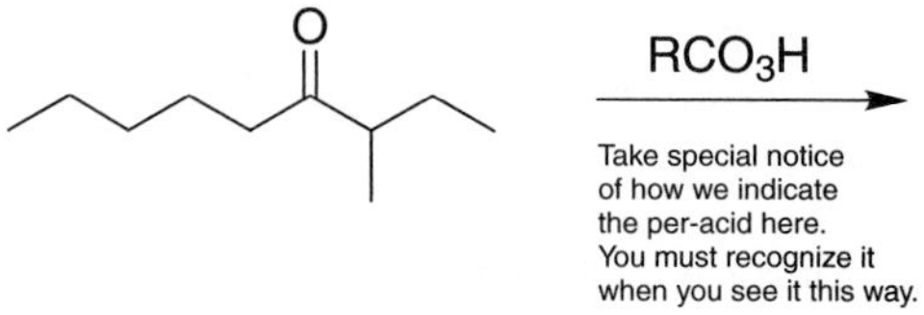

RCO₃H

Take special notice
of how we indicate
the per-acid here.
You must recognize it
when you see it this way.

5.79.

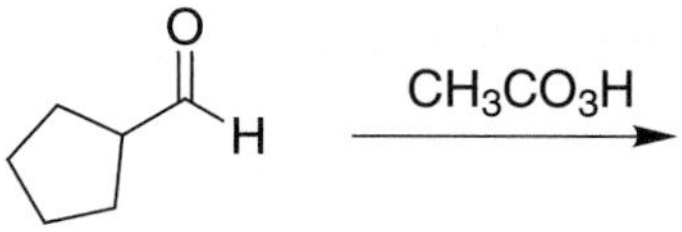

CH₃CO₃H

5.9 HOW TO APPROACH SYNTHESIS PROBLEMS

In this chapter, we have seen many reactions which you will need to have at your fingertips in order to solve synthesis problems. In the beginning of this chapter, we saw a few ways to make aldehydes and ketones. Do you remember those reactions? If you don't, then you are in trouble. This is why organic chemistry can get tough at times. It is not sufficient to be a master of mechanisms. That is an excellent start, and it builds an excellent foundation for understanding the material. But at the end of the day, you have to be able to solve synthesis problems. And in order to do that, you must have all of the reactions organized in your mind.

Let's start with a short review of everything we saw:

We started with a few ways of making ketones and aldehydes (two ways to oxidize, and then an ozonolysis). Then we looked at hydrogen nucleophiles (NaBH₄ and LAH) and at oxygen nucleophiles (making ketals), and we saw that ketals can be used to protect ketones. Then we saw sulfur nucleophiles (to form thioketals), and we saw how to use them to either *create* ketones and aldehydes or to *reduce* ketones and aldehydes. Next we saw nitrogen nucleophiles (primary amines and secondary amines), we examined special primary amines (hydroxylamine and hydrazine), and we saw how hydrazones could be used to reduce ketones to alkanes. Then we moved on to three kinds of carbon nucleophiles—Grignard reagents, phosphorus ylides, and sulfur ylides. Finally, we took a close-up look at the Baeyer-Villiger oxidation, focusing on its utility for synthesis. That is everything we saw in this chapter.

In this last section of this chapter, we will bring everything together to solve synthesis problems. The first step is to make sure that you know these reactions well enough to claim that you have them at your fingertips. So to guarantee that you get there, try to do the following. Take a separate piece of paper and try to write down all of the reactions listed in the paragraph above (without looking back in the chapter, if possible). Check that you can draw all of the reagents and all of the products. If you cannot do so, then you are not ready to even START thinking about synthesis problems. Students often complain that they just don't know how to approach synthesis problems. The difficulty is usually due NOT to the student's poor abilities, but rather to the student's poor study habits. You CAN do synthesis problems. You might even enjoy them, believe it or not. But you have to walk before you can run. If you try to run before you learn how to walk, you will trip and you will get frustrated. Many students make this mistake with synthesis problems.

So take my advice, and focus right now on mastering the individual reactions. Try to fill out a blank sheet of paper with everything that we have done in this chapter. If you find that you have to look back into the chapter to get the exact reagents (or to see what the exact products are), then that is fine. It is part of the studying process. But don't trick yourself into thinking that you are ready for synthesis problems once you have filled out the sheet. You are not ready until you can fill out the entire sheet of paper, start-to-finish, without looking back even once into the chapter to get the fine details. Keep filling out a new sheet, again and again, until you can do it all without looking back. Ideally, you should get to a point where you do not even need to look at the short summary that we just gave. You should get to a point where you can reconstruct the summary in your head, and then based on that summary, you should be able to write out all of the reactions.

It sounds like a lot of work, and it is. It will take you a while, but when you are done, you will be in an excellent position to start tackling synthesis problems. If you get lazy and you decide to skip this advice, then don't complain later if you are frustrated with synthesis problems. It would be your own fault for trying to run before you have mastered walking.

Once you get to the point where you have all of the reactions at your fingertips, then you can come back to here, and try to prove it, by doing some simple problems. These problems are designed to test you on your ability to list the reagents that you would need in order to do simple one-step transformations. Once you have all of the reactions down cold, then we will be able to move on and conquer some multistep synthesis problems.

EXERCISE 5.80.   What reagents would you use to do the following transformation?

Answer: This transformation involves a ketone being converted into an enamine. Clearly, we will need a nitrogen nucleophile; we just need to decide what kind of nitrogen nucleophile. Because our product is an enamine, we will need a secondary amine. When we look at the product, we can determine that we would need the following secondary amine:

Finally, we just need to decide whether any special conditions should be mentioned. And we did learn that there are special conditions for the reaction between a ketone and secondary amine. Specifically, we need to have acid-catalysis and Dean-Stark conditions. So our answer is:

The following problems are designed to be simple, so that you can prove to yourself that you can do these reactions cold. I highly recommend that you photocopy the next page *before* filling it out. You might find that you get stuck on a few problems, and it might be helpful for you to come back in the near future to fill it out again. The following problems are not listed in the order they appear in this chapter.

5.81.    **5.82.**

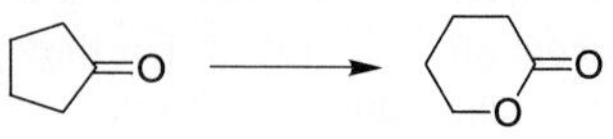

5.83.    **5.84.**

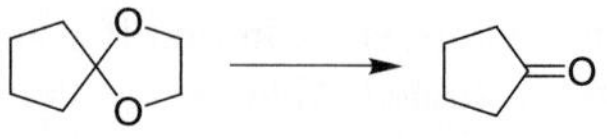

5.85.    **5.86.**

5.87.    **5.88.**

5.89.    **5.90.**

5.91. 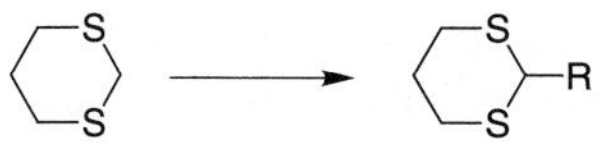   **5.92.**

5.93.    **5.94.**

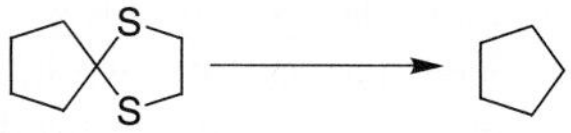

5.95. 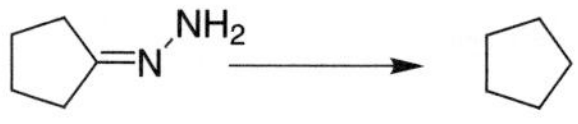   **5.96.**

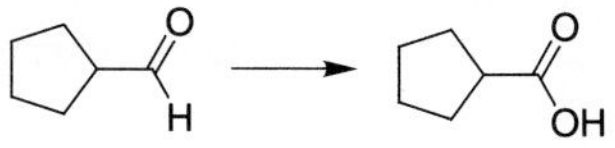

5.97. 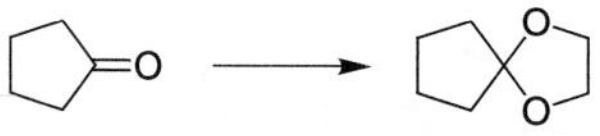   **5.98.**

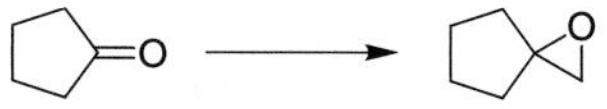

5.99. 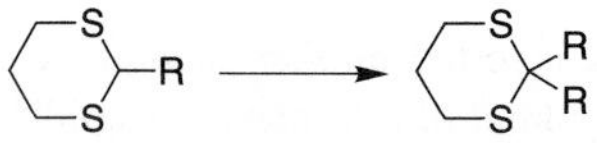   **5.100.**

If you felt comfortable with those problems, then you should be ready to move on to solving some multistep synthesis problems. Let's see an example:

EXERCISE 5.101. Propose an efficient synthesis for the following transformation:

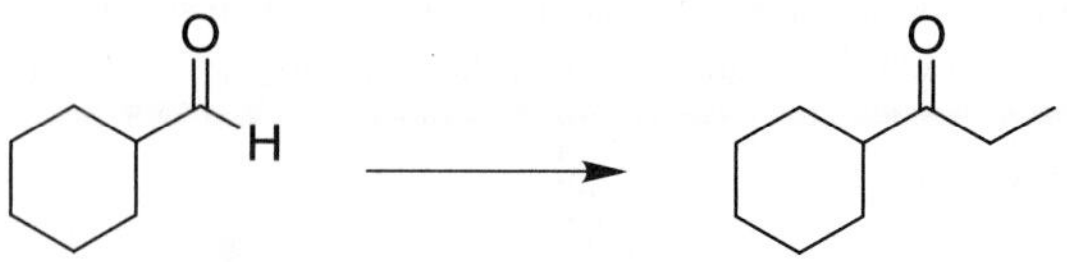

Answer: This problem is a bit trickier than the problems on the previous page because this problem cannot be done in one step. We need to hook on an ethyl group, while maintaining the carbonyl group. If we use a Grignard to reagent, we can hook on the ethyl group, but we will reduce the ketone to an alcohol in the process:

This issue can be easily overcome because we can oxidize the alcohol back up to a ketone:

So, we have a two-step synthesis to accomplish this transformation.

There is a completely different way to do this problem, also using reactions that we learned in this chapter. Remember that when we learned about thioacetals we saw that you can convert an aldehyde into a thioacetal, then alkylate it, and finally pull off the thioketal to give a ketone:

This example illustrates an important aspect of synthesis problems. Notice that we just came up with two completely different solutions for this problem, and that both solutions are perfectly correct answers. Here is the take-home message: rarely is there only one answer to a synthesis problem. As we learn more and more reactions, you will find that there are even more possible ways to do the same problem. Don't get stuck into thinking that you have to find THE answer. You may even find that you might come across a perfectly acceptable answer that no one else in the class thought of. Those are the most exciting moments. There is room for you to express some creativity when you solve synthesis problems.

Before you try to solve some problems yourself, we need to make one more important point about synthesis problems. Very often, it is helpful to work "backwards." We call this *retrosynthetic analysis*. Let's see an example:

EXERCISE 5.102.   Propose an efficient synthesis for the following transformation:

Answer: If we look at the product first, we notice that it is an enamine. We have only seen one way to make an enamine—from the reaction between a ketone and a secondary amine. So we can work our way backwards:

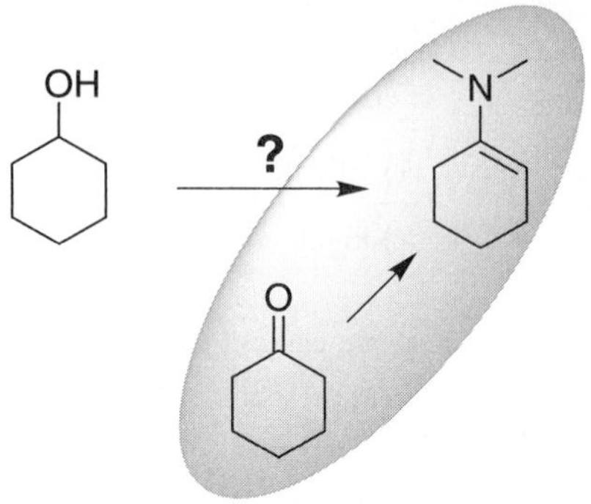

All we need to do is find a way to convert the starting compound into a ketone. And we have seen how to do that. We can convert an alcohol into a ketone, using an oxidation reaction. So our synthesis is:

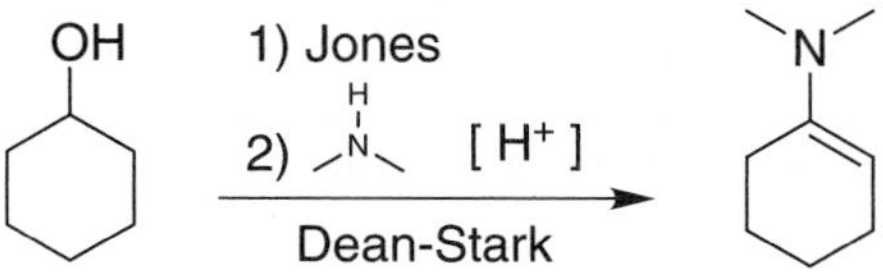

Of course, this last problem was not very difficult because it had a two-step solution. As you solve problems that require more steps, this approach will become increasingly important (retrosynthetic analysis). But don't worry: you won't have to solve any problems that require 10 steps because that is way beyond the scope of this course. You will generally not have to deal with problems that require more than three or four steps. So, with a lot of practice, it is definitely realistic to become a master of solving synthesis problems. Once again, it all depends on how well you know all of the reactions.

Now let's get some practice:

For each of the following problems, suggest an efficient synthesis. Remember that there will not be just one answer for each of these problems. If you propose an answer and it does not match the answer in the back of the book, do not be discouraged. Carefully analyze your answer because it might also be correct.

5.103.

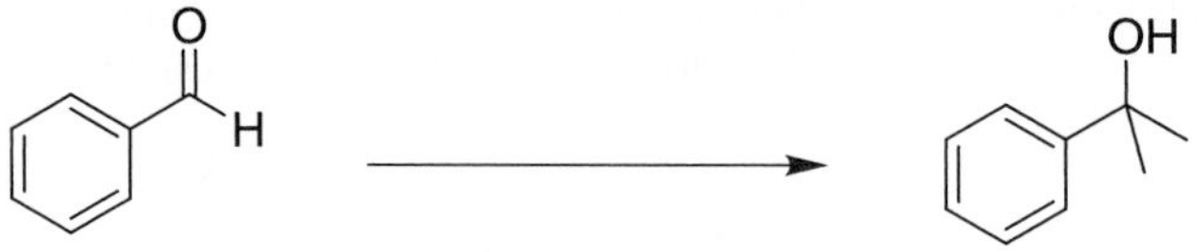

5.104.

5.105.

5.106.

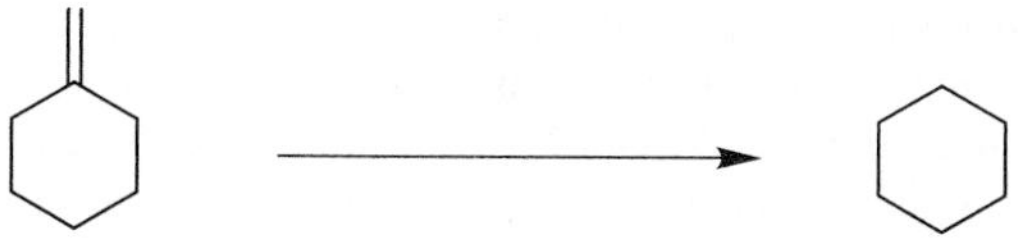

5.107.

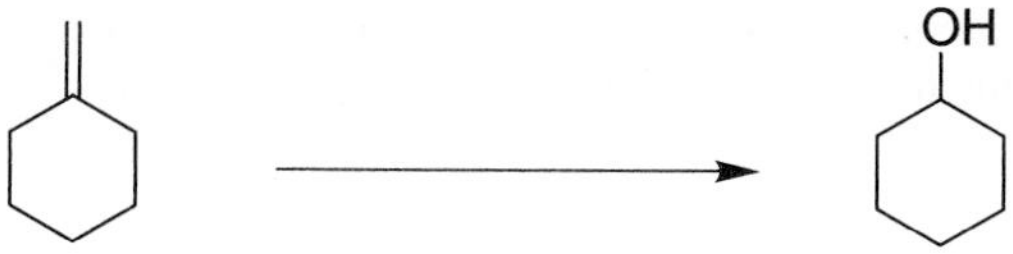

5.108.

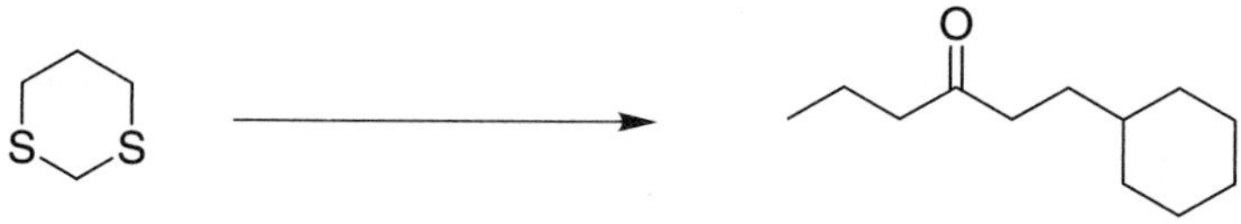

5.109.

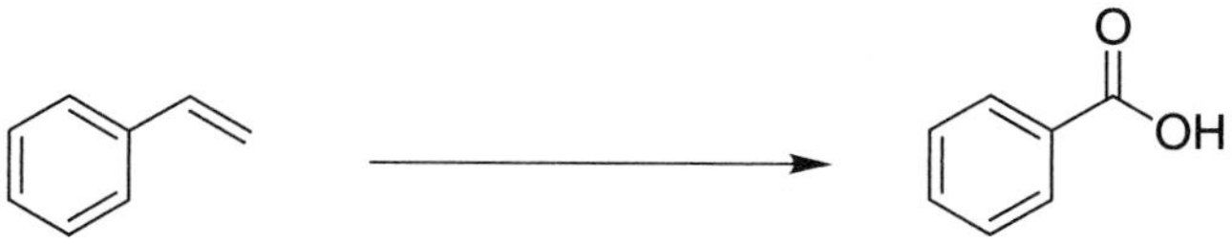

5.110.

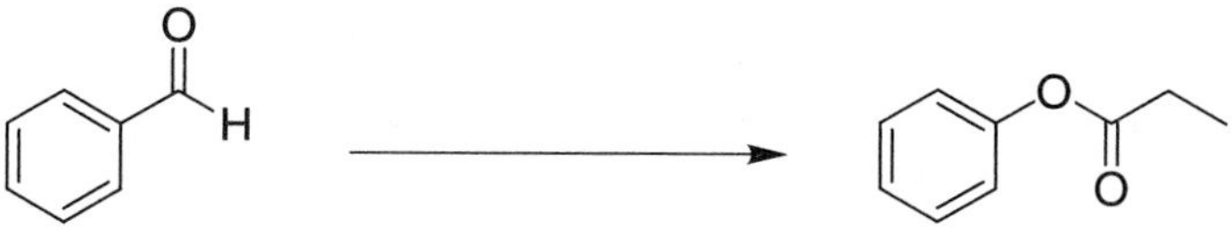

5.111.

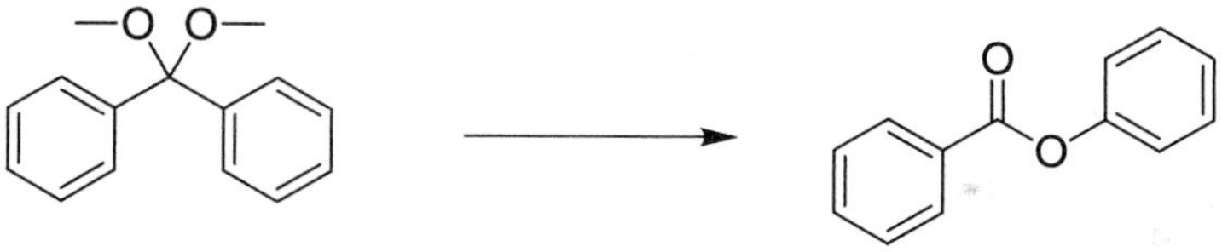

And now let's end up with a couple of tougher problems:

5.112.

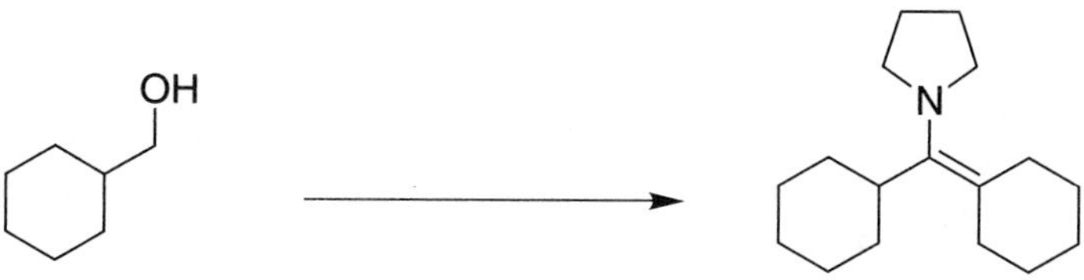

5.113.

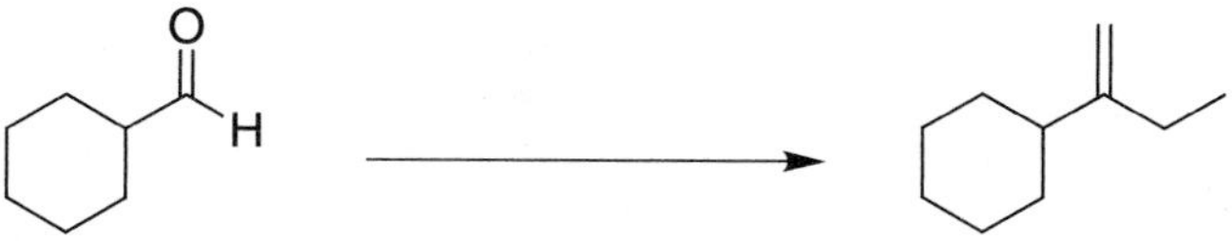

Before we end this chapter, we must make one last note. In this chapter, we did **not** see EVERY reaction that is in your textbook or lecture notes. We covered the basic, core reactions (probably 90 or 95% of the reactions you need to know). The goal of this chapter was NOT to cover every reaction; rather, it was to lay a foundation for you when you are reading your textbook and lecture notes. We saw the similarities between mechanisms, and we saw a simple way of categorizing all nucleophiles (hydrogen nucleophiles, oxygen nucleophile, sulfur nucleophiles, etc.).

Now you can go back through your textbook and lecture notes, and look for those reactions that we did not cover here in this chapter. With the foundation we have built in this chapter, you should be in good shape to fill in the gaps and study more efficiently.

And make sure to do ALL of the problems in your textbook. You will find more synthesis problems there. The more you practice, the better you will get it. Good luck . . .

CHAPTER *6*

CARBOXYLIC ACID DERIVATIVES

Carboxylic acid derivatives are similar to carboxylic acids because they also have a heteroatom (an atom other than C or H) connected to a carbonyl group:

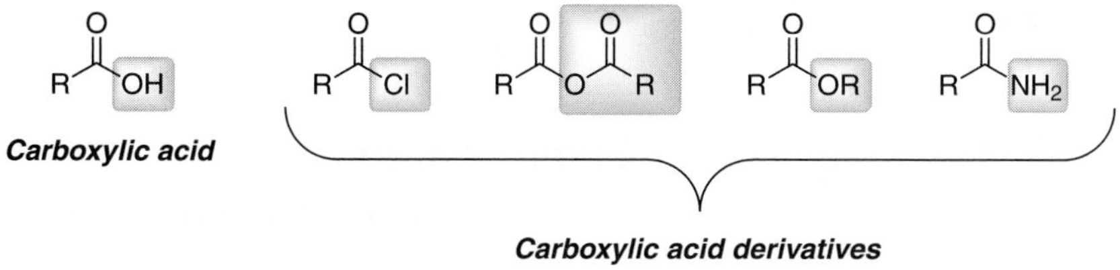

Carboxylic acid

Carboxylic acid derivatives

The difference is that we are replacing the OH group (of the carboxylic acid) for some other group. The chemistry of carboxylic acids is slightly different from the chemistry of ketones and aldehydes because carboxylic acid derivatives have a leaving group that can leave after the carbonyl is attacked:

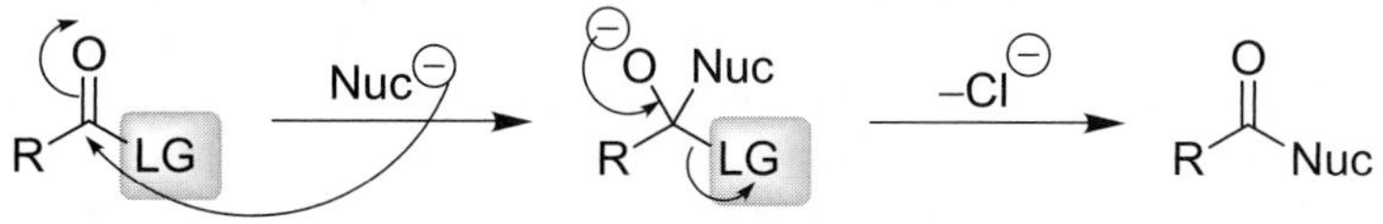

The nature of the leaving group will determine how reactive the compound is. For example, acyl halides are extremely reactive:

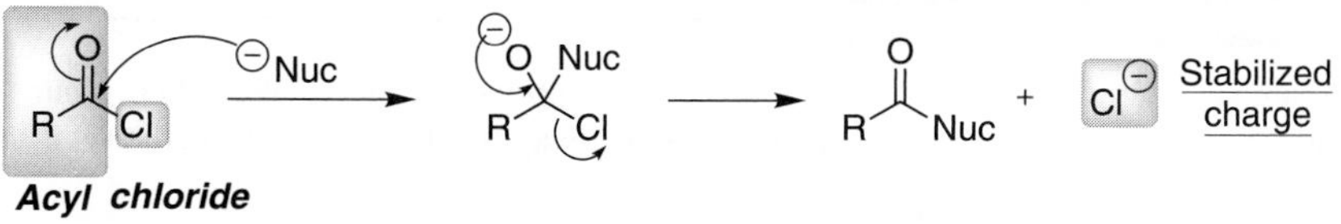

Acyl chloride

Acyl halides are the most reactive because they have the best leaving groups. We can therefore use acyl halides in synthesis problems to form any of the other carboxylic acid derivatives.

Here is the way to think about it: Carboxylic acid derivatives have a "wild card" next to the carbonyl group. I am calling it a "wild card" because we can easily exchange it for a different group:

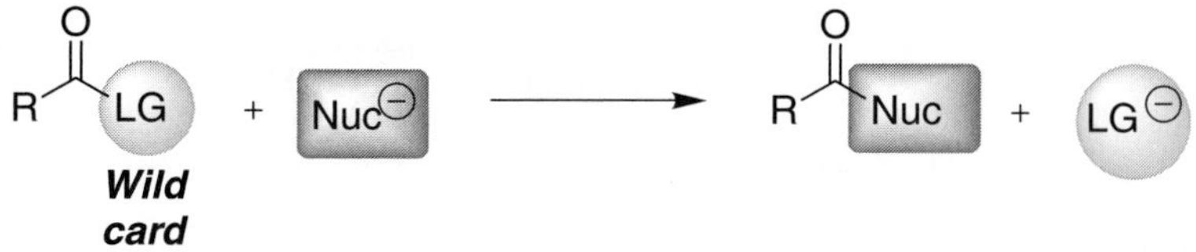

Wild card

In this chapter we will learn how to exchange the groups so that we can convert from one carboxylic acid derivative to another. In order to do so, we will have to know something about the order of

159

reactivity of carboxylic acid derivatives. We have already said that acyl halides are the most reactive. After acyl halides, the second most reactive carboxylic acid derivatives are anhydrides. Here is the order of reactivity that you need to know:

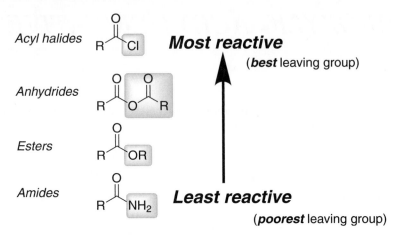

As we learn how to exchange the "wild card," we will see that just a few simple rules will determine everything. We will see dozens of reactions, but all of these reactions are completely predictable and understandable if you understand how to apply just a few simple rules.

In this chapter, we will master these rules first, and we will then use those rules for predicting products, proposing mechanisms, and proposing syntheses. We will NOT cover every reaction in your textbook or lecture notes. Rather, we will focus on the core skills you need. When you are finished with this chapter, you must go through your textbook and lecture notes to learn any reactions that we did not cover here in this chapter. This chapter will arm you with the skills you need in order to master the material in your textbook.

6.1 GENERAL RULES

We have already learned the most important rule in the previous chapter. We called it our "golden rule," and it went like this: after attacking a carbonyl, always try to re-form the carbonyl if you can, but never kick off H⁻ or C⁻.

With this rule, we can now appreciate the fact that the outcome of a reaction involving H⁻ or C⁻ (as a nucleophile) will be very different from the outcome of a reaction involving any other type of nucleophile. When we use a hydrogen nucleophile or a carbon nucleophile, we find that the carboxylic acid derivative gets attacked *twice*. Let's see why.

The first attack takes place as we would expect:

Then, we apply our golden rule: if you can re-form the carbonyl, then do it, but don't kick off H⁻ or C⁻. Since we started with a carboxylic acid derivative, we will have a leaving group. So, the leaving group leaves:

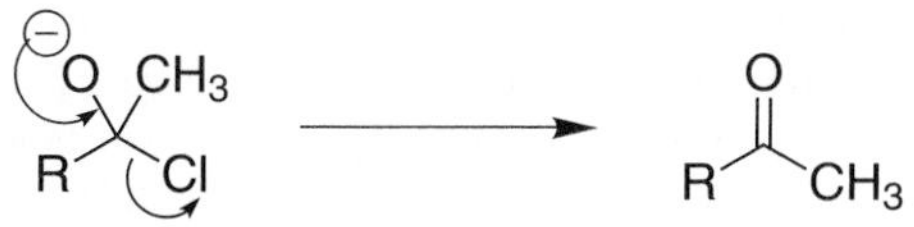

But notice that we now have a ketone. (It would have been an aldehyde if we had attacked with a hydrogen nucleophile.) This compound can be attacked a *second* time, as follows:

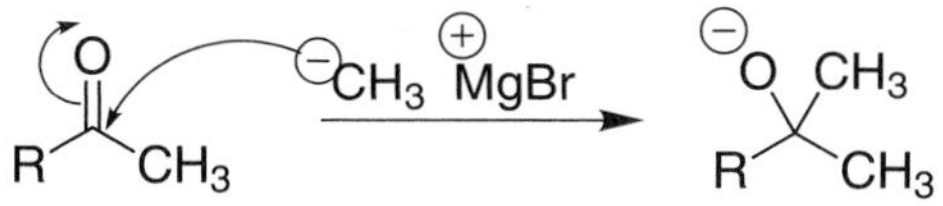

This intermediate is now incapable of re-forming the carbonyl because there is no leaving group. We have already said again and again that we cannot kick off H$^-$ or C$^-$ to re-form the carbonyl. So the reaction is over, and all we can do is provide a source of protons to get our product (we add the proton source AFTER the reaction is complete):

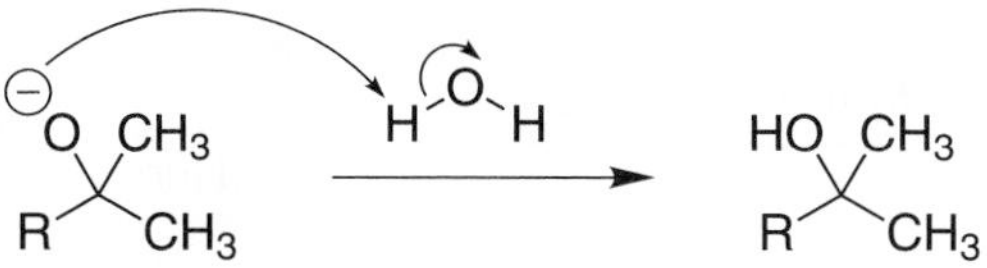

In the end, our product is an alcohol because the nucleophile attacked *twice*.

The situation is very different when we use any other nucleophile (not H$^-$ or C$^-$). Suppose, for example, that we attack with RO$^-$:

We re-form the carbonyl to give an ester,

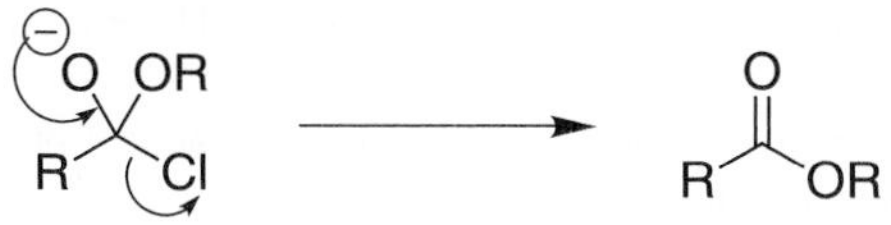

and then we try to attack the carbonyl a second time:

This intermediate CAN re-form because we can just kick off the second RO$^-$ that just attacked. This brings us right back to the ester:

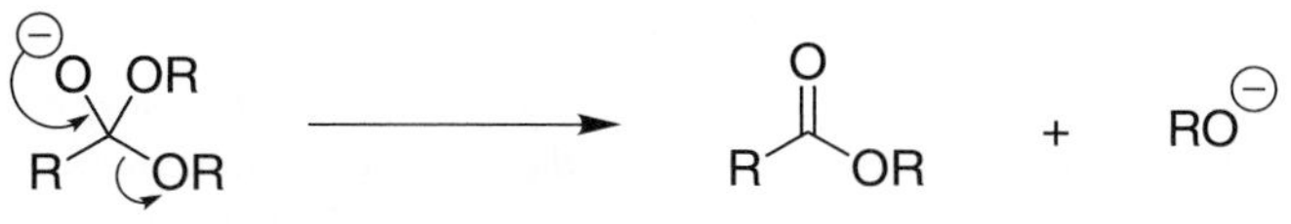

So attacking the ester with a second nucleophile did not get us anywhere. When we tried to attack the ester with a second nucleophile, it just shot that nucleophile right back out to re-form the carbonyl. And once again, we are left with an ester.

The second attack is only permanent when we use H⁻ or C⁻. This should make sense based on our golden rule: if the second attack is by H⁻ or C⁻, then the carbonyl will not be able to re-form.

So, we have seen that there is a difference in the types of products we get when we use H⁻ or C⁻, versus when we use any other nucleophile. Keep this in mind: H⁻ or C⁻ will attack twice, but all other nucleophiles will only attack once:

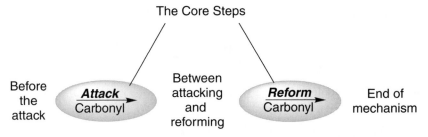

We have seen that all nucleophiles (except for H⁻ or C⁻) will attack only once, and in those reactions, the outcome will be to exchange one type of carboxylic acid derivative for another. For example:

The basic structure of this mechanism (and all others like it) goes like this: Attack and then re-form the carbonyl. That's it—just two core steps. Often, however, we must include proton transfers in our mechanism. So, if you want to become a master of mechanisms in this chapter, you will have to know when to use proton transfer and when not to use them. There are just a few simple rules, and we will go over them now.

To understand when to use proton transfers, let's first realize that proton transfers can only happen at three different times in the mechanism:

The transfers can happen *before the attack*, or immediately *after the attack* (before the leaving group leaves), or after the leaving group leaves at the *end of the mechanism*. Sometimes, you won't need to do any proton transfers at all (the examples we have seen so far did not require any proton transfers). Sometimes, you will need to use one or perhaps two proton transfers. And other times, you will have to use proton transfers at all three times in the mechanism. For example, consider the following mechanism:

PROTONATE   **ATTACK**

PROTON TRANSFERS

DEPROTONATE   **REFORM**

If you follow the steps and describe them out loud, you would say the following:

Proton transfer, *attack*, proton transfer, *re-form*, proton transfer

Notice once again that we have our two major steps (attack and re-form), but we also have three proton transfers here. This is the maximum number that we could possibly have in this mechanism: one before the attack, one between the attacking and re-forming, and the last one after the leaving group leaves. In order to master the mechanisms in this chapter, we will therefore need some rules to help us decide whether or not to use a proton transfer at each of the three possible times when proton transfers can occur.

Let's break this discussion down into three different parts:

Proton transfer, *attack*, proton transfer, *re-form*, proton transfer
1   2   3

Let's start by seeing how to decide whether we need a proton transfer *before* the attack. In the previous chapter we saw that protonating a carbonyl group makes that carbonyl group even more electrophilic. But some carbonyl groups simply don't need that. Acyl halides and anhydrides, which are very reactive, do not need to be any more electrophilic. Therefore, you will not need to protonate them. But an ester is not so reactive, so, if we are attacking with a weak nucleophile (like water), then we will need a proton transfer in the first step. This explains why the mechanism we just saw has a proton transfer in the first step:

Proton transfers

When we try to hydrolyze an acyl halide, we do *not* need to use a proton transfer in the first step:

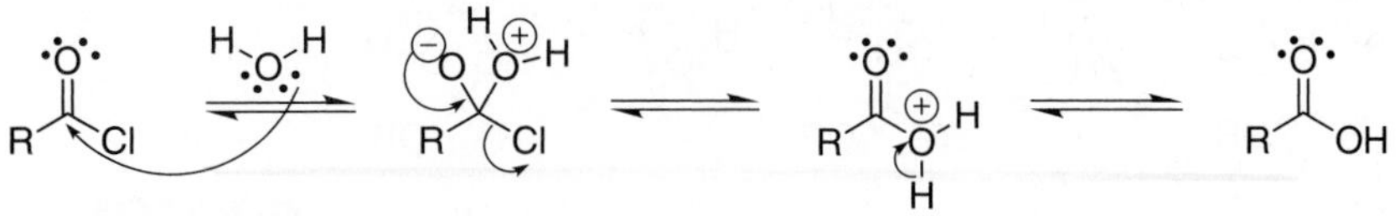

So we look at the reactivity of the carboxylic acid derivative, and we do not need to protonate the carbonyl if it is an acyl halide or an anhydride.

Now let's look at the second place where we might need a proton transfer:

<u>Proton transfer</u>, *attack*, <u>proton transfer</u>, *re-form*, <u>proton transfer</u>
    1   *2*   3

At this point in a mechanism (in between attacking and re-forming), you would need a proton transfer to ensure that the leaving group is compatible with the conditions. For example, consider the step highlighted below in the hydrolysis of an ester:

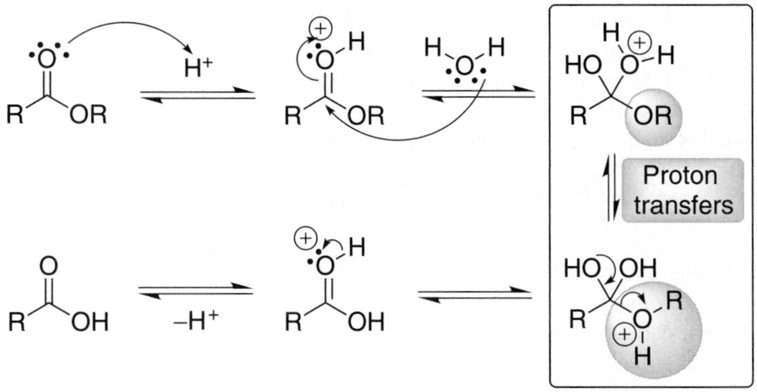

We need to protonate the OR group, so that it can leave without a negative charge. We don't want to kick off RO⁻ because the reaction takes place in acidic conditions. Never kick off RO⁻ into acidic conditions. Rather, protonate it first and then kick it off as ROH (neutral), as shown above. If we were in basic conditions (instead of acidic conditions), then it would be fine to kick off RO⁻.

Finally, let's consider the last place where we might need to do a proton transfer:

<u>Proton transfer</u>, *attack*, <u>proton transfer</u>, *re-form*, <u>proton transfer</u>
    1   2   **3**

This point represents the end of a mechanism. The purpose of a proton transfer at the very end is to form our final product, for example:

We must pull this proton off in the end of the mechanism in order to get our product.

Now we have seen everything that we need in order to decide when and where to do proton transfers. To recap, we have seen that the core steps of the mechanism will always be the same (attack and then re-form). In addition to these core steps, there are three distinct places where you will have to decide whether or not to do a proton transfer: before the attack, in between attacking and re-forming, and at the end of the mechanism.

Let's get some practice, and then we will be able to move on and apply what we have learned:

EXERCISE 6.1. Propose a plausible mechanism for the following reaction:

Answer: This reaction involves the conversion of one carboxylic acid derivative into another. Therefore, our mechanism should have at least two steps: attack and re-form. We just need to decide if we need to do any proton transfers. Because the starting material is an acyl halide, we don't need to protonate it. In fact, the reaction does not indicate the presence of protons anyway. So, we attack the carbonyl with MeOH:

Now we need to ask ourselves if we need to do any proton transfers before we re-form the carbonyl. We start by looking at the leaving group that we are about to kick off, and we ask whether it is OK to kick off this leaving group under these conditions. There is no problem with kicking off Cl⁻ as a leaving group. So we do not need to do any proton transfers before we re-form the carbonyl. Our next step is to just re-form the carbonyl:

Finally, we deprotonate to get our product:

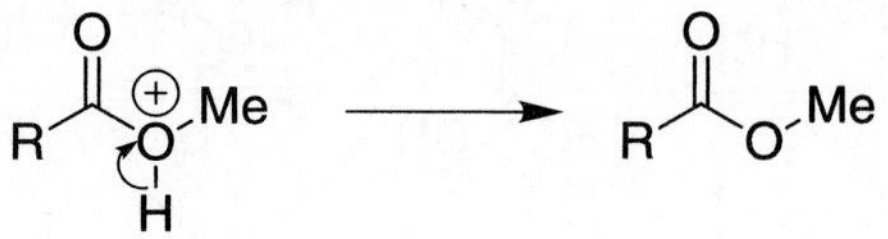

Overall, our mechanism looks like this:

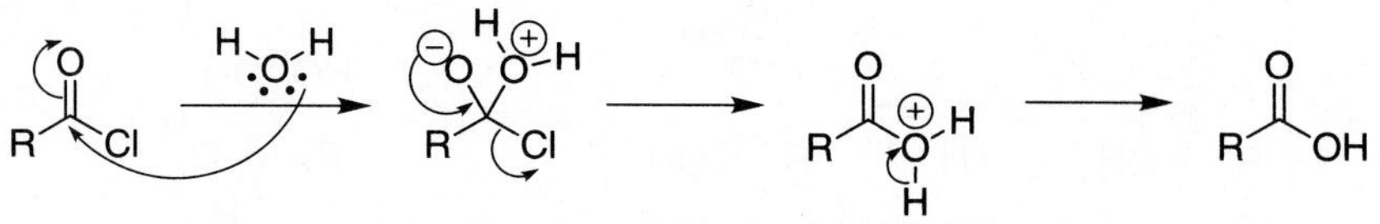

Propose a plausible mechanism for each of the following problems:

6.2.

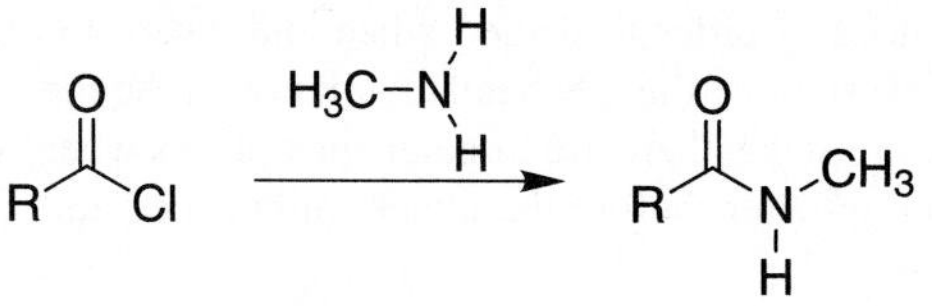

6.3.

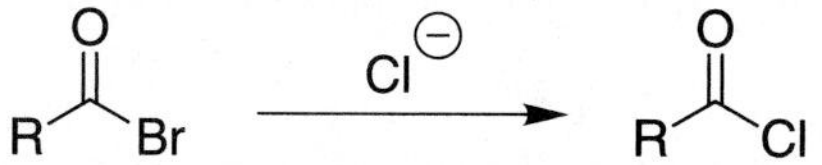

6.4.

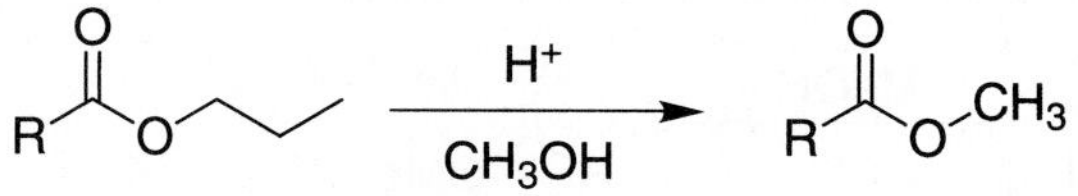

6.5.

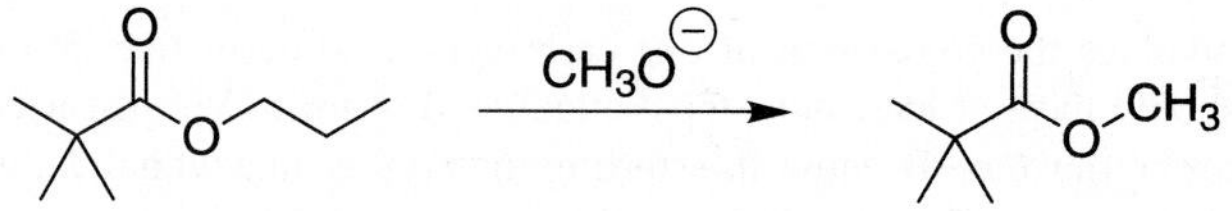

Earlier we saw that H⁻ and C⁻ are special nucleophiles because they will attack the carbonyl twice, yielding an alcohol as a product. Thus, there are more than two core steps: (1) attack, (2) re-form the carbonyl, and (3) then attack again. In these kinds of reactions, we generally do not need any proton transfers until the very end, for example:

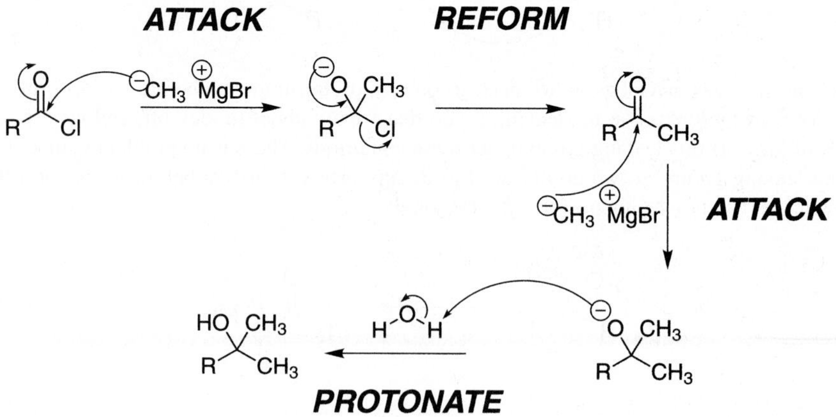

ATTACK   **REFORM**

ATTACK

PROTONATE

We just have to give a proton at the very end in order to get our product, which will be an alcohol. **But** no protons are to be found when we are dealing with H⁻ or C⁻. (As we argued in the previous chapter, any protons would destroy these reagents). When we attack a carboxylic acid with H⁻ or C⁻, we will therefore need to add a source of protons after the reaction is over, *as a separate step*:

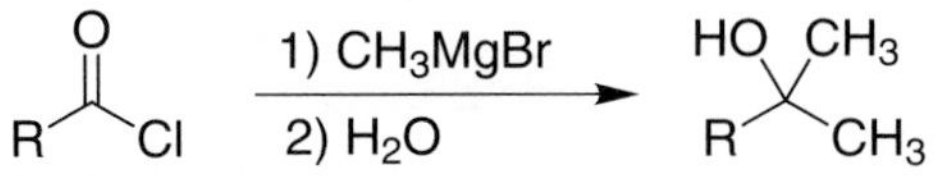

Propose a plausible mechanism for the following reactions:

6.6.

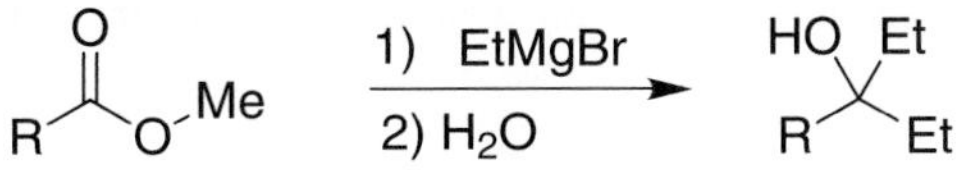

6.7.

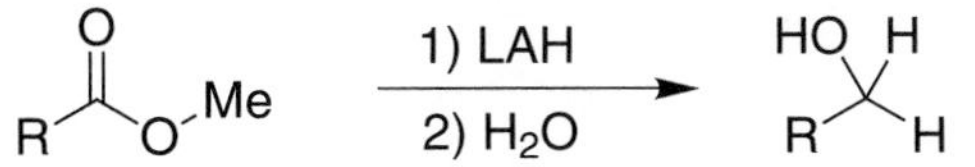

6.8.

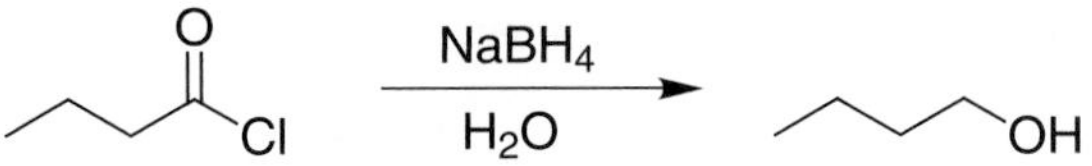

6.2 ACYL HALIDES

As we have mentioned, acyl halides are the most reactive of the carboxylic acid derivatives because they produce the most stable leaving groups. Therefore, we can make any of the other carboxylic acid derivatives from acyl halides (by just kicking off Cl⁻). Accordingly, it is critical that you know how to make an acyl halide. It is very common to encounter a synthesis problem where you will need to make an acyl halide at some point in the synthesis. You will make heavy use of the following reactions.

To make an acyl halide, it would be nice if Cl⁻ could be used to just kick off the OH group:

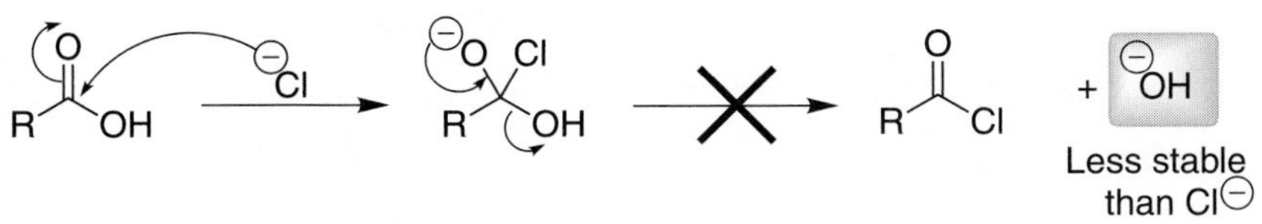

Less stable
than Cl⊖

But OH⁻ is less stable than Cl⁻, so this would be an uphill battle. Cl⁻ is not going to kick off OH⁻. We must therefore convert the OH group into a different group that CAN be kicked off by Cl⁻. Here is our strategy:

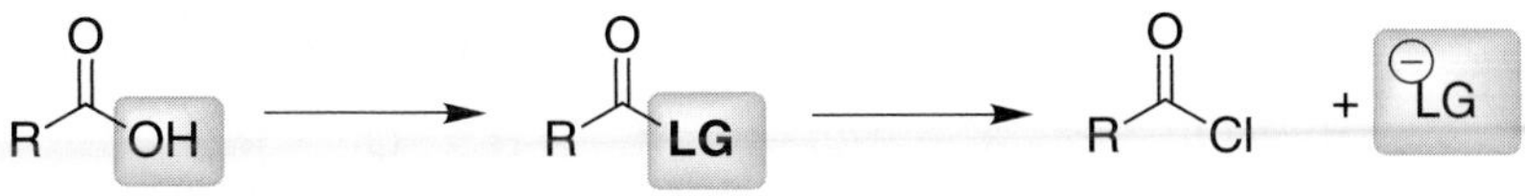

And there are two common ways to do this:

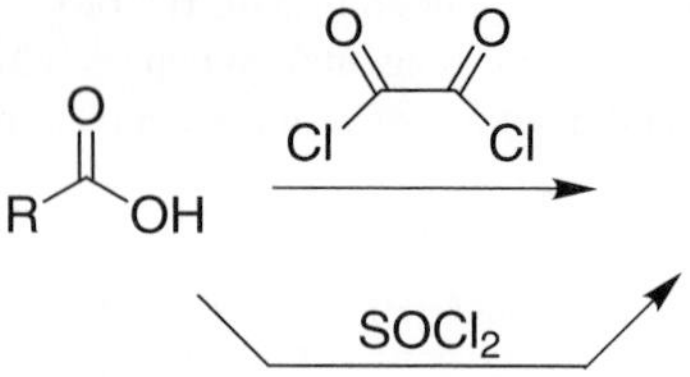

Let's explore the mechanisms for each of these reactions. We will see that these mechanisms utilize all of the rules that we have seen so far, and there is no new information here. Let's take a closer look at the first reaction.

In the first step, the carboxylic acid functions as the nucleophile, attacking the carbonyl of oxalyl chloride:

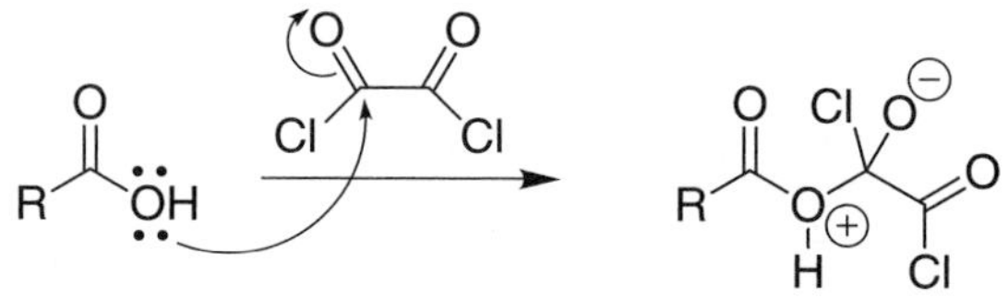

Then, we re-form the carbonyl to kick off Cl⁻ as a leaving group:

These steps should seem familiar to you (attack carbonyl and then re-form the carbonyl). This gives us an intermediate that is positively charged, so we deprotonate to make it neutral:

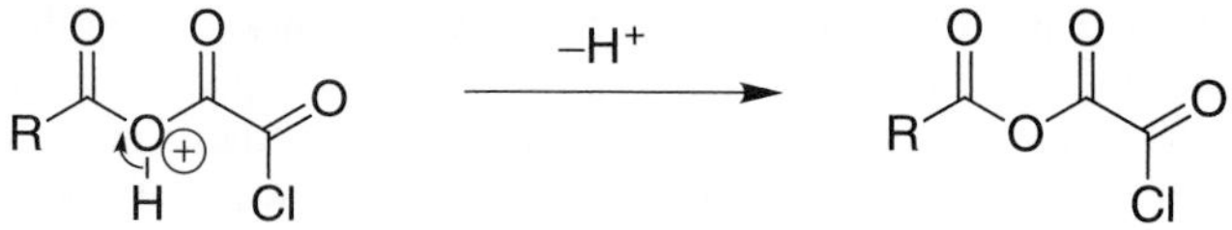

Now, we have a better leaving group. Compare this leaving group to what we had before:

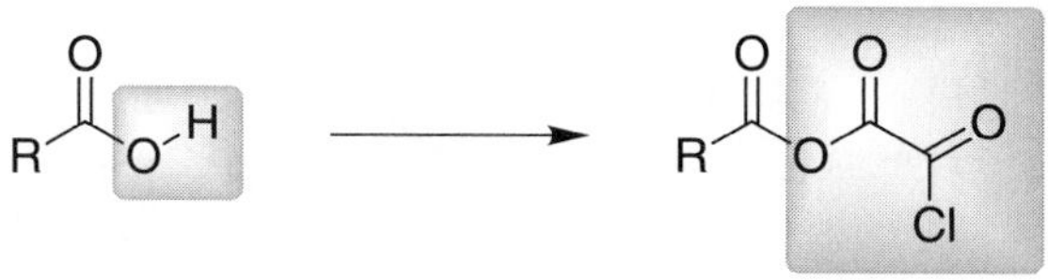

Better leaving group

To understand why this is a better leaving group now, we need to see what happens when Cl⁻ attacks. (Remember that we kicked off Cl⁻ as a leaving group in the previous step, and now it comes back to attack.) So, we attack with Cl⁻ at the carbonyl group all the way on the left:

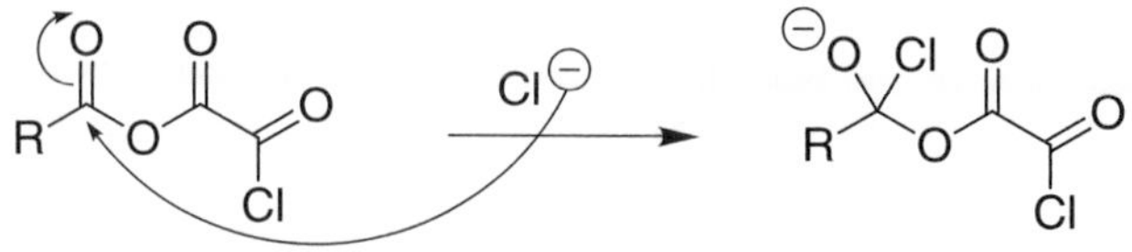

You might wonder why we are attacking this carbonyl rather than the other two. If you try to attack the middle carbonyl group, you will find that you go right back to what you started with. And if you attack the carbonyl on the right side, you will only regenerate the same intermediate when you re-form the carbonyl. So, attacking the carbonyl on the left side is the only path that leads to the formation of something new. After the chloride attacks this carbonyl, we can re-form the carbonyl as follows:

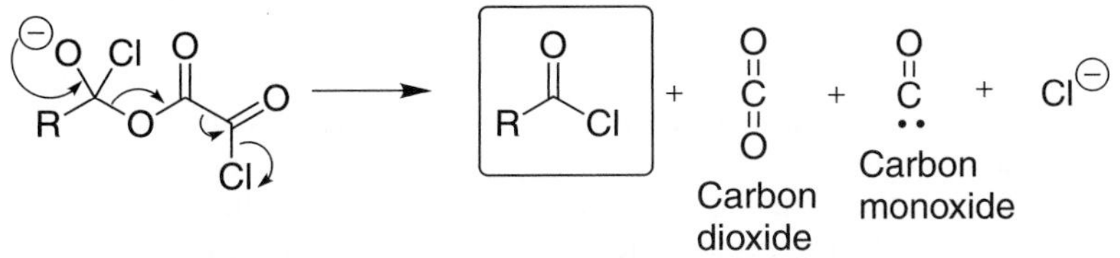

Notice that we kick off two gases here (CO_2 and CO). This essentially drives the reaction to completion. (We saw in the previous chapter that the production of gases can push a reaction to completion.)

Now let's look at our second reagent for converting a carboxylic acid into an acyl halide:

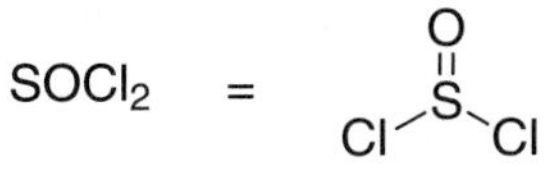

Thionyl chloride

Once again, the mechanism is very familiar. We begin by using the carboxylic acid as a nucleophile. We attack, re-form the S=O double bond, and then deprotonate (just as we did before):

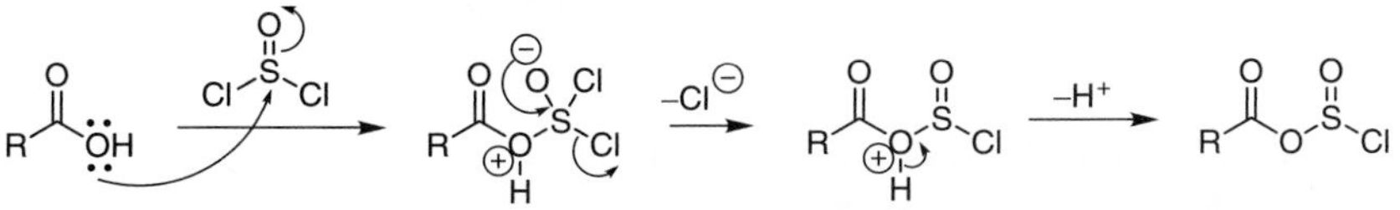

Now we have a much better leaving group than we did before:

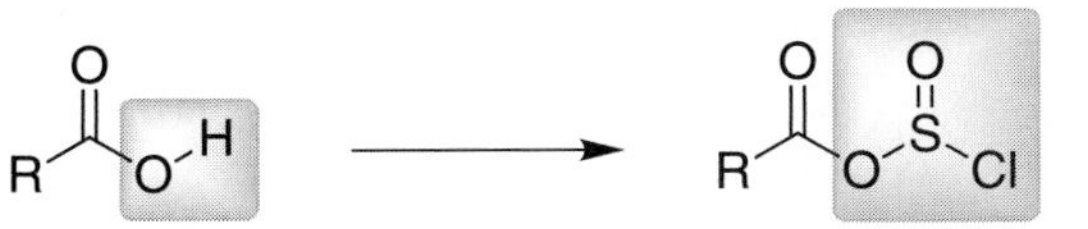

Better
leaving group

So, Cl^- attacks, and then we re-form the carbonyl to give our acyl chloride:

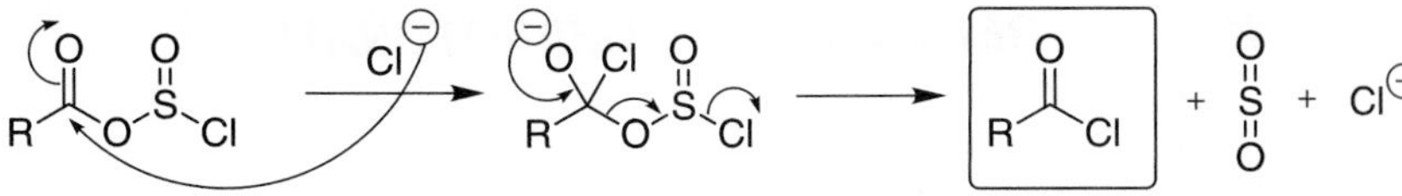

Once again, the production of a gas (SO_2) will force the reaction to go to completion.

We have thus far seen two ways to prepare acyl halides. When we took a close look at the mechanisms of these two reactions, we were guided by just a few general rules.

Now we are ready to see the reactions of acyl halides. We will see innumerable reactions, BUT don't try to memorize them. Instead, try to appreciate that they all follow the same general rules. As we saw in the previous section, there are two core steps (attack and re-form the carbonyl). Then, we just need to be careful about proton transfers in the three potential times where we might need to use them:

$$\underline{\text{Proton transfer}}, \textbf{\textit{attack}}, \underline{\text{proton transfer}}, \textbf{\textit{re-form}}, \underline{\text{proton transfer}}$$

| 1 | 2 | 3 |

But with acyl halides, it gets much easier because you will generally only need to do one proton transfer at the end of the mechanism (you can skip "1" and "2" on the diagram above). We don't need a proton transfer in the beginning ("1" above) because an acyl halide is electrophilic enough (we don't need to protonate it to get an attack). Then, we also don't need a proton transfer in the middle of the reaction ("2" above) because the leaving group (Cl$^-$) has no problem leaving (under any conditions). We will therefore have to do a proton transfer only at the very end to get our product.

Therefore, whenever we attack an acyl halide, we will generally get the following mechanism: attack, re-form, deprotonate. Say that out loud 10 times real fast (attack, re-form, deprotonate). You will find that this order of events keeps repeating itself in all of the reactions we are about to see. Let's go through these reactions, one-by-one:

When **water** attacks an acyl halide, we get a carboxylic acid:

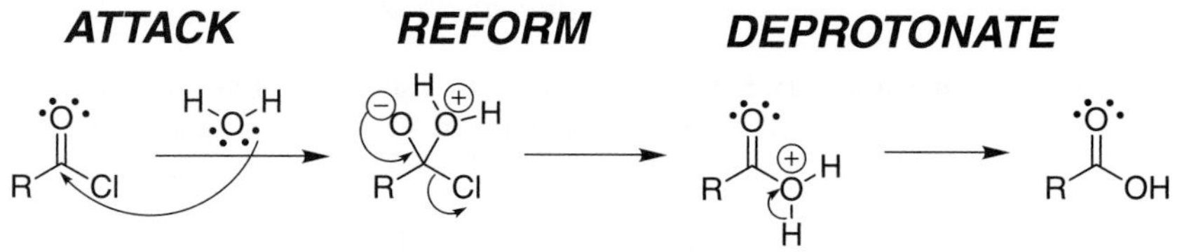

Notice that the mechanism is attack, re-form, deprotonate.

When an **alcohol** attacks, our product is an ester:

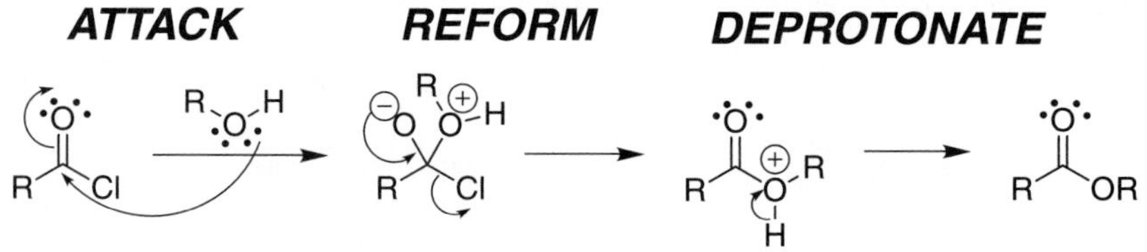

Once again, the mechanism is: attack, re-form, deprotonate.

When an **amine** attacks, we get an amide:

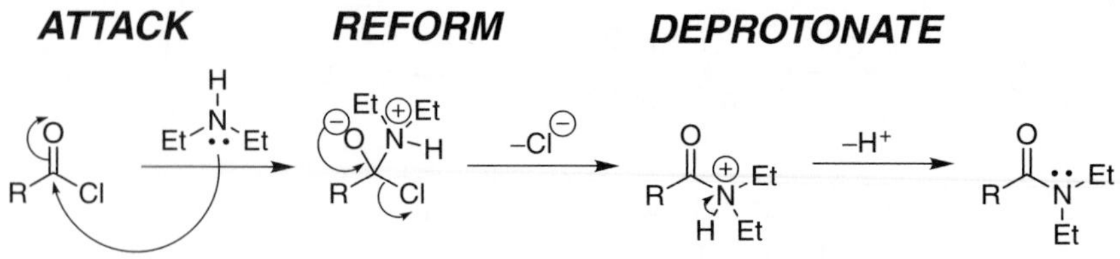

But now let's consider what happens when we use H⁻ or C⁻ to attack an acyl halide. We have already seen that H⁻ and C⁻ are special because they will attack twice:

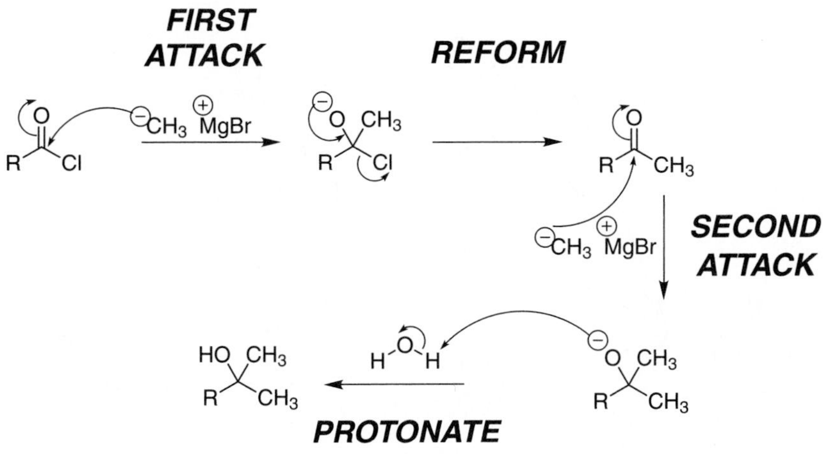

**FIRST
ATTACK**   **REFORM**

**SECOND
ATTACK**

PROTONATE

This leaves us with an obvious question: what if we wanted to attack with C⁻ just once? In other words, what if we wanted to do the following?

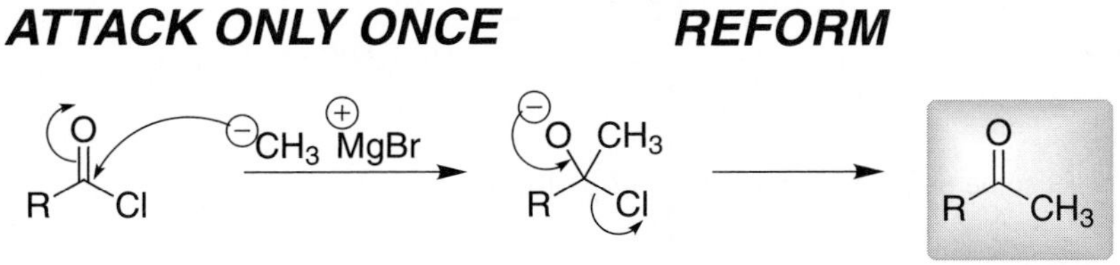

ATTACK ONLY ONCE   **REFORM**

What if we want our product to be the ketone? We have a problem here because the Grignard reagent will attack twice. We can't stop it from attacking twice. If we just try to put in one equivalent of the Grignard, we will get a mess of products. (Some acyl halides will get attacked twice, and others will not get attacked at all.) In order to stop at the ketone, we need a carbon nucleophile that will only react with an acyl halide but *not* with a ketone. And we are in luck because there is a class of compounds that will do this, namely, lithium dialkyl cuprates (R_2CuLi).

You might have been introduced to these compounds in the first semester of organic chemistry. Lithium dialkyl cuprates are carbon nucleophiles, but they are less reactive than Grignard reagents. Lithium dialkyl cuprates will react with acyl halides, but not with ketones. Therefore, we can use these reagents to attack acyl halides just once, stopping at the ketone:

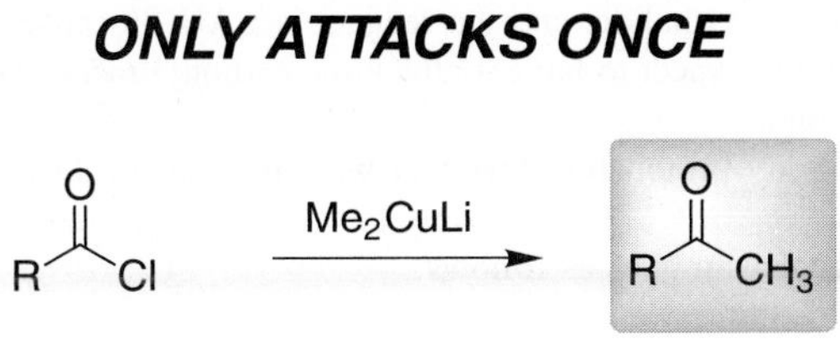

ONLY ATTACKS ONCE

We have seen a lot of reactions, so let's just quickly review them. Look at each reaction carefully and check that you are familiar with the mechanism for each of these reactions:

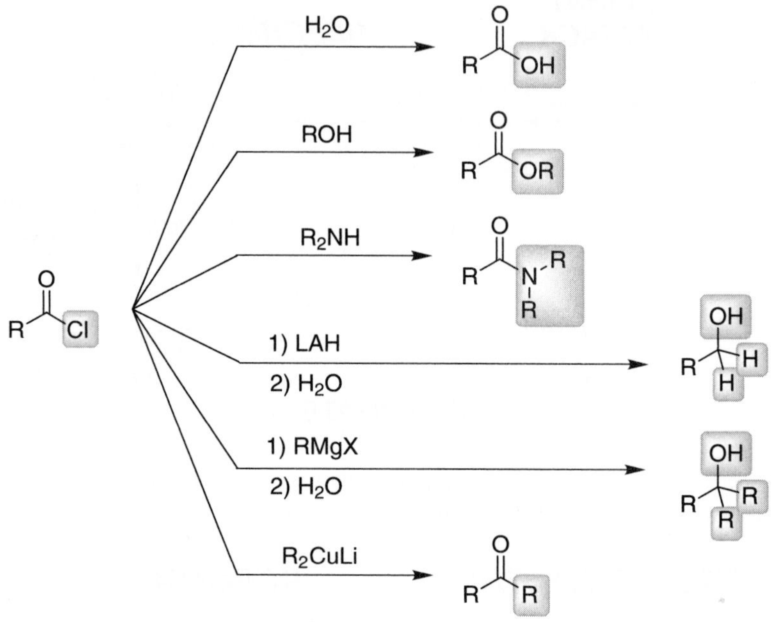

Now let's get some practice with a variety of problems.

EXERCISE 6.9. Propose a mechanism for the following reaction:

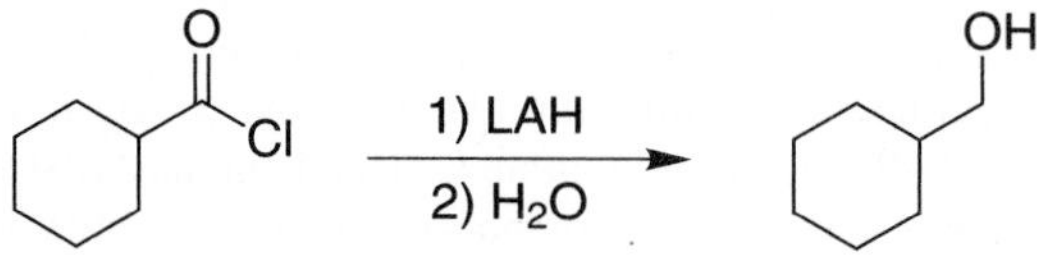

Answer: First let's decide if the nucleophile attacks once or twice, and then we will decide when to do proton transfers.

We are starting with an acyl halide, and the reagent is LAH. Because LAH is a hydrogen nucleophile (source of H^-), we expect to have it attack the carbonyl *twice*. That means that our final product should be an alcohol.

Next, we think about any proton transfers that we will need to do to facilitate this reaction. Whenever we attack with H^- or C^-, we will only have one proton transfer at the very end of the mechanism. And our mechanism goes as follows:

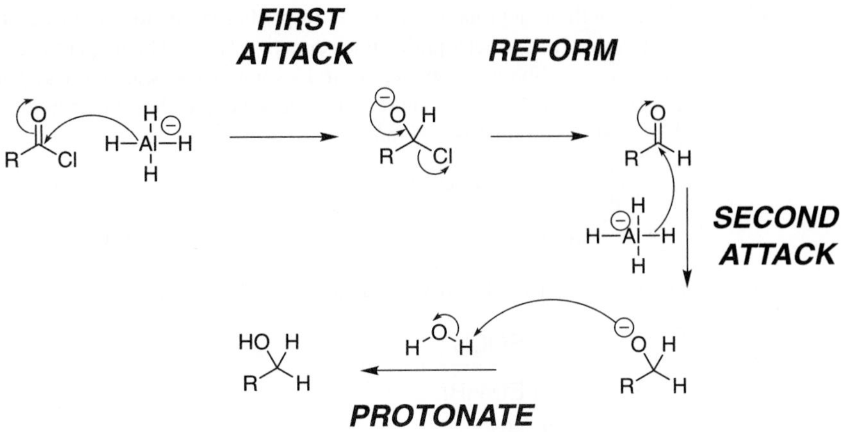

**FIRST
ATTACK**   **REFORM**

**SECOND
ATTACK**

PROTONATE

Propose a plausible mechanism for each of the following reactions:

6.10.

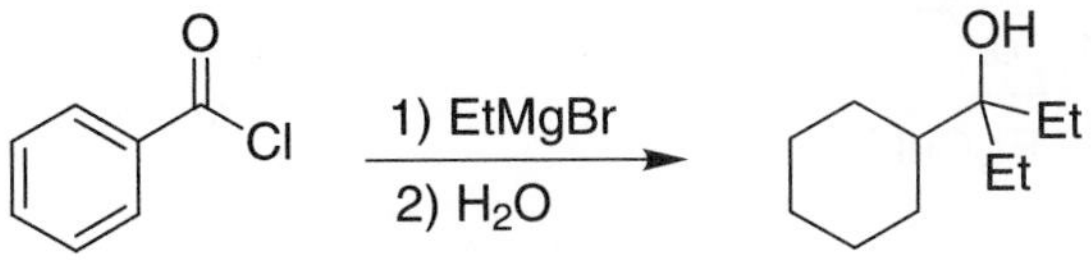

1) EtMgBr

2) H_2O

6.11.

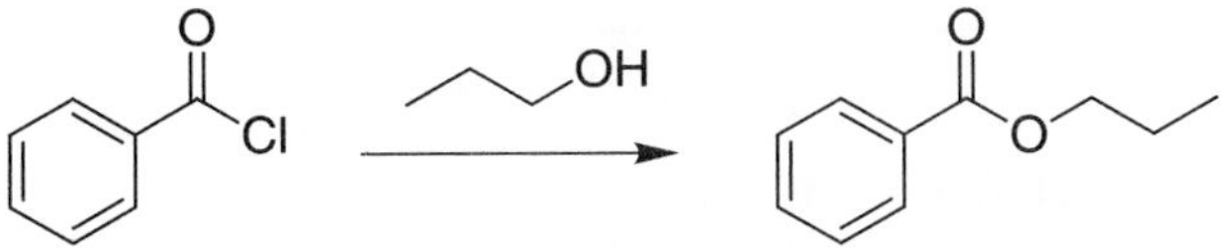

6.12.

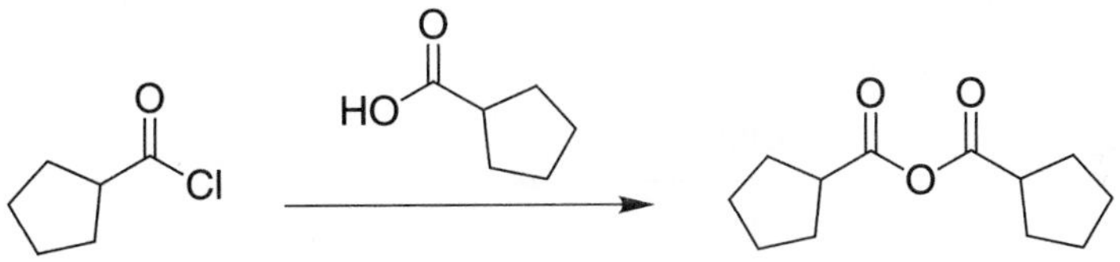

6.13.

$SOCl_2$

6.14.

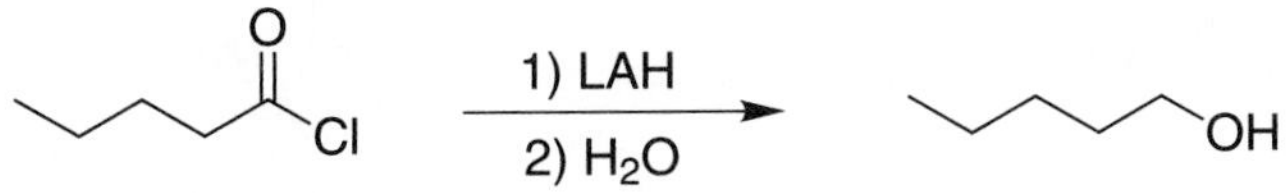

1) LAH

2) H_2O

EXERCISE 6.15.   Predict the major product of the following reaction:

Me_2CuLi

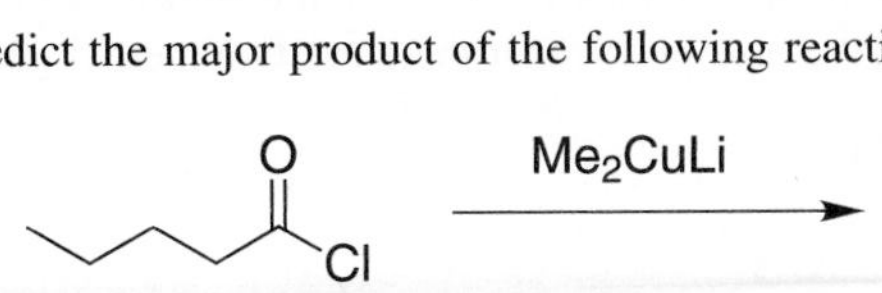

Answer: We are starting with an acyl halide. In order to predict the major product of this re-action, we will have to determine if the nucleophile attacks once or twice. The reagent is a lithium dialkyl cuprate. This reagent is a carbon nucleophile, but it does not attack twice the way Grignard reagents do. Instead, it will only attack once because it is a very *tame* carbon nucleophile, as we saw earlier in this section. And the product is a ketone:

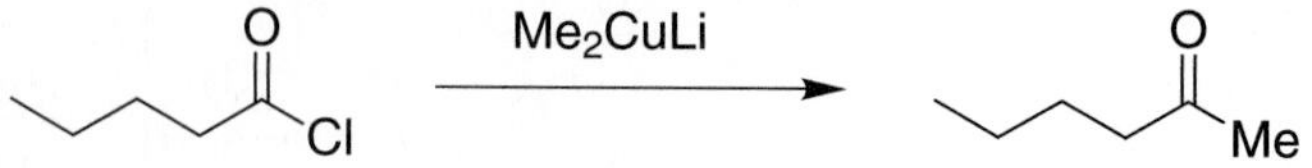

Predict the major product for each of the following reactions:

6.16.

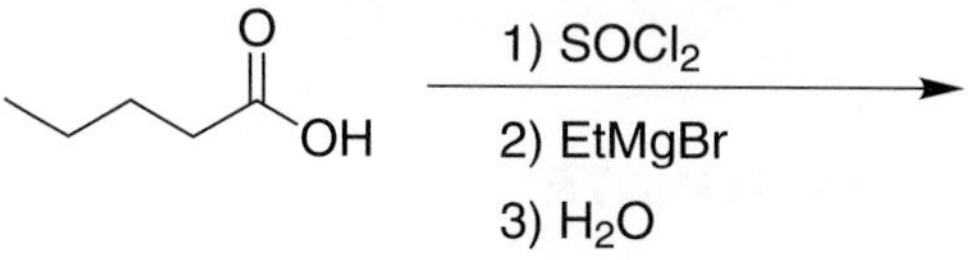

1) $SOCl_2$
2) EtMgBr
3) H_2O

6.17.

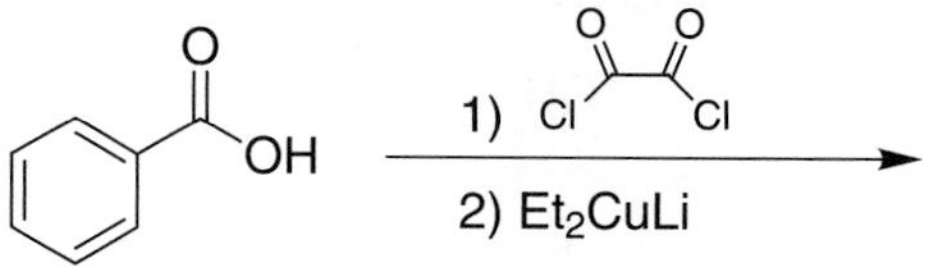

1)
2) Et_2CuLi

6.18.

NH_3

6.19.

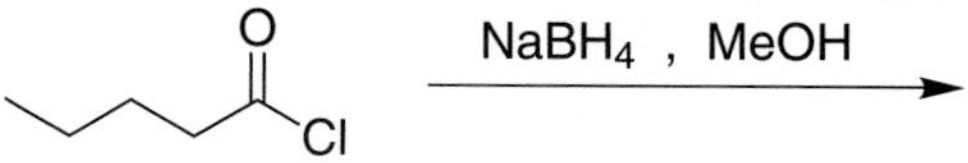

$NaBH_4$, MeOH

6.20.

EtOH

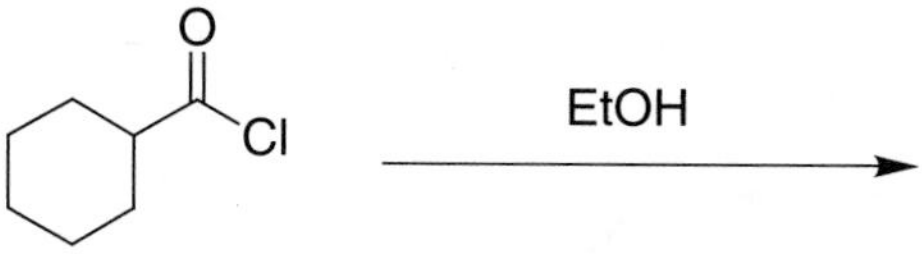

EXERCISE 6.21. What reagents would we use to carry out the following transformation?

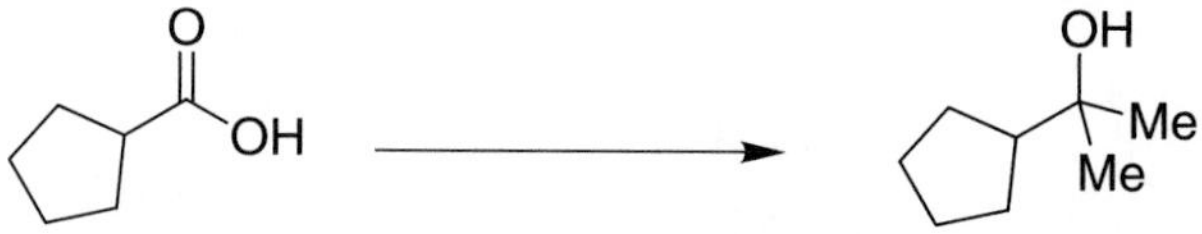

Answer: Let's carefully look at what we have. We are starting with a carboxylic acid, and the final product is an alcohol. We also notice that there are two methyl groups in the product:

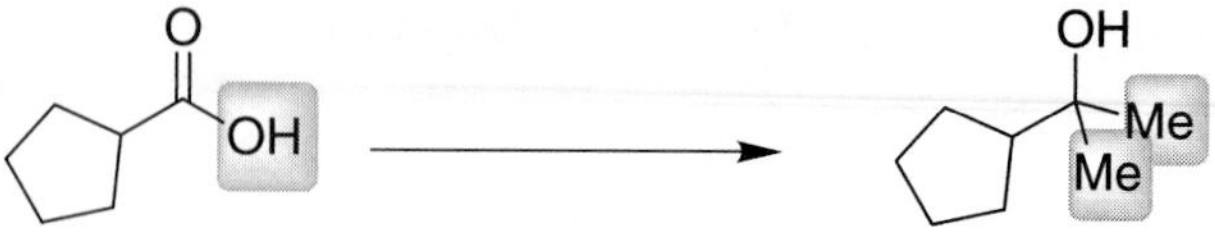

We therefore need to hook on *two* methyl groups, which means that we need to attack the carbonyl *twice*. In the process, the carbonyl must be reduced to an alcohol. This sounds like a Grignard reaction.

But we have to be careful. We cannot run a Grignard directly on a carboxylic acid. Remember that a Grignard reagent is sensitive to its conditions: if there are any protons around, they will destroy the Grignard reagent. Since our starting compound (a carboxylic acid) has an acidic proton, we must therefore first convert the carboxylic acid into an acyl halide. So our strategy goes like this:

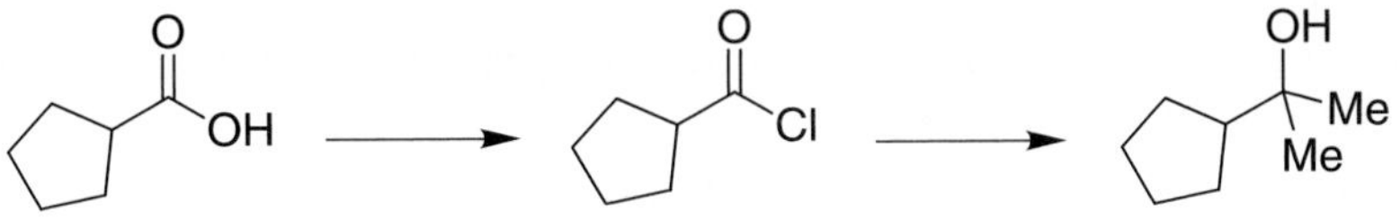

And to do this, we would use the following reagents:

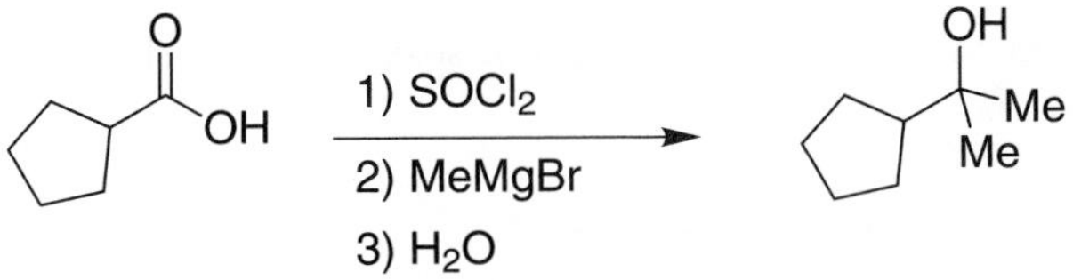

1) $SOCl_2$
2) MeMgBr
3) H_2O

Identify the reagents you would use to carry out the following transformations:

6.22.

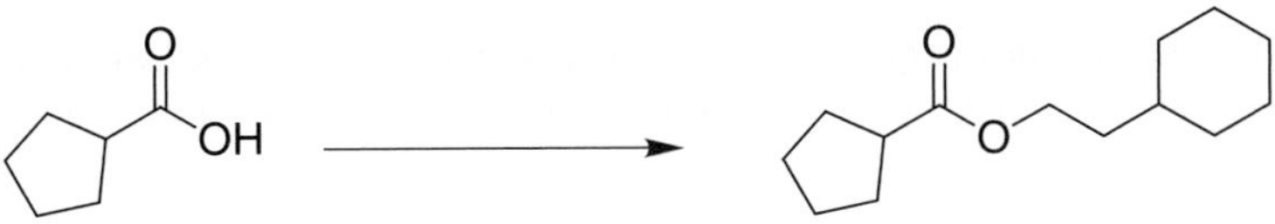

6.23.

6.24.

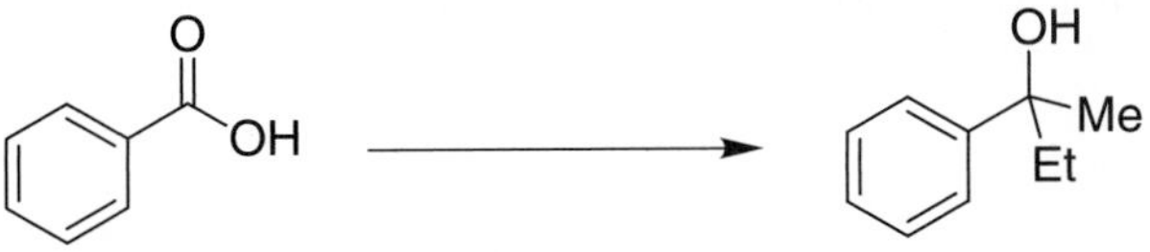

6.25.

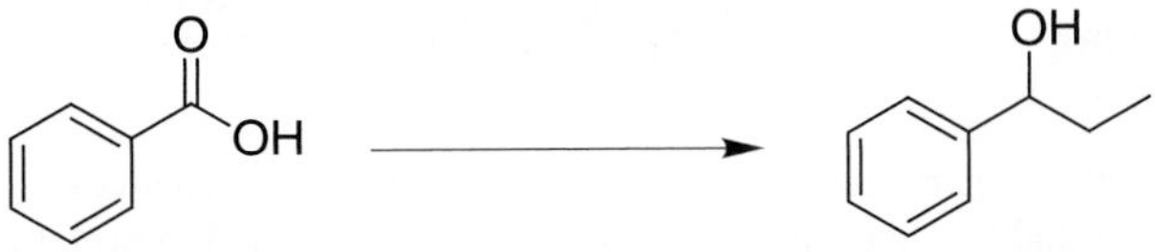

6.26.

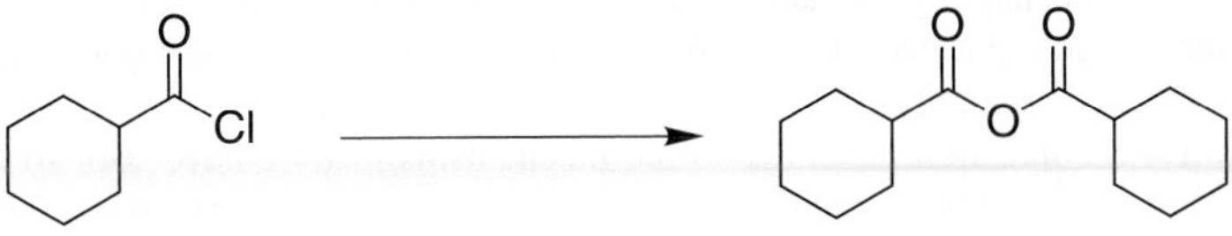

6.3 ANHYDRIDES

Anhydrides can be prepared by the reaction between a carboxylic acid and an acyl halide:

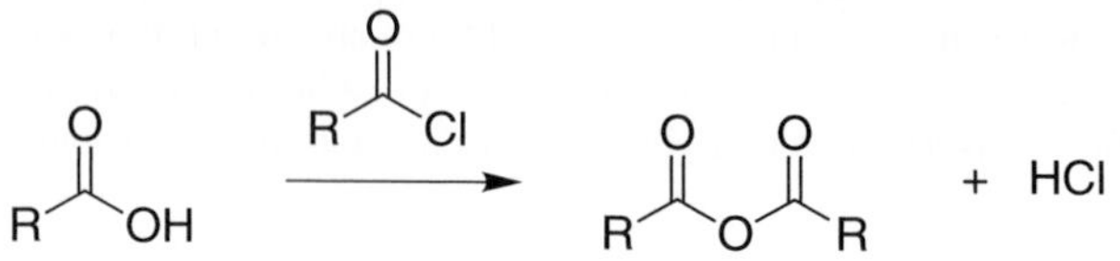

Notice that a byproduct of this reaction is HCl. To soak up this HCl as it is being formed, we generally use a compound called pyridine:

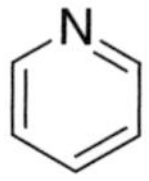

Pyridine functions as an "acid sponge" because it soaks up the HCl as it is being formed, as follows:

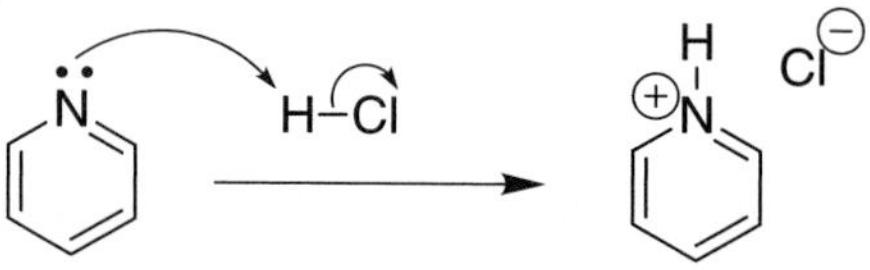

Pyridine actually serves another purpose in this reaction (which your textbook may or may not discuss). But in general, when you see pyridine, you should think of it as an acid sponge. You will often see it expressed as "py":

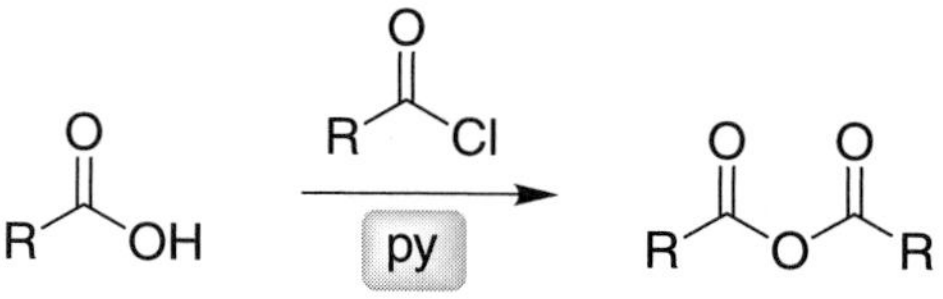

We can avoid the need for pyridine if we just use a carboxylate ion (a deprotonated carboxylic acid) as our nucleophile:

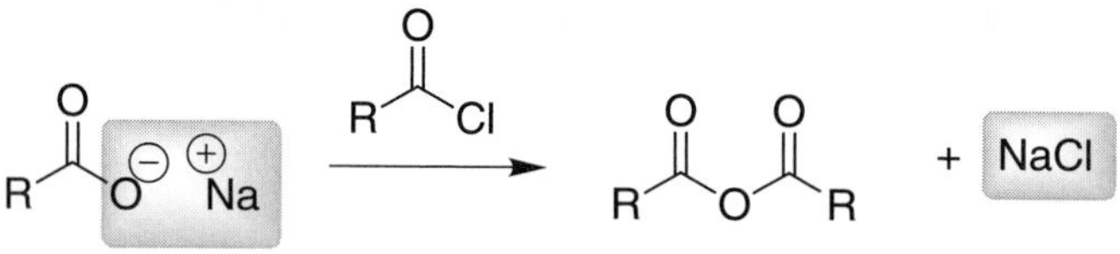

When we do it this way, we do not generate HCl, so we don't need the acid sponge.

One other common way to make anhydrides is to use a compound, called phosphorus pentoxide (P_2O_5). This compound has a poor name because the structure of the compound is actually P_4O_{10}. But chemists used to think that the structure was P_2O_5, and that is how it got its name. And old habits die hard, so we commonly call this compound phosphorus pentoxide.

The mechanism for this reaction is similar to the mechanisms that we have seen so far. There are a few subtle differences, but the mechanism makes perfect sense when we apply the rules that we have seen. For some reason, however, most textbooks do not explain the mechanism of this reaction. I cannot explain why they skip it because I think this mechanism would further reinforce all of the rules we have seen. But since your textbook probably doesn't cover it, I, too, will skip this mechanism. Perhaps if you are feeling adventurous, you could look up the structure of P_4O_{10} on the Internet and then try to work through the mechanism yourself.

Anhydrides are almost as reactive as acyl halides; therefore, the reactions of anhydrides are identical to the reactions of acyl halides. You just have to train your eye to see the leaving group:

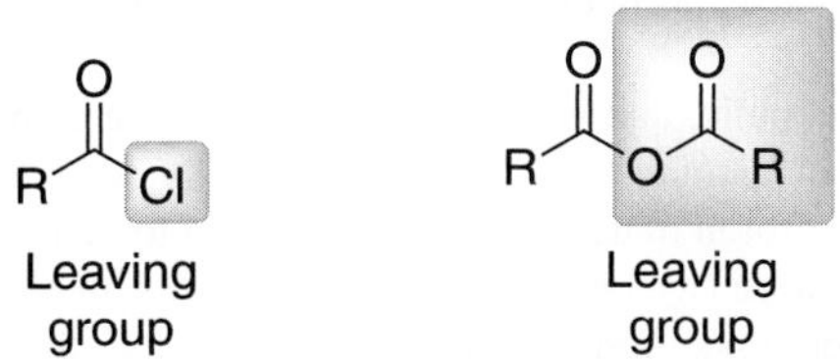

Leaving
group

Leaving
group

When a nucleophile attacks, the carbonyl can re-form to give a leaving group that is stabilized by resonance:

Stabilized by
resonance

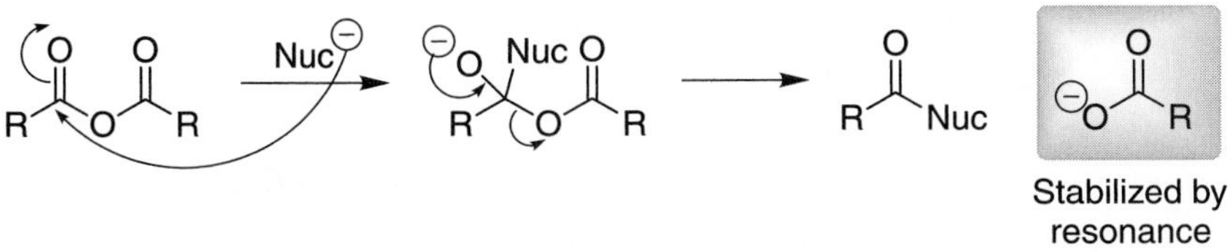

So, we can react an anhydride with any of the nucleophiles that we saw in the previous section, and we will get the same products that we got in the previous section:

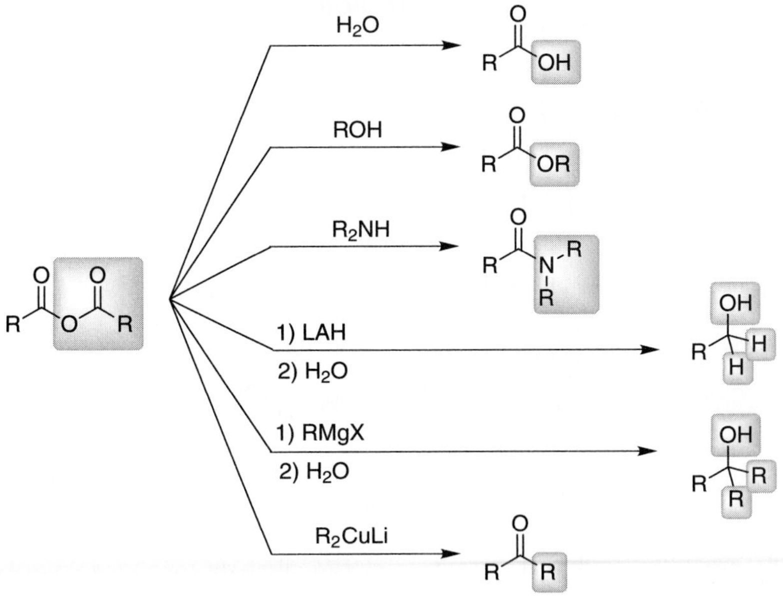

EXERCISE 6.27. Propose a mechanism for the following reaction:

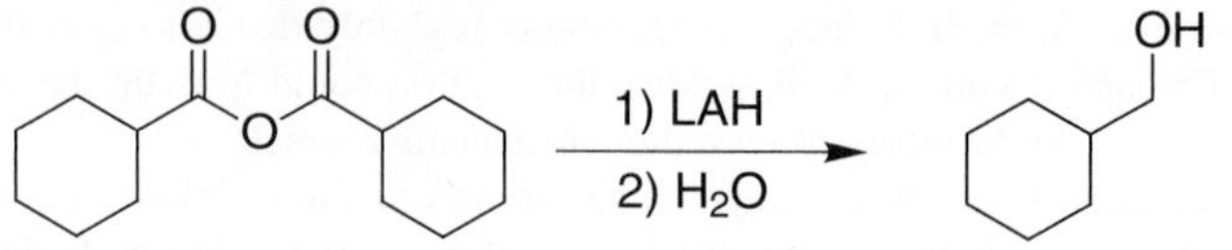

Answer: First let's decide if the nucleophile attacks once or twice, and then we will decide when to do proton transfers.

LAH is a hydrogen nucleophile, and we expect it to attack the carbonyl *twice*. That means that our final product should be an alcohol.

Next, we think about any proton transfers that we will need to do to facilitate this reaction. Since we are starting with an anhydride (which is almost as reactive as an acyl halide), we will only have one proton transfer at the very end. And our mechanism goes like this:

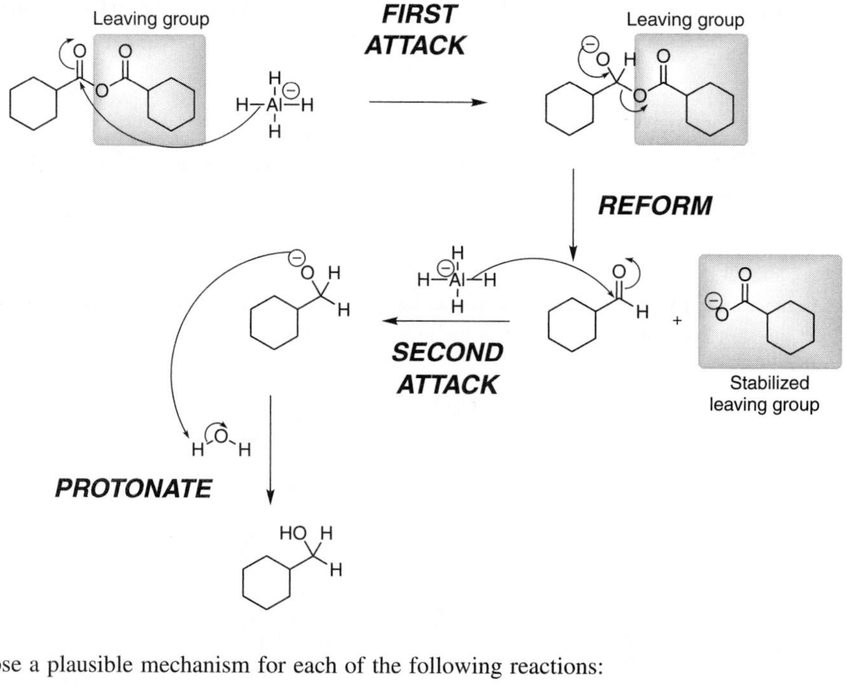

Propose a plausible mechanism for each of the following reactions:

6.28.

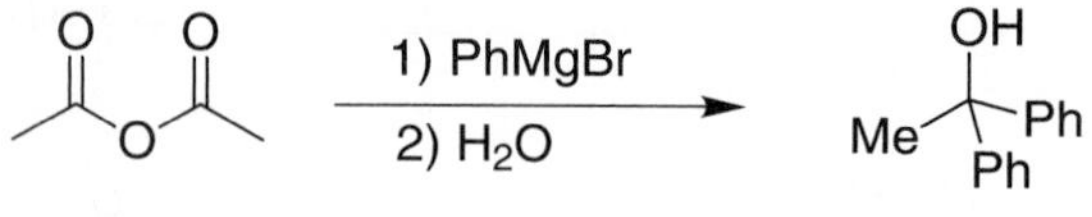

6.29.

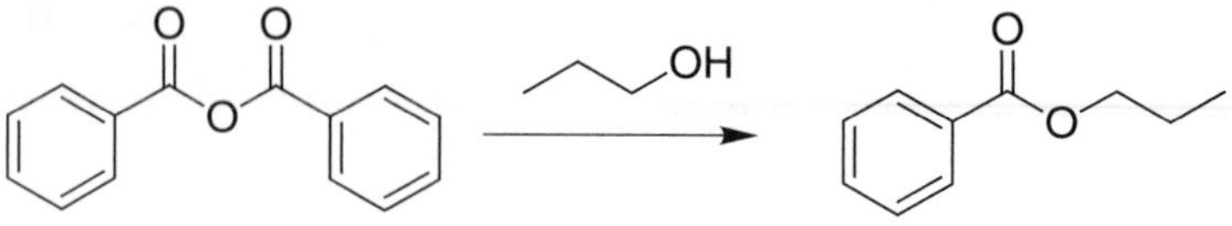

EXERCISE 6.30.   Predict the major products of the following reaction:

1) PhMgBr

2) H₂O

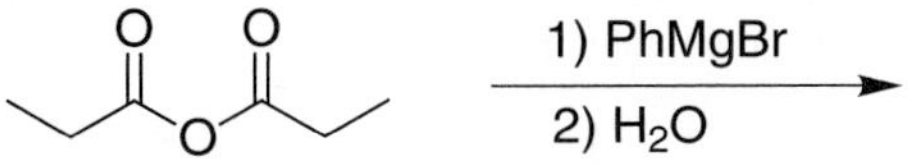

Answer: In order to predict the major products of this reaction, we will have to determine if the nucleophile attacks once or twice. The reagent is a Grignard, which is a strong carbon nucleophile. We therefore expect it to attack twice, which will produce an alcohol:

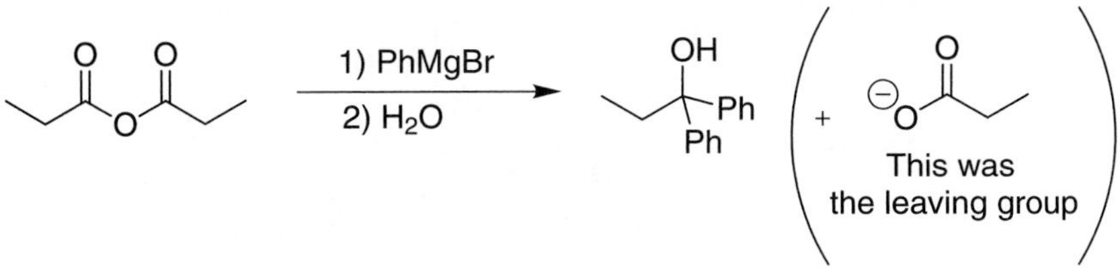

1) PhMgBr

2) H₂O

OH

Ph
Ph

+

This was
the leaving group

Predict the major products for each of the following reactions:

6.31.

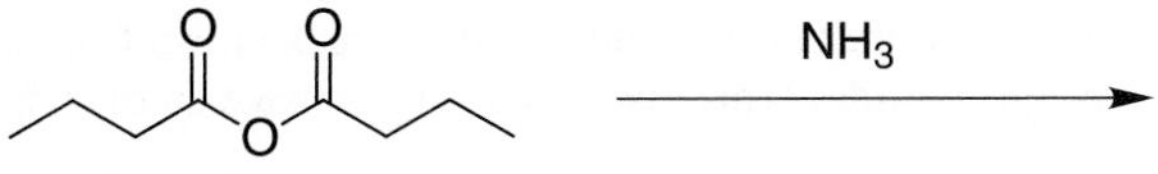

NH₃

6.32.

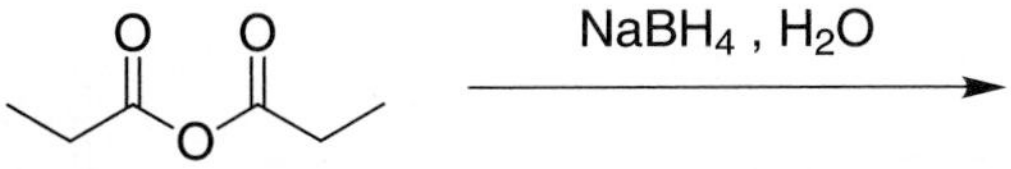

NaBH₄ , H₂O

6.33.

H₃O⁺

6.4   ESTERS

Let's quickly review the order of reactivity for the carboxylic acid derivatives:

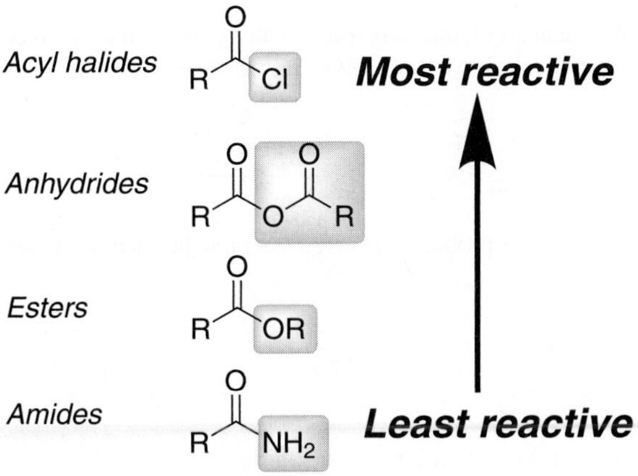

Acyl halides

R   Cl

Most reactive

Anhydrides

R   O   R

Esters

R   OR

Amides

R   NH₂

Least reactive

It is helpful to remember this order. Esters can be made from any carboxylic acid derivatives that are more reactive than an ester (anything above the ester on the chart above). In other words, we can make an ester from an acyl halide or from an anhydride:

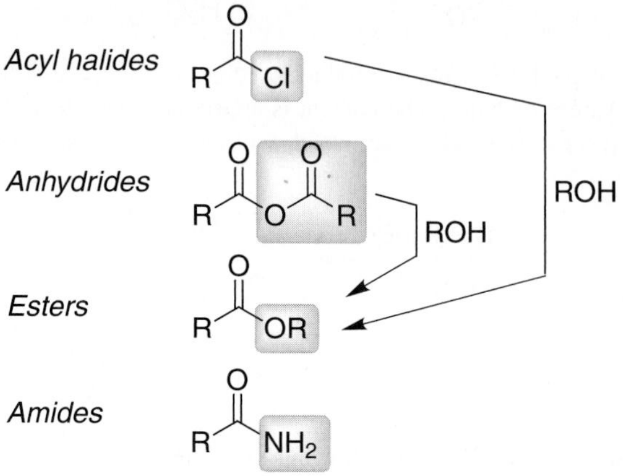

We have already seen a couple of ways to make acyl halides: we can make them from carboxylic acids (using thionyl chloride or oxalyl chloride). This gives us a way to make an ester from a carboxylic acid in two steps:

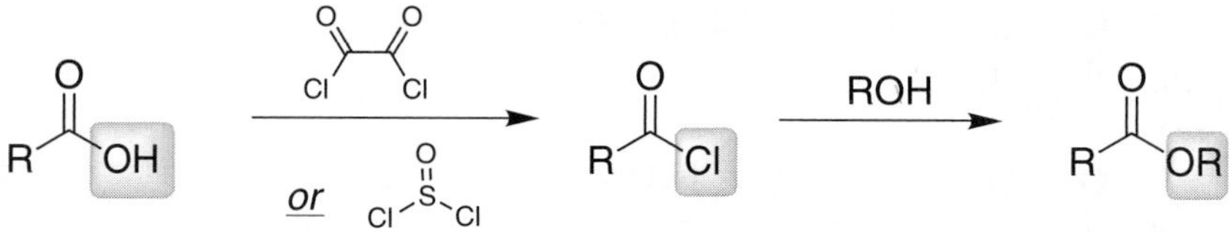

But this begs the question: can we make an ester directly from a carboxylic acid *in one step*? Can we skip the need for making the acyl halide? If we simply try to mix an alcohol and a carboxylic acid, we do ***not*** get a reaction:

Let's therefore see what we can do to force this reaction along. We can try to make the nucleophile more nucleophilic. In other words, we can try to use RO⁻ instead of ROH:

But that would create a separate problem. The RO⁻ would just act as a base and deprotonate the carboxylic acid:

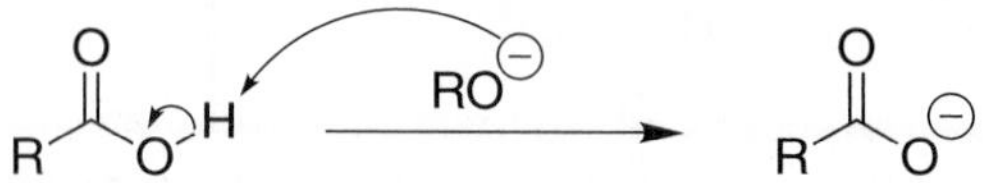

Once again, then, we would not get an ester as our product.

But there is one more thing we can try. Rather than making the nucleophile more nucleophilic, we can try to make the electrophile more electrophilic. Do you remember how to do that? We just protonate the carbonyl, as follows:

More electrophilic

H^+ can function as a catalyst by making the electrophile more electrophilic. Under these conditions (acidic conditions), we *do* get the reaction we want (a one-step synthesis of an ester from a carboxylic acid):

This reaction is incredibly useful and important (and it is considered to be a staple of any organic chemistry course), so let's go over the mechanism in detail.

Notice that only two core steps are involved here: attack and then re-form. Everything else is just proton transfers (which we discussed at length at the beginning of this chapter):

PROTONATE   ATTACK

PROTON TRANSFERS

DEPROTONATE   REFORM

We have three proton transfers here. This exactly follows the pattern we described in the beginning of this chapter:

Proton transfer, *attack*, proton transfer, *re-form*, proton transfer

The first proton transfer is used to make the electrophile more electrophilic. The proton transfers in the middle of the mechanism are used to protonate the leaving group (so that we do not kick off OH^- in acidic conditions). And finally, the last proton transfer is to deprotonate our product to get rid of the charge.

This mechanism is called a Fischer Esterification. The position of equilibrium is very sensitive to the concentrations of starting materials and products. Excess ROH favors formation of the ester, whereas excess water favors the carboxylic acid:

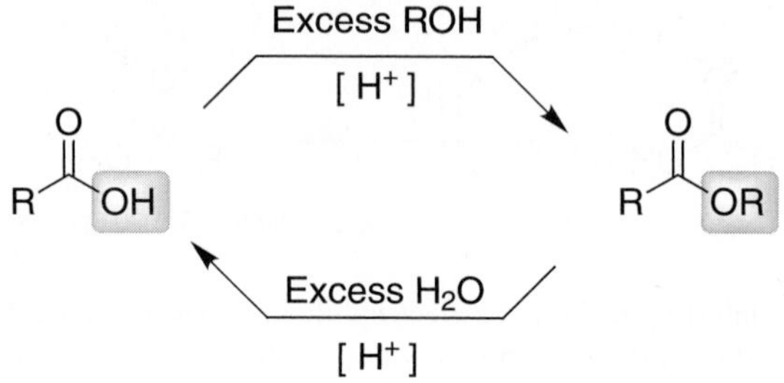

Excess ROH
[H⁺]

Excess H₂O
[H⁺]

This is very helpful because it gives us a way to convert an acid into an ester *or* an ester into an acid.

For now, let's focus on converting an acid into an ester:

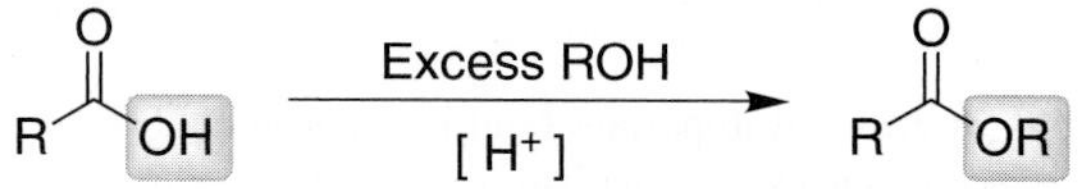

Excess ROH
[H⁺]

We will spend some time on the reverse process very soon.

As we have seen, a Fischer Esterification is the reaction between a carboxylic acid and an alcohol (with acid catalysis). In the examples we have seen, the carboxylic acid and the alcohol were different molecules. But it is possible for both functional groups to be present in one compound, for example:

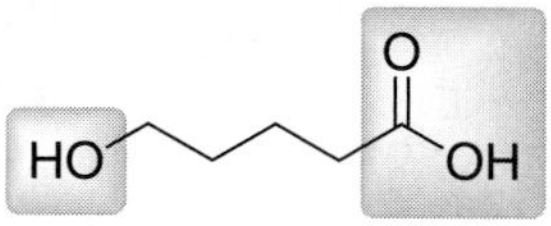

This compound has both a COOH group and an OH group. So, in this case, it is possible to have an *intramolecular* reaction:

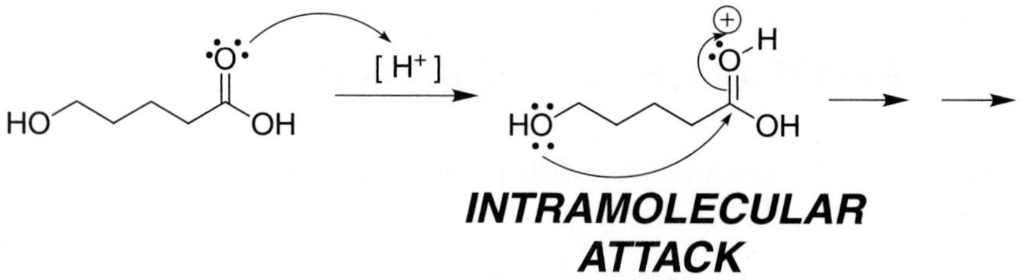

[H⁺]

INTRAMOLECULAR ATTACK

The rest of the mechanism is exactly the same as what we have already seen. The mechanism has two core steps (attack and re-form). The rest of the mechanism is just proton transfers: in the beginning, in the middle, and then in the end.

PROBLEM 6.34.   Draw the mechanism of the following reaction:

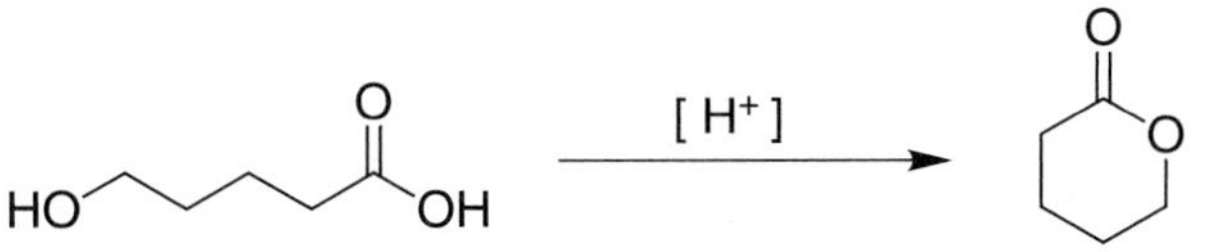

This reaction is just an intramolecular Fischer Esterification. Try to get some practice drawing a Fischer Esterification. Remember that the general steps are:

   Proton transfer, *attack*, proton transfer, *re-form*, proton transfer

Answer:

EXERCISE 6.35.   Suppose you wanted to make the following compound with a Fischer Esterification:

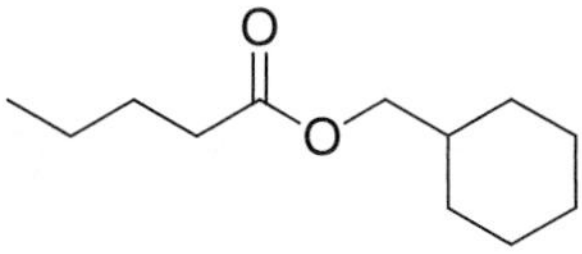

What reagents would you use?

Answer: To make an ester using a Fischer Esterification, we need to start with a carboxylic acid and an alcohol. The question is: how do we decide which carboxylic acid and which alcohol to use? To do this, we must realize the bond that will be formed during the reaction:

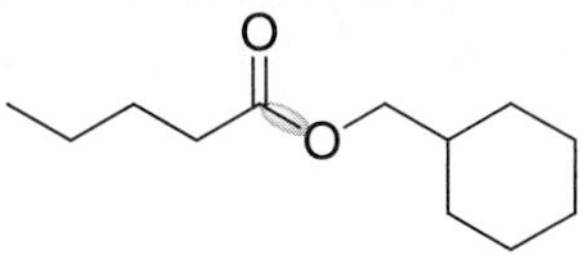

Therefore, we will need the following reagents:

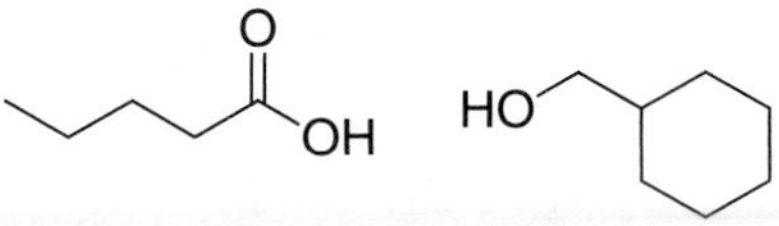

And don't forget that we need acid catalysis. So our synthesis would look like this:

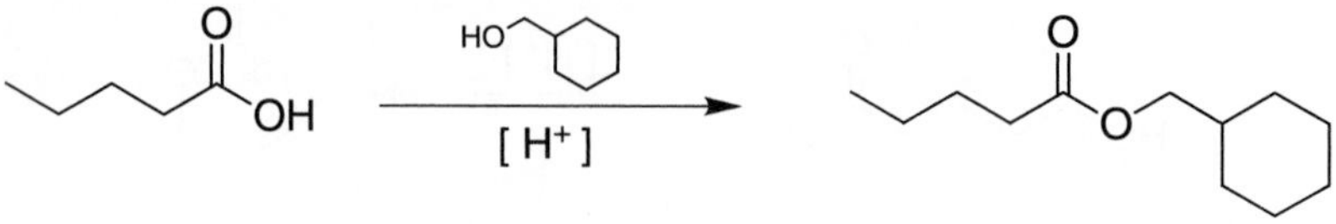

For each of the following problems, identify the reagents you would use to make the ester that is shown:

6.36.

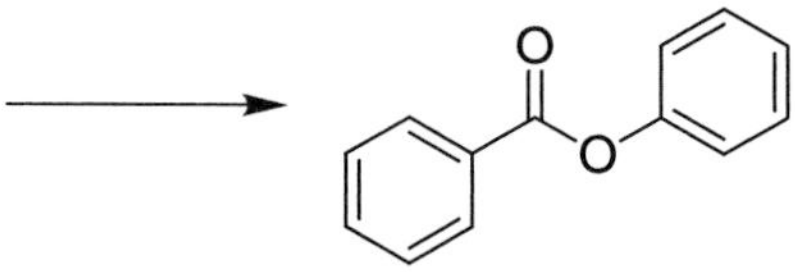

6.37.

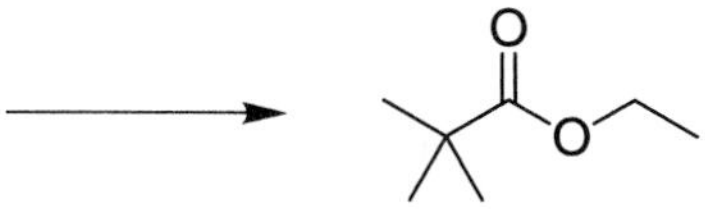

6.38.

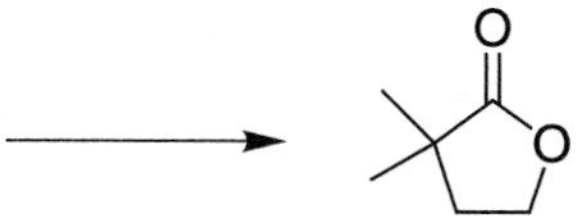

6.39.

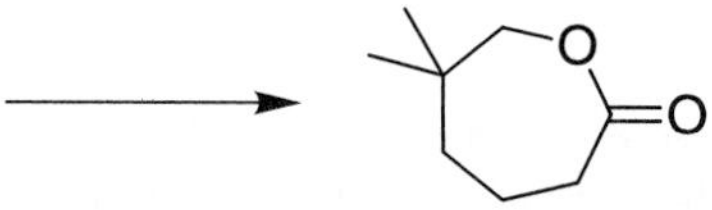

Now that we have seen how to make esters, let's focus our attention on the reactions of esters. We will center on two reactions in particular. Esters can be hydrolyzed to give carboxylic acids, under two different sets of conditions:

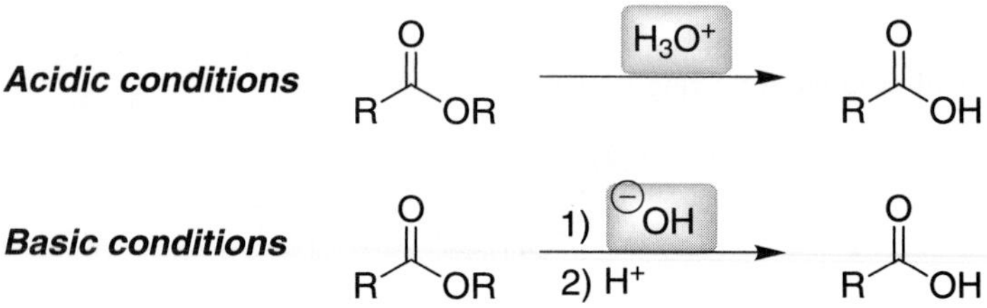

Acidic conditions

Basic conditions

The first set of conditions above (acidic conditions) should seem very familiar to you. This reaction is simply the reverse of a Fischer Esterification. When we learned about the Fischer Esterification, we mentioned that the reverse of a Fischer Esterification is possible. Now we will explore that reverse reaction in more detail. The mechanism goes like this:

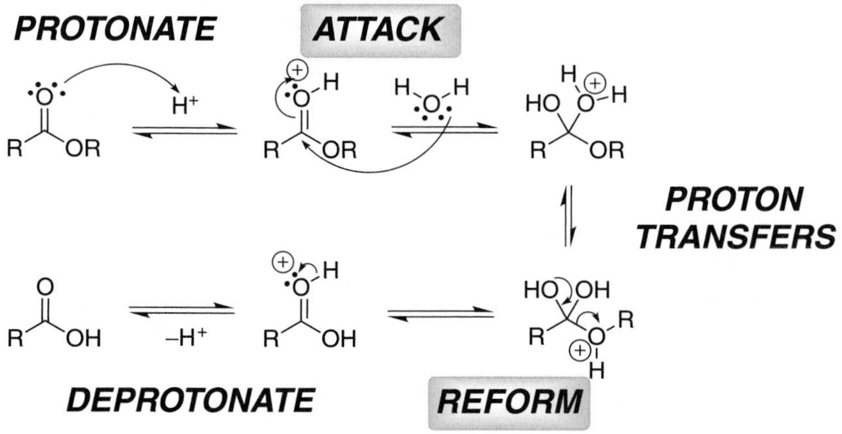

Once again, we see the same pattern again and again. Look closely at this mechanism. There are two core steps: attack and re-form to lose a leaving group. The rest of the mechanism is just proton transfers that facilitate the reaction. We have one proton transfer in the beginning (to protonate the carbonyl), proton transfers in the middle (so that the leaving group can leave as a neutral species), and proton transfer at the end (to deprotonate and form the product).

But this reaction was under acidic conditions. We can also hydrolyze an ester under basic conditions through the following mechanism:

Once again, we have two core steps (attack and re-form), followed by a deprotonation. This proton transfer at the end is unavoidable under these conditions. In basic conditions, a carboxylic acid will get deprotonated. In fact, that is the driving force for this reaction to take place: we are forming a more stable anion:

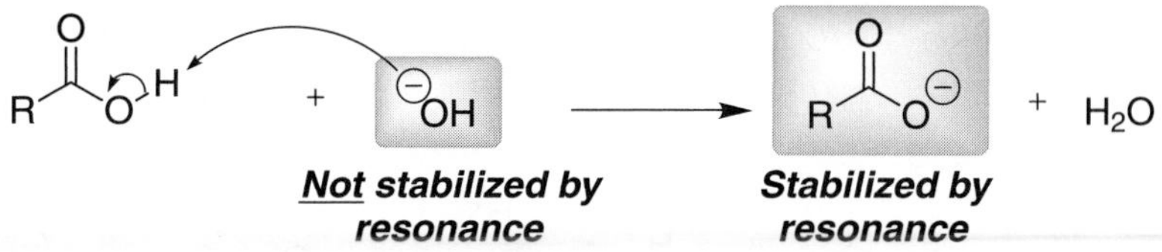

In order to isolate our product, we need to give back a proton, but since we are in basic conditions, we don't have any protons around. We will therefore need to add a source of protons when the reaction is complete:

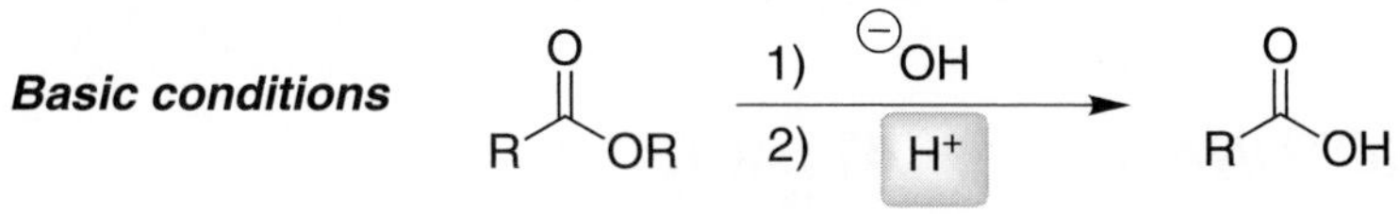

Notice that we need to use H^+ rather than H_2O because a carboxylate ion will not pull a proton off of water:

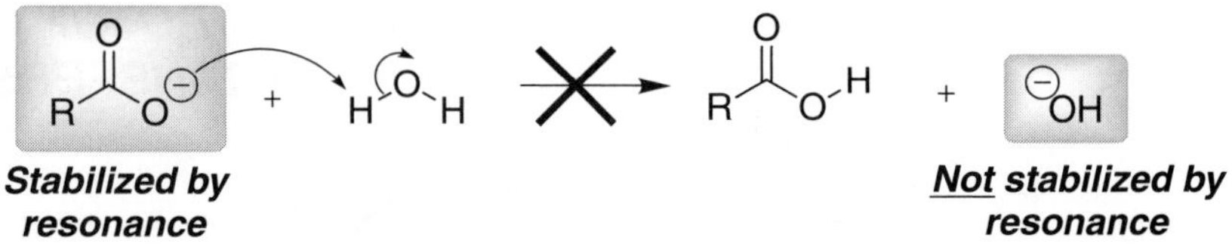

Stabilized by resonance   **Not stabilized by resonance**

This process (hydrolyzing an ester under basic conditions) has a special name: *saponification.*

EXERCISE 6.40. Propose a mechanism for the following transformation:

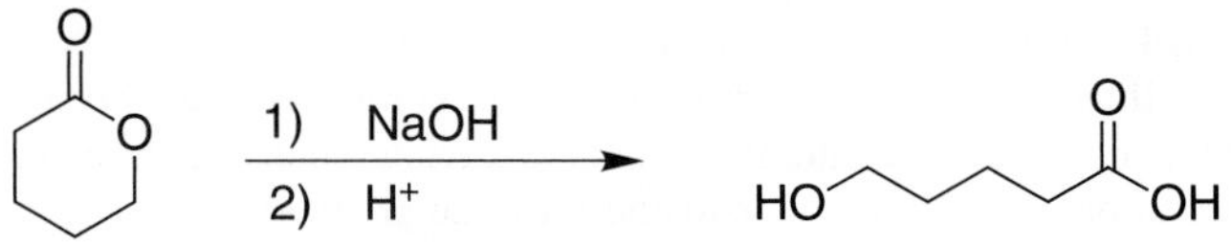

Answer: This reaction utilizes basic conditions to convert an ester into a carboxylic acid and an alcohol. We call this a saponification. The only twist here is that the reaction is intramolecular, but that does not change the mechanism at all. It is the same as the mechanism we just saw. There are no proton transfers until the very end of the mechanism. Thus, we start by attacking the carbonyl and then re-forming it. Then, under basic conditions, the carboxylic acid is deprotonated, which is why we add H^+ at the end (to give the proton back):

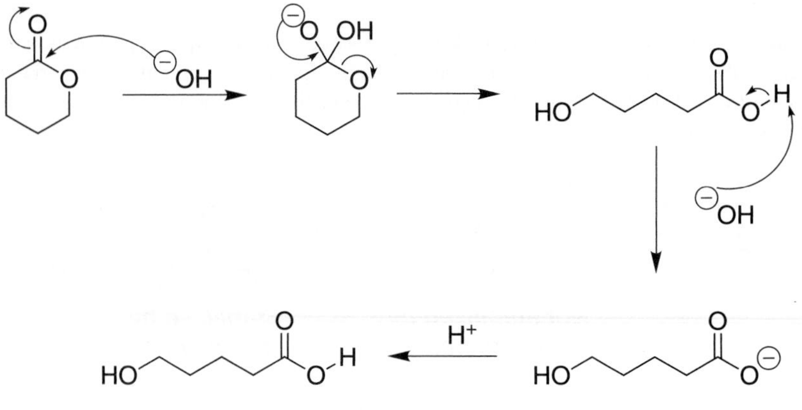

Propose a plausible mechanism for each of the following reactions:

6.41.

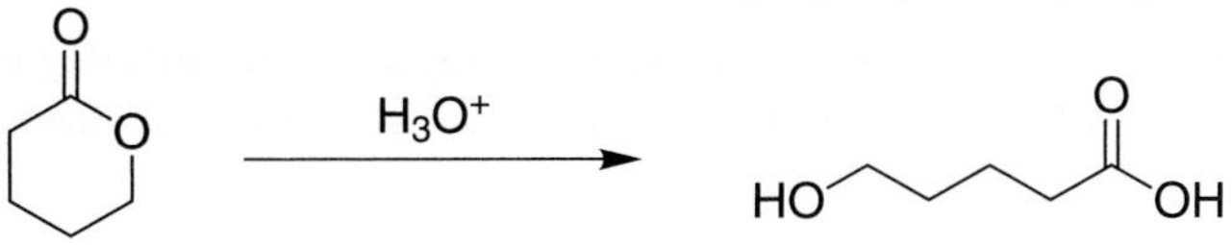

6.42.

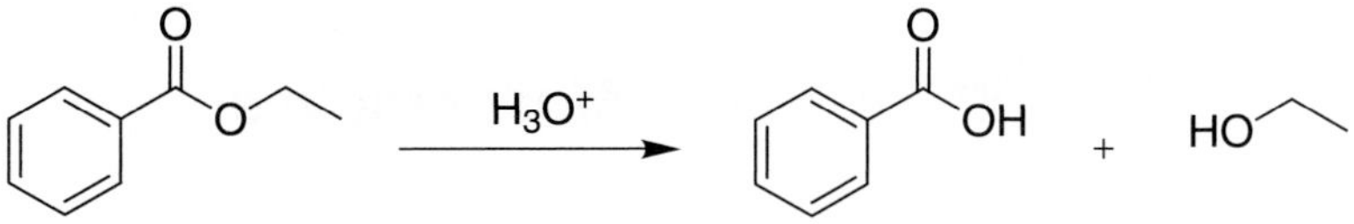

EXERCISE 6.43.   Predict the products of the following reaction:

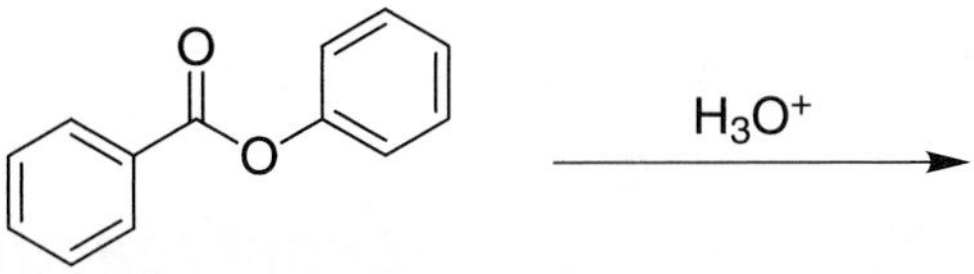

Answer: We are starting with an ester, and we are subjecting it to acidic conditions. This will convert the ester into a carboxylic acid and an alcohol (the reverse of a Fischer Esterfication). This will give us the following products:

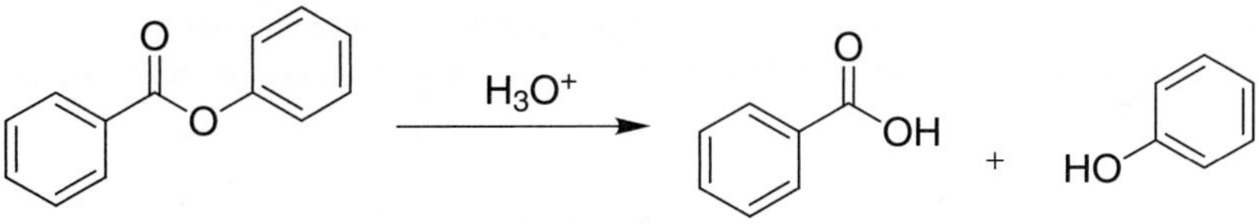

Predict the products of each of the following reactions:

6.44.

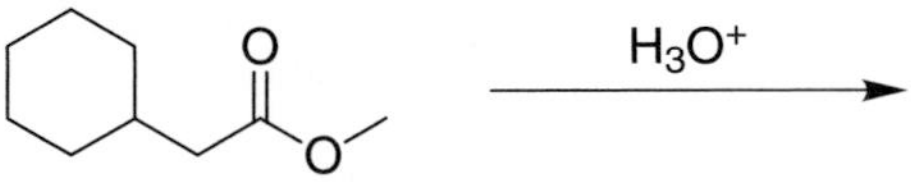

6.45.

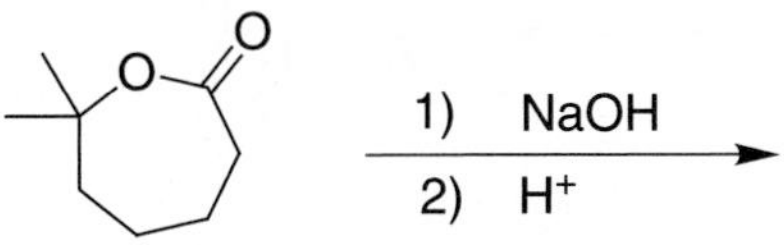

1)   NaOH
2)   H⁺

6.46.

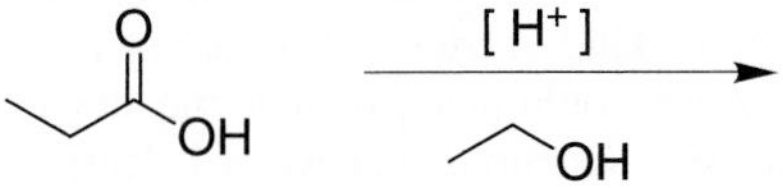

[H⁺]

6.47.

1)   NaOH
2)   H⁺

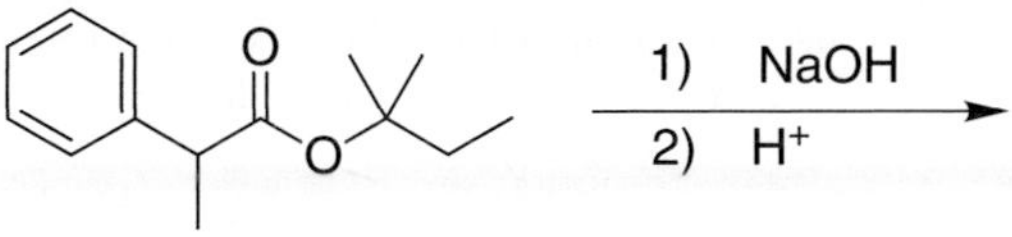

6.5 AMIDES AND NITRILES

We have said before that a carboxylic acid derivative can be prepared from any other carboxylic acid derivative that is more reactive. Let's go back to our reactivity chart to see what this means practically:

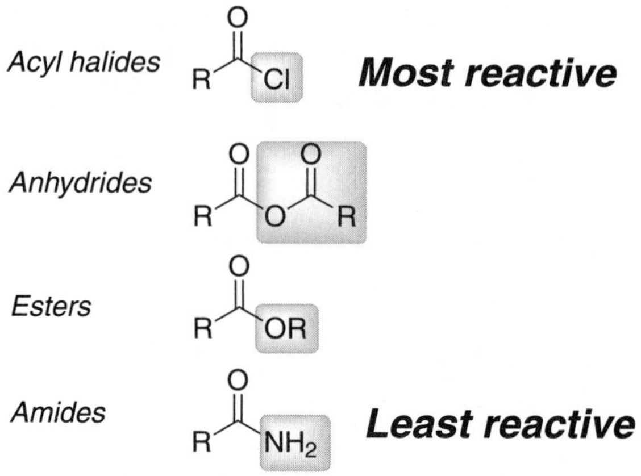

Since amides are the least reactive of the carboxylic acid derivatives shown on this chart, we can therefore make amides from any carboxylic acid derivatives that are higher on the chart. In other words, we can make amides from acyl halides, from anhydrides, or from esters.

Earlier in this chapter, we saw how to make amides from acyl halides or anhydrides:

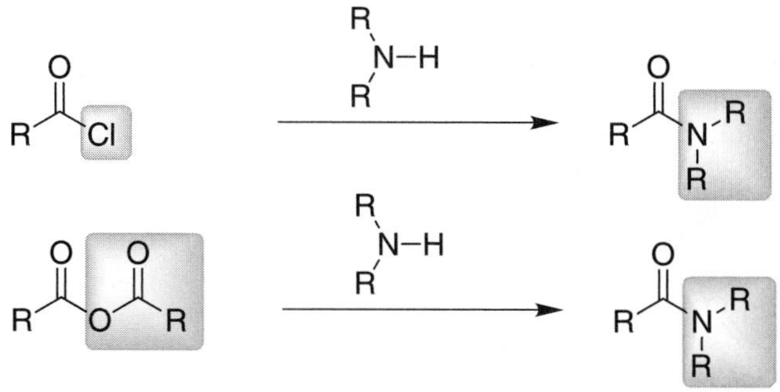

But now the question is: how do we make amides from esters? Esters are less reactive than acyl halides or anhydrides. We therefore have to use some kind of trick to get the reaction to go. We cannot use acid or base to help this reaction go. (Acid would just protonate the attacking amine, rendering it useless; and base would cause other side reactions that we will learn in the next chapter.) So we cannot make our nucleophile more nucleophilic, and we cannot make our electrophile more electrophilic. Instead, we will use the simplest trick there is: brute force. We just heat the reaction for a very long time, and we can get a reaction, which follows the following mechanism:

ATTACK   **REFORM**   **PROTON TRANSFER**

Notice that we kick off RO⁻ to re-form the carbonyl. That might seem strange because there is a much better leaving group available to leave:

Much better leaving group

But this gets back to something we have said many times before. Of course, it is possible for the amine to leave; in fact, it happens all of the time. The amine attacks, and then it gets kicked off; it attacks, and then it gets kicked off again. Every time this happens, there is no change for us to observe, but every once in a while, something else can happen. We can kick off RO⁻, which then immediately grabs a proton, as shown in the mechanism above. We are allowed to kick off RO⁻ when re-forming a carbonyl because the tetrahedral intermediate is so high in energy (negative charge on an oxygen).

The equilibrium that gets established favors the products (amide + alcohol) over the reactants (ester + amine):

So, we can use this as another way to make amides (and when we use this method, we form an alcohol as a by-product).

In this section, we have seen that we can make amides from acyl halides, from anhydrides, or from esters. Now that we know how to make amides, let's see a couple of the important reactions of amides. Much of biochemistry is dependent on how, when, and why amides will undergo a reaction. Consequently, if you plan to take biochemistry, you should know some of the basics of amide chemistry. Some organic chemistry textbooks will go into significant detail on this topic, whereas other textbooks will only show two generic reactions. You should take special note to look in your textbook and lecture notes to see how much you are responsible for. In this book, we will focus on just the two most common reactions that you will find in every organic chemistry textbook.

In both of these reactions, we are converting an amide into a carboxylic acid. The only difference is the conditions (basic conditions or acidic conditions):

Let's begin with the acid-catalyzed mechanism. This reaction is no different than the other acid-catalyzed reactions we have seen. Take a close look at the mechanism:

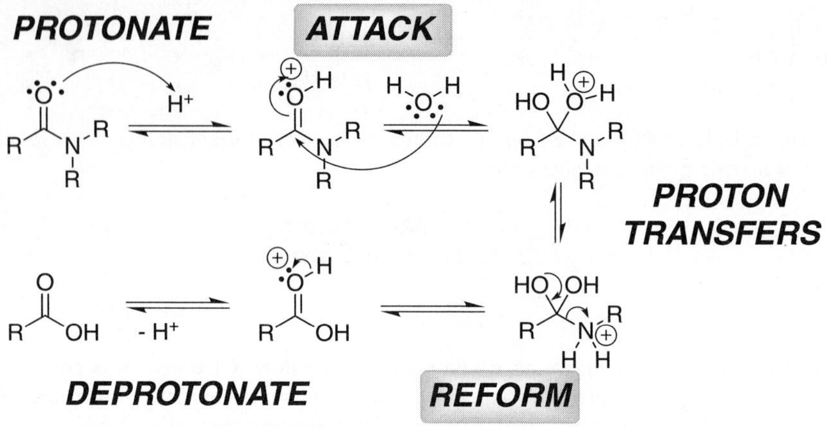

PROTONATE   **ATTACK**

PROTON TRANSFERS

DEPROTONATE   **REFORM**

Notice that it follows the exact same pattern that we have seen again and again:

Proton transfer, **attack**, proton transfer, **re-form**, proton transfer

Now let's look at how it works under basic conditions. Here is the mechanism:

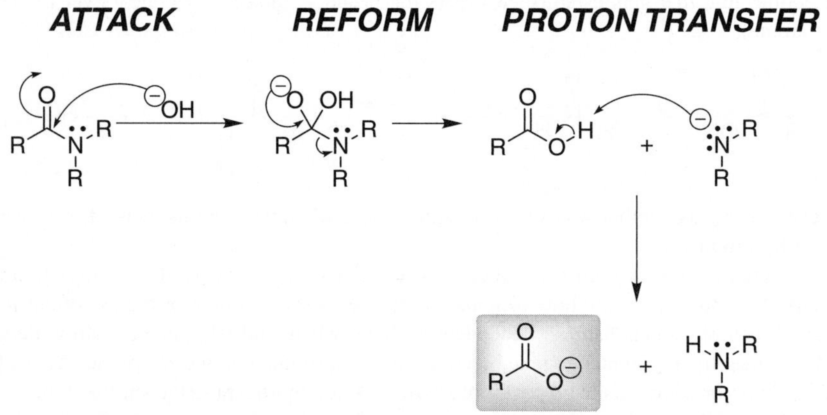

ATTACK   **REFORM**   **PROTON TRANSFER**

Under basic conditions, the product we get is a carboxylate ion (highlighted above); we must then add a source of protons in order to get our product:

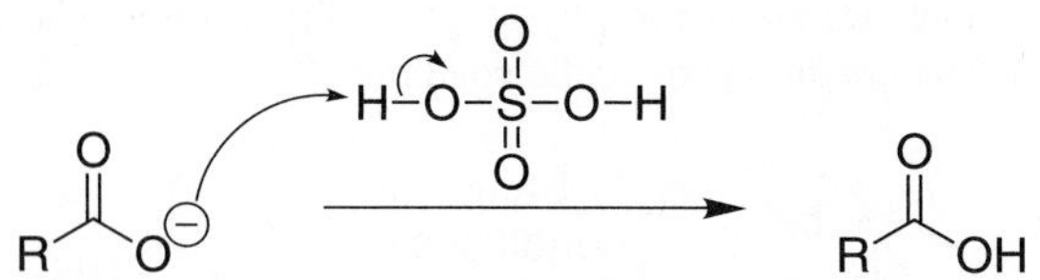

Before we do some problems, let's look at one last carboxylic acid derivative that we have not yet seen. Compounds containing a cyano group are called nitriles:

$$R-C \equiv N$$ *Cyano group*

You might be wondering why nitriles are carboxylic acid derivatives. After all, a nitrile looks very different from the other carboxylic acid derivatives. To make sense of this, we need to look at oxidation states. Each of the carboxylic acid derivatives has three bonds to electronegative atoms:

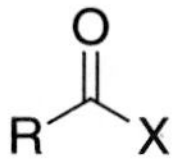

The carbon atom of the carbonyl group has two bonds with an oxygen, and it also has one more bond with some heteroatom, X (depending on which carboxylic acid derivative you are talking about). That gives a total of three bonds to heteroatoms. The carbon atom of a cyano group also has three bonds to a heteroatom. Thus, nitriles are at the same oxidation level as the other carboxylic acid derivatives.

Nitriles can be prepared using cyanide as a nucleophile to attack an alkyl halide:

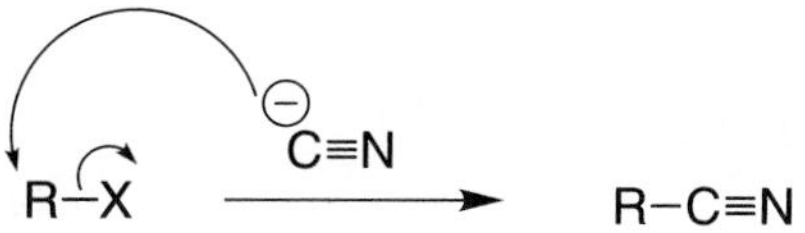

This is an S_N2 reaction, so you can only use this method when you are dealing with primary or secondary alkyl halides (primary halides are much better). Don't use this method on tertiary alkyl halides. Some textbooks cover other methods to prepare nitriles. You should look through your textbook (and your lecture notes) to see if you are responsible for knowing any other ways to make nitriles.

Nitriles are at the same oxidation level as other carboxylic acid derivatives because hydrolysis (which is **not** an oxidation-reduction reaction) produces an amide:

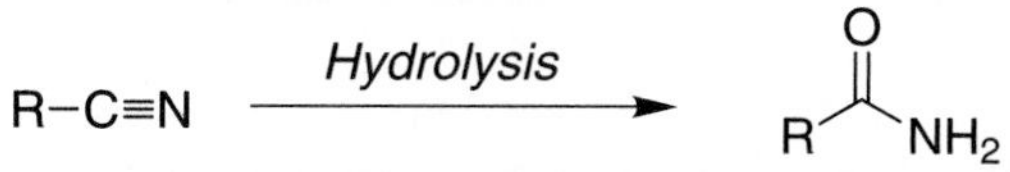

We can hydrolyze under acidic conditions or under basic conditions.

Whether we do an acid-catalyzed hydrolysis or a base-catalyzed hydrolysis, the core steps are slightly different from the core steps of the mechanisms we have seen in this chapter. So far, all of our reaction mechanisms have had at least **two** core steps (attack the carbonyl and then kick off a leaving group to re-form the carbonyl), with all other steps being proton transfers. But now we will see a mechanism that has just **one** core step (attack the carbonyl). With regard to the hydrolysis of nitriles, we don't need to kick off a leaving group. We can actually re-form the carbonyl simply through proton transfers:

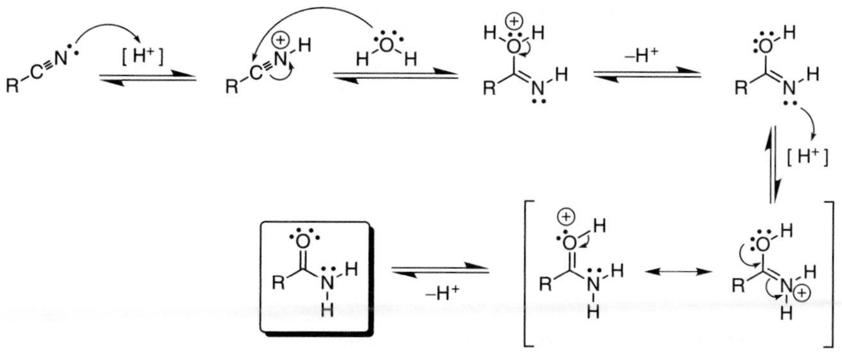

This is the mechanism for the hydrolysis of a nitrile under **acidic** conditions. Notice that the second step in the mechanism is attacking the cyano group (in much the same way we attack a carbonyl group). All other steps in the mechanism are just proton transfers. When you think of it in this way, it greatly simplifies the mechanism. We need to do so many proton transfers to avoid forming any negative charges in our intermediates. Notice that in acidic conditions, all of the intermediates are either positively charged or neutral.

Now let's consider what happens when we hydrolyze a nitrile under **basic** conditions. The mechanism is actually very similar to the mechanism above. There is also only **one** core step (attacking the cyano group), and all the other steps are just proton transfers. But that's where we have our difference between acidic conditions and basic conditions. For example, in basic conditions, you would *not* protonate the cyano group first. Rather, you would just attack with hydroxide first:

The rest of the mechanism is all just proton transfers. In order to do the proton transfers properly, we must keep one thing in mind: stay consistent with the conditions. When we were in acidic conditions, all of our intermediates were either positively charged or neutral. That was consistent with acidic conditions. In basic conditions, however, all of the intermediates should either be negatively charged or neutral.

With this in mind, let's see if you can propose the mechanism for the hydrolysis of a nitrile under basic conditions.

PROBLEM 6.48. Based on everything we have just seen, try to propose the mechanism for the hydrolysis of a nitrile under basic conditions:

Remember: there is just one core step (attacking the cyano group with hydroxide). After that, everything is just proton transfers. When you have finished, you can look in the back of the book (or in your textbook) to see if you got it right.

Answer:

Now let's get some more practice with amides and nitriles

EXERCISE 6.49.   Propose a plausible mechanism for the following reaction:

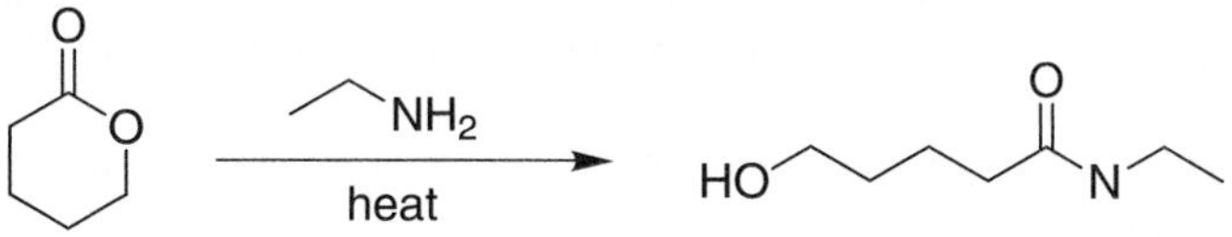

Answer: This reaction involves an ester reacting with an amide under conditions of heating. We have seen these conditions before. The first step of the mechanism involves the amine attacking the ester:

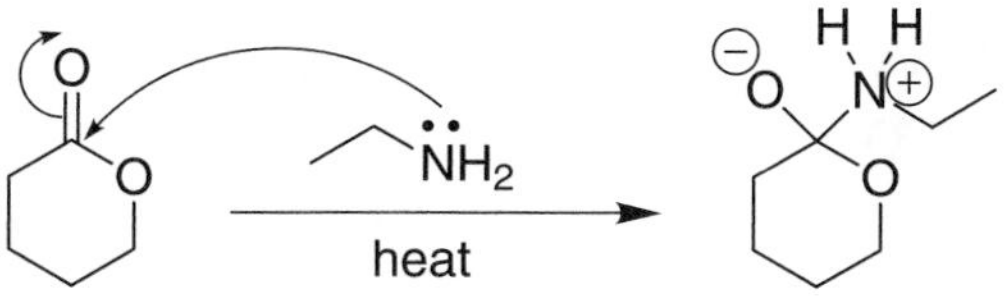

Then we re-form the carbonyl to kick off an alkoxide (RO⁻) as a leaving group:

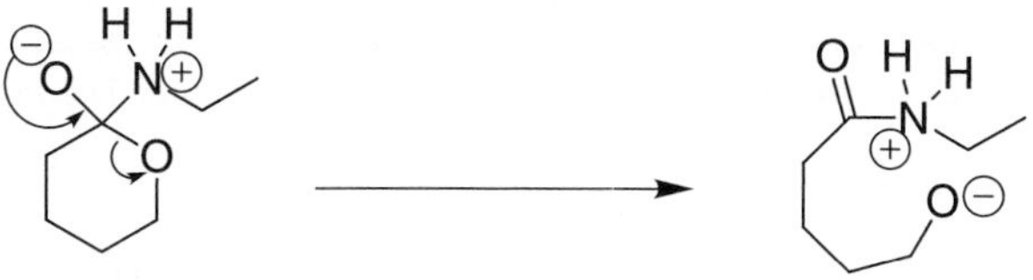

This negative charge then grabs a proton to give us our product:

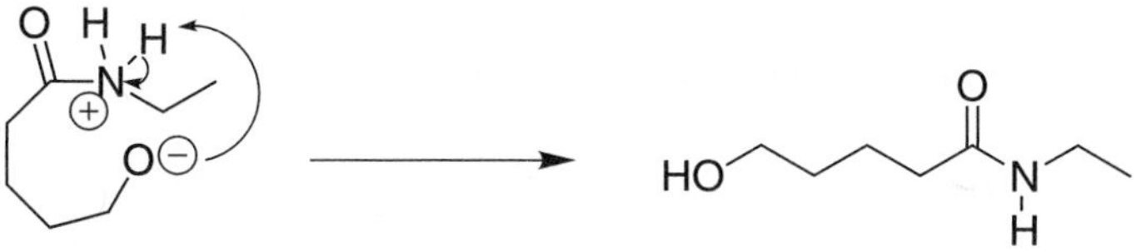

Propose a plausible mechanism for each of the following reactions. If you recognize the reaction but can't remember the mechanism, try to propose the mechanism anyway without looking back to find the answer. See if you can apply all the rules we have learned to rediscover the mechanisms:

6.50.

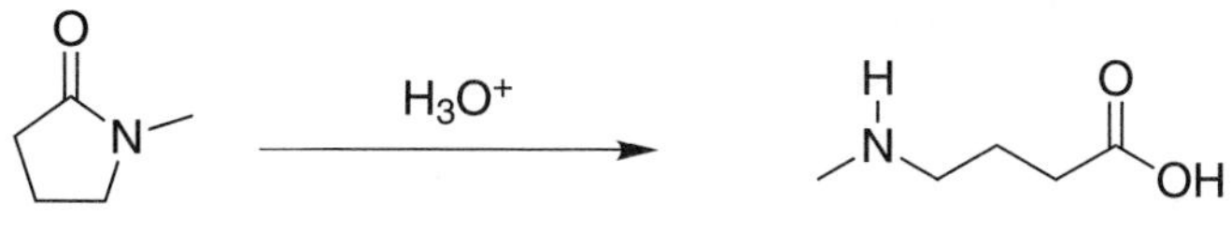

6.51.

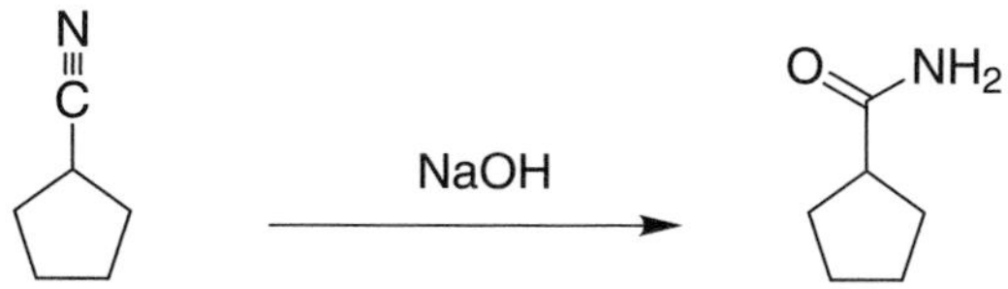

6.52.

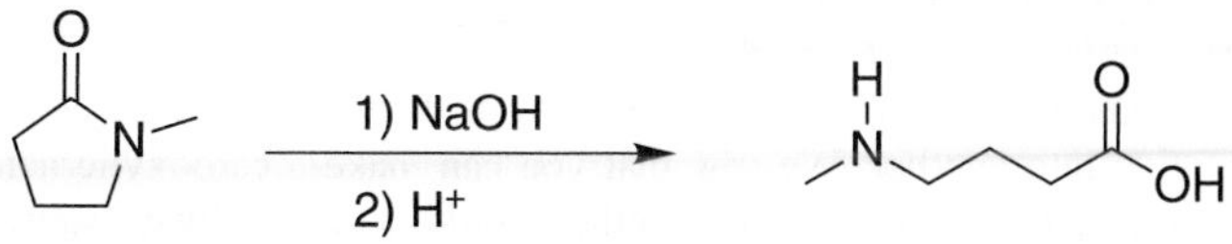

And now let's just do one more challenging mechanism. I say "challenging" not because it is difficult but because you have not seen this exact mechanism before. Rather, you should be able to work your way through the mechanism, using all of the skills we have developed in this chapter:

PROBLEM 6.53.   On a separate piece of paper, propose a mechanism for the following reaction:

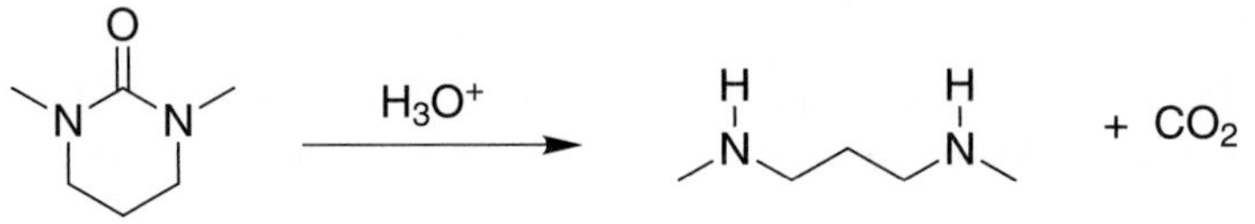

EXERCISE 6.54.   Predict the products of the following reaction:

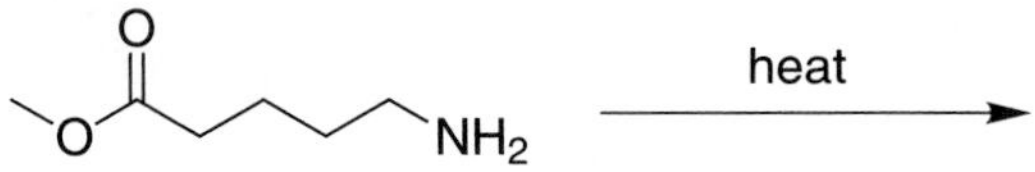

Answer: We were not given any reagents here (just conditions of heat), and so we look carefully at the starting material to see if we can have an intramolecular reaction. We notice that there are two functional groups in our starting compound. We have an ester and an amine. And we have seen that an ester can react with an amine under conditions of heating. The products should be an amide and an alcohol:

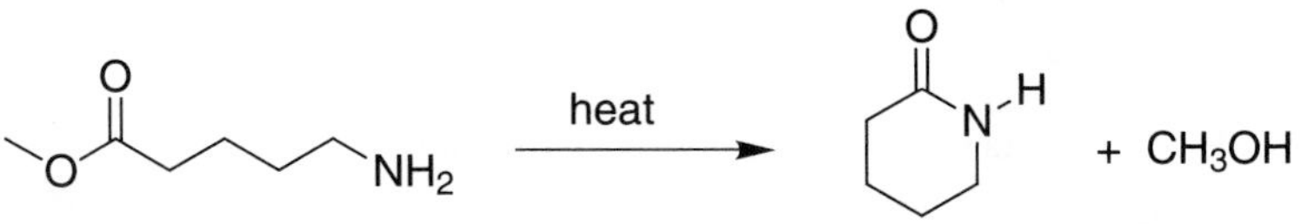

Predict the products for each of the following reactions:

6.55.

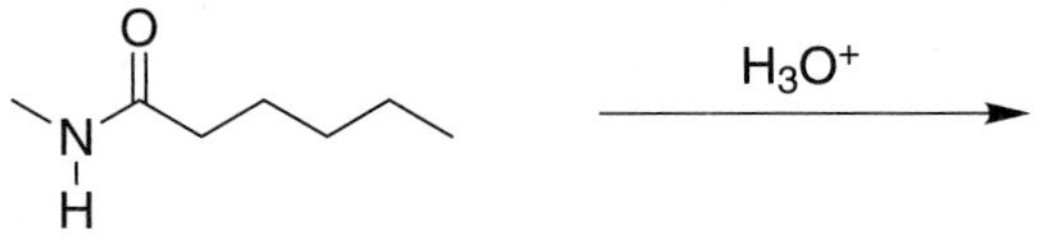

6.56.

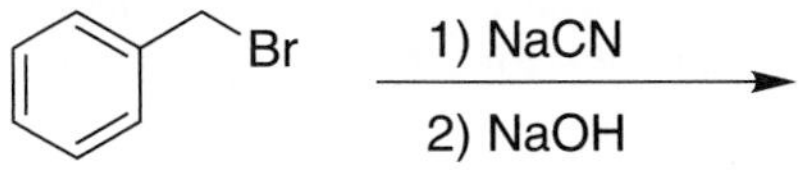

6.57.

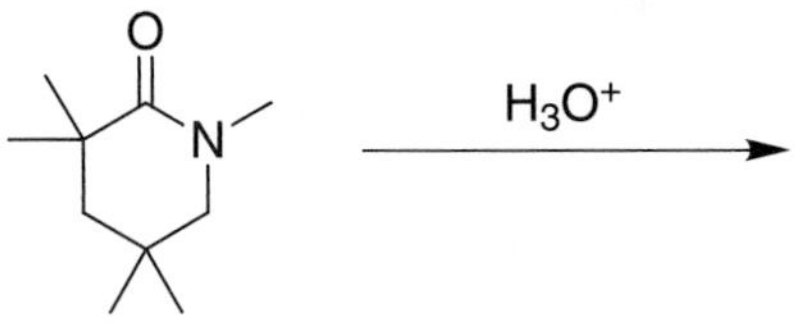

6.6 SYNTHESIS PROBLEMS

We have seen a lot of reactions in this chapter, and almost all of them involved the conversion of one carboxylic acid derivative into another. We saw that you can make a carboxylic acid derivative from any other derivative that is more reactive. In other words, you can always step your way *down* the following chart:

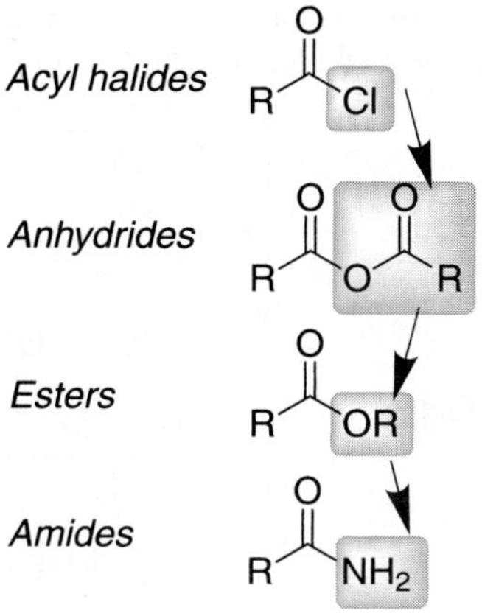

Acyl halides

Anhydrides

Esters

Amides

You can even jump down the chart if you want:

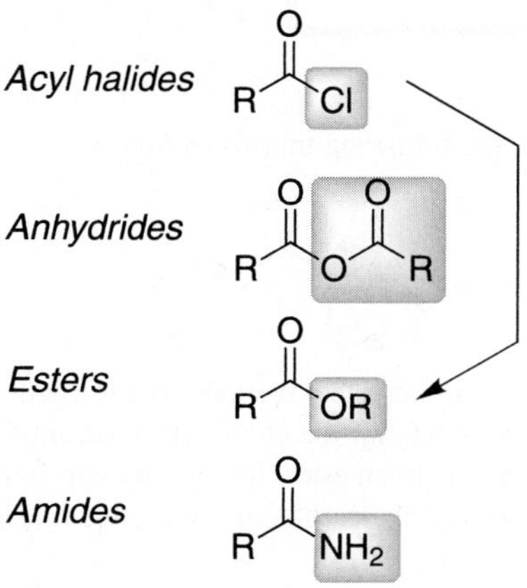

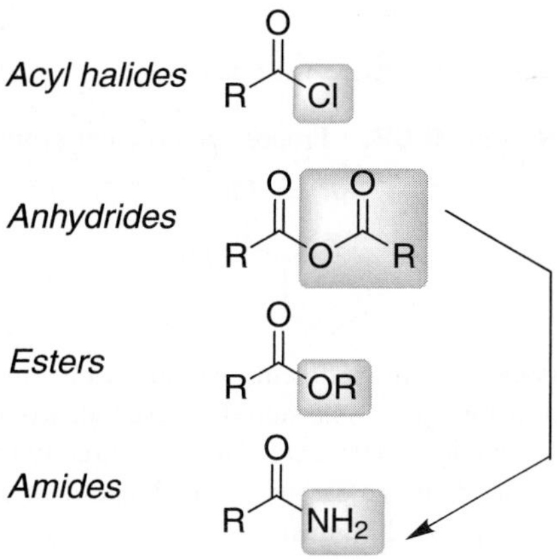

Acyl halides

Anhydrides

Esters

Amides

Acyl halides

Anhydrides

Esters

Amides

But you *cannot* travel *up* the chart in one step:

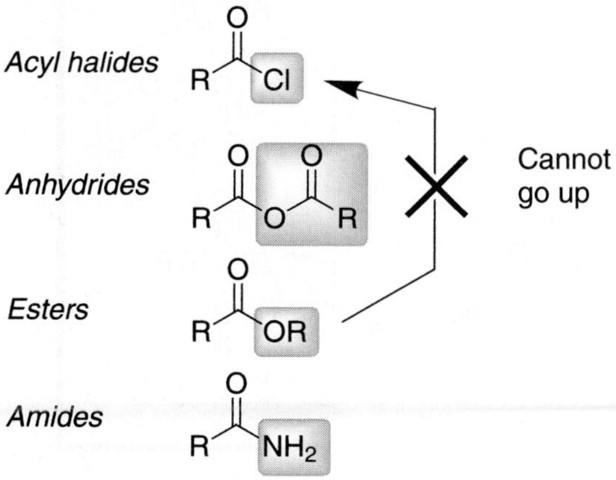

Acyl halides

Anhydrides

Esters

Amides

Cannot
go up

So, *how* do you travel *up* the chart? Here is the way to do it: you can exit this chart by converting into a carboxylic acid, and then come back into the chart, as follows:

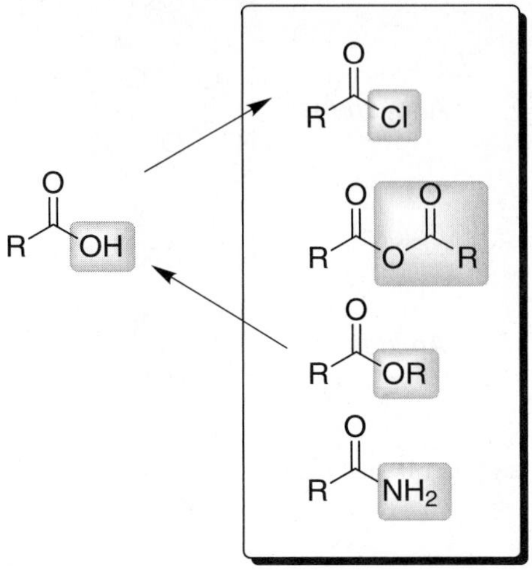

Let's get some practice with this:

EXERCISE 6.58. Propose an efficient synthesis for the following transformation:

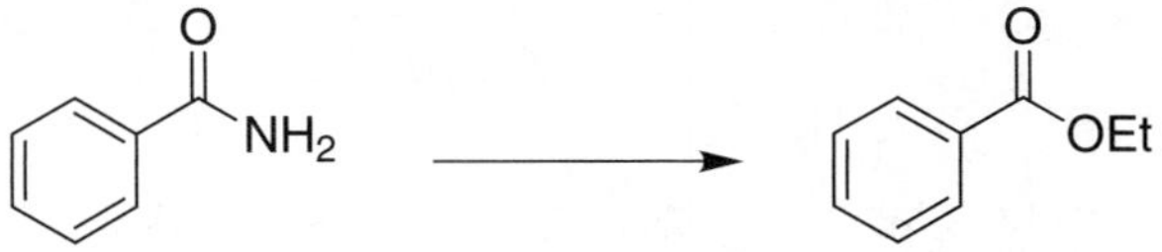

Answer: In this problem, we must convert an amide into an ester. We have not learned a way to do this directly in one step (because that would involve going *up* the chart). Because amides are less reactive than esters, we cannot go directly from an amide to an ester. Instead, we can first convert the amide into a carboxylic acid, and then we can convert the carboxylic acid into an ester. So our strategy goes like this:

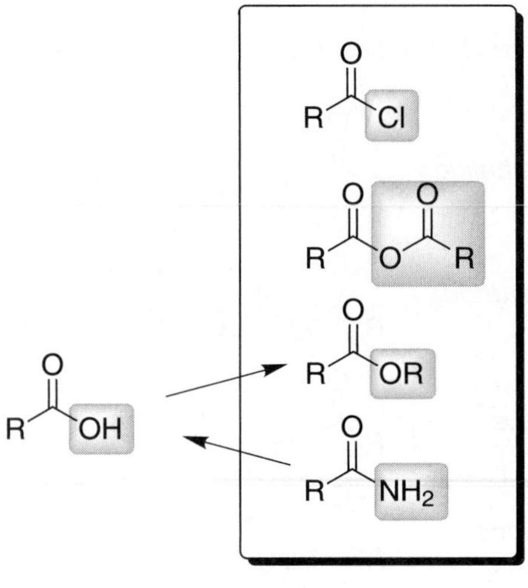

So, our answer is:

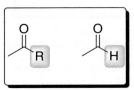

1) H₃O⁺

2) Excess EtOH, [H⁺]

Propose an efficient synthesis for each of the following problems:

6.59.

6.60.

6.61.

6.62.

6.63.

6.64.

You need to keep one other important strategy in mind when you are solving synthesis problems. In this chapter, we studied the chemistry of carboxylic acid derivatives; and in the previous chapter, we studied the chemistry of ketones/aldehydes. These two chapters form two different realms:

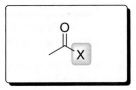

Realm of
carboxylic acid derivatives

Realm of
ketones and aldehydes

These realms are not completely isolated from one another because we have seen ways to convert from one realm into another. In this chapter, we saw how to convert an acyl halide into a ketone:

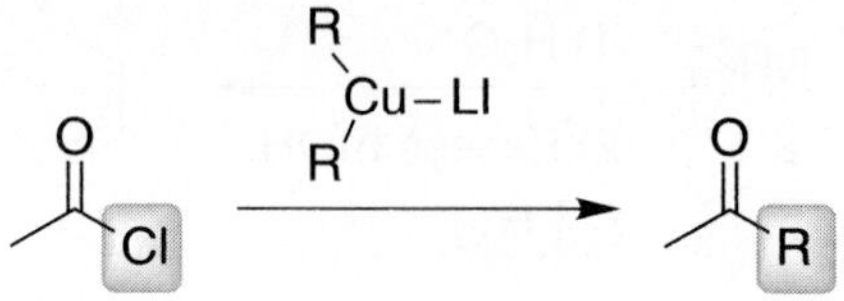

There is also another way to cross over from the realm of carboxylic acids into the realm of ketones and aldehydes. Rather than making a ketone (like above), we can make an aldehyde using the following two steps:

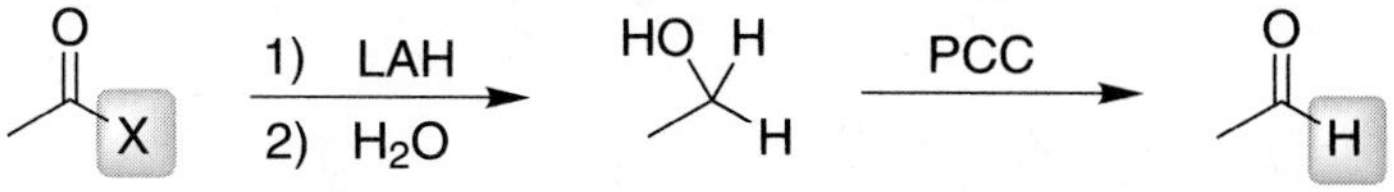

Some textbooks and instructors will teach you a way to do this in one step (converting from an acyl halide into an aldehyde). Many hydride reagents are tame enough to convert an acyl halide into an aldehyde (very much the way lithium dialkyl cuprates are tame enough to convert an acyl halide into a ketone, without attacking the carbonyl a second time). You should look through your textbook and lecture notes to see if you have learned about a tame hydride nucleophile. If you haven't, you can always use the two-step method (shown above) for converting an acyl halide into an aldehyde.

With the reactions above, we have seen how to "cross over" from the realm of carboxylic acid derivatives into the realm of ketones and aldehydes:

Realm of
carboxylic acid derivatives

Realm of
ketones and aldehydes

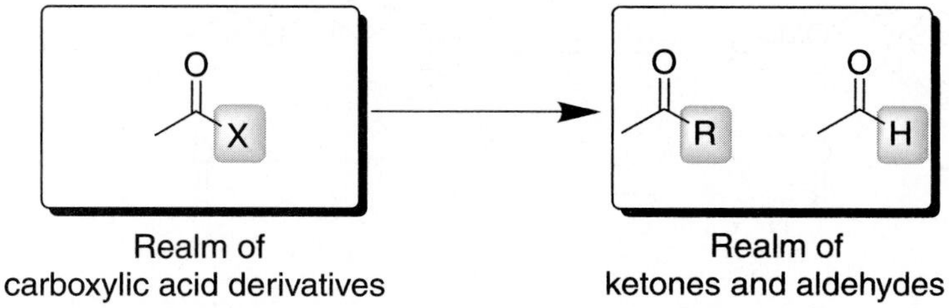

But what about the reverse direction? Do we have a way to "cross over" from the realm of ketones and aldehydes into the realm of carboxylic acid derivatives?

Realm of
carboxylic acid derivatives

Realm of
ketones and aldehydes

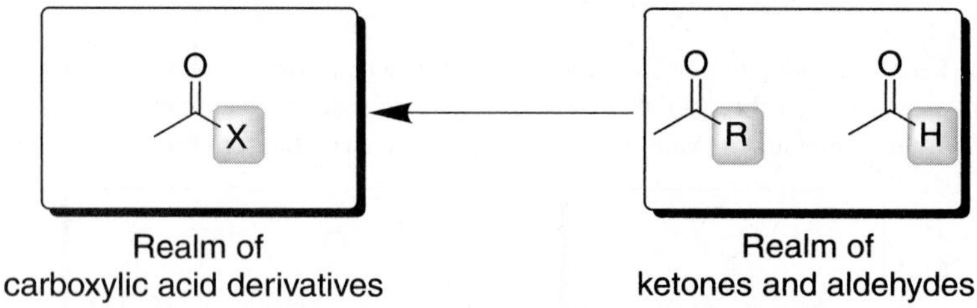

We have seen a way to do this also. Do you remember the Baeyer-Villiger oxidation from the previous chapter? This reaction can be used to convert a ketone into an ester:

We can also use a Baeyer-Villiger oxidation to convert an aldehyde into a carboxylic acid (remember migratory aptitude?):

We now have reactions that allow us to "cross over" from one realm into the other (in either direction). Let's see a concrete example of how to use this:

EXERCISE 6.65.   Propose an efficient synthesis for the following transformation:

Answer: Our final product is an alcohol. At first glance, it is difficult to see how we will do this transformation. But don't get discouraged. You are not expected to know how to solve problems like this one instantly. Synthesis problems require a lot of thought and strategizing. Remember that you always want to try to go backwards as much as possible (retrosynthetic analysis). So, let's work our way backwards.

Did we learn any simple ways to make alcohols? In the previous chapter, we learned how to make an alcohol from a ketone, using LAH:

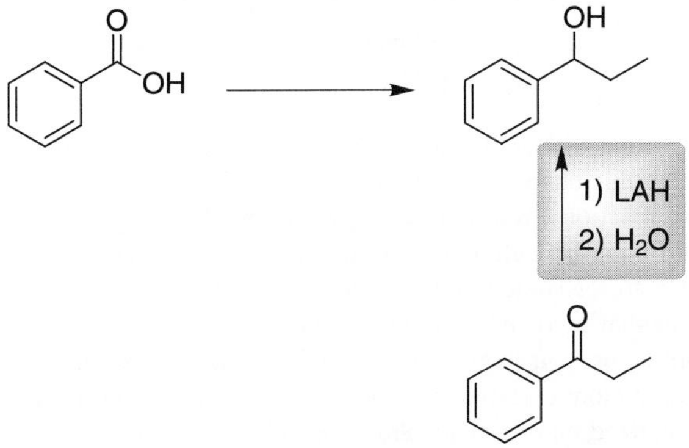

With this one important step, we are now in a position to realize that this problem can be thought of as a "crossover" problem. The starting material is a carboxylic acid, and we need to turn it into a ketone:

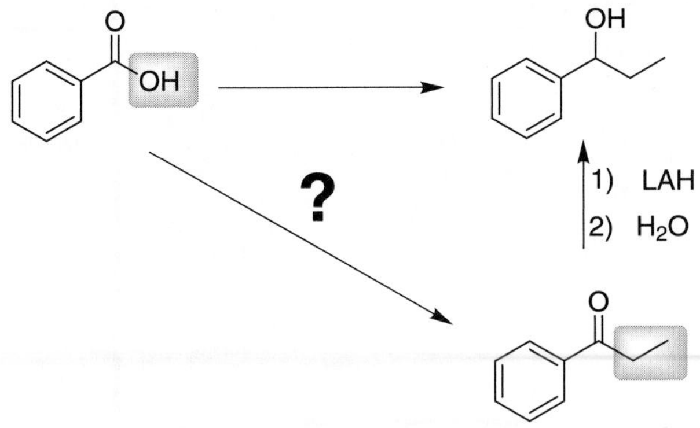

In other words, we need to cross over from the realm of carboxylic acids into the realm of ketones. And we did see one reaction that allows us to do that. We can make a ketone from an acyl halide, using a lithium dialkyl cuprate. So, now we have worked backwards again:

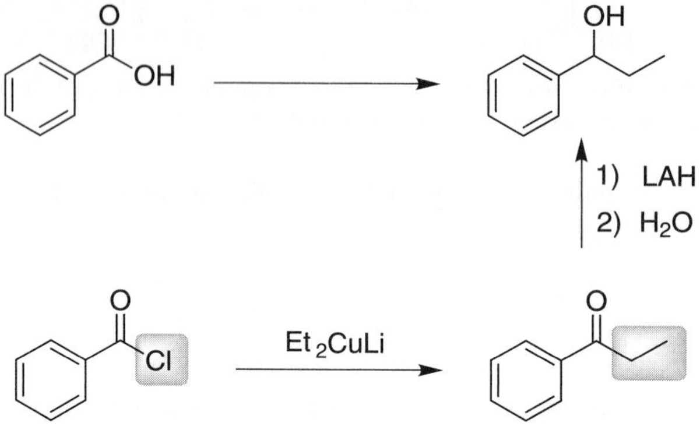

Finally, we need to convert the carboxylic acid into an acyl halide, and we can do that in one step with either thionyl chloride or oxalyl chloride.

So, our answer is:

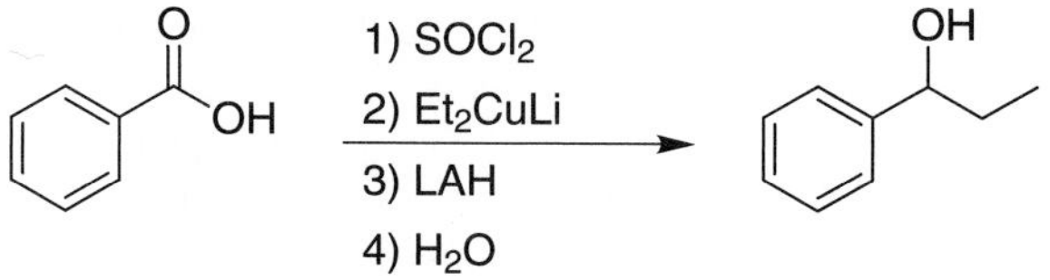

Now let's get some practice with some more "crossover" problems. In order to do these problems, you will need to review this chapter *and* the previous chapter (ketones and aldehydes). You will need to have all of the reactions from both chapters at your fingertips.

At first, you might find it difficult to identify the following problems as crossover problems, but hopefully, you will start to see some trends as you solve them. The hope is that you will train your eyes to locate problems that involve "cross over" reactions.

In solving these problems, you should be familiar with the ways we have seen for crossing over. Although there are many more crossover reactions, we have seen four of these reactions. All four reactions can be summarized on one chart. Study this chart carefully:

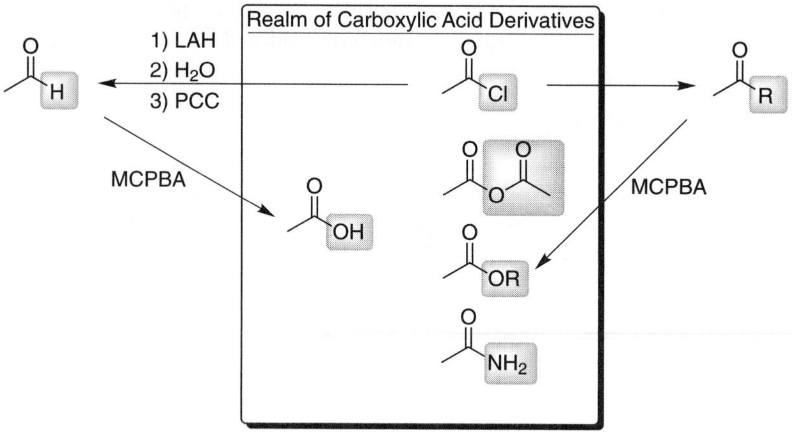

You will never have to cross over twice; that would be silly. If you start with a carboxylic acid de-rivative and you need to convert it into another carboxylic acid derivative, it would make no sense to do two crossover reactions. You would just want to stay in the realm of the carboxylic acid de-rivatives, using the methodology we saw in the beginning of this section. Don't be confused by the fact that all four reactions above are shown on one chart. You will rarely use more than one of these four reactions in a single problem.

Unless you know these four reactions cold, you will be unable to do "crossover" problems. To help you remember them, you should notice one thing that all four reactions have in common: they are all reduction-oxidation reactions. This should make sense because carboxylic acid derivatives are at an oxidation state that is different from ketones and aldehydes.

Each of the following problems will take you a while, so don't sit down to do them when you only have five minutes to study. That would just frustrate you. These problems are tough, and it will take a long time to work through all of them.

Propose an efficient synthesis for each of the following transformations. In each case, remem-ber to work backwards (retrosynthetic analysis) and try to determine which crossover reaction to use. When you compare your answers to the answers in the back of the book, keep in mind that a synthesis problem can often be solved in more than one way. Note, too, that if your answer is dif-ferent from the answer in the back of the book, you should not necessarily conclude that your an-swer is wrong.

6.66.

6.67.

6.68.

6.69.

6.70.

6.71.

6.72. ⟶

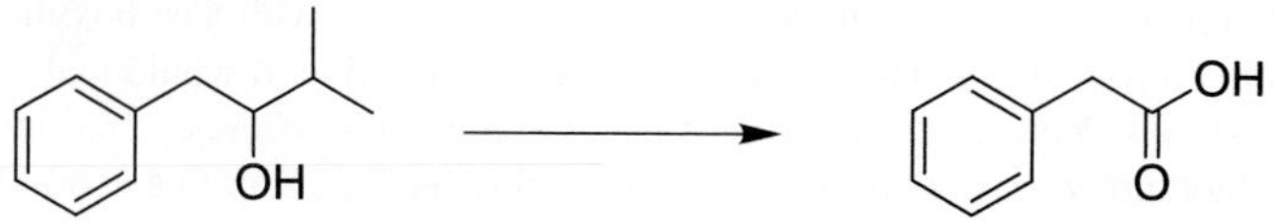

6.73. ⟶

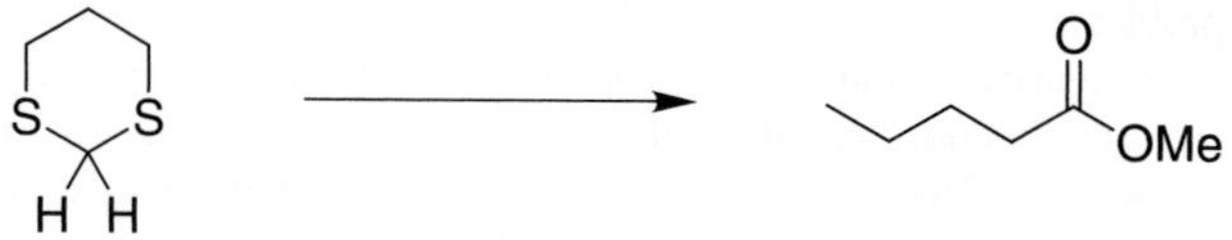

6.74. ⟶

6.75. ⟶

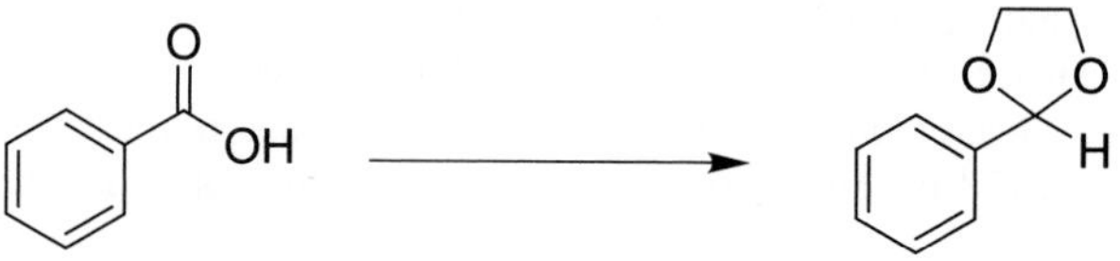

6.76. ⟶

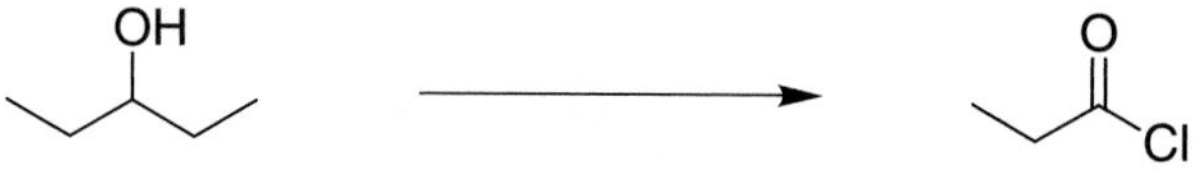

6.77. ⟶

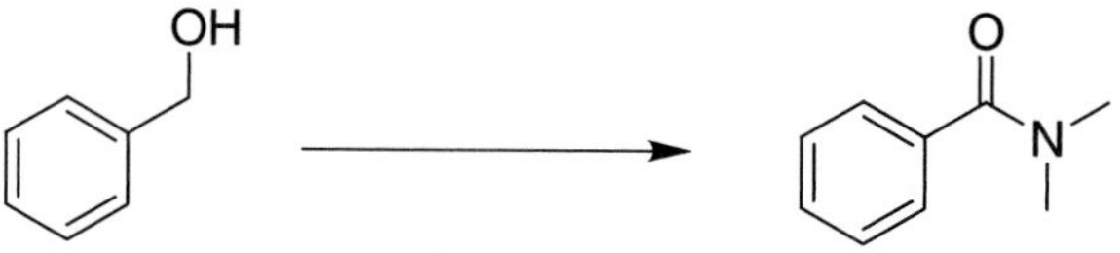

The goal of this chapter was to lay a foundation for enabling you to study your textbook and lecture notes more efficiently. We saw the simple rules behind all of the mechanisms, and we learned a bunch of synthesis strategies.

Now you can go back through your textbook and lecture notes, and look for those reactions that we did not cover here in this chapter. With the foundation we have built in this chapter, you should be in good shape to fill in the gaps and study more efficiently.

And make sure to do ALL of the problems in your textbook. You will find more synthesis problems there. The more you practice, the better you will get it. Good luck . . .

CHAPTER 7

ENOLS AND ENOLATES

In the previous two chapters, we focused on the reactions that can take place when a nucleophile attacks a carbonyl group:

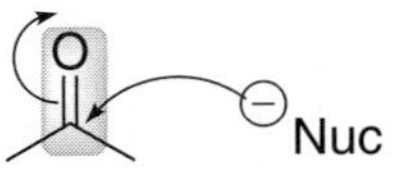

We first learned about nucleophilic attacks on ketones and aldehydes in Chapter 5. Then, in Chapter 6, we looked at carboxylic acid derivatives. Now, we are ready to move away from the carbonyl group, and we will look at the chemistry that can take place at the alpha (α) carbon:

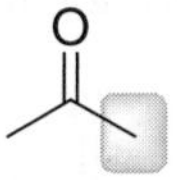

We call this the alpha carbon because it is the carbon atom directly connected to the carbonyl group. We use the Greek alphabet to label carbon atoms, moving away from the carbonyl group, in either direction:

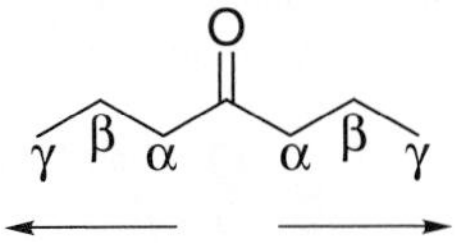

Notice that in this compound, there are two alpha positions. In this chapter, we will focus on the chemistry that can take place at the alpha carbon.

Before we get started, we should discuss one more piece of terminology. When an alpha carbon has protons connected to it, we call them alpha protons:

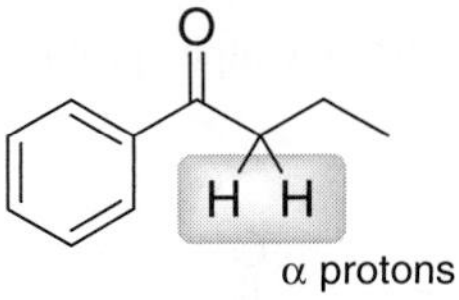

α protons

Not all alpha carbon atoms will have alpha protons. For example, consider the following compound:

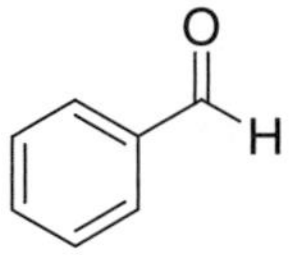

This compound has no alpha protons. If you look just to the right of the carbonyl, you will see that there is no alpha carbon (it is just an aldehyde). That aldehydic H is NOT an alpha proton because

203

it is not connected to an alpha carbon. And if you look just to the left of the carbonyl, you will see that there IS an alpha carbon, but this carbon has no protons.

It is important to recognize the presence or absence of alpha protons. We will see a lot of reactions in this chapter, most of which will be based on the presence of alpha protons. It turns out that alpha protons are somewhat acidic, and when you pull off one of these protons, you get an anion that is very reactive. We will see this in greater detail very soon. For now, let's just make sure that we can identify alpha protons when we see them.

EXERCISE 7.1. Consider the following compound:

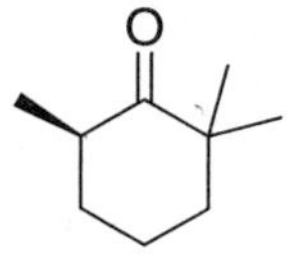

Identify any alpha protons that you see.

Answer: To see if there are any α protons, we must look at the alpha carbon atoms:

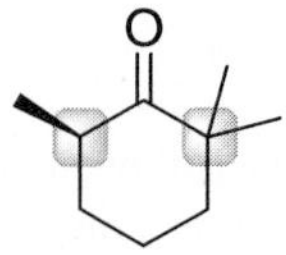

The alpha carbon on the right does *not* have any protons on it. The alpha carbon on the left *does* have a proton that was not shown in the drawing above:

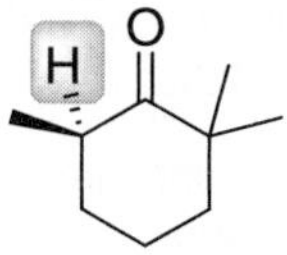

So, there is just one alpha proton in this compound.

For each of the compounds below, identify the alpha protons (some compounds may not have any alpha protons).

7.2.

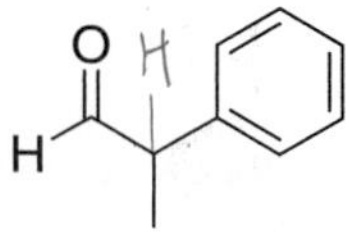

7.3.

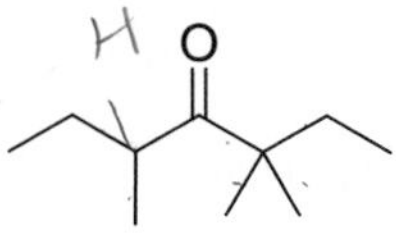

7.4. nune

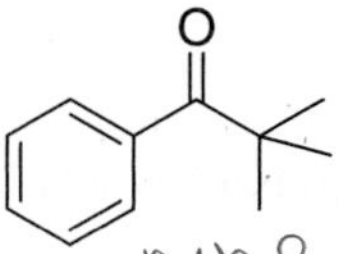

7.5.

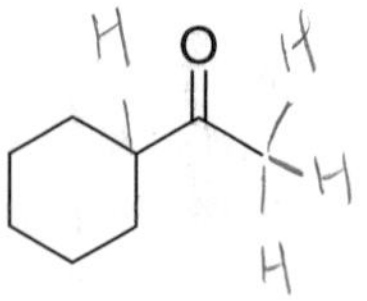

7.6.

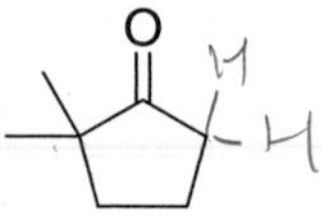

7.7. nune

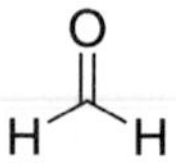

7.1 KETO-ENOL TAUTOMERISM

When a ketone has an alpha proton, an interesting thing can happen. In either acidic *or* basic conditions, the ketone exists in equilibrium with another compound:

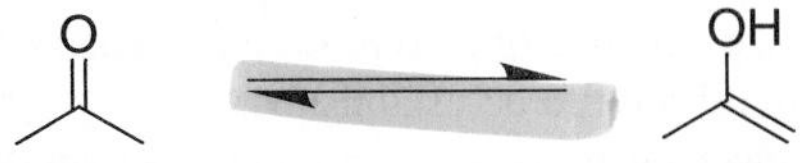

Ketone   **Enol**

This other compound is called an ***enol*** because it has a $C=C$ double bond ("ene") and an OH group ("ol"). The equilibrium shown above is actually very important because you will see it in many mechanisms. So let's take a closer look.

If we focus on the connections of atoms, we will find that the two compounds differ from each other in the placement of one proton. The ketone has the proton attached to an alpha carbon, and the enol has the proton connected to the oxygen:

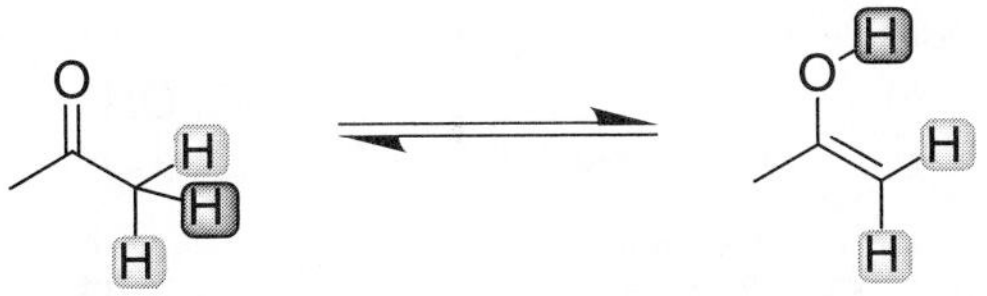

The π bond is also in a different location. But when we just focus on the atoms (which atoms are connected to which other atoms), we find that the difference is in the placement of just one proton. We have a special name to describe the relationship between compounds that differ from each other in the placement of just one proton. We call them *tautomers*. So, the enol above is said to be the *tautomer* of the ketone, and similarly, the ketone is called the *tautomer* of the enol. The equilibrium shown above is called *keto-enol tautomerism*.

Keto-enol tautomerism is ***not*** resonance. The two compounds shown above are *not* two representations of the same compound; they are, in fact, different compounds. These two compounds are in equilibrium with each other.

In most cases, the equilibrium favors the ketone greatly:

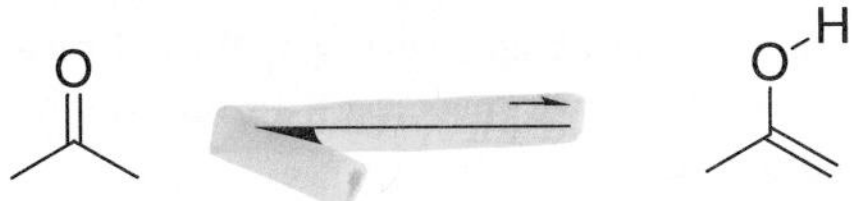

This should make sense to us because the last two chapters focused on the formation of $C=O$ double bonds as a driving force for reactions. A ketone has a $C=O$ double bond, but an enol does not. So we should not be surprised that the equilibrium favors the ketone.

There are some situations where the equilibrium can favor the enol. For example:

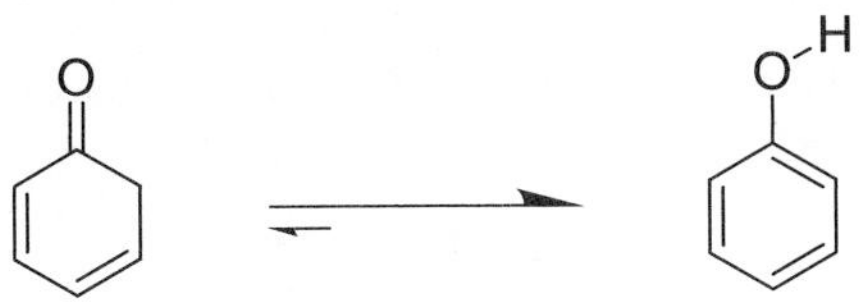

In this case, the enol is an aromatic compound, and it is much more stable than the ketone (which is not aromatic). In many other situations an enol can be more stable than its tautomer. You will probably find some of these examples in your textbook (such as 1,3-diketones). But in most cases (other than these few exceptional cases), the equilibrium will favor a ketone over an enol.

It is nearly impossible to avoid the equilibrium. Imagine that you have a pure ketone and that you make great efforts to remove all traces of acid or base. Your hope is that you can avoid the equilibrium, so as to avoid the production of small amounts of the enol. But you will find that you cannot do this easily. Even trace amounts of acid or base adsorbed on the glassware (that you cannot remove) will allow the equilibrium to be established.

We will now explore the mechanism for keto-enol tautomerism. We have said that compounds will tautomerize in the presence of either acid or base, so we will need to see two mechanisms: one under *acidic* conditions and one under *basic* conditions. The truth is that these two mechanisms are VERY similar to each other. The core steps are identical, and the only difference is the order in which we do the core steps.

We saw that, by definition, tautomers will differ in the position of one proton. In order to convert a ketone into an enol, we will therefore need to do two things: (1) give a proton, and (2) remove a proton:

Similarly, to convert the enol back into a ketone, we will again need to give a proton and remove another proton:

You might wonder why we are talking about *two* steps. Why do we need to give a proton *and* remove a proton? Why can't we just move the proton in one step (in an intramolecular proton-transfer reaction), as in the following:

This doesn't work because the oxygen is just too far away (in space) from the proton it wants to grab:

So, we will have to do this in two steps. One proton comes *off*, and then another proton comes back *on*. But these two steps can be in either order: it can be *on* and *then off*, or it can be *off* and *then on*. If we first pull a proton off and then put one back on, our mechanism will look like this:

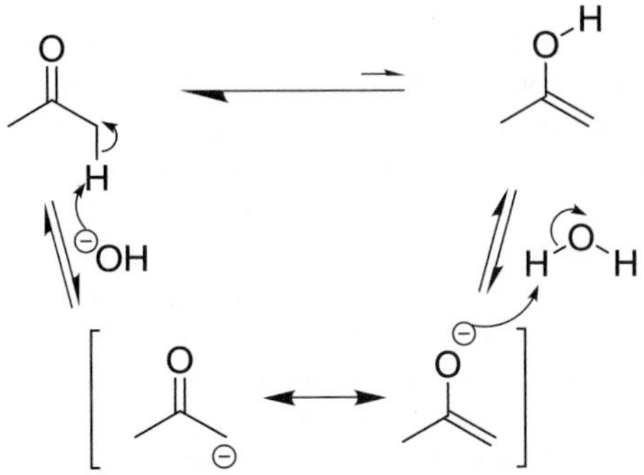

Notice that there is one intermediate (for which we must draw resonance structures), and this intermediate is negatively charged. If you look at the second resonance structure, it looks like an enol that is missing one proton. So we call this intermediate an ***enolate***. Don't be fooled into thinking there are more than two steps here. Resonance (of the intermediate) is NOT a step. Our mechanism only has two steps, like this:

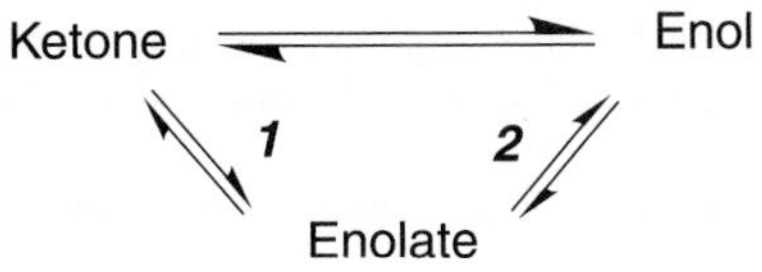

The enolate is very important for the rest of this chapter because it can function as a nucleophile. We will see many examples in the coming sections. For now, let's finish our discussion of keto-enol tautomerism.

In the mechanism above, we first pulled a proton *off*, and *then* we put a proton back *on*. But we could have done these two steps in a different order. What if we *first* put a proton *on*, and *then* pull a proton back *off*. In that case, our mechanism would look as follows:

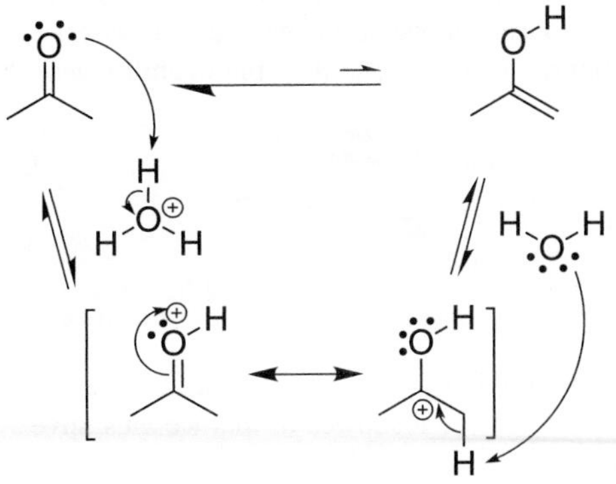

Once again, there are just two steps here. Don't be fooled by the resonance of the intermediate. Resonance is not a step; rather, resonance is just our way of dealing with the fact that we cannot draw the intermediate with only one drawing. We need two drawings to capture its essence. And if you look at these resonance structures, you will see that this intermediate is positively charged.

Notice the difference between these two mechanisms. The first mechanism has a negatively charged intermediate, and the second mechanism has a positively charged intermediate. Other than that, the difference between these two mechanisms is pretty small. Each mechanism has only two steps, and both steps are just proton transfers. The only question is the sequence of events. Is it: deprotonate, then protonate? Or is it: protonate, then deprotonate?

In order to determine which mechanism to use, we must look carefully at the conditions. In acidic conditions, we must protonate first, and then we deprotonate. This gives us a positively charged intermediate, which is consistent with the conditions (do not form a negatively charged intermediate in acidic conditions). But in basic conditions, we must deprotonate first, and then we protonate. This gives us a negatively charged intermediate, which is consistent with the conditions (do not form a positively charged intermediate in basic conditions).

EXERCISE 7.8. Consider the following compound:

It is not possible to isolate and purify this compound because it will rapidly tautomerize to form a ketone. Show the mechanism for formation of the ketone under acidic conditions.

Answer: This compound is an enol, and its tautomer will be the following ketone:

To convert the enol into a ketone, our mechanism will have two steps: *give* a proton, and *remove* a proton. But we must decide what order to use. Do we first *remove* a proton and *then give* a proton? Or do we first *give* a proton and *then remove* a proton? To answer this question, we look at the conditions. Since we are in acidic conditions, we should first *give* a proton (forming a positively charged intermediate), and only then do we remove the other proton.

Now we know what order to use, but we also have to decide **where** to give the proton, and **where** to remove the proton. To figure this out, we look at the overall reaction:

When we analyze the reaction in this way, it is easy to see **where** to give and **where** to remove a proton. This step was important because it showed us that we must give a proton to the double bond (not to the oxygen), as follows:

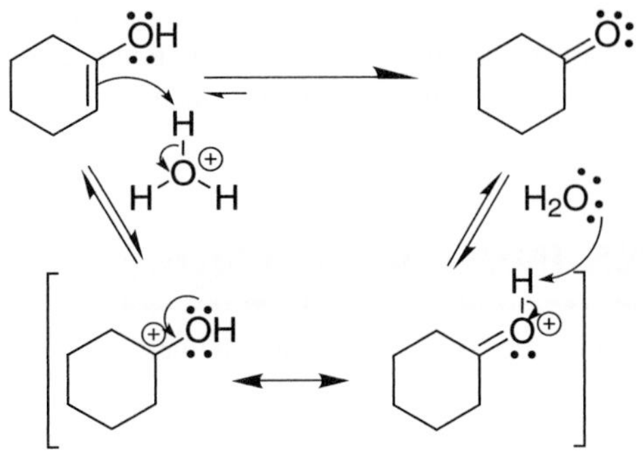

So often, students will start this problem by protonating the OH group. Although that might make sense at first, you will find that this step will NOT lead to formation of the ketone. The first step is to protonate the double bond, *not* the OH group.

Here is one last piece of advice before you try to propose a mechanism yourself. Never use OH⁻ and H_3O^+ in the same mechanism. When you are in acidic conditions, use H_3O^+ to give the proton and use H_2O to remove the proton. We can't use hydroxide to remove a proton because we are in acidic conditions.

Similarly, whenever you are in basic conditions, use OH⁻ to remove the proton and use H_2O to give back a proton. We can't use H_3O^+ to give back a proton, because we are in basic conditions. Here is the take-home message: always stay consistent with your conditions.

In summary, we need to think about three things in order to correctly draw the mechanism of a keto-enol tautomerization: (1) what *order* to use (first give and then remove, or first remove and then give), (2) *where* to protonate and where to deprotonate, and (3) *what reagents* to use for the proton transfers (stay consistent with the conditions).

For each of the following reactions, propose a mechanism that is consistent with the conditions indicated (you will need a separate piece of paper to record your answers):

7.9.

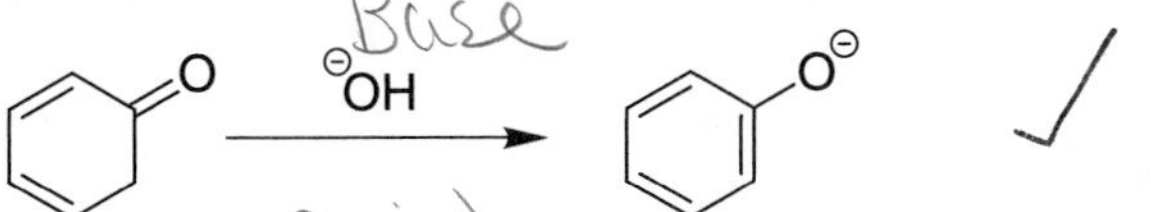

7.10.

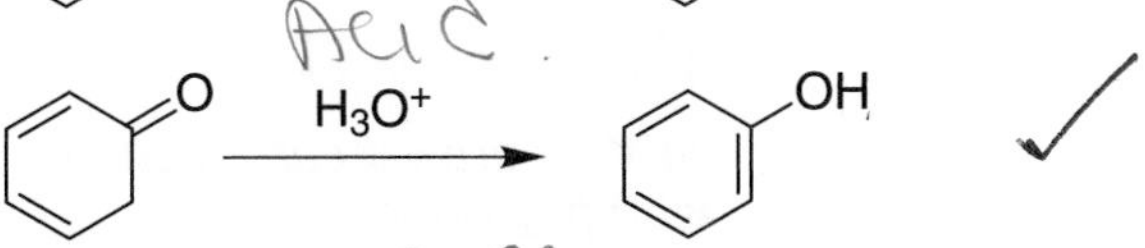

7.11.

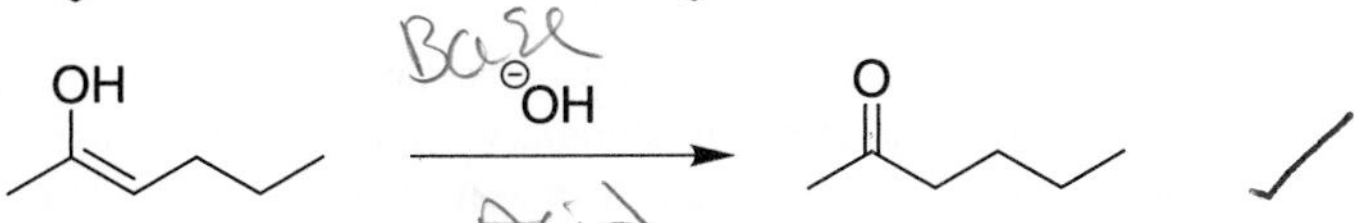

7.12.

7.13. The following reaction is also a keto-enol tautomerization:

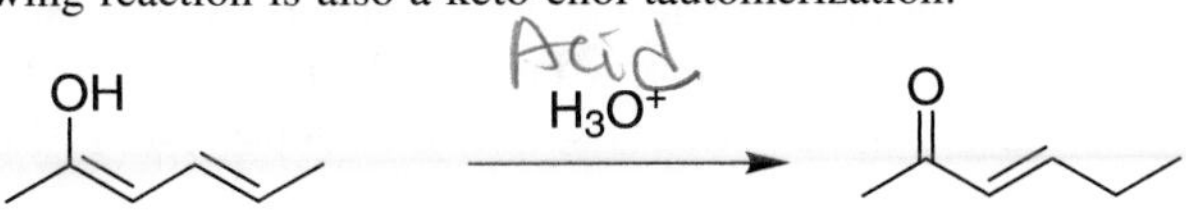

Try to propose a mechanism for this reaction, using the strategy that we developed. Remember to ask three important questions: (1) what *order* to use (first protonate or first deprotonate), (2) *where* to protonate and deprotonate, and (3) *what reagents* to use. You will need a separate piece of paper to record your answer.

7.2 REACTIONS INVOLVING ENOLS

It is hard to see how the alpha carbon of a ketone can be nucleophilic:

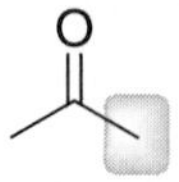

The alpha carbon does *not* have a lone pair or a π bond that can attack something. However, when we examine the structure of the enol (that is in equilibrium with this ketone), we get a different picture:

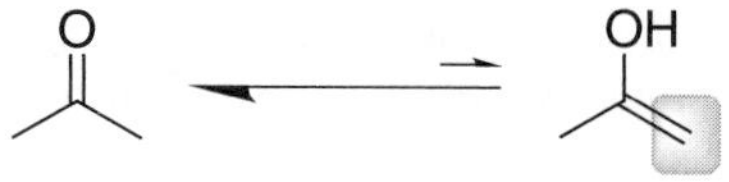

The enol has a π bond on the alpha carbon, and therefore, the alpha carbon can function as a nucleophile to attack some electrophile:

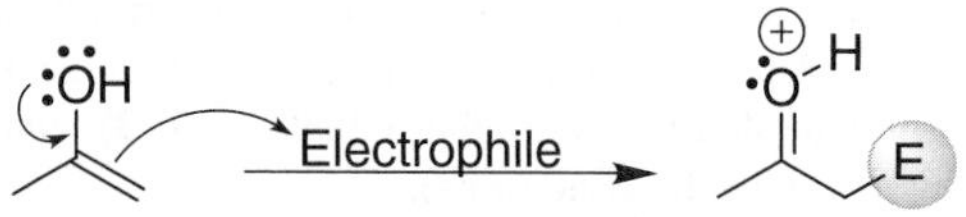

In order for the attack to take place, we are relying on the ability of the ketone to tautomerize. But not every ketone will exist in equilibrium with an enol. A ketone that lacks alpha protons will *not* tautomerize to form an enol:

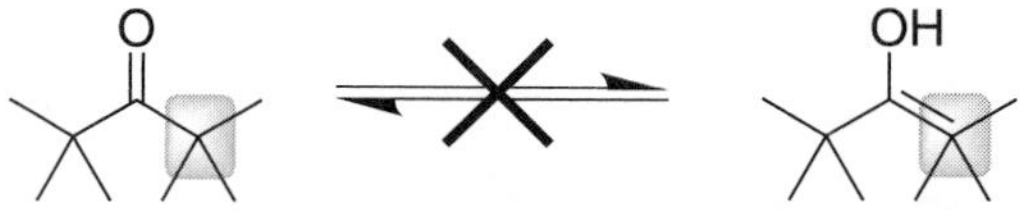

**No proton
to pull off here**

**NEVER draw a carbon atom
with _five_ bonds**

But most ketones that you will see do in fact have alpha protons, and therefore, a typical ketone will exist in equilibrium with an enol. In the previous section, we saw some rare cases where the equilibrium can actually favor the enol, but in general, the equilibrium favors the ketone. Therefore, you will generally only have trace amounts of the enol present in equilibrium with the ketone.

This small amount of enol is able to react as a nucleophile and attack some electrophile. After the enol attacks the electrophile, the keto-enol equilibrium shifts to produce some more enol (to account for the enol that "disappeared" as a result of the reaction). Slowly but surely, most of the ketone molecules end up converting into enols and reacting with the electrophile. The most common example is alpha-halogenation, where the electrophile is a halogen (giving us an alpha-halo ketone):

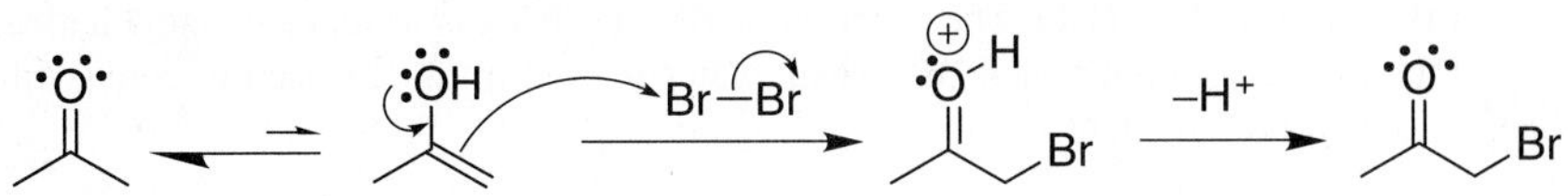

In the first step of this mechanism, the ketone tautomerizes from a small amount of enol. Then comes the critical step: the enol functions as a nucleophile to attack Br_2 (the electrophile). In the last step, we just deprotonate to get our product. Notice that most of the steps in this mechanism are just proton transfers. Our mechanism follows the following pattern: tautomerize, attack, deprotonate. But "tautomerize" is just a new name for a special kind of proton transfer. There is really only one step where an attack takes place (when the enol attacks the electrophile).

In the end, this provides us with a way to place a halogen at the alpha position of a ketone:

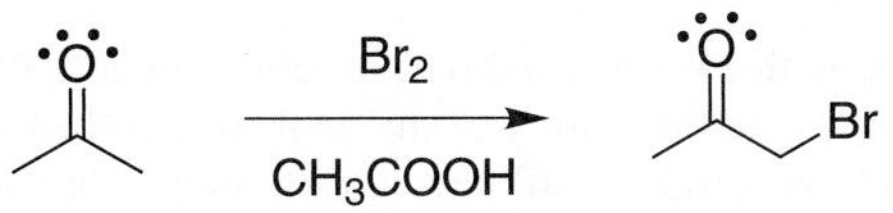

We use a mild acid (such as acetic acid, CH_3COOH) to facilitate the tautomerization. We don't have to worry about this acid getting halogenated (at its alpha position), like this:

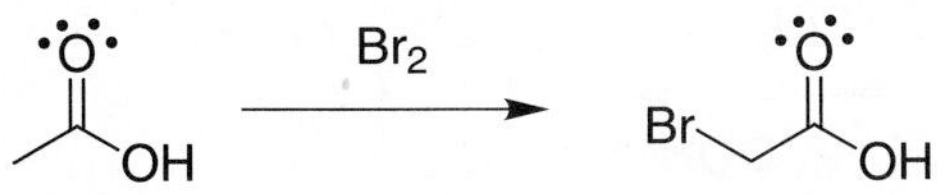

CH₃COOH   **THIS REACTION IS SLOW**

We don't have to worry about this because carboxylic acids are much slower to react in this kind of reaction.

If we **want** to halogenate the alpha position *of a carboxylic acid*, it is possible, but it will require some extra steps. First, we must convert the carboxylic acid into an acyl halide. We do this because the enol of an acyl halide is very fast at attacking a halogen. Then, in the end, we just need to convert the acyl halide back into a carboxylic acid:

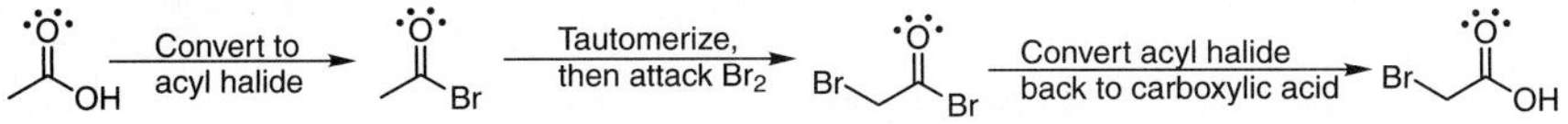

This strategy (for halogenating carboxylic acids) is called the Hell–Volhard–Zelinsky reaction.

Here is the bottom line: in the section, we have seen two reactions that exploit the nucleophilic nature of enols. These reactions can be used to place a halogen at the alpha position of a ketone or at the alpha position of a carboxylic acid:

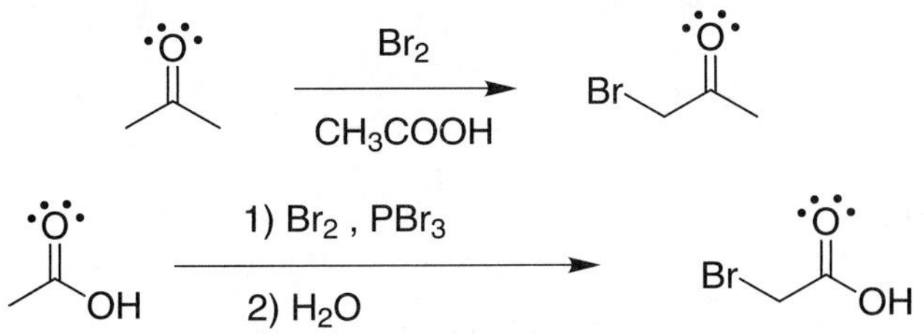

Notice the reagents that we used. We saw the reagent in the first reaction (Br_2 and some mild acid—to halogenate a ketone). But to halogenate a carboxylic acid, we use a different set of reagents. We use Br_2 and PBr_3, followed by H_2O. The function of Br_2 and PBr_3 is to make the acyl halide, form the enol, and then have the enol attack Br_2. Then, water is used in the last step to convert the acyl halide back into a carboxylic acid.

EXERCISE 7.14. Predict the product of this reaction:

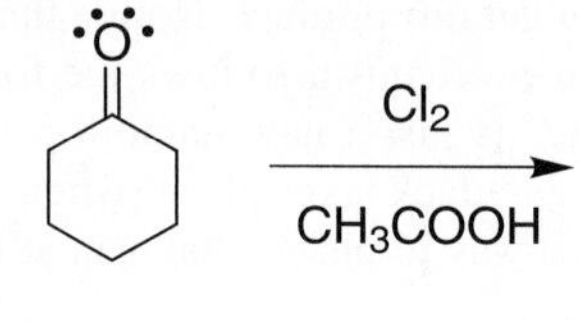

Answer: We are starting with a ketone, and we are subjecting it to Cl_2 in the presence of a mild acid. The mild acid promotes tautomerization to the enol, which then attacks the Cl_2 in an alpha halogenation. So, in the end, our product will have a Cl at one of the alpha positions. Because either side is the same, we can just pick a side:

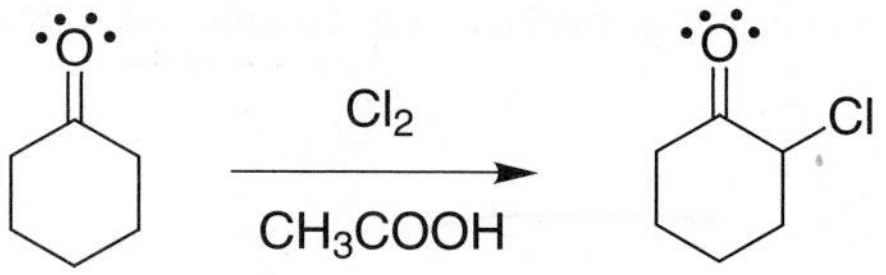

Predict the products of each of the following reactions. Remember that you can only halogenate an alpha position that has protons.

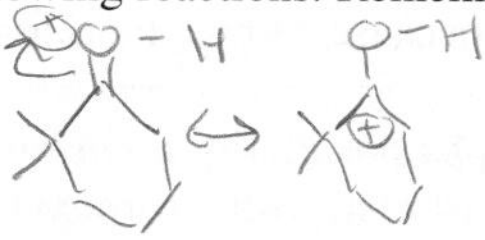

7.15.

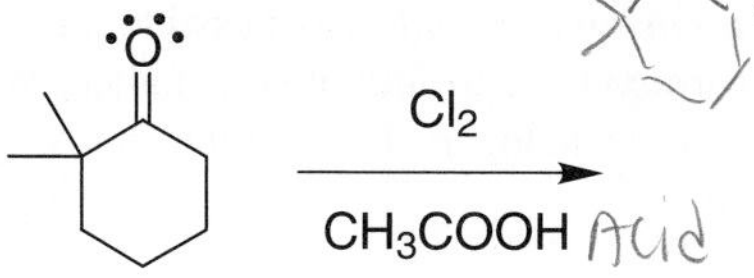

7.16.

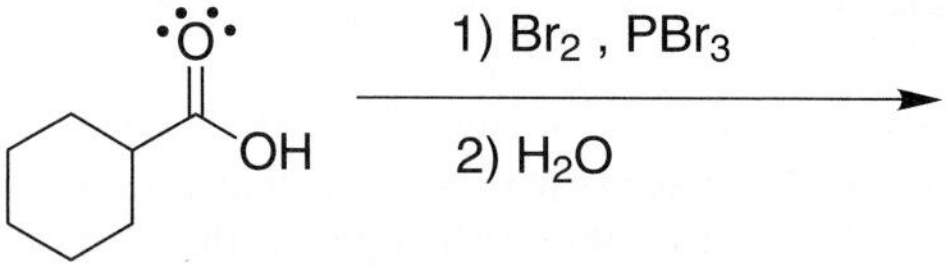

7.17.

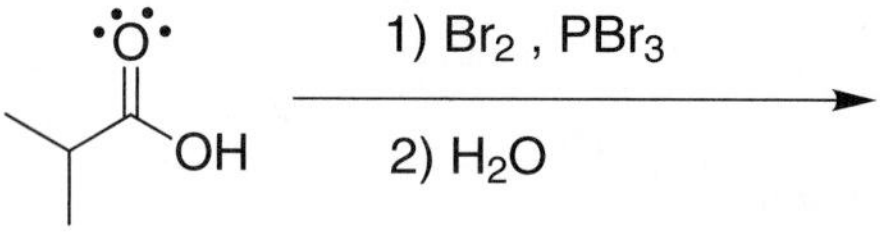

7.18.

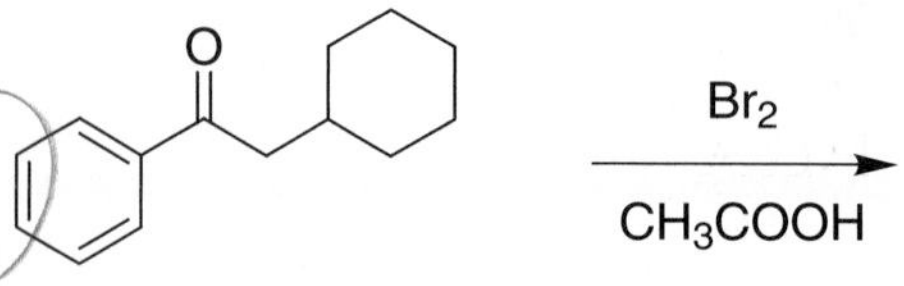

7.3   MAKING ENOLATES

In the previous section, we saw that enols can be nucleophilic. But enols are only mild nucleophiles. So, the question is: how can we make the alpha position even more nucleophilic (so that we can get a broader range of possible reactions)? There is a way to do this. We just need to give the alpha position a negative charge. To see how we do this, let's quickly review the mechanism we saw for tautomerization under basic conditions, and let's focus on the intermediate (highlighted below):

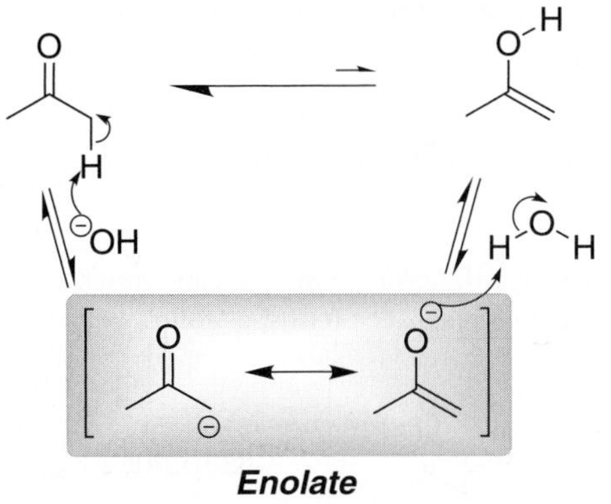

Enolate

The intermediate is negatively charged, and we mentioned before that we call it an enolate. In order to capture the essence of the enolate, we have to draw resonance structures for it. Remember what resonance structures are for: we cannot draw this *one* intermediate with any single drawing. So, we draw two drawings, and we must meld these two images together in our minds in order to get a true picture of this intermediate. And that picture shows the enolate as being electron rich in two locations: the alpha carbon *and* the oxygen:

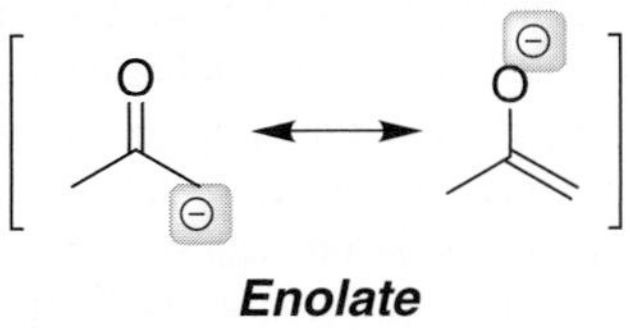

Enolate

So, we expect *both* of these locations to be very nucleophilic. However, we won't see any reactions that have the oxygen functioning as a nucleophile. In some conditions we can get O-attack, rather than C-attack, but you probably won't learn about those reactions in this course. Most textbooks and instructors do not teach the conditions for O-attack because it is considered to be a more advanced topic. So, from now on, everything we see will be examples of a C-attack (where the alpha carbon acts as the nucleophile, attacking some electrophile):

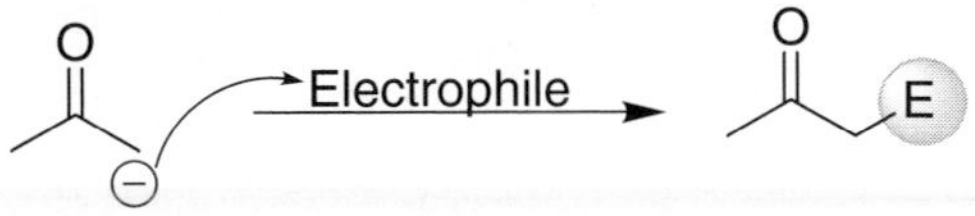

Notice that we have used only one resonance form of the enolate. If we had used the other resonance form, it would have looked like this:

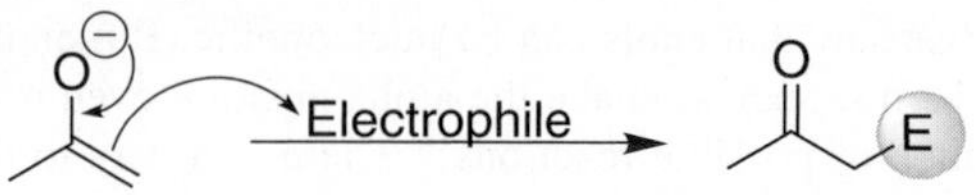

This is just another way of showing the same reaction. Many textbooks will show it the second way (starting with the resonance form that has the negative charge on the oxygen). Perhaps this is more appropriate because this resonance form contributes more to the overall character of the enolate. However, in this book, we will use the resonance structure where the negative charge is on the carbon:

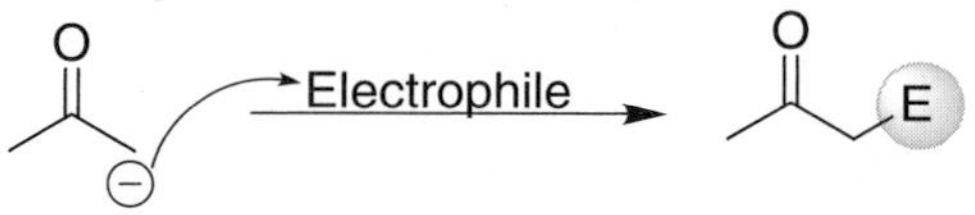

We will do it this way because it will make the mechanisms easier to follow. To be absolutely correct, we should actually draw *both* resonance forms, as follows:

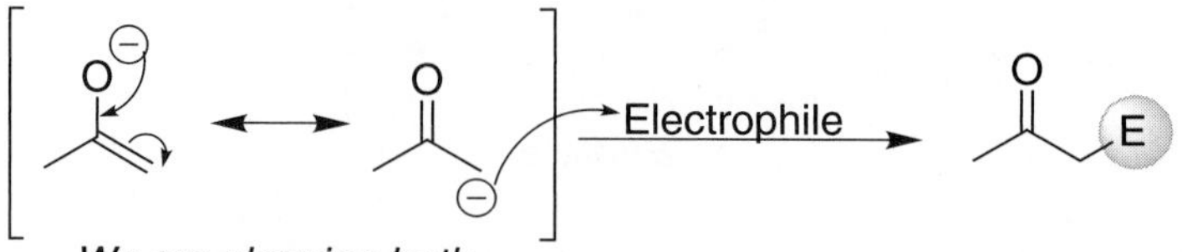

We are showing both
resonance structures
of the enolate

For simplicity, however, we will just show one resonance structure for the enolate (in most of the mechanisms that we will see in this chapter).

Now let's think about the base that we use *to make* enolates. If we use bases such as HO⁻ or RO⁻ (bases with a negative charge on oxygen), we find that these bases are *not* strong enough to completely convert the ketone into an enolate. Rather, there is an equilibrium that gets established between the ketone, the enolate, and the enol. This equilibrium only produces very small amounts of the enolate, but that doesn't matter. Once an enolate reacts with an electrophile, the equilibrium produces more enolate to replenish the supply. Over time, all of the ketone can convert into the enolate and then react with some electrophile. This is very similar to the situation we saw with enols. Once again, we are relying on the equilibrium to continuously produce more of the enolate. The major difference here is that enolates are so much more reactive than enols. Therefore, the chemistry of enolates is more robust than the chemistry of enols.

As an example for the richness of enolate chemistry, consider this: some enolates are much more stabilized than other enolates. These "super-stabilized" enolates are more "tame" nucelophiles (more selective in what they react with). For example, a compound with two carbonyl groups (separated by one carbon) can be deprotonated to form an intermediate that is like a *double* enolate:

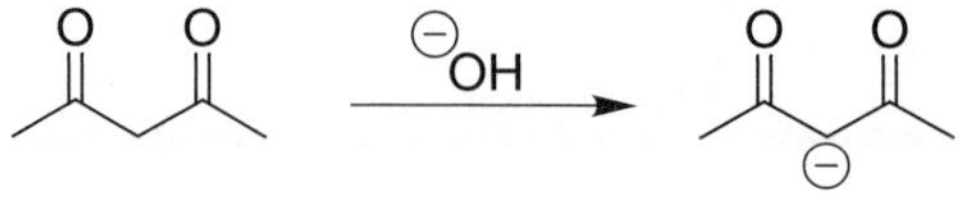

The negative charge in this intermediate is delocalized over both carbonyl groups:

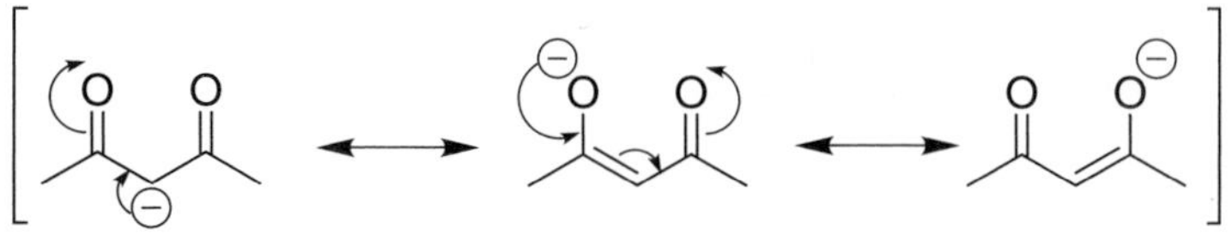

Therefore, it is extremely stable; it is even more stable than HO⁻ or RO⁻. So, when we use bases such as HO⁻ or RO⁻, the equilibrium lies greatly toward the side of this very stabilized enolate:

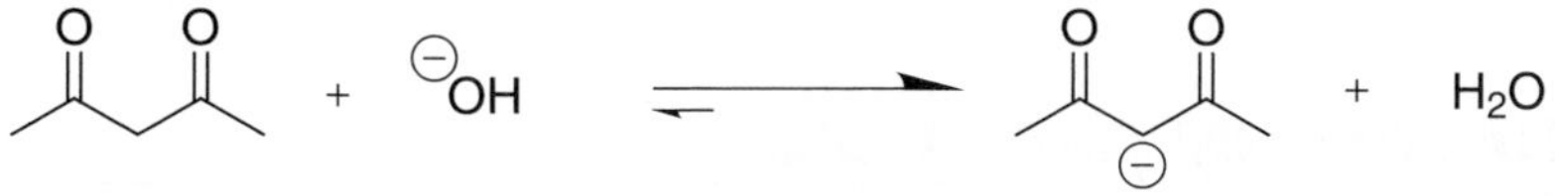

We will soon see that the position of this equilibrium will be a driving force in the Claisen Condensation (later in this chapter).

EXERCISE 7.19. Consider the following compound:

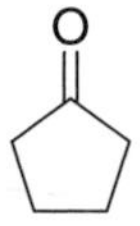

Draw the enolate that is formed when you deprotonate this compound. Make sure to draw all resonance structures.

Answer: We just need to identify the alpha proton, and then pull it off:

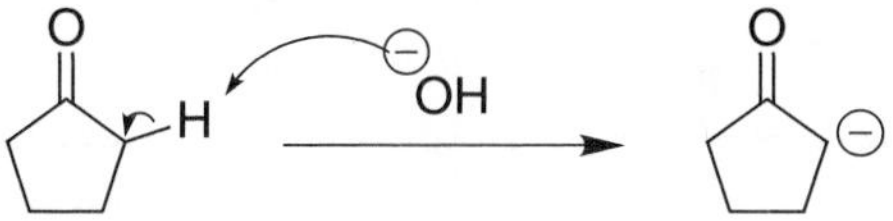

And then we draw the resonance structures:

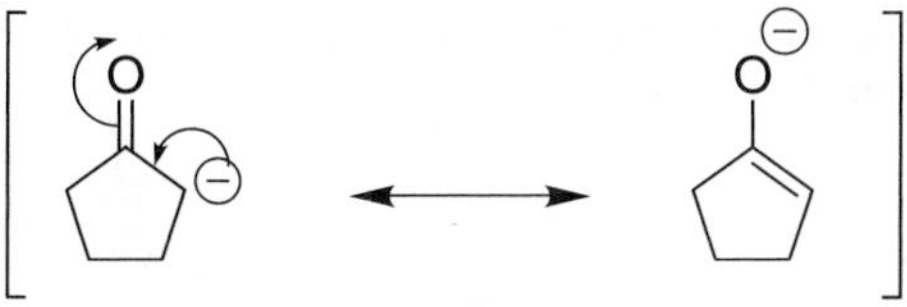

For each compound below, draw the enolate that would form in the presence of hydroxide. Make sure to draw all resonance structures.

7.20.

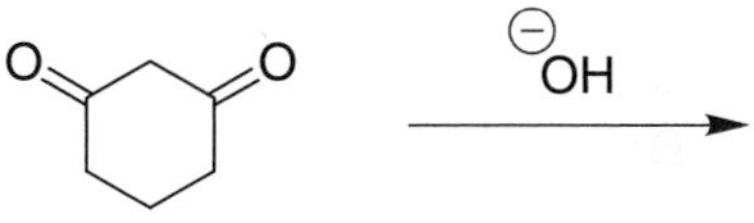

7.21.

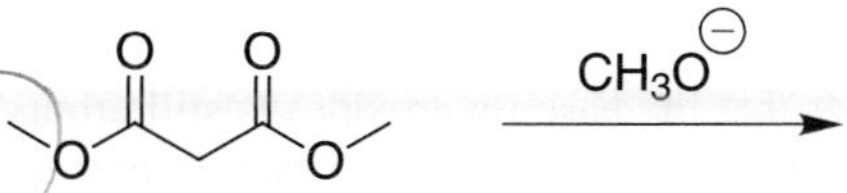

7.22.

7.23.

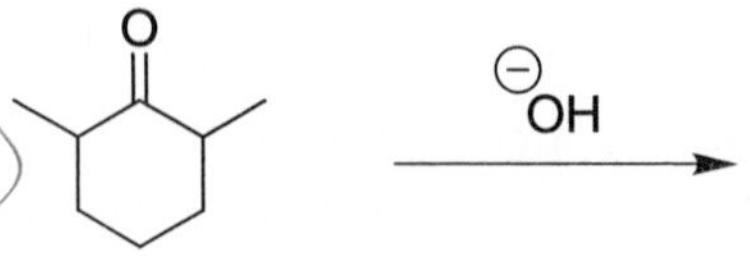

7.4   HALOFORM REACTIONS

In the previous section, we learned how to make enolates. Now, we will begin to see what an enolate can attack. In this section, we will attack a halogen (such as Br, Cl, or I), and in the following sections, we will see what happens when an enolate attacks different electrophiles.

Consider what might happen under the following conditions:

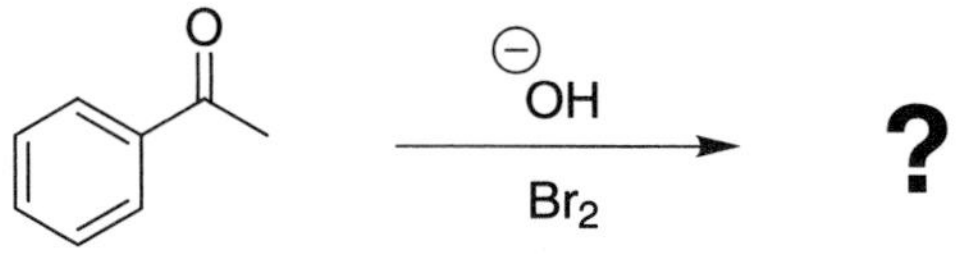

We have a ketone and hydroxide, which means that we will have an equilibrium that produces a small amount of enolate:

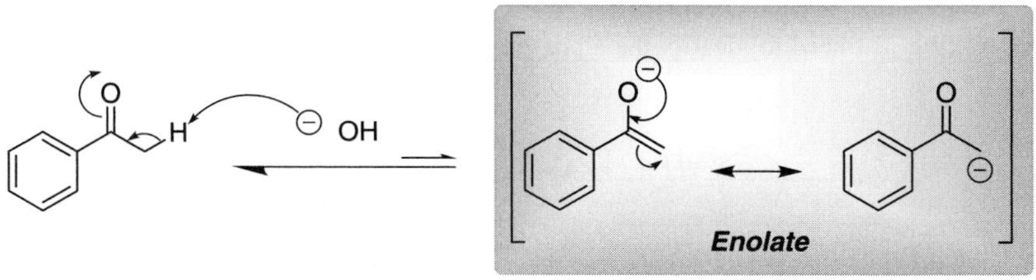

Enolate

This enolate is formed in the presence of Br_2, so the enolate has an electrophile to attack. The initial product is not surprising:

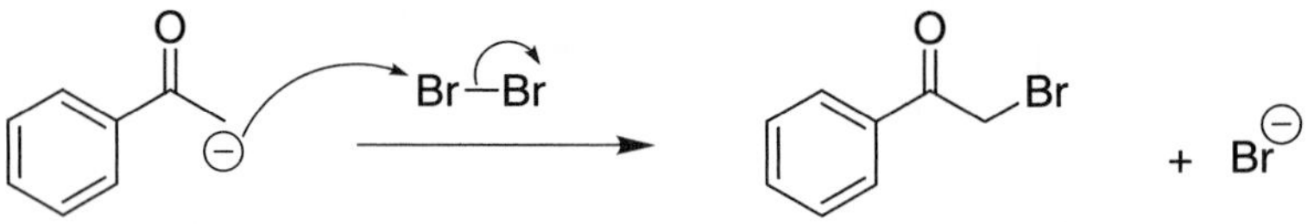

The enolate attacks Br_2 and kicks off Br^- as a leaving group. The result is that we have placed a Br at the alpha position:

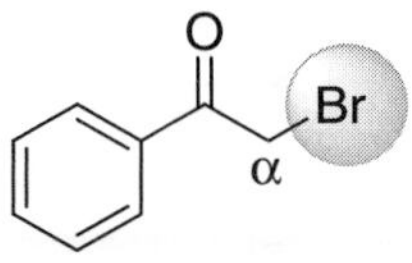

But the reaction doesn't stop there. Remember that the base (hydroxide) is still around in solution. So hydroxide can pull off another alpha proton. In fact, it is even easier to pull off this proton because the inductive effect of the bromine atom serves to further stabilize the enolate we will get. This enolate can then attack Br$_2$ again:

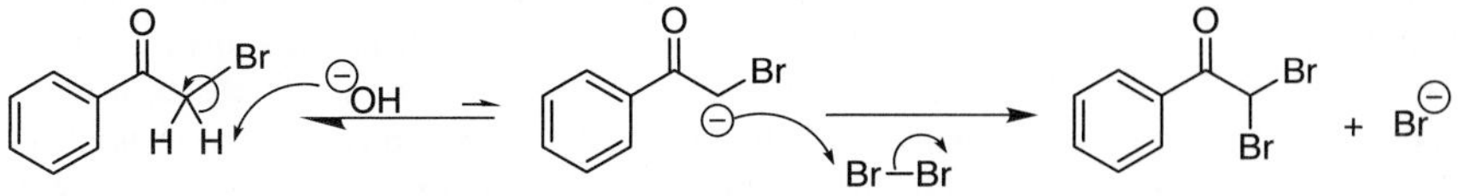

Now we have **two** Br atoms on our compound. And then, it happens again:

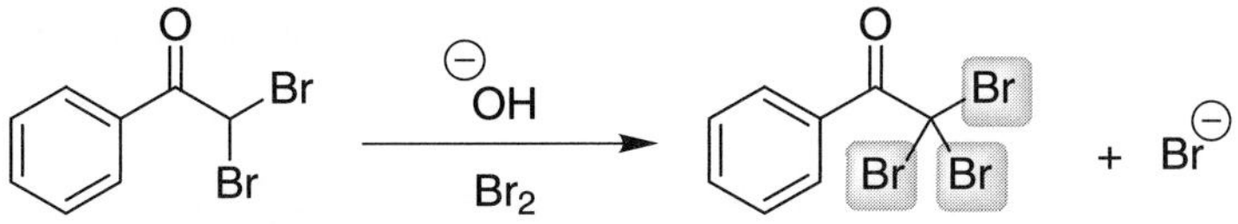

Think about what we have done so far. We have converted a CH$_3$ group into a CBr$_3$ group. This transformation is very significant because a CBr$_3$ group is able to function as a leaving group:

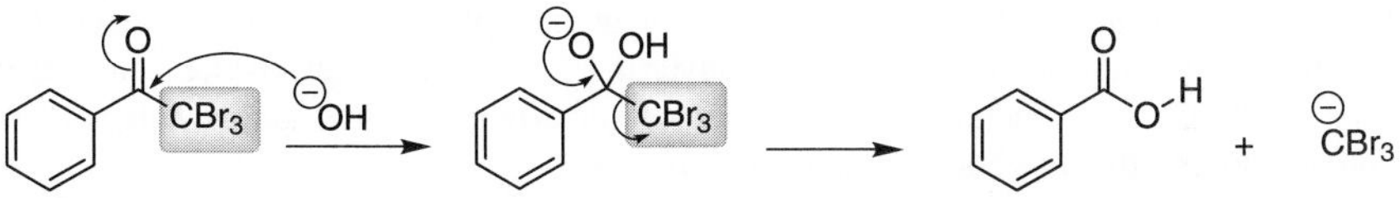

At this point, you should be feeling uncomfortable. You probably remember our golden rule from the previous chapters (don't kick off H$^-$ or C$^-$), and it seems as though we are breaking our golden rule. Aren't we kicking off a C$^-$ here? Yes, we are. This is actually one of the rare exceptions to the golden rule. In general, the golden rule holds true **most** of the time because C$^-$ is generally too unstable to serve as a leaving group. But in some cases a C$^-$ can be stabilized enough for it to serve as a leaving group, and this is one of those rare situations. CBr$_3$ (with a negative charge on the carbon) is actually a pretty good leaving group because of the combined electron-withdrawing effects of all three bromine atoms. But even though it can leave, it is not the most stable anion on the planet. In fact, it is not even as stable as a carboxylate ion (that's the ion you get when you deprotonate a carboxylic acid). So, our reaction finishes off with a proton transfer to form a more stable carboxylate anion and bromoform:

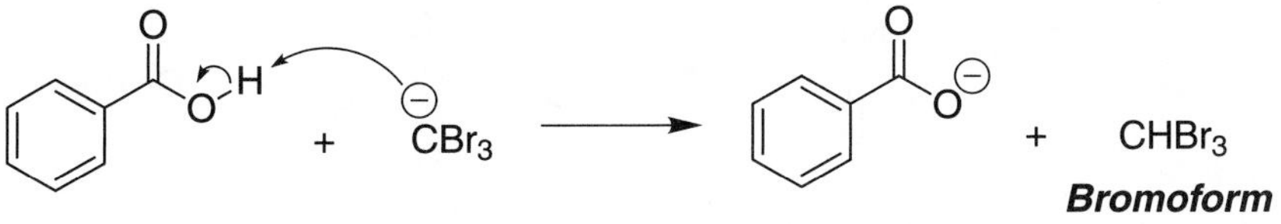

Bromoform

This forms a carboxylate anion and CHBr$_3$ (called bromoform). And this is the end of our mechanism. If we want to isolate the carboxylic acid, we will have to add a mild source of protons to protonate the carboxylate anion.

When the same reaction is done with iodine instead of bromine, we get iodoform as a byproduct (instead of bromoform):

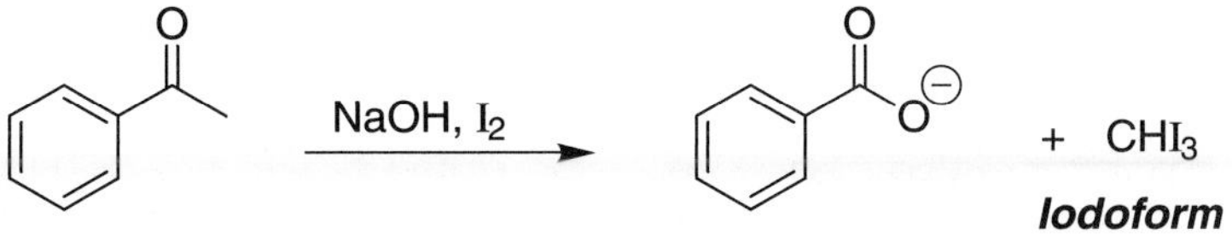

Iodoform

Iodoform is a yellow solid that will precipitate out of solution. Therefore, this reaction was used as one way of probing the identity of unknown compounds. If your unknown compound is a methyl ketone, then it will produce iodoform under these conditions (NaOH and I_2). This iodoform test is not really used anymore (we now have spectroscopy techniques that give us this information and much, much more). So, this chemical test is a relic of the past. But for some reason, it is still used in textbook problems. You will usually see it like this: "An unknown compound tests positive for iodoform, and. . . . " The beginning of this problem is telling you that you have a methyl ketone. If you see this in a problem in your textbook, you should know what it means.

But there is a much more important use for this reaction. You can use it when solving synthesis problems. This reaction gives us a way to convert a methyl ketone into a carboxylic acid:

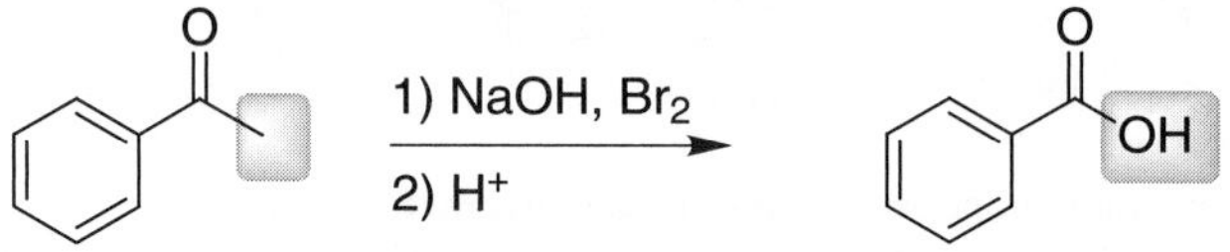

This should stick out in your mind because it is a new example of a "crossover" reaction. In the previous chapter, we talked about ways of converting ketones into carboxylic acid derivatives (crossover reactions). Our reaction here is a way of converting a methyl ketone into a carboxylic acid. You should add this to your synthetic toolbox.

EXERCISE 7.24. Predict the products of the following reaction:

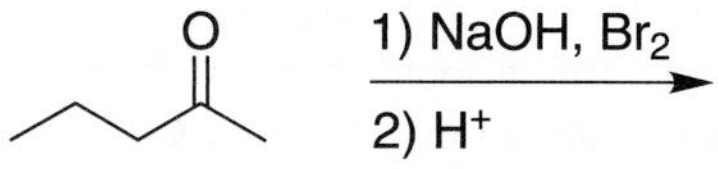

Answer: The trick here is to recognize that we are dealing with a methyl ketone and we have the conditions that will turn a methyl ketone into a carboxylic acid (with a by-product of bromoform):

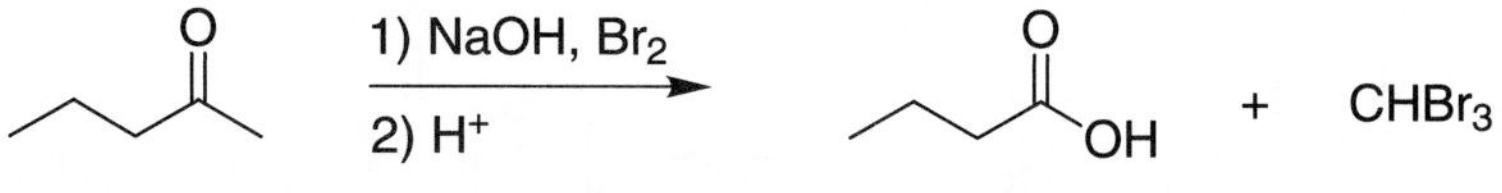

Predict the products for each of the following reactions:

7.25.

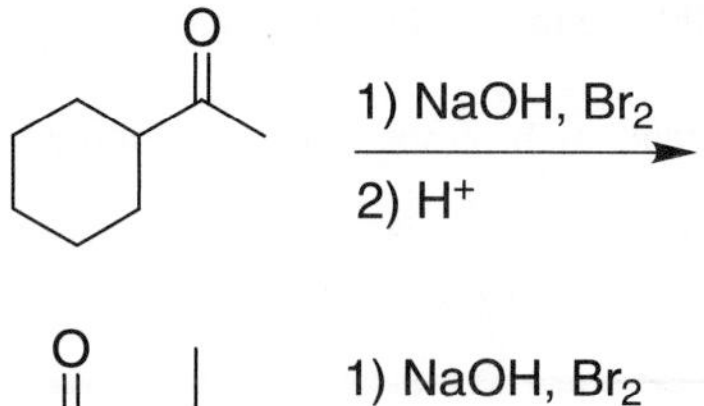

7.26.

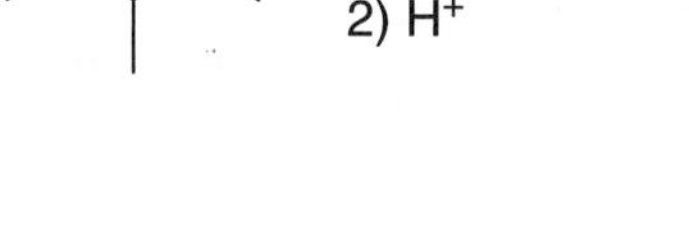

7.27. Without reviewing the mechanism that we developed in this section, try to draw the mechanism of the previous problem (7.26):

7.5 ALKYLATION OF ENOLATES

In this section, we will continue our discussion on the kinds of compounds that enolates can attack. We will learn how to hook an alkyl group onto an alpha carbon:

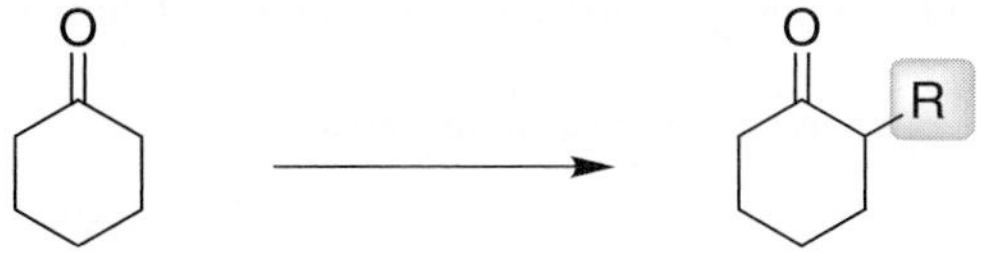

In order to alkylate the alpha position, it makes sense to use an enolate to attack an alkyl halide, for example:

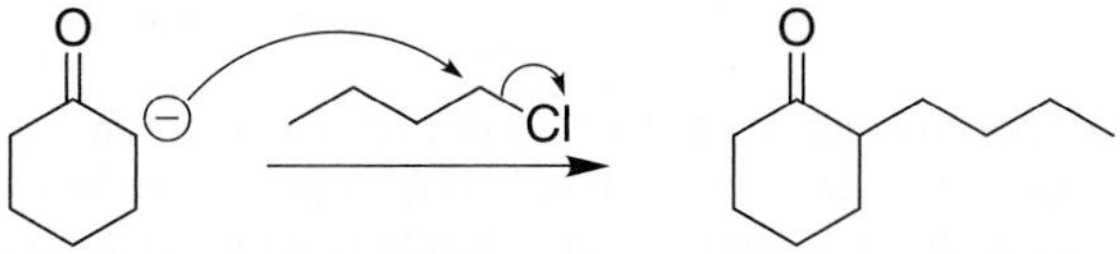

This is just an S_N2 reaction, so it will work best with ***primary*** alkyl halides.

But we run into a major obstacle when we try to make the enolate by mixing the ketone with hydroxide. Remember that when we use hydroxide as the base to form our enolate, we find that the equilibrium lies very far to the side of the ketone:

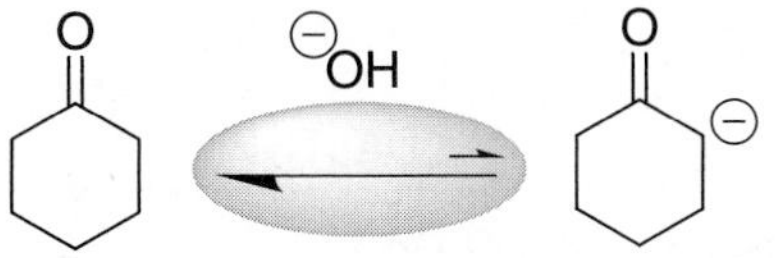

At equilibrium there is a very small amount of enolate, and there is a lot of ketone and a lot of hydroxide floating around. So, if we put some alkyl halide into the reaction flask, we run into a major obstacle. The excess hydroxide can attack the alkyl halide, which creates competing side reactions. And this gives us a mess of products.

In order to avoid this problem, we will need to form our enolate under conditions where most of the ketone molecules are converted into enolates. If we are able to do this, we will have very little base left over, and therefore, we won't have to worry about the base reacting with the alkyl halide. It is possible to do this, but we will need to use a base that is much stronger than the oxygen bases we have been using so far (HO^- and RO^-). We will use a nitrogen base instead:

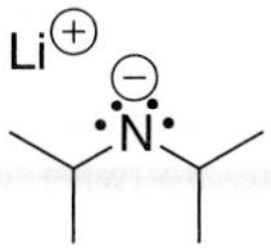

The name of this compound is lithium diisopropylamide (LDA). LDA is a very strong base, because the negative charge is on a nitrogen atom (which is less stable than a negative charge on an oxygen atom). The two isopropyl groups are sterically bulky, so LDA is *not* a good nucleophile. LDA is used primarily as a strong, sterically hindered base, which is exactly what we need in our situation. By using LDA, we get a very good conversion of the ketone into the enolate. The equilibrium lies very far to the side of the enolate:

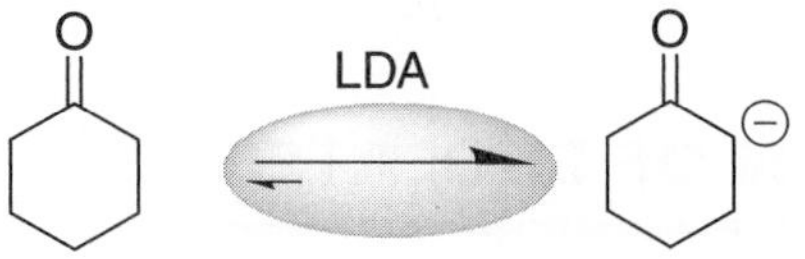

So, we will have mostly enolates in our reaction flask (and very little ketone or base). Now when we toss in some alkyl halide into our reaction flask, we don't have to worry about competing side reactions.

So, to alkylate a ketone, we use the following reagents:

In step 1, we use LDA to deprotonate the ketone, to form an enolate. When you see THF (tetrahydrofuran) in the reagents above, don't get confused. THF is just the solvent that is typically used to dissolve LDA. In step 2 above, we use an alkyl halide (RX) to hook on the alkyl group. R is some primary (or secondary) alkyl group, and X is a halogen (Cl, Br, or I).

This works very well when we start with symmetrical ketones, as in the case above with cyclohexanone. But what happens when we start with an unsymmetrical ketone? For example, consider the following situation:

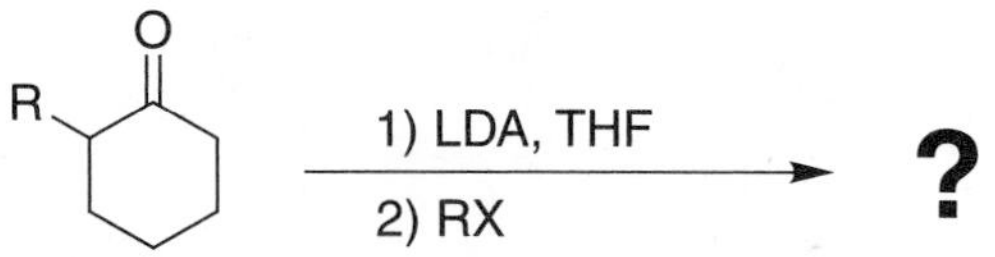

Where should we put the alkyl group? on the left side, or on the right? In order to answer this question, we need to take a close look at the two possible enolates that can form here:

This enolate is more stable   **This enolate forms faster**

The more substituted enolate (above left) is the more stable enolate. However, the less substituted enolate (above right) can form faster. There are twice as many protons to grab from the less substituted side:

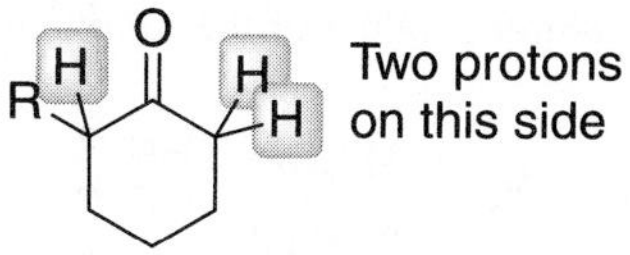

Two protons on this side

So from a probability point of view, we expect to form more enolate on the less substituted side. Also, we expect the sterically hindered base to have a much easier time grabbing one of these protons. So, we have two competing arguments:

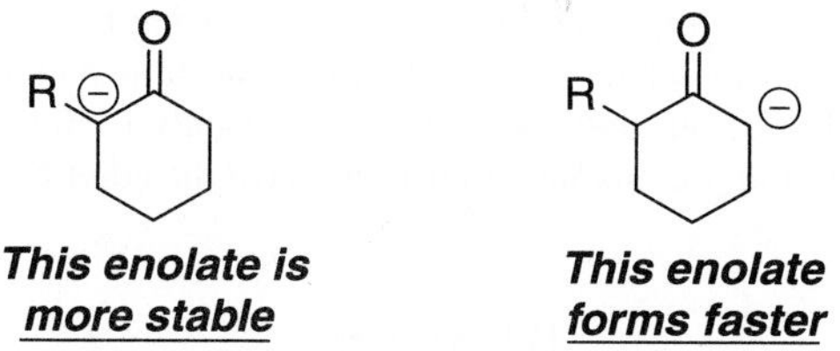

This enolate is more stable

This enolate forms faster

This is a classic example of thermodynamics vs. kinetics. Thermodynamics is all about stability and energy levels. So a thermodynamic argument states that we should predominantly form the more stable enolate. However, a kinetic argument tells us to expect the other enolate simply because it forms faster. Which argument wins? The truth is that we will get a mixture of products. But with LDA at low temperature, there is a clear preference to form the kinetic enolate:

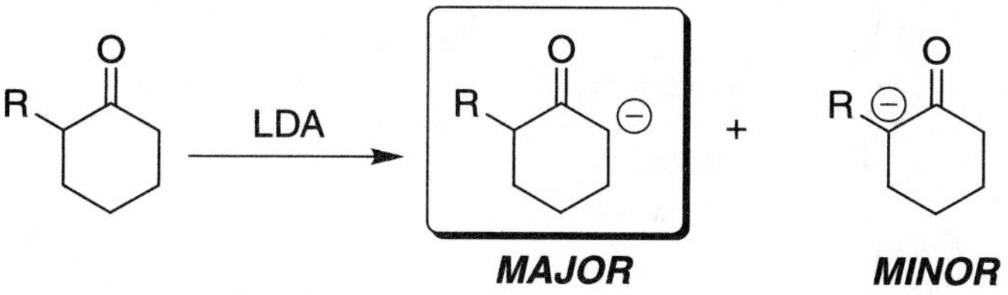

MAJOR   **MINOR**

When we add the alkyl halide to the reaction flask, alkylation will occur primarily at the less substituted alpha position:

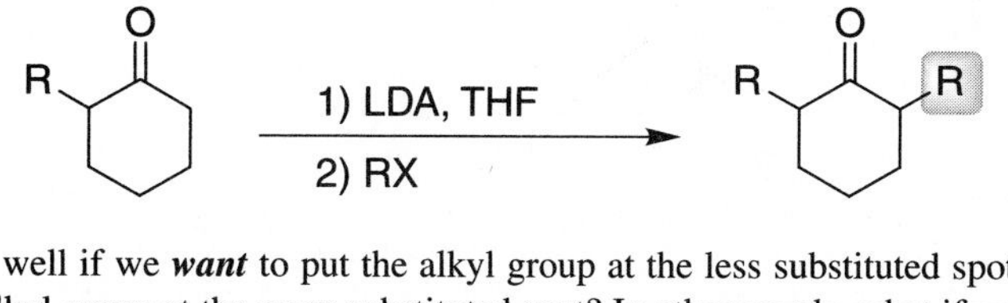

1) LDA, THF

2) RX

That works very well if we **want** to put the alkyl group at the less substituted spot. But what if we want to put the alkyl group at the more substituted spot? In other words, what if we want to do this:

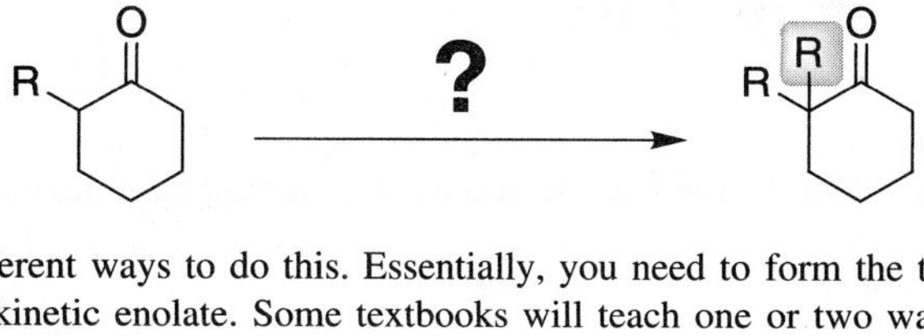

?

There are many different ways to do this. Essentially, you need to form the thermodynamic enolate rather than the kinetic enolate. Some textbooks will teach one or two ways to do this, while other textbooks will skip it altogether. You should look through your textbook and lecture notes to see if you are responsible for knowing how to alkylate on the more substituted side.

EXERCISE 7.28.   Predict the *major* product of the following reaction:

1) LDA, THF

2) EtCl

?

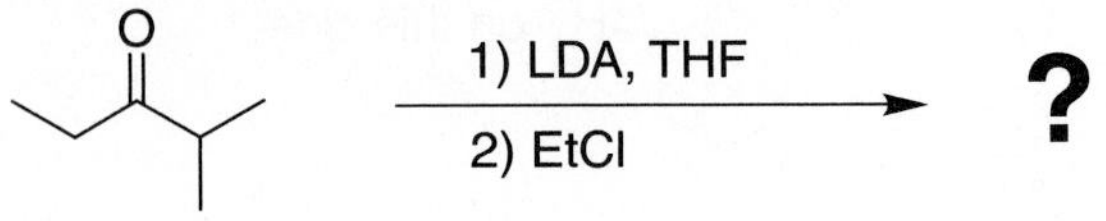

Answer: This is an alkylation reaction. In step 1, we are using LDA to form an enolate. And then in step 2, we are using an alkyl halide to alkylate.

Because our alkyl halide is ethyl chloride, we will be placing an ethyl group on an alpha carbon. The only question is: which alpha carbon? the more substituted carbon or the less substituted carbon? When we use LDA as our base, we will predominantly form the kinetic enolate (the less substituted enolate). Therefore, our *major* product will have the ethyl group on the less substituted alpha position:

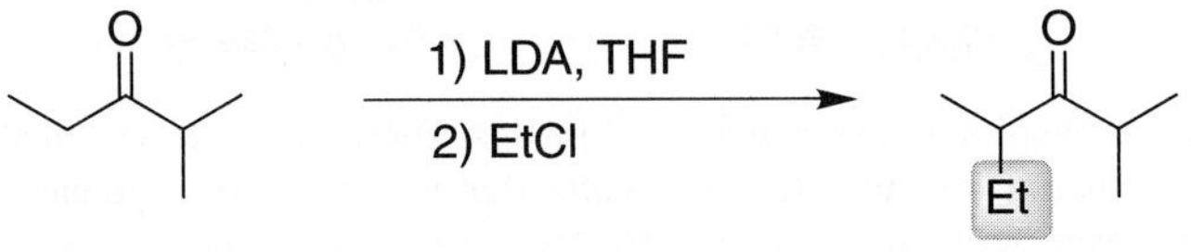

1) LDA, THF

2) EtCl

Predict the *major* product for each of the following reactions:

7.29.

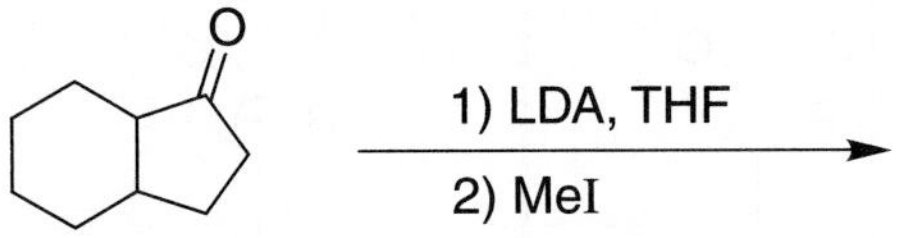

1) LDA, THF

2) MeI

7.30.

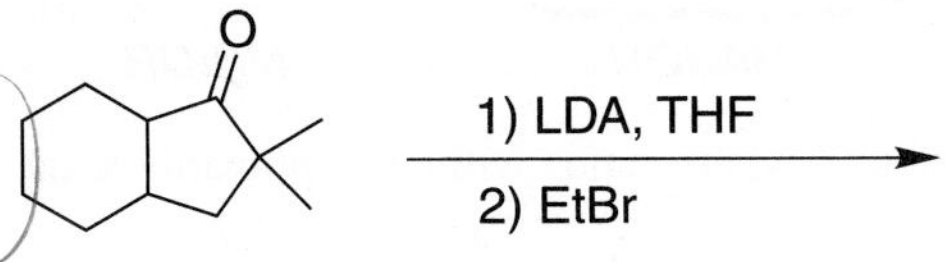

1) LDA, THF

2) EtBr

7.31.

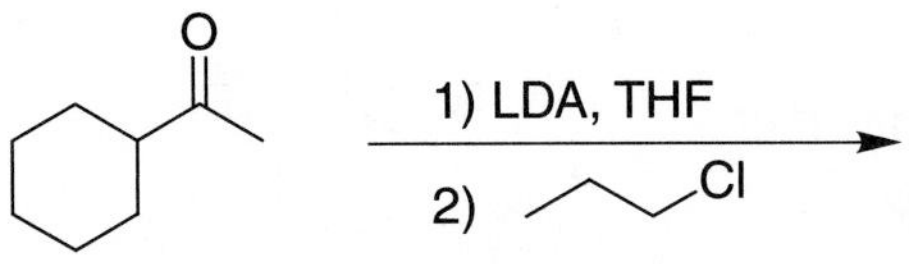

1) LDA, THF

2)   ⌃⌃Cl

7.32.

1) LDA, THF

2) MeI

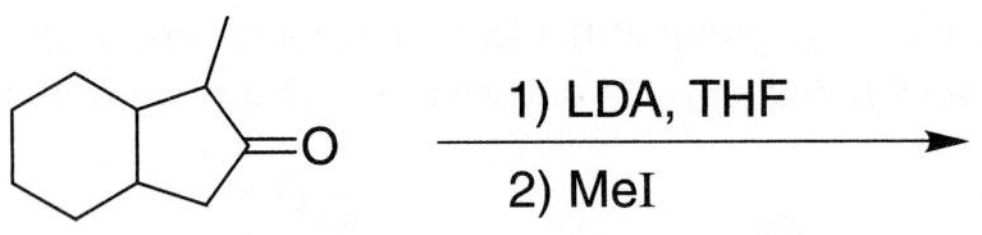

EXERCISE 7.33.   What reagents would you use to carry out the following transformation:

?

Answer: If we look at the difference between the starting material and the product, we will see that an extra methyl group was introduced. This methyl is placed on the less substituted side, so we will need to use LDA and a methyl halide:

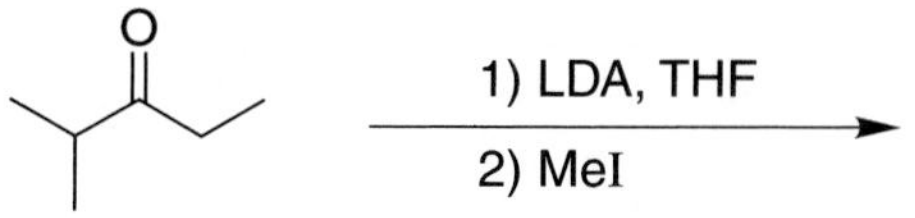

1) LDA, THF

2) MeI

What reagents would you use to carry out each of the following transformations:

7.34.

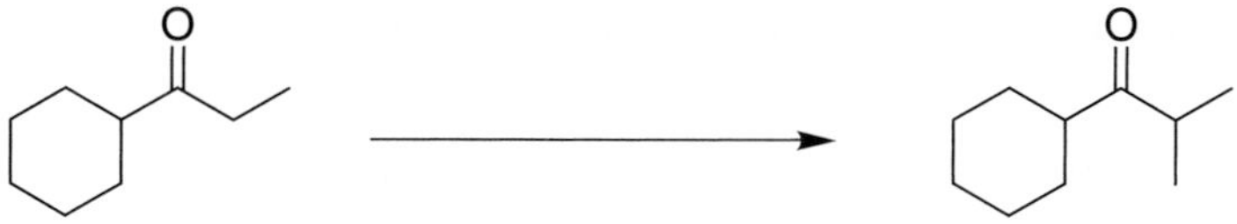

7.35.

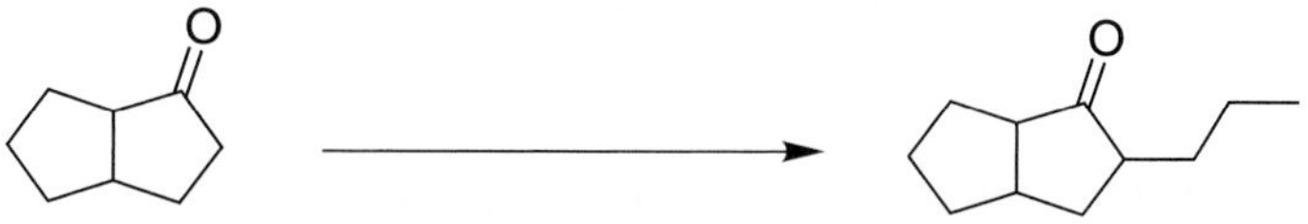

7.36.

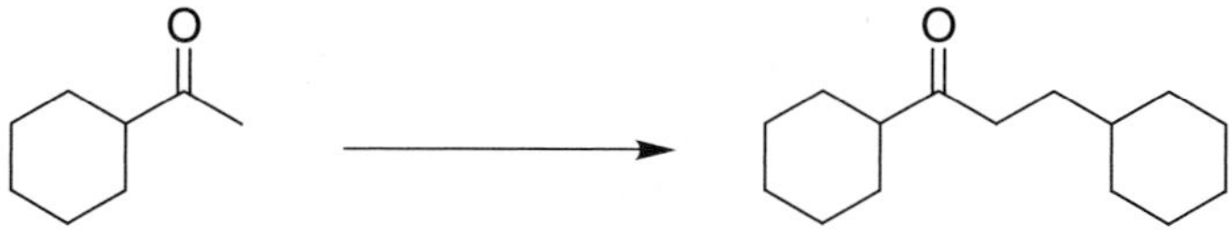

7.6   ALDOL REACTION AND ALDOL CONDENSATION

So far in this chapter, we have learned how to make enolates, and we have used them to attack various electrophiles. We attacked halogens, and we attacked alkyl halides. In this section, we will see what happens when an enolate attacks a ketone or aldehyde.

Suppose we start with a simple ketone, and we subject it to basic conditions, using hydroxide as our base. We have already seen that often an equilibrium will be established between the ketone and the enolate:

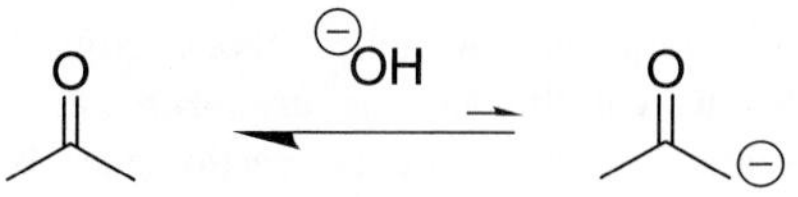

If we do this in the presence of an electrophile, the enolate that forms will attack the electrophile. Then the equilibrium will produce more enolate to replenish the supply. But what if we do not add any other electrophiles to the reaction mixture? What if we just stir the ketone and hydroxide together?

It turns out that there actually *is* an electrophile present. We said that the enolate is in equilibrium with the ketone (and there is a lot of ketone around). Well, ketones are electrophilic, aren't they? We devoted an entire chapter to the reactions that take place when ketones get attacked. So, what happens when an enolate attacks a ketone? We get the following reaction:

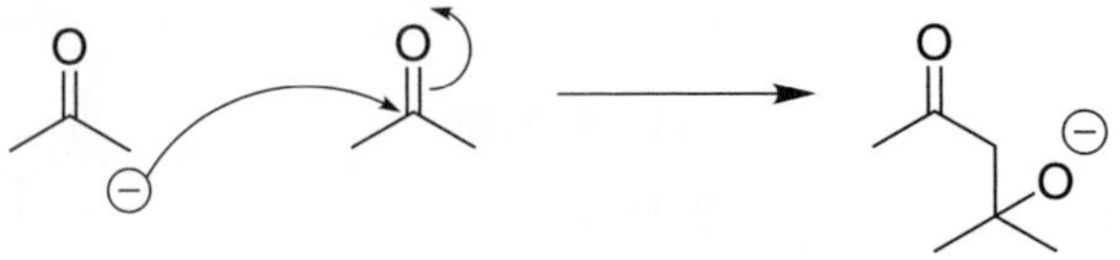

The enolate attacks the ketone, kicking the negative charge up onto an oxygen. Now, our golden rule tells us to try and re-form the carbonyl, but don't kick off H^- or C^-. In this case, we have no leaving groups that we can kick off. So, the only way to get rid of the charge is to grab a proton. In basic conditions, we will have to grab a proton from water (not H^+, because there isn't any around):

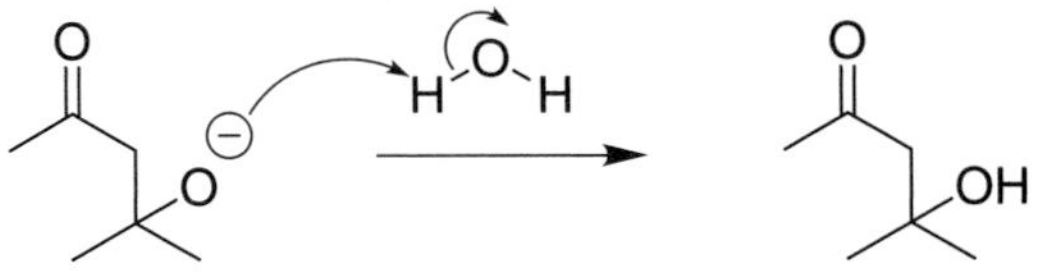

This is the initial product of this reaction. Notice that the OH group is at the β position relative to the one carbonyl group:

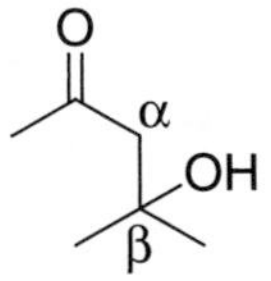

This will always be the case whenever an enolate attacks a carbonyl, regardless of the structure of the starting ketone and the structure of the enolate. The alpha carbon of the enolate is directly attacking the carbonyl of the ketone. That will always place the OH group in the beta position. Always. This product is called a *β-hydroxy ketone*, and the reaction is called an ***aldol*** reaction.

In general, the reaction doesn't stop there (at the β-hydroxy ketone). With heating, we will generally get one more step where we eliminate water to form a double bond:

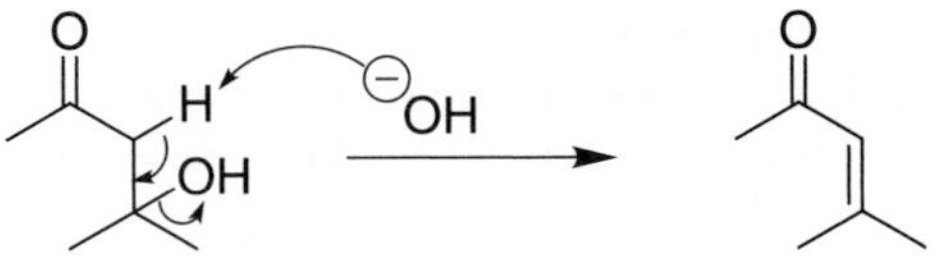

This product has a double bond in conjugation with the carbonyl group. The double bond is located between the α and β positions. So, we call this an α, *β-unsaturated ketone*.

In the laboratory, we can usually control how far the reaction goes. By carefully controlling the conditions of the reaction (temperature, concentrations, etc.), we can usually control whether the reaction stops at a *β-hydroxy ketone* or whether it continues to form an *α,β-unsaturated ketone*. So, you can use an aldol to form either product.

But you should be familiar with the proper terminology. When we go all the way to the *α, β-unsaturated ketone*, we call the reaction an aldol ***condensation***. By definition, a condensation is any reaction where two molecules come together, and in the process, they liberate a small molecule. The small molecule can be N_2 or CO_2 or H_2O, and so on. In this case, we have two molecules of ketone coming together, and in the process, a molecule of water is liberated:

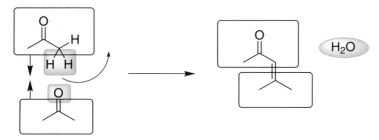

Therefore, we call this reaction an aldol ***condensation***. But what if we control the reaction conditions so that we stop at the β-*hydroxy ketone*?:

Stop here

If we stop here, then we cannot call it a condensation reaction anymore because we are not losing a water molecule in the process. Instead, we call it an aldol ***reaction***. The difference between an aldol condensation and an aldol reaction is how far we go in the process:

Aldol *condensation*

Aldol *reaction*

This distinction (between the aldol *reaction* and the aldol *condensation*) is often absent in textbooks, and you might find the terms being used interchangeably in your textbook. I am taking the time to point out the distinction because I believe that it will help you to remember and master the mechanism (by dividing it in your mind into two distinct steps, where each step has a specific name).

The mechanism of an aldol condensation is fairly straightforward. But sometimes, it can get hard to see what reagents to start with when you are solving a synthesis problem. So try to think of it the way we showed it just a few moments ago:

We are pulling off two alpha protons from one ketone, and we are pulling off the oxygen from the other ketone. Do **not** confuse the drawing above with a mechanism. A mechanism is when you show all of the curved arrows and intermediates. But this way of thinking about the reaction might come in handy during problem solving. Let's get a bit of practice with this way of thinking about it:

EXERCISE 7.37. Predict the major product of the following reaction:

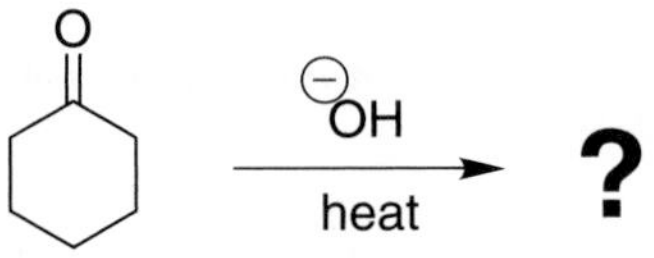

We are using heat, so you can assume that you get an aldol *condensation*.

Answer: We can work through the mechanism to get the answer, and we will soon get practice with that. But for now, let's just make sure that we can use our simple method for figuring out what the answer is.

We start by drawing two molecules of ketone, and we draw them so that the oxygen of one ketone is pointing directly at the alpha protons of the other ketone:

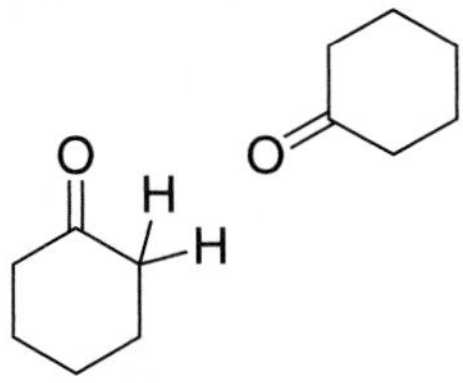

Then, we erase the two alpha protons and the oxygen atom, and we push the fragments together (connecting them with a double bond):

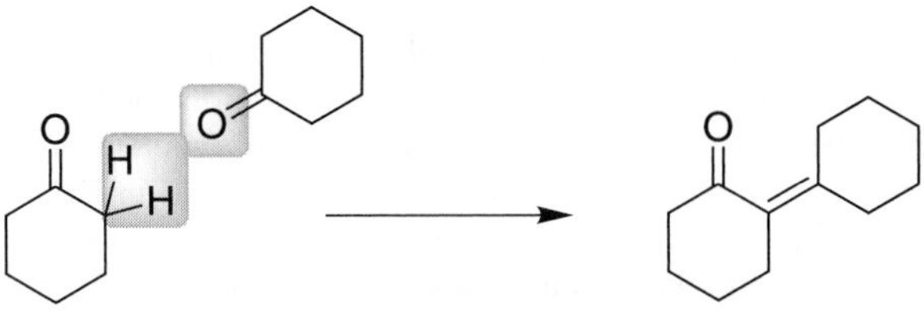

That's all there is to it. It is a simple, but powerful, way of thinking about this reaction.

Predict the major product for each of the following reactions. In each case, assume that a condensation takes place, and draw the α, β-unsaturated ketone that is produced.

7.38.

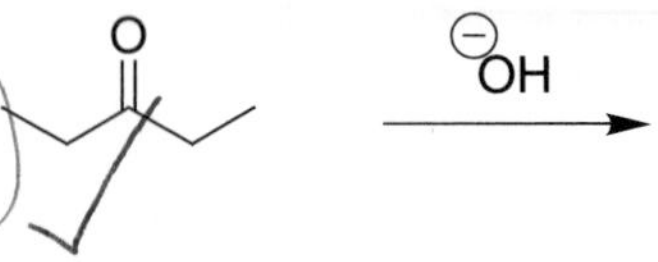

7.39.

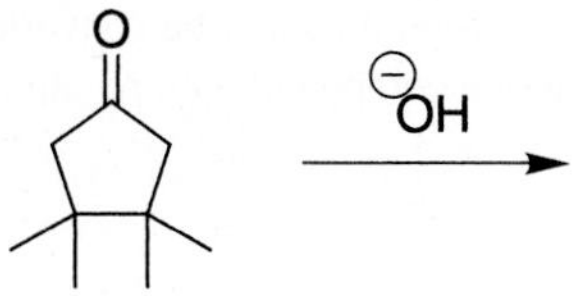

7.40.

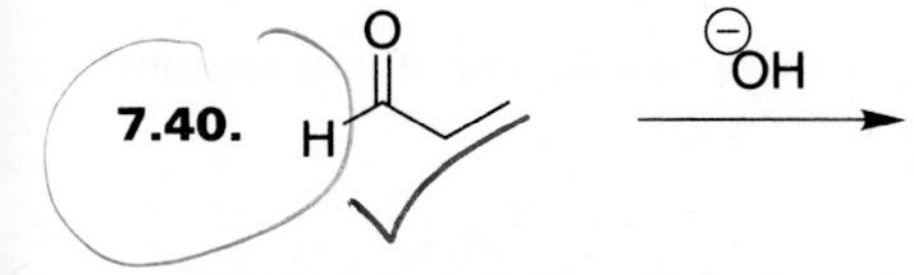

In all of the aldol condensations that we have seen so far, we used two molecules *of the same ketone*. We would just deprotonate one molecule of ketone (to give an enolate), and then we would use that enolate to attack another molecule of the same ketone. But what if we use two very different ketones? For example, what if we try to do this:

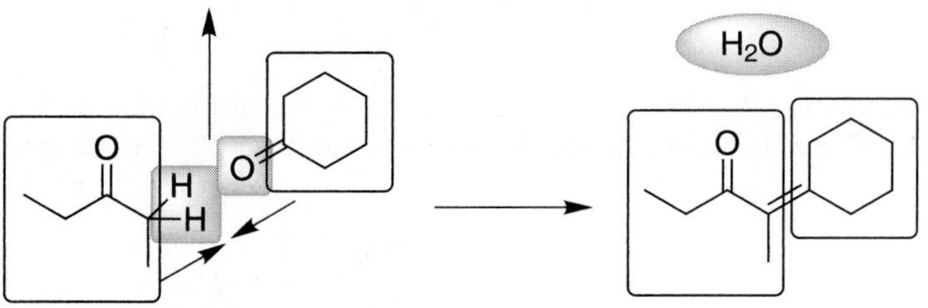

Notice that the ketones are different from each other. We call this a *crossed*-aldol. This can work, but we will get a lot of different products. To see why this is the case, we must realize that it is possible for enolates and ketones to exchange protons:

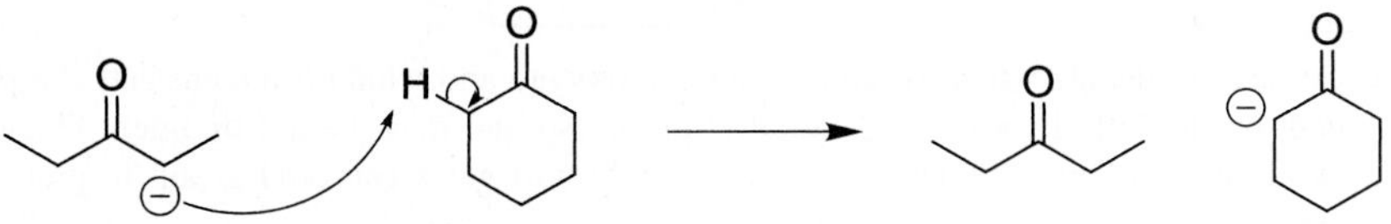

So, you can't really control which ketone will be converted into the enolate. This means that there will be more than one type of enolate and more than one type of ketone floating around in solution (and you can't prevent that from happening). So a number of possible reactions can take place, and this will give us a mess of products.

Practically, then, it is important to try to avoid these types of situations. There is one very easy way to avoid this issue. If one of our ketones has no alpha protons, then it cannot form an enolate. For example, consider the following compound

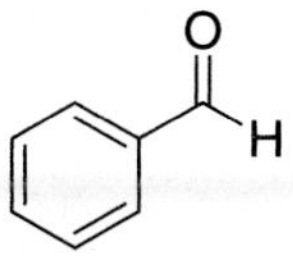

This compound, called benzaldehyde, has no alpha protons. Therefore, it cannot be converted into an enolate; it will just wait to be attacked. Here is another example of a compound with no alpha protons:

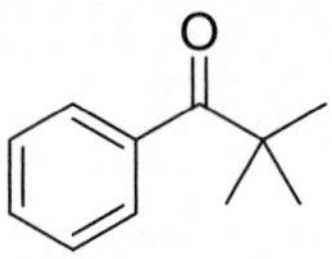

So, if you really want to do a crossed aldol, you should try to make sure that one of your reagents has no alpha protons. That will minimize the number of potential products.

EXERCISE 7.41.   What reagents would you use to make the following compound, using an aldol condensation:

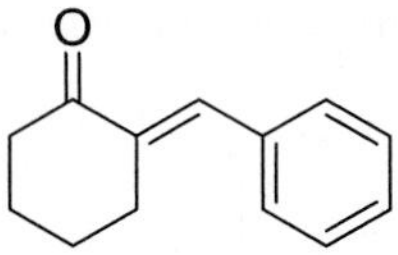

Answer: We can use the same method we used earlier. We just need to do it in reverse. We break the molecule apart into two fragments in order to insert water. We break it apart at the C=C double bond:

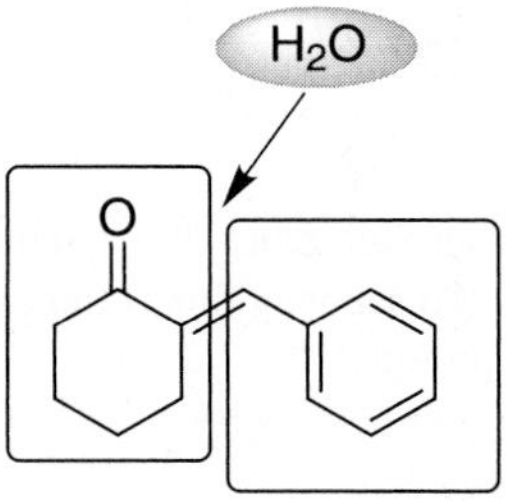

And we just have to decide which fragment gets the oxygen and which fragment gets the protons. The fragment on the left already has a carbonyl group, so that must be the fragment that we will give the two alpha protons. The fragment on the right will get a carbonyl group in place of the C=C double bond:

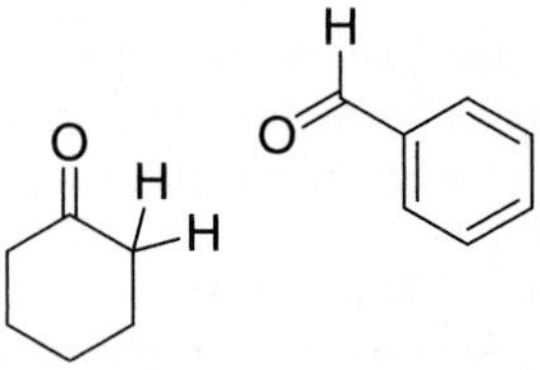

Identify what reagents you would use to make each of the following compounds (using an aldol condensation):

7.42.

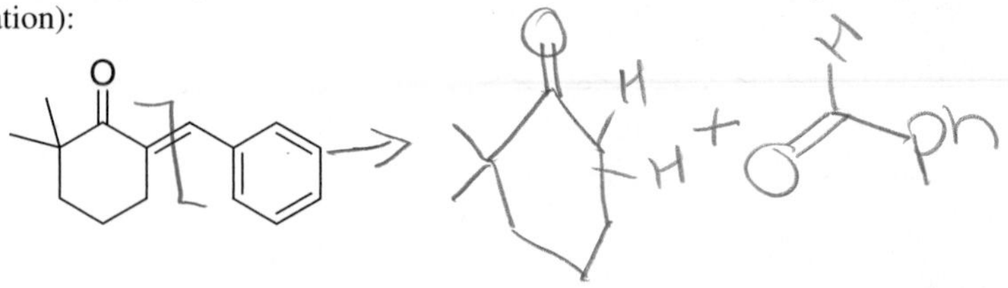

7.43.

7.44.

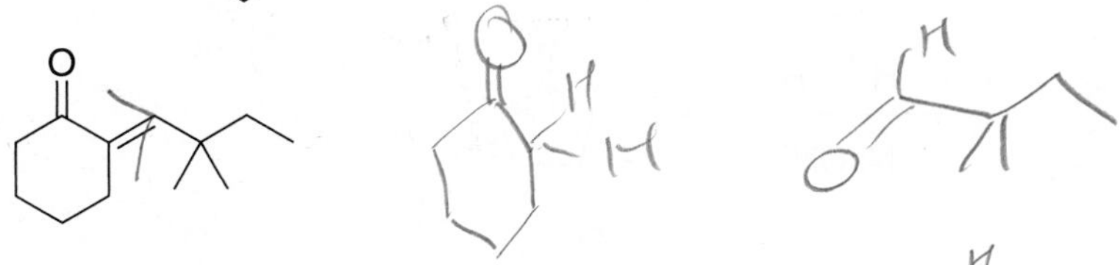

7.45.

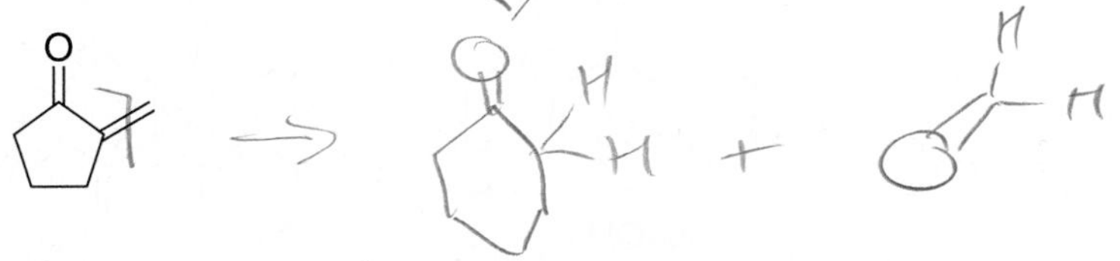

EXERCISE 7.46. Propose a plausible mechanism for the following reaction:

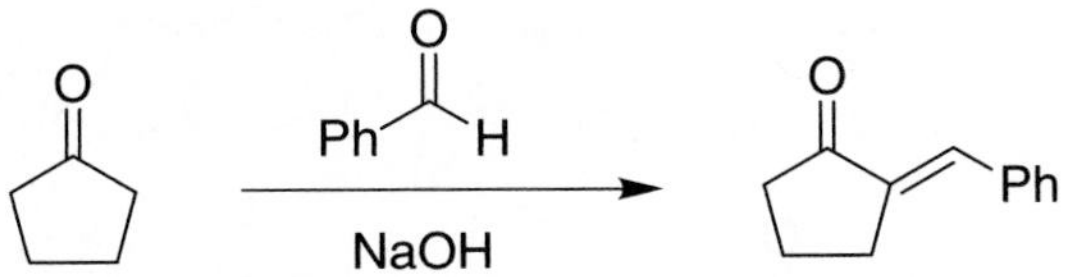

Answer: In the first step, hydroxide is used to generate an enolate:

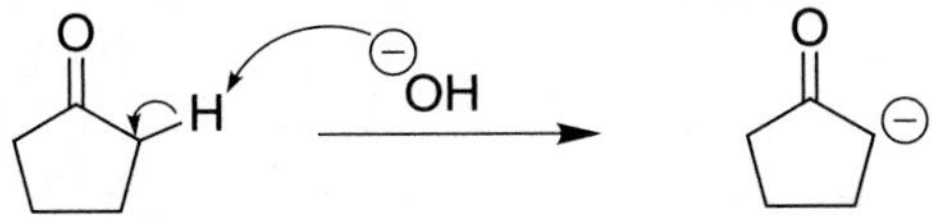

Then, this enolate attacks benzaldehyde:

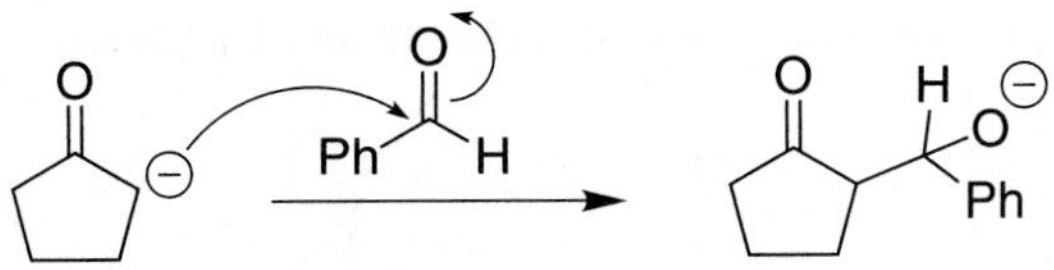

We then grab a proton from water, to form the β-hydroxy ketone:

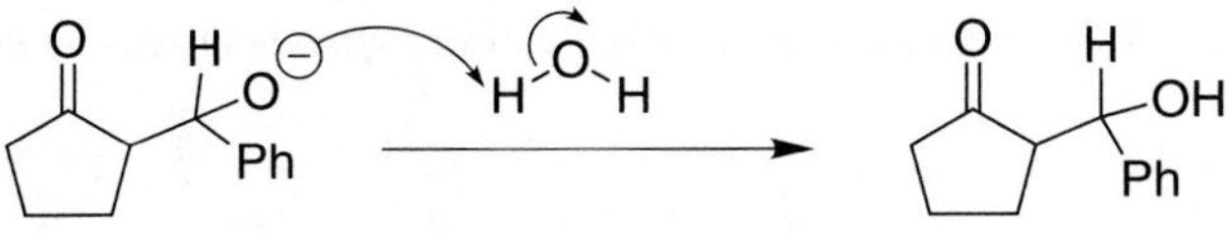

Finally, we eliminate water to form the α,β-unsaturated ketone:

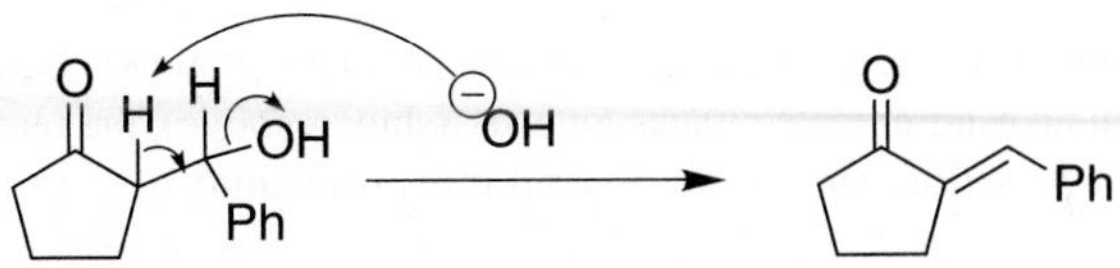

Now let's get some practice with the mechanism of an aldol condensation. Draw the mechanism for each of the following transformations. You will need a separate piece of paper to record your answers:

7.47.

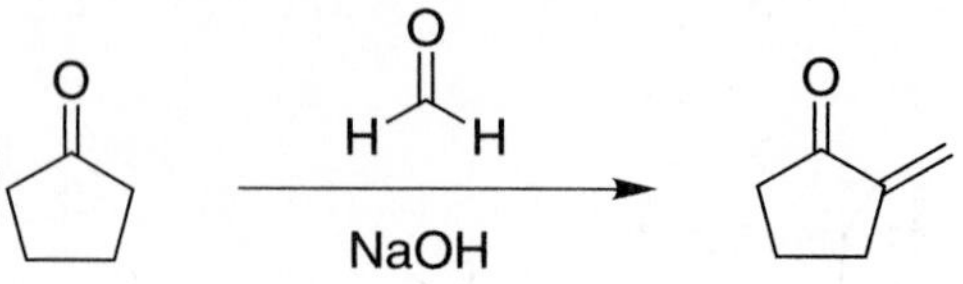

7.48.

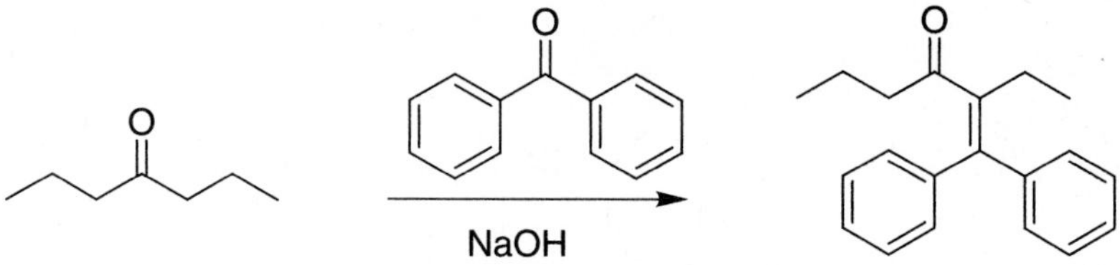

7.49.

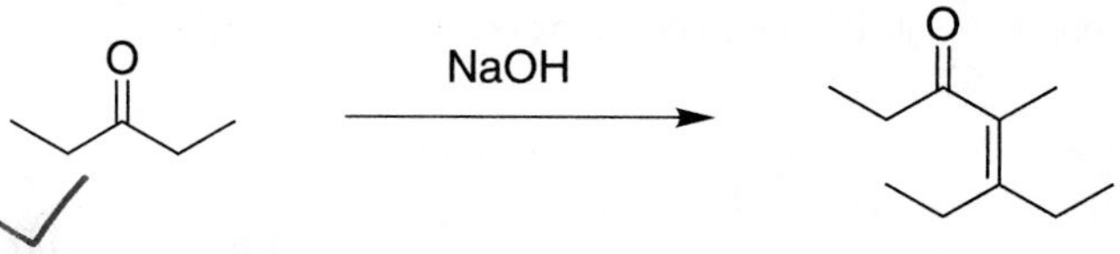

NaOH

7.50.

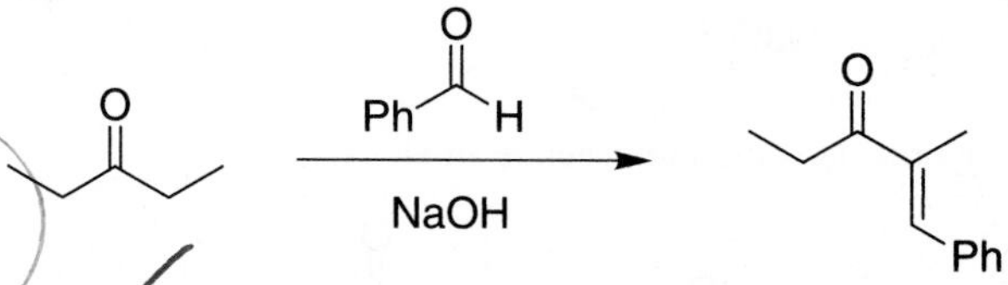

7.7 CLAISEN CONDENSATION

In the previous section, we saw how to use an enolate to attack a ketone:

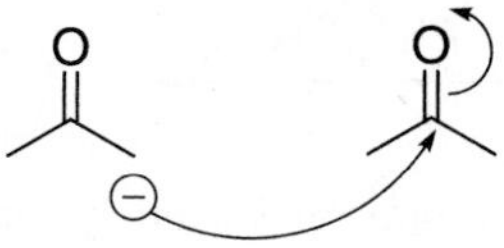

In this section, we will explore what happens when an **ester enolate** attacks an ester:

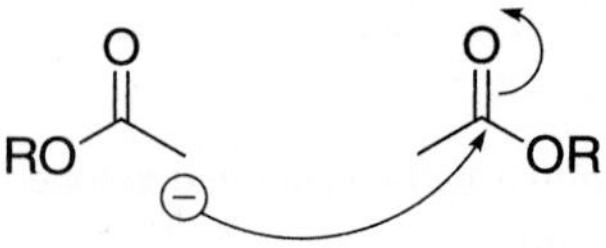

An ester enolate is similar to a regular enolate: an ester enolate is nucleophilic, and it will also attack a carbonyl. When an ester enolate attacks an ester (shown above), the reaction that takes place is called a Claisen condensation. Here is the overall transformation:

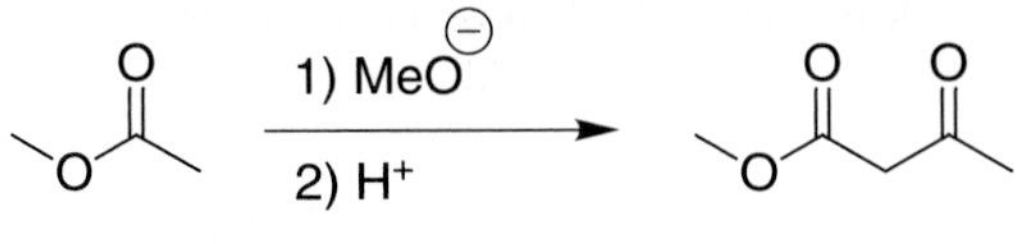
1) MeO⊖
2) H+

This product is called a β-keto ester:

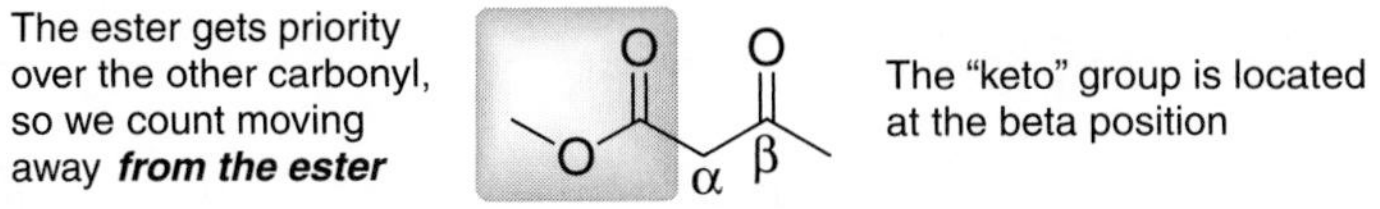
The ester gets priority over the other carbonyl, so we count moving away ***from the ester***
α
β
The "keto" group is located at the beta position

At first glance, this product seems very different from the α,β-unsaturated ketones that we got from aldol condensations. But when we explore this mechanism in detail, we will see the parallel between the aldol and the Claisen condensations.

Let's start with the first step: preparing the enolate:

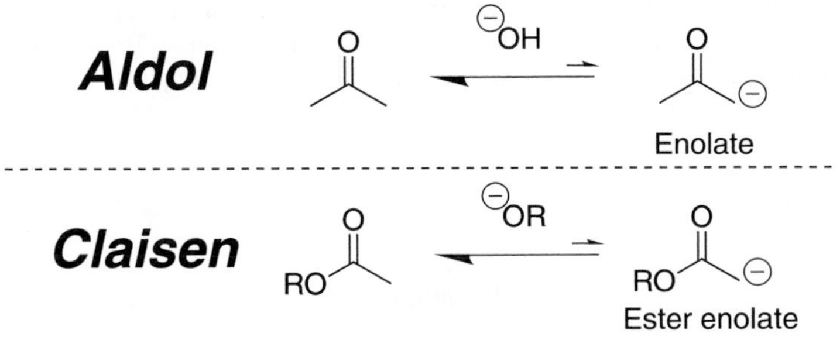
Aldol
OH
Enolate
Claisen
OR
RO
Ester enolate

So far, both mechanisms are almost identical. The only difference is the choice of base (the aldol used hydroxide, and the Claisen used alkoxide), and we will discuss the reason for this shortly. For now, let's just finish through the mechanism.

In the next step, the enolate attacks:

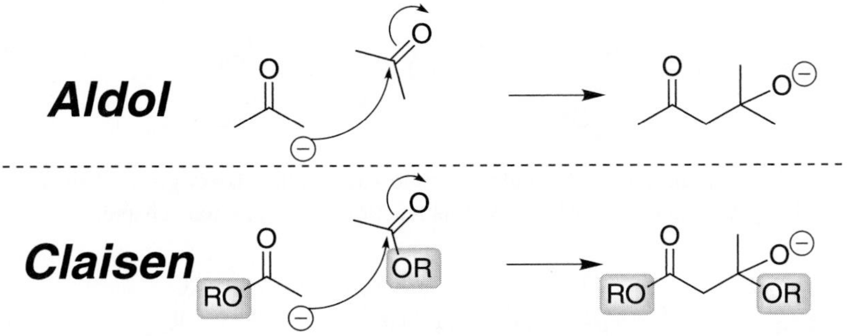
Aldol
Claisen
RO
OR

Once again, we see that both mechanisms are essentially identical. In the Claisen condensation, the alkoxy groups seem to just come along for the ride.

But now the two mechanisms take different routes. And we can use our golden rule to understand why. In the aldol reaction, the carbonyl cannot re-form, so the oxygen can only grab a proton. But in a Claisen condensation, we CAN re-form the carbonyl because we have a group that can leave:

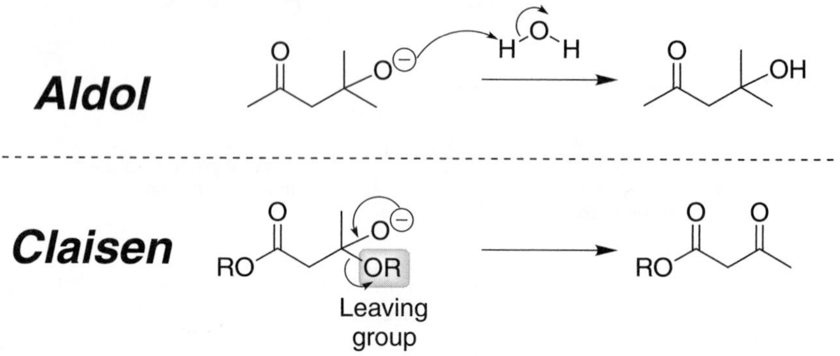

Aldol

Claisen

Leaving
group

And this is why the product of a Claisen condensation looks very different from the product of an aldol condensation. But when you understand the mechanisms, you can appreciate that these reactions are very similar. The difference between these two reactions stems from the fact that Claisen condensations involve esters, and esters have a "built-in" leaving group:

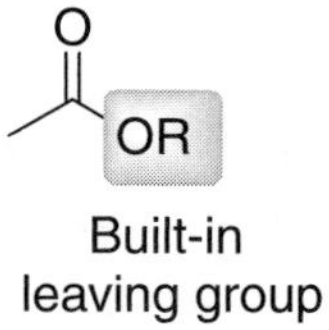

Built-in
leaving group

Now that we have seen the whole mechanism, let's go back and explore it a bit more. In the first step, we made an ester enolate. To do this, we used a strong base. But we pointed out at the time that we did NOT use hydroxide. Instead, we used an alkoxide. Let's try to understand why.

If we had used hydroxide, then we could have gotten a competing reaction. Instead of the hydroxide acting as a base to grab a proton, it is possible for the hydroxide to function as a nucleophile, attacking the carbonyl of the ester:

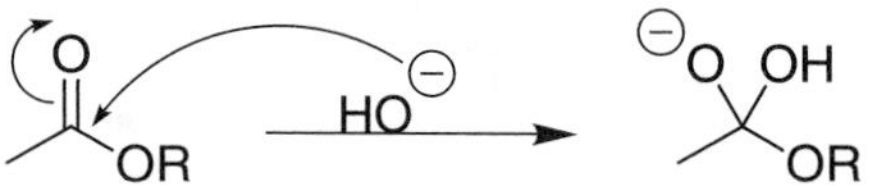

After the initial attack, the carbonyl could re-form to kick off the alkoxy group. This unwanted side reaction would hydrolyze the ester (we saw this reaction in the previous chapter):

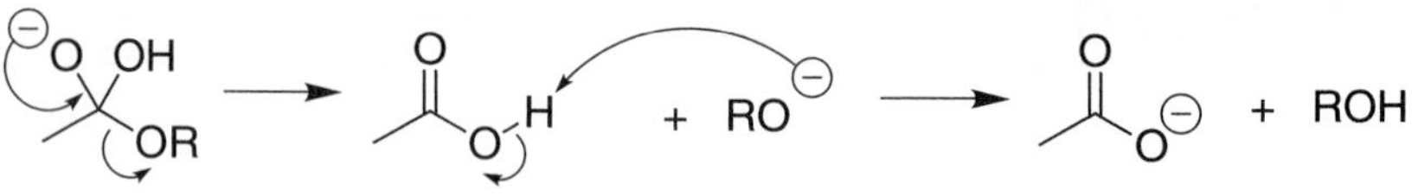

In order to avoid this unwanted side reaction, we use an alkoxide as our base. It is true that alkoxides can *also* function as nucleophiles, but think about what happens if the alkoxide attacks as a nucleophile:

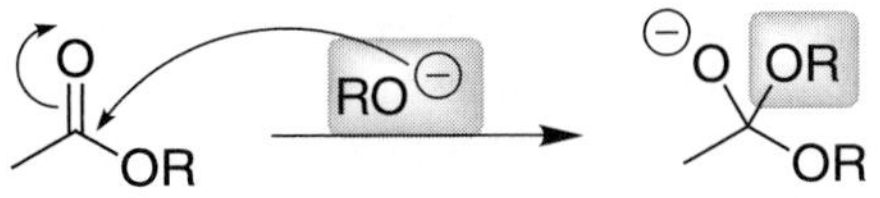

When the carbonyl re-forms, it doesn't matter which alkoxy group gets kicked off. Either way, we will regenerate the same ester that we started with:

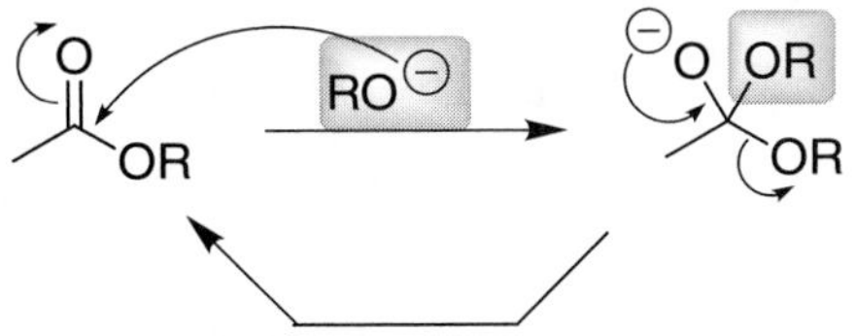

Although the alkoxide **can** attack the carbonyl, we don't have to worry about it because it does not actually lead to new products. So, we can avoid unwanted side reactions by using an alkoxide as the base for a Claisen condensation.

And we don't just use *any* alkoxide. We choose the alkoxide carefully. If we are dealing with a methyl ester, then we will use methoxide:

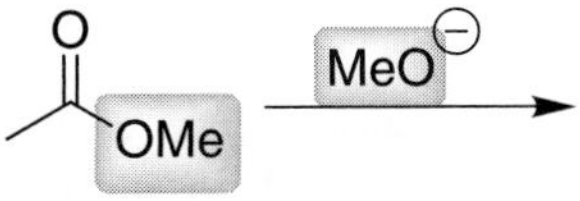

The reason for this is simple. If we were to use ethoxide in this case, this would actually change some of our ester:

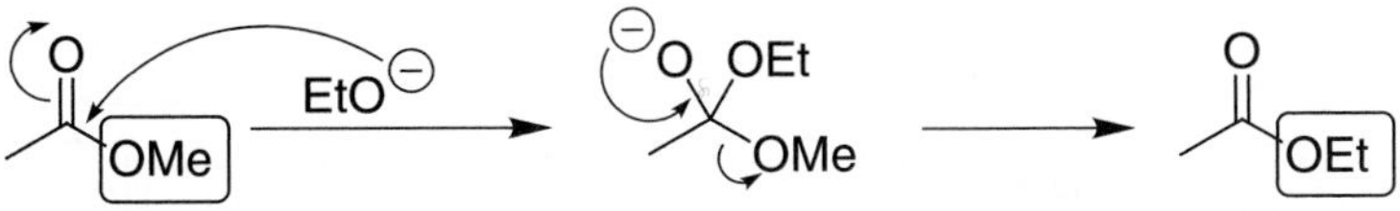

This is called trans-esterification, and we can avoid it by choosing our base to match the alkoxy group in the ester. That way we avoid unwanted side reactions. If we are dealing with an ethyl ester, then we just use ethoxide as our base:

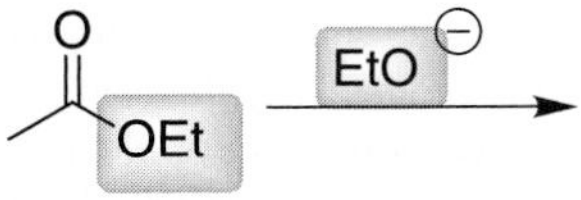

Now that we know what base to choose for a Claisen condensation, let's talk about another special role that the base plays in a Claisen condensation. We said that the final product is a β-keto ester. But remember that this reaction is run under basic conditions (in the presence of some alkoxide acting as a base). Under these conditions, the β-keto ester gets deprotonated to form an enolate that is especially stabilized:

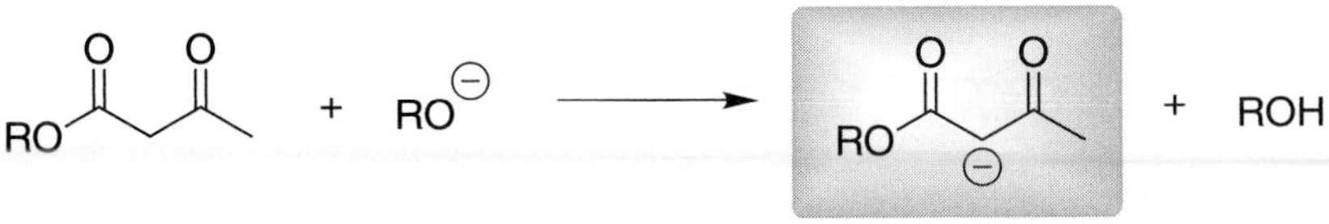

In the beginning of this chapter, we talked about the extra stability of this kind of "double" enolate. This enolate is even more stable than an alkoxide ion. That is an important point because it means that the reaction will favor formation of the product. Why? Because the reaction is converting alkoxide ions into enolate ions (which are more stable):

More stabilized negative charge

The formation of this stabilized enolate is a strong driving force pushing this reaction toward formation of products.

So, when the reaction is finished, we must protonate the enolate in order to get our product:

The Claisen condensation is important because it gives us a way to make β-keto esters. And we will soon see that there is a clever synthetic trick that you can do with β-keto esters. So, let's make sure that we have mastered the Claisen condensation.

Overall, here is what is happening:

We are removing *one* proton from an ester, and we are removing an alkoxy group from the other ester. The remaining fragments are then joined together. Notice that a small molecule is liberated in the process (ROH). That is why we call this reaction a Claisen *condensation*.

Now let's see if we can use this to predict some products. We will soon come back and master the mechanism. But for now, let's make sure that you train your eye to see the products of a Claisen in an instant. You will need that skill in order to master synthesis problems.

EXERCISE 7.51.   Predict the products of the following reaction:

Answer: We slam together two fragments, while kicking off EtOH, like this:

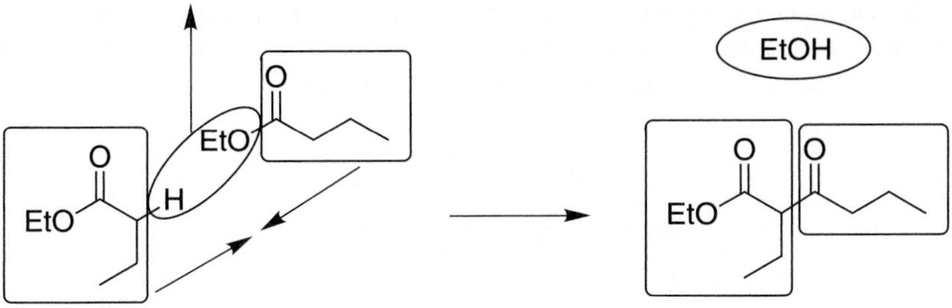

EtOH

Predict the products of the following reactions:

7.52.

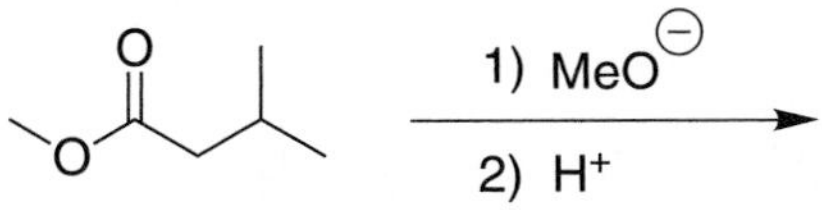

1) MeO$^{\ominus}$

2) H$^+$

7.53.

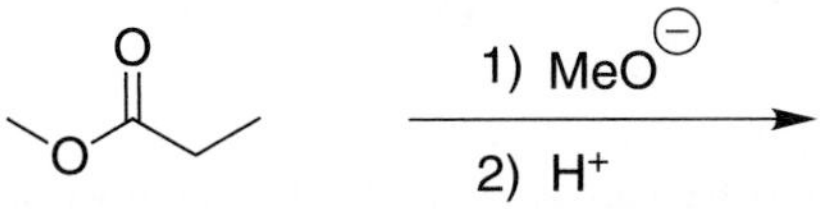

1) MeO$^{\ominus}$

2) H$^+$

7.54.

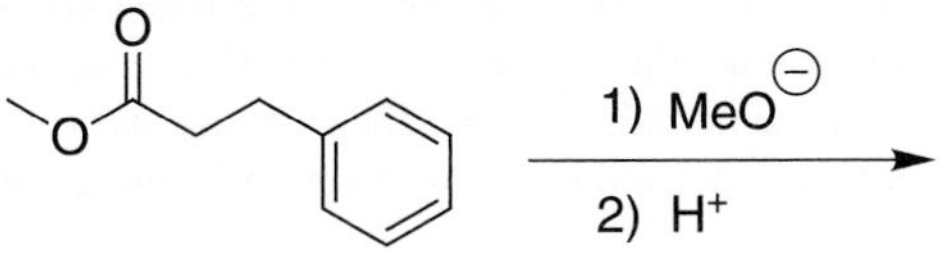

1) MeO$^{\ominus}$

2) H$^+$

7.55.

1) EtO$^{\ominus}$

2) H$^+$

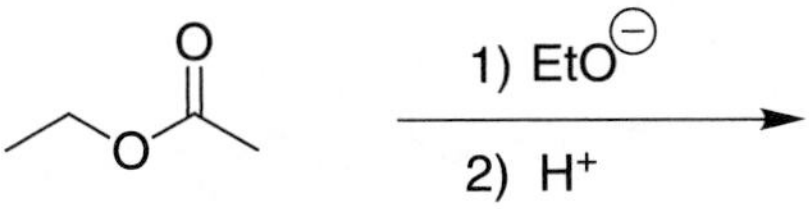

EXERCISE 7.56.   Consider the following compound:

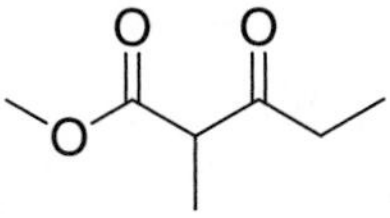

Using a Claisen condensation, what reagents would you use to prepare this compound?

Answer: We break the molecule apart into two fragments in order to insert MeOH. We break it apart between the α and β positions:

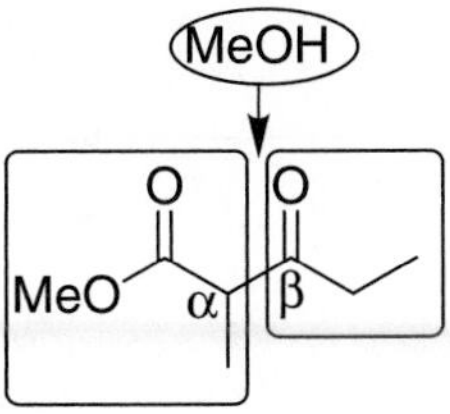

MeOH

And we just have to decide which fragment gets the methoxy group and which fragment gets the proton. The fragment on the left already has its alkoxy group, so that must be the fragment that gets the proton. The fragment on the right will get the alkoxy group:

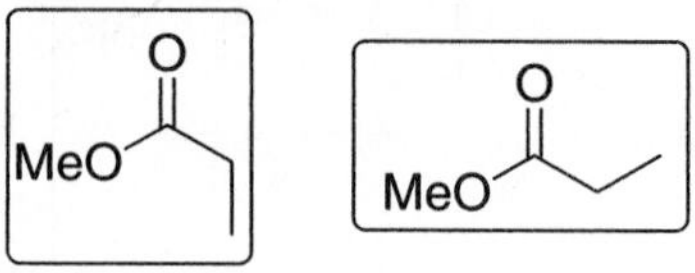

These two esters are identical, which is good. That means that we just need one kind of ester. We choose our base to match the alkoxy group (methoxy in this case), so our synthesis would look like this:

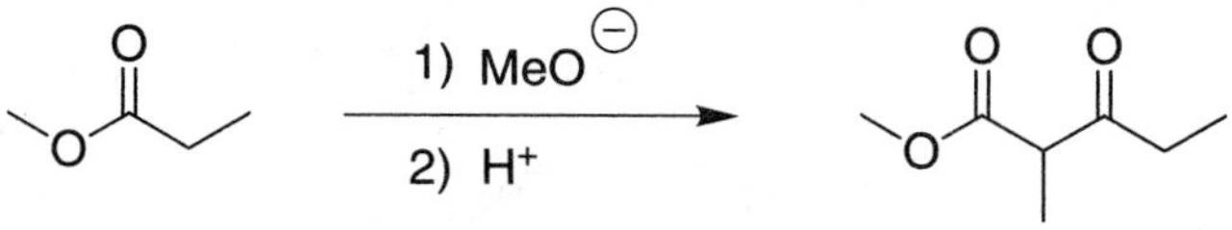

We can do crossed Claisen condensations (just as we can do crossed aldols), but we would have the same concerns as before. We would have to worry about potential side reactions. A crossed Claisen will work best when one of the esters has no alpha protons. You will see that some of the problems below are the products of crossed Claisen condensations. Keep an eye out for them.

For each of the compounds below, what reagents would you use to prepare the compound using a Claisen condensation:

7.57.

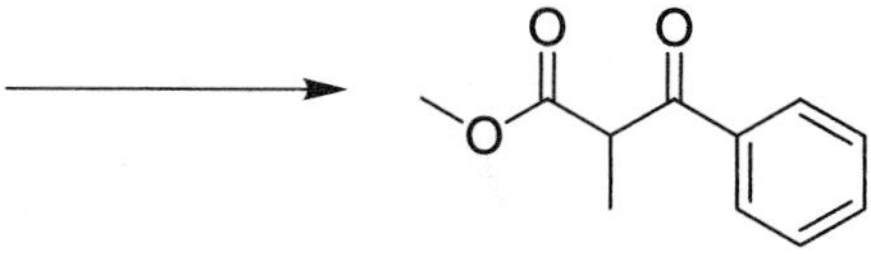

7.58.

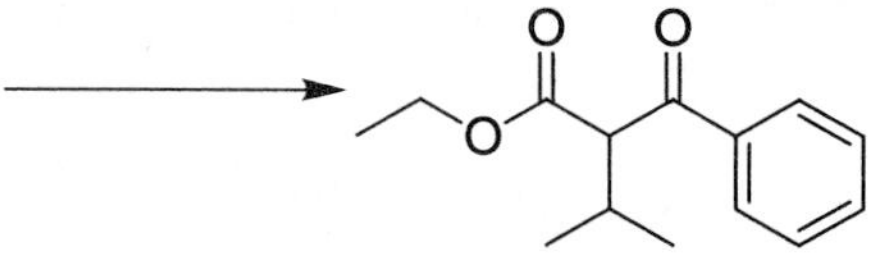

7.59.

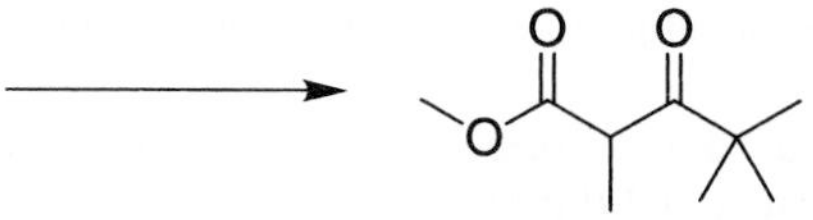

7.60.

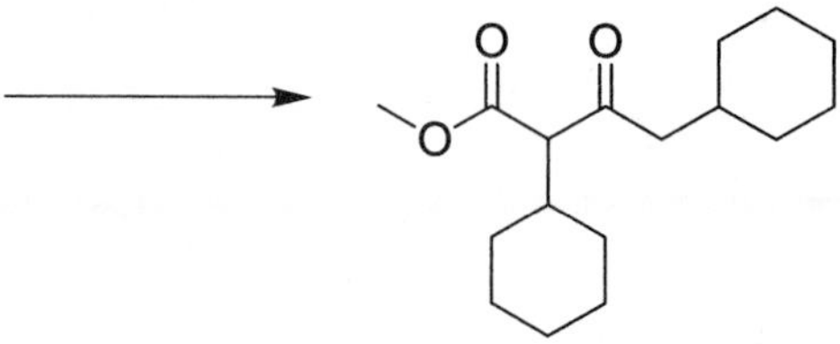

EXERCISE 7.61. Propose a mechanism for the following transformation:

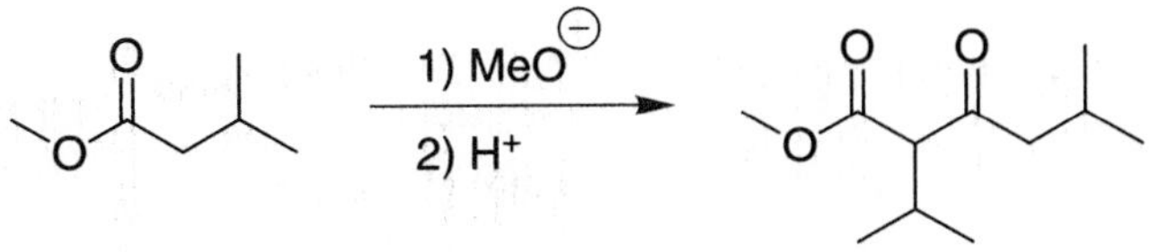

Answer: Methoxide functions as a base and deprotonates the ester:

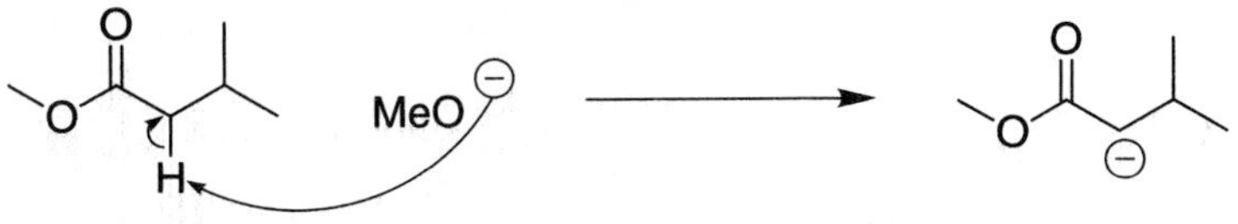

This ester enolate then attacks an ester to give the following intermediate:

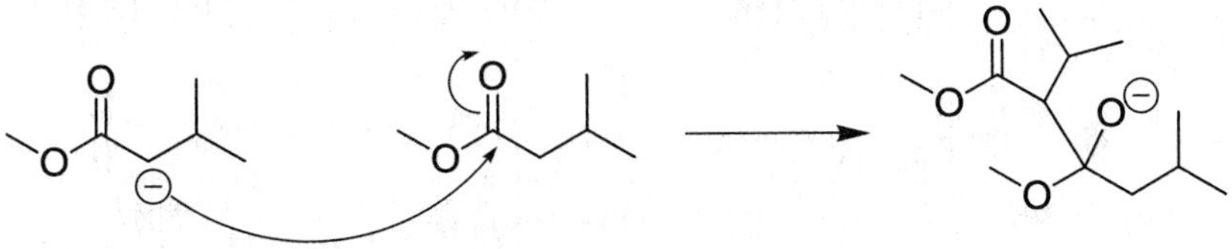

This intermediate then can re-form the carbonyl to kick off methoxide:

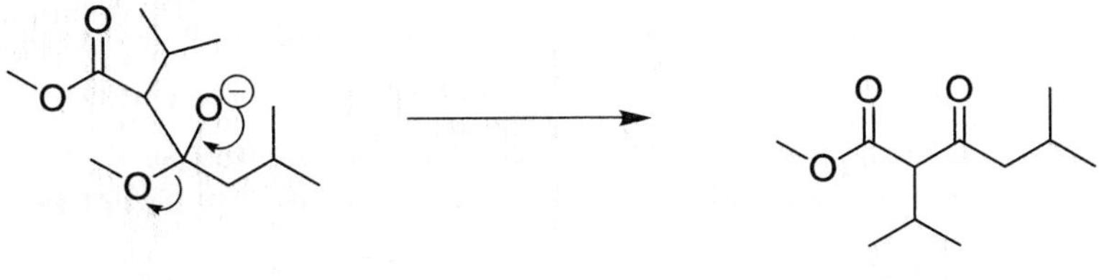

Under these basic conditions (methoxide), the β-keto ester is deprotonated:

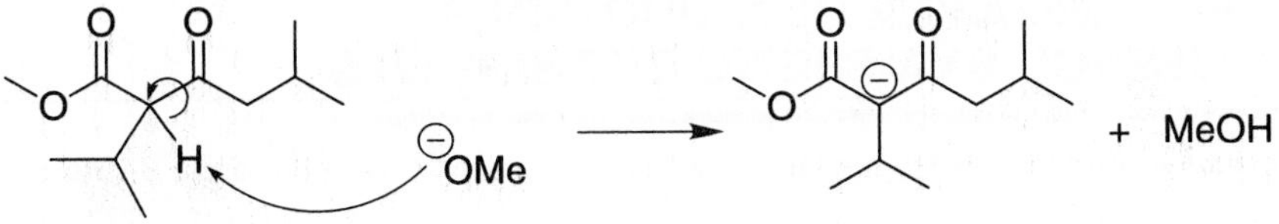

+ MeOH

This deprotonation step is important (even though we are about to give a proton right back in just a second) because the formation of this stabilized anion is the driving force for the reaction. That is why we must show this step. That explains why the reagents indicate that acid is added to the flask at the end of the reaction. We need a source of protons to form our final product:

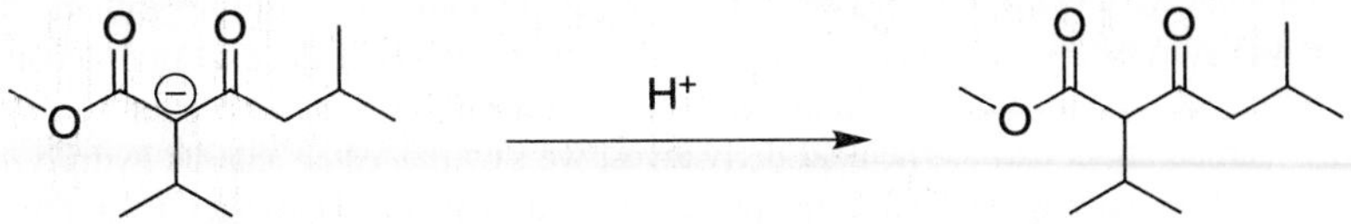

H+

Propose a mechanism for each of the following reactions. You will need a separate piece of paper to record your answers:

7.62.

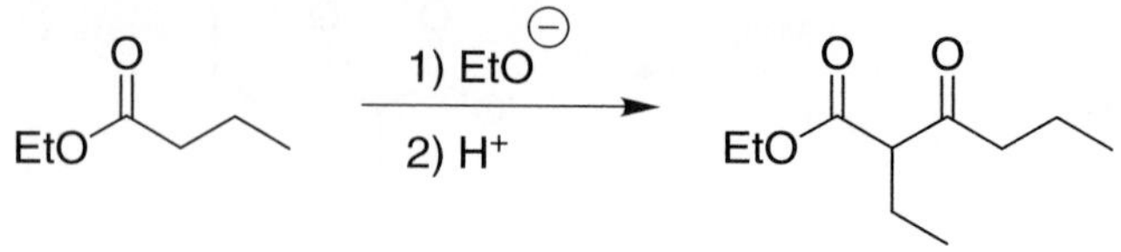

7.63.

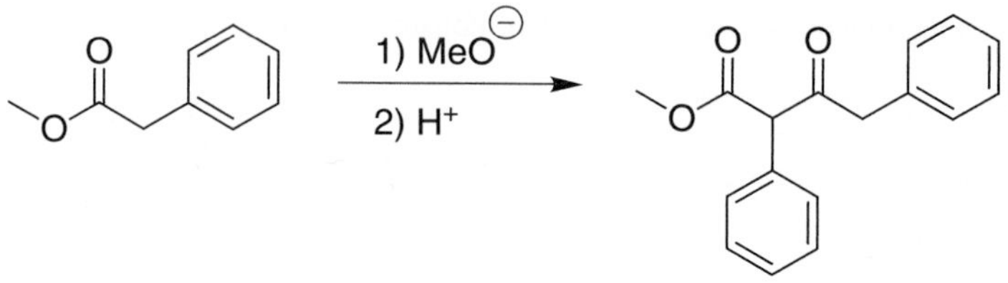

When you start with a di-ester as your starting material, it is possible to get an *intramolecular* Claisen condensation:

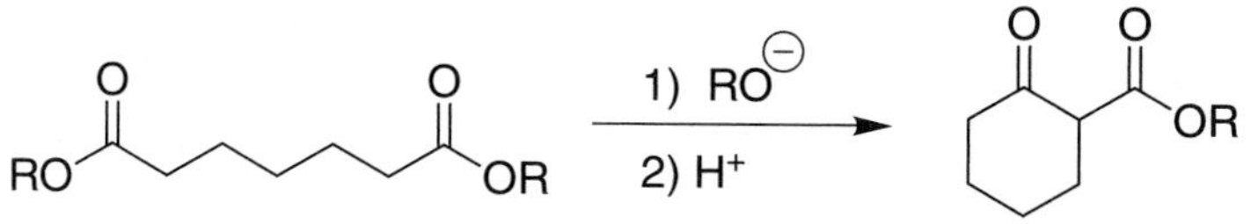

Notice once again that the product is just a β-keto ester. This reaction has its own name (the Dieckmann condensation), but it is really just an intramolecular Claisen condensation. Therefore, the steps of this mechanism are identical to the steps of a regular Claisen condensation. See if you can do it on your own.

7.64.   Propose a mechanism for the Dieckmann condensation (shown above). Try to do it without looking back at your previous work. You will need a separate piece of paper to record your answer.

7.8   DECARBOXYLATION PROVIDES SOME USEFUL SYNTHETIC TECHNIQUES

In the previous section we learned how to use a Claisen condensation to prepare a β-keto ester:

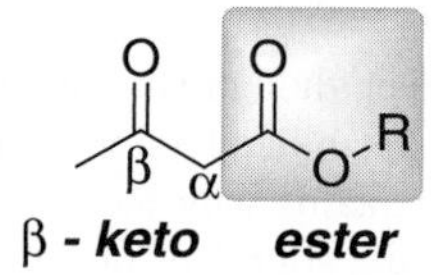

β - *keto*   *ester*

Now let's see what we can do with β-keto esters. There are some very useful synthetic techniques that start with β-keto esters. In order to see how they work, we will need to remind ourselves of one reaction that we saw a couple of chapters ago.

When we learned about carboxylic acid derivatives, we saw that esters can be hydrolyzed to give carboxylic acids. We can use the exact same process to hydrolyze a β-keto ester, as in the following:

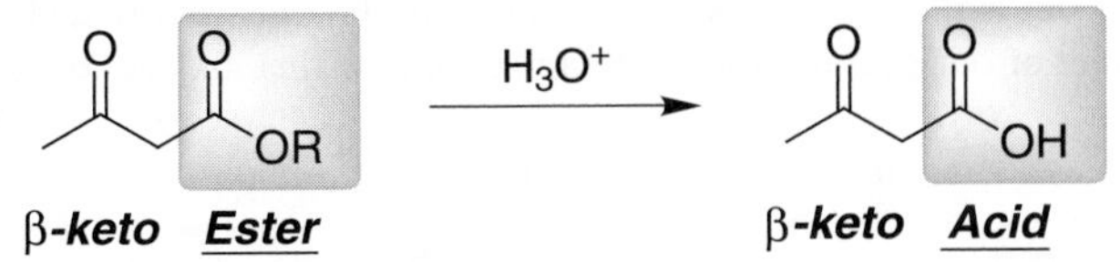

β-*keto* *Ester*   β-*keto* *Acid*

And the product we get is a β-keto acid.

β-keto acids do something very interesting when you heat them. The carboxyl group gets blasted away:

This carboxyl group
gets blasted away

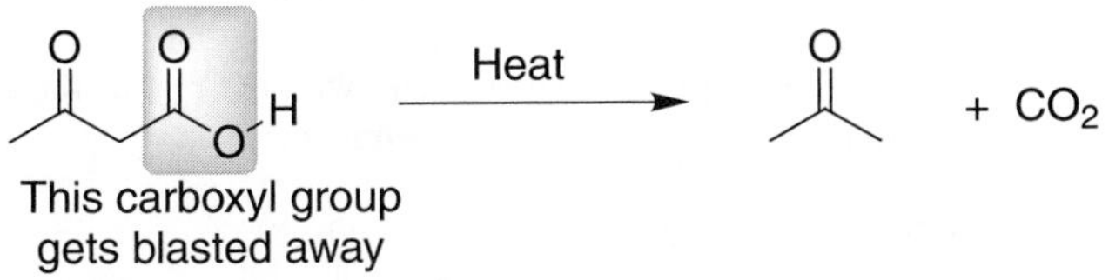

We call this a *decarboxylation*. This reaction is the basis for the synthetic techniques we will learn in this section, so let's make sure we understand how a decarboxylation happens. Let's take a close look at the mechanism.

In the first step, we get a pericyclic reaction, which liberates CO_2 as a gas:

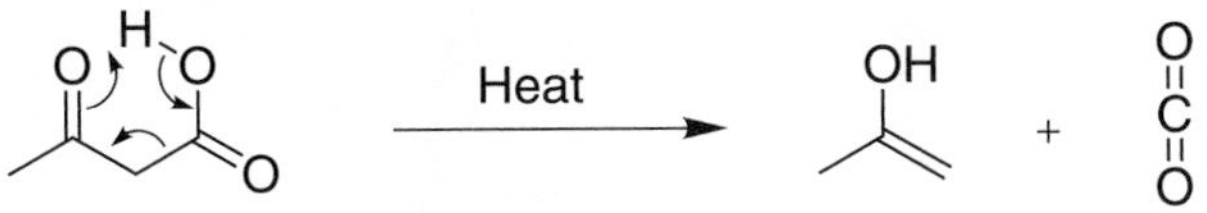

Pericyclic reactions are characterized by a ring of electrons moving around in a circle. There are many kinds of pericyclic reactions (including the Diels-Alder reaction, which you probably saw last semester). Pericyclic reactions truly deserve their own chapter, and unfortunately, most textbooks do not devote an entire chapter to pericyclic reactions (references to them are just scattered throughout the text). Perhaps your instructor will spend some time on pericyclic reactions. We will not cover them right now, because we must continue with the topic at hand.

In the reaction above, we liberate CO_2 gas (that's how the carboxyl group gets blasted away), and we generate a compound that is an enol. And we know that enols will quickly tautomerize to give ketones:

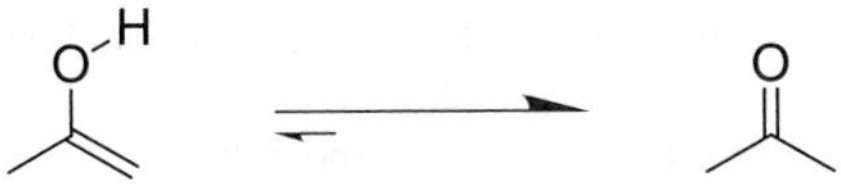

So, when we heat a β-keto acid, we blast away the carboxyl group, and we end up with a ketone.

Now consider what we have just done. We took a β-keto ester (which is the product of a Claisen condensation), and we hydrolyzed it to produce a β-keto *acid*. Then, we heated this compound, and we blasted away the carboxyl group:

β-*keto* *Ester*   β-*keto* *Acid*   *Ketone*

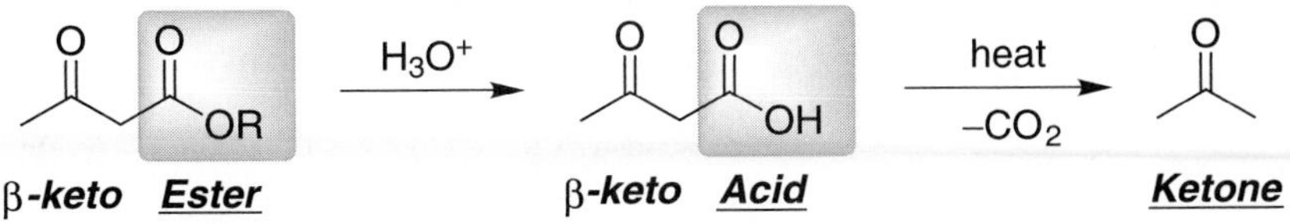

In the end, our product is a ketone. To see why this so useful, we need to add just one more step at the very beginning of our overall process. Imagine that we first alkylate our β-keto ester:

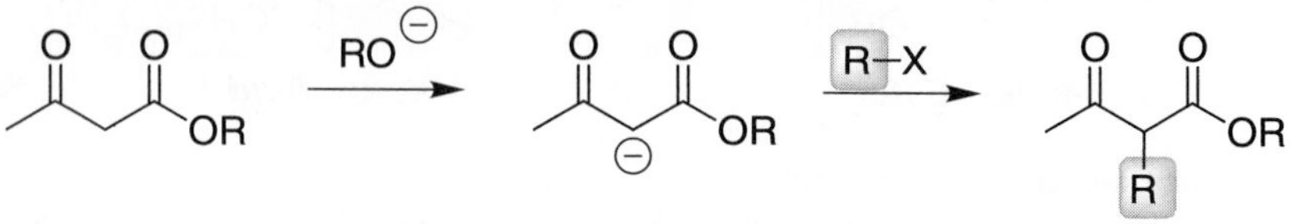

We have already seen this kind of reaction before (section 7.5). It is just an alkylation. We used an alkoxide ion to produce a very stabilized enolate, which then attacks the alkyl halide in an S_N2 reaction.

So, if we start with an alkylation, and then we continue with the rest of the strategy (hydrolysis, followed by decarboxylation), we get the following sequence:

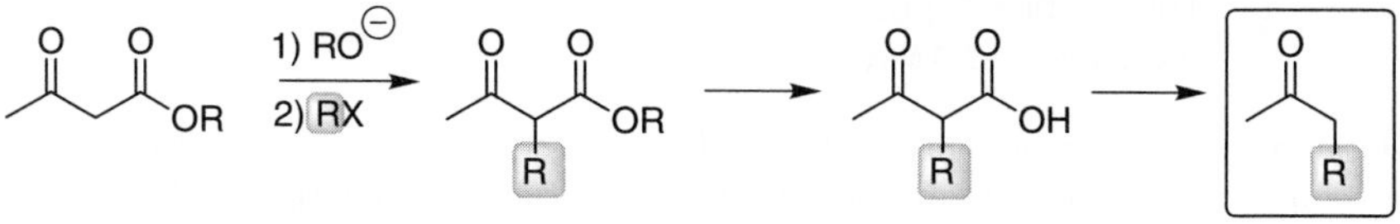

Now take a close look at the product. This compound is a substituted derivative of acetone (acetone is the common name for dimethyl ketone):

Acetone   **A substituted
derivative of acetone**

This provides us with a way to make a wide variety of substituted derivatives of acetone.

This is useful because we run into trouble if we try to alkylate acetone directly:

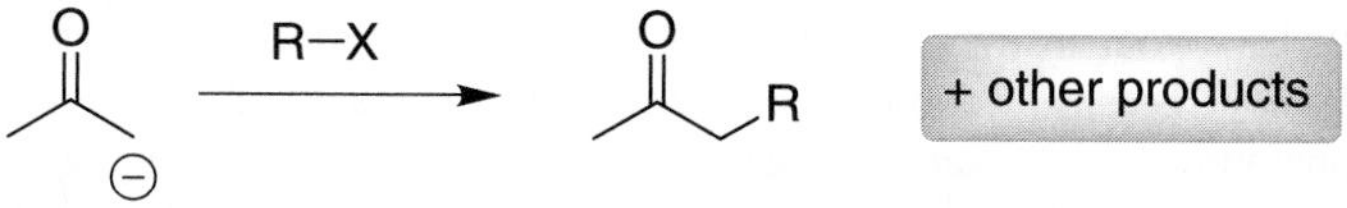

+ other products

We get our desired product together with a mixture of other side products (from polyalkylation and from elimination reactions). The strategy we have learned provides a clean way to make substituted derivatives of acetone. But be careful—remember that the alkylation step is an S_N2 reaction, which means you can only use primary or secondary alkyl halides (primary is better). In other words, you *could* use this strategy to prepare the following compound:

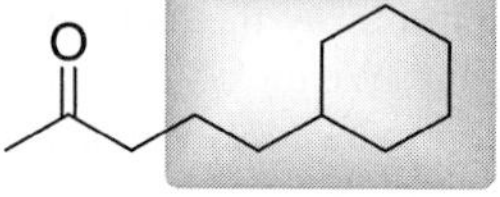

But you could *not* use this synthetic strategy to make this compound:

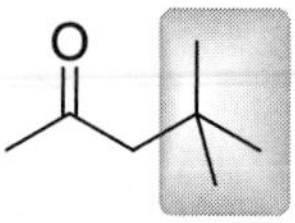

because that would have required an alkylation step involving a tertiary alkyl halide, which would not work.

In order to use this synthetic strategy, we will always have to start with the following compound:

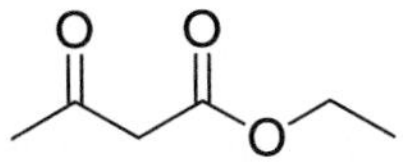

This compound, called ethyl acetoacetate, belongs to a class of compounds called acetoacetic esters. Therefore, we call our strategy the ***acetoacetic ester synthesis***.

To summarize what we have seen, the acetoacetic ester synthesis has three main steps: alkylate, hydrolyze, and then decarboxylate. Say that ten times real fast.

Now let's get some practice using this synthetic strategy.

EXERCISE 7.65. Starting with ethyl acetoacetate, show how you would prepare the following compound:

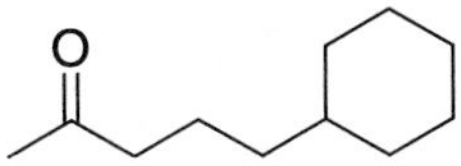

Answer: Remember that the acetoacetic ester synthesis has the following steps: alkylate, hydrolyze, and then decarboxylate. So, in order to solve this problem, we only have to determine the alkyl group that is needed. To do that, we look at the compound like this:

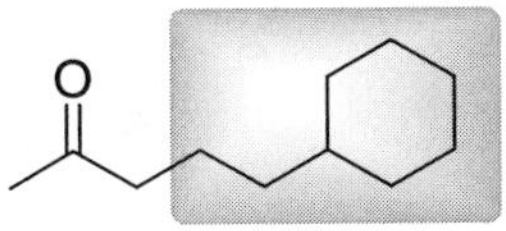

So, we will need the following alkyl halide:

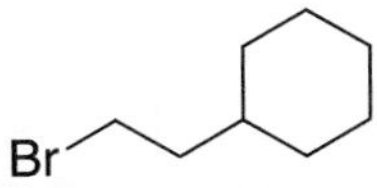

Now that we have determined what alkyl halide to use, we are ready to propose our synthesis:

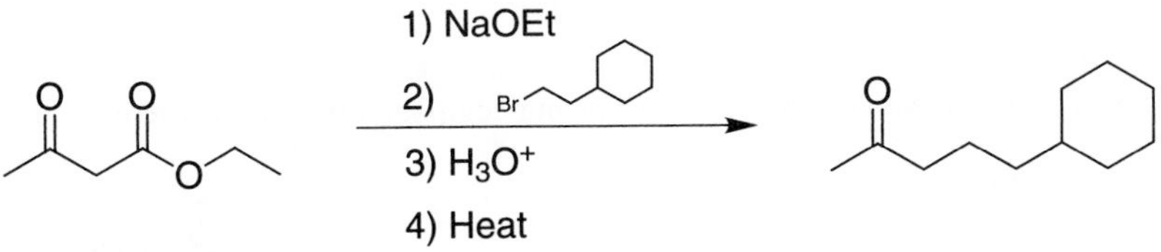

1) NaOEt
2) Br~
3) H₃O⁺
4) Heat

$$\text{1) NaOEt} \quad \text{2) } Br \quad \text{3) } H_3O^+ \quad \text{4) Heat}$$

Show how you would prepare each of the following compounds from ethyl acetoacetate:

7.66.

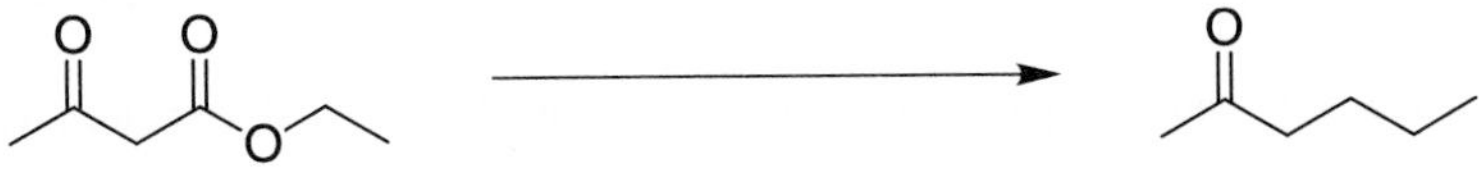

7.67.

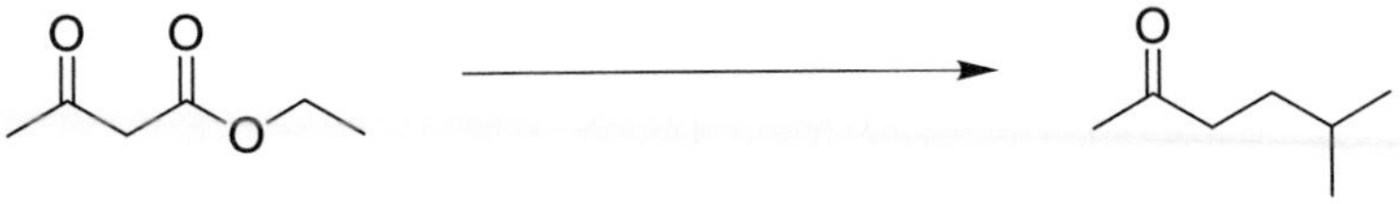

7.68.

7.69.   The following compound ***cannot*** be made with an acetoacetic ester synthesis.

Why not? (*Hint:* Think about what alkyl halide you would need)

In all of the problems we have done so far, we have focused on alkylating ***once.*** But it is also possible to alkylate twice, which would give us a product with two alkyl groups:

And the R groups don't even have to be the same. We would do it like this:

7.70.   Show how you would prepare the following compound from ethyl acetoacetate:

7.71.   Show how you would prepare the following compound from ethyl acetoacetate:

7.72. Propose a synthesis for the following transformation:

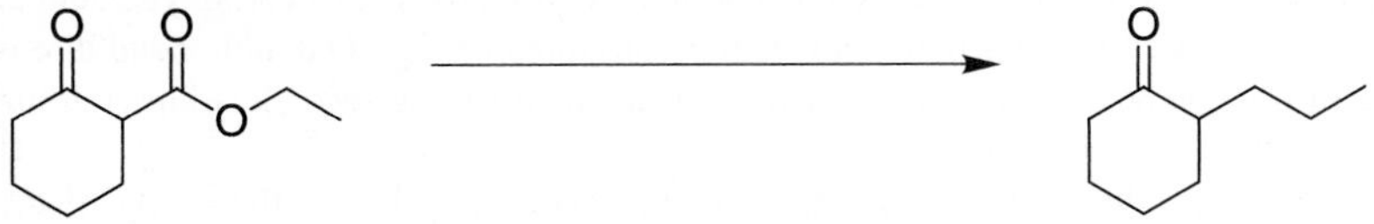

Hint: This reaction is similar to an acetoacetic ester synthesis,
but we are just starting with a different β-keto ester.

There is another common synthetic strategy that utilizes the same concepts as the acetoacetic ester synthesis. So, let's now focus on this other strategy. It is called the *malonic ester synthesis* because the starting material is a malonic ester (called diethyl malonate):

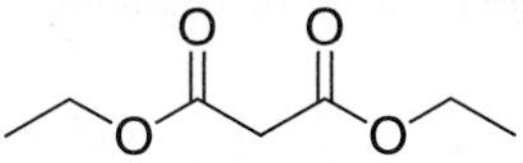

We follow the same three steps that we followed in our previous strategy: alkylate, hydrolyze, and then decarboxylate. The only difference is that we start with a slightly different starting material (malonic ester, instead of acetoacetic ester), and therefore, our product will be slightly different. Compare the structures of ethyl acetoacetate and diethyl malonate:

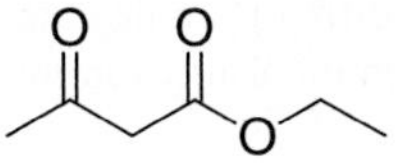

Ethyl acetoacetate 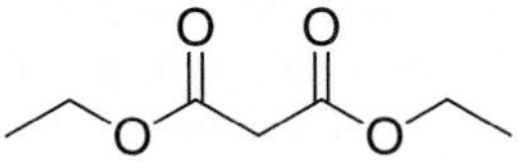   **Diethyl malonate**

Notice that diethyl malonate has *two* carboxyl groups (as opposed to ethyl acetoacetate, which has one carboxyl group and one carbonyl group). To see how this extra carboxyl group affects the structure of our end-product, let's go through the three steps: alkylate, hydrolyze, and then decarboxylate. We start with an alkylation:

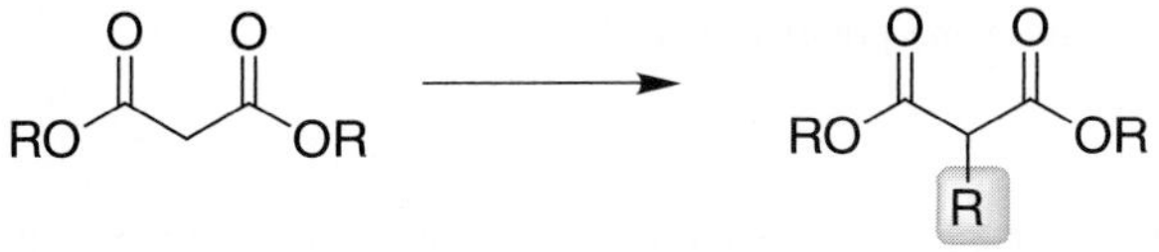

Then, we hydrolyze:

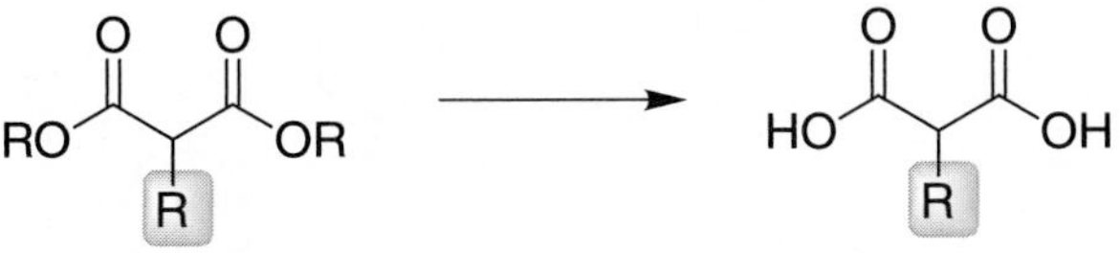

Notice that *both* sides get hydrolyzed.
Then, finally, we decarboxylate:

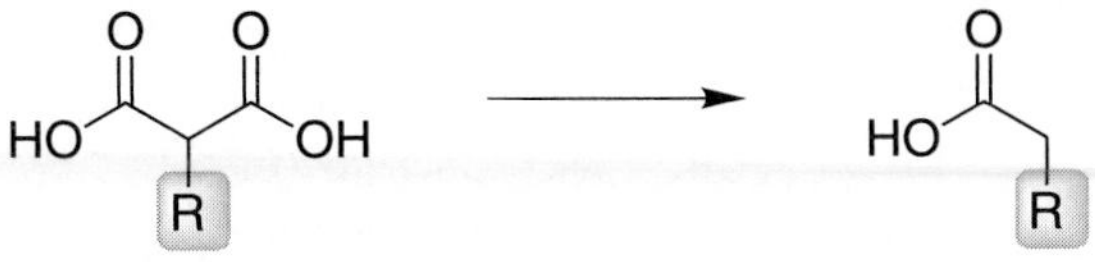

Only one side gets decarboxylated. Remember how a decarboxylation works. It was a pericyclic reaction that you get when you have a C=O double bond that is β to a carboxylic acid group. After you blast the first carboxyl group away, there is no longer a C=O double bond that is β to the remaining carboxylic acid group. Try to draw a mechanism for the second carboxyl group leaving, and you will find that you can't do it.

Notice that the product is now a substituted carboxylic acid. This is the power of the malonic ester synthesis: it allows you to make a wide variety of substituted carboxylic acids:

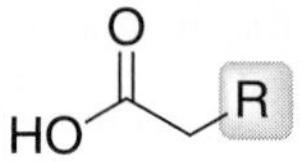

We can also use this synthesis to put on *two* alkyl groups (just as we did with the acetoacetic ester synthesis). We would just alkylate twice at the beginning of our procedure:

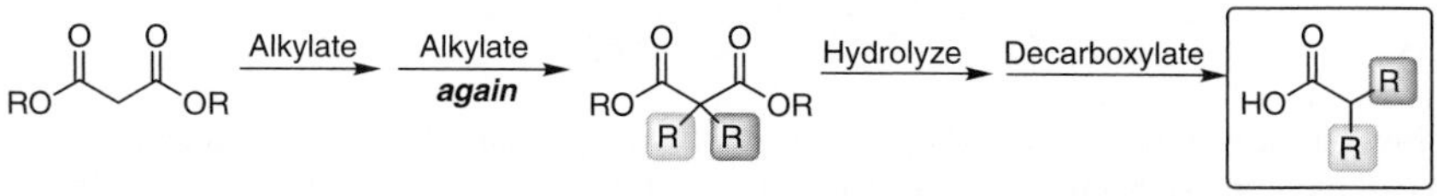

But once again, the R groups must be primary or secondary (primary is best) because alkylation is an S_N2 process.

This strategy is very useful because it would be very difficult to alkylate a carboxylic acid directly. If we try to alkylate a carboxylic acid directly, we immediately run into an obstacle because we cannot form an enolate of a carboxylic acid:

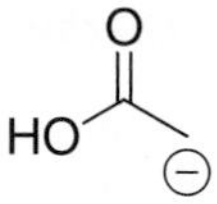

Cannot form
this enolate

You cannot form an enolate in the presence of an acidic proton (from the carboxylic acid). So the malonic ester synthesis gives us a way around this obstacle. It gives us a way to make substituted carboxylic acids. Let's get some practice with this:

EXERCISE 7.73. Starting with diethyl malonate, show how you would prepare the following compound:

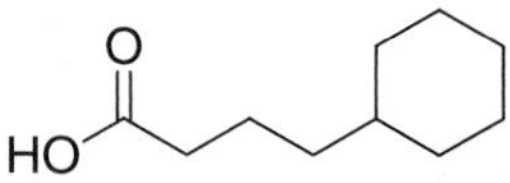

Answer: Remember that the malonic ester synthesis has the following steps: alkylate, hydrolyze, and then decarboxylate. In order to solve this problem, we only have to determine the alkyl group that is needed. And to do that, we look at the compound like this:

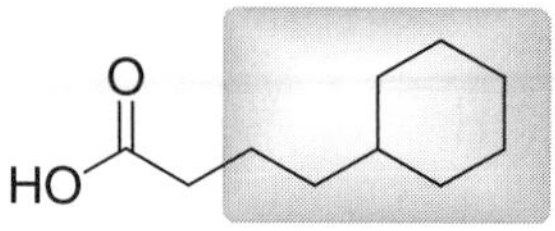

So, we will need the following alkyl halide:

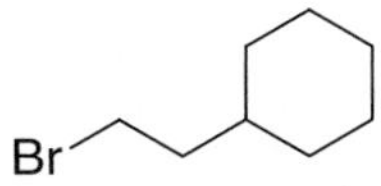

Now that we have determined what alkyl halide to use, we are ready to propose our synthesis:

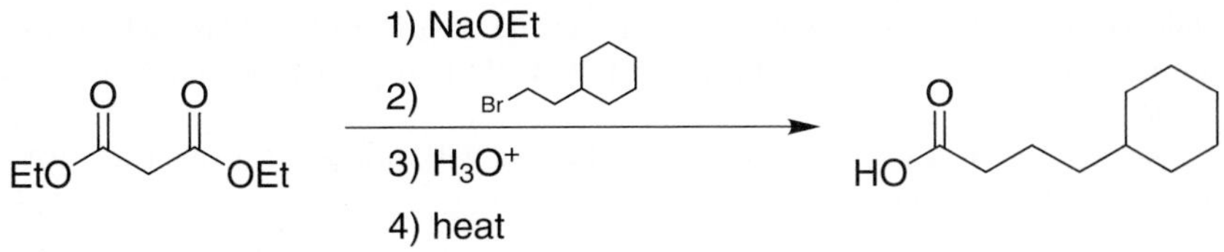

1) NaOEt
2) (alkyl halide shown)
3) H₃O⁺
4) heat

Show how you would prepare each of the following compounds from diethyl malonate:

7.74.

7.75.

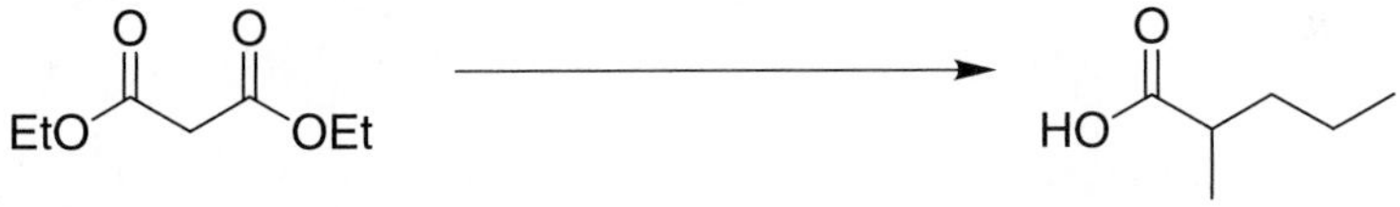

7.76.

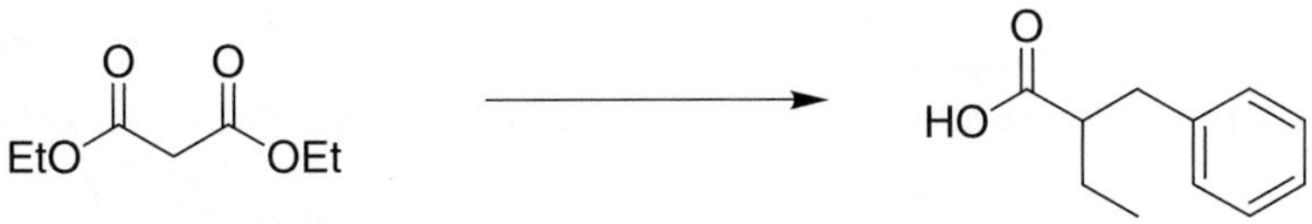

7.9 MICHAEL REACTIONS

In this chapter, we have seen that enolates can attack a wide variety of electrophiles. We started the chapter with enolates attacking halogens, and then we looked at enolates attacking alkyl halides. We also saw that enolates can attack ketones or esters. In this section, we will conclude our discussion of enolates by looking at a special kind of electrophile that enolates can attack. Consider the following compound:

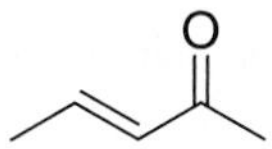

This compound is an α,β-unsaturated ketone (it is the product of an aldol condensation, remember?). This compound is a special kind of electrophile. To understand why it is special, let's take a close look at the resonance structures:

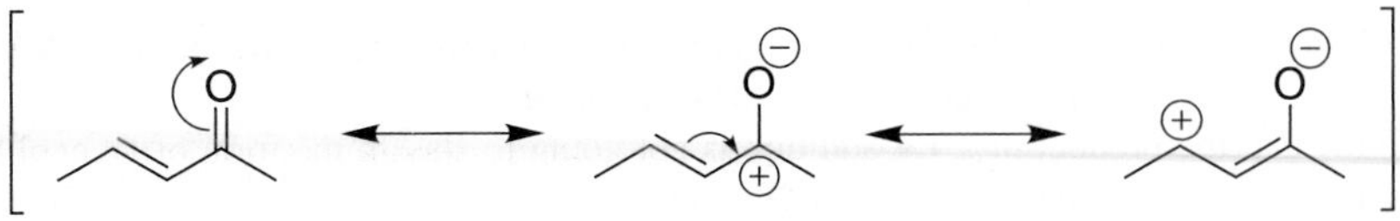

These resonance structures paint the following picture for us:

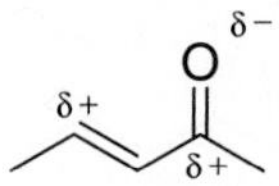

We see that there are **two** electrophilic centers. We already knew that the carbonyl itself was electrophilic, but now we can appreciate that the β position is also electrophilic. That means that any nucleophile has two choices when it attacks. (1) It can attack at the carbonyl (as we have seen many times already), **or** (2) it can attack at the β position. Let's look at both possibilities, and we will compare the products.

If the nucleophile attacks the carbonyl group, then we will form a negative charge that grabs a proton to give us our product:

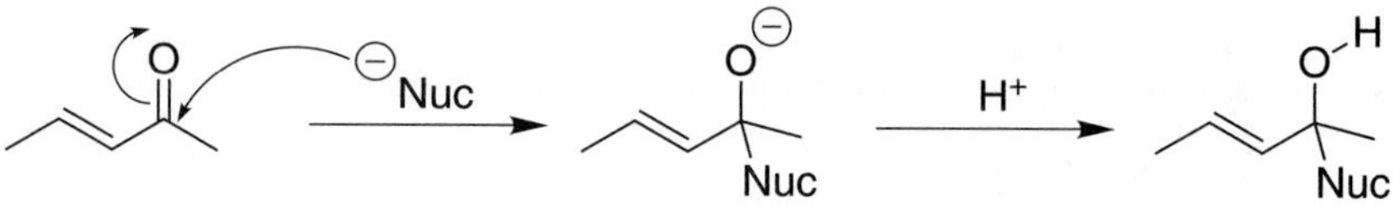

Notice that we had a π system that spanned four atoms, and we added the nucelophile and H to positions 1 and 2:

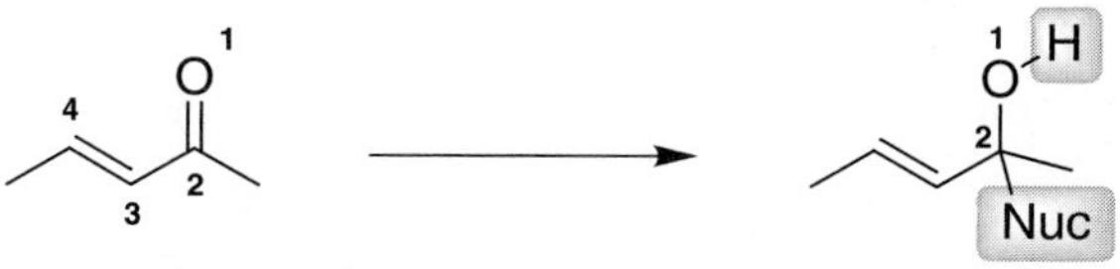

Therefore, we call this a **1,2-addition**.

But what happens when we attack the β position? The initial intermediate is an enolate:

An enolate

Then, when this enolate grabs a proton, it forms an enol:

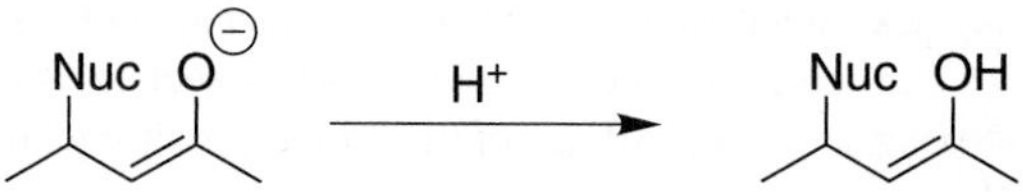

An enol

Once again, we have added the nucleophile and H across the π system. But this time, we have added them across the ends of this system:

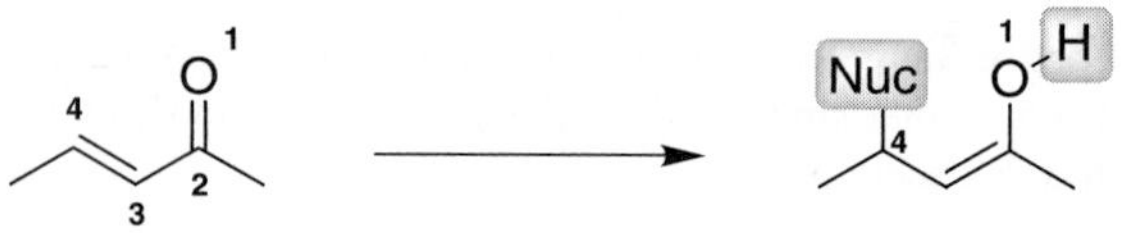

So, we call this a **1,4-addition**. Chemists have given this reaction other names as well. A 1,4-addition is often called a **conjugate addition**, or a **Michael addition**.

We know that the product of a 1,4-addition is not going to stay in the form of an enol because we know that an enol will tautomerize to form a ketone:

When you look at this ketone, it is hard to see why we call it a 1,4-addition. After all, it looks like the nucleophile and the H have added across the C=C double bond:

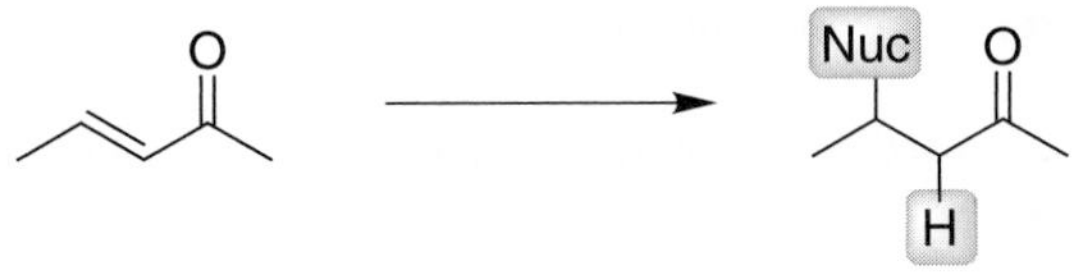

You need to draw the entire mechanism (as we did just a moment ago) in order to see why we call it a 1,4-addition.

Now that we know the difference between a 1,2 addition and a 1,4-addition, let's take a look at what happens when our attacking nucleophile is an enolate.

If you mix an α,β-unsaturated ketone together with an enolate in a flask, you will get a mixture of products. Not only do we get both possible attacks (at the carbonyl or at the β position), but it gets even more complicated. The product of the 1,4 addition is a ketone, which can get attacked again by an enolate. You can get crossed aldol condensations and all sorts of unwanted products. So, we can't use an enolate to attack an α,β-unsaturated ketone. The enolate is simply too reactive, and we get a mess of products.

The way around this problem is to create a more stabilized enolate. A more stable enolate will be less reactive, and therefore, it will be more selective in what it reacts with. But how do we make a more stabilized enolate? We have actually already seen this earlier in this chapter. Consider the following enolate:

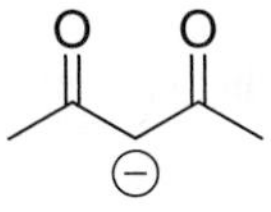

We argued that this enolate is more stable than a regular enolate because the negative charge is delocalized over *two* carbonyl groups. If we use this enolate to attack an α,β-unsaturated ketone, we find that the predominant reaction is a 1,4-addition:

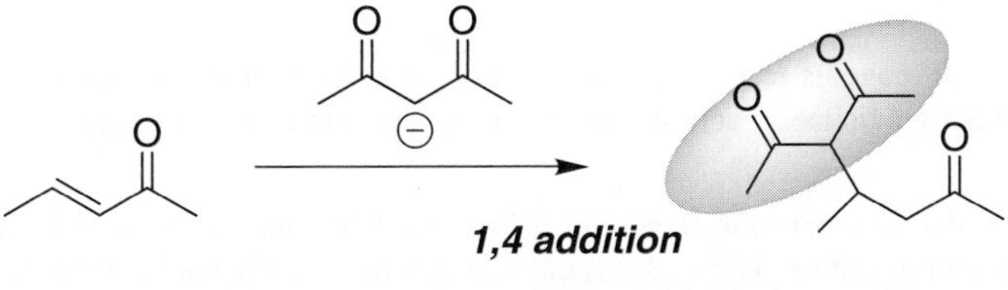

1,4 addition

We said earlier that a 1,4-addition is also called a Michael addition. In order to get a Michael addition, you need to have a stabilized nucleophile, like the stabilized enolate shown in the reaction above. This stabilized enolate is called a *Michael donor*. There are many other examples of Michael donors. Look at them carefully because you will need to recognize them as being Michael donors when you see them in problems:

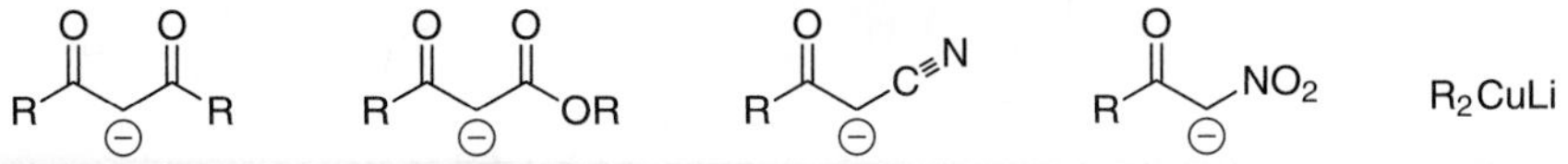

All of these nucleophiles are stabilized enough to be Michael donors.

In any Michael reaction, there is always a Michael donor *and* a Michael acceptor:

Nuc⊖

Michael donor   **Michael acceptor**

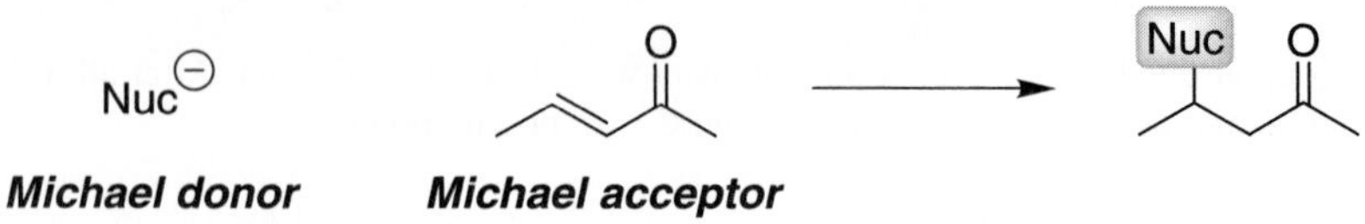

In this reaction, the Michael acceptor is an α,β-unsaturated ketone. But other compounds can also function as Michael acceptors:

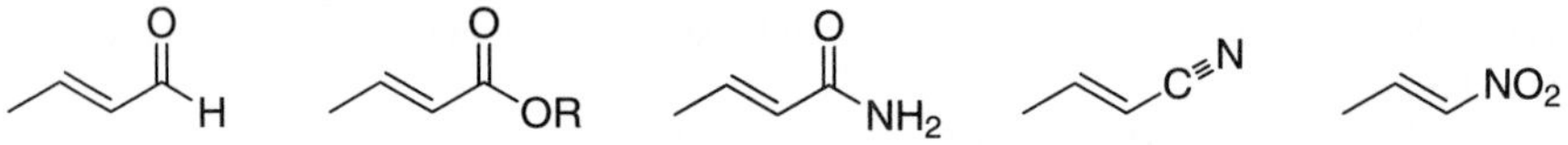

You can use any Michael donor to attack any Michael acceptor. For example, the following reaction would also be called a Michael reaction:

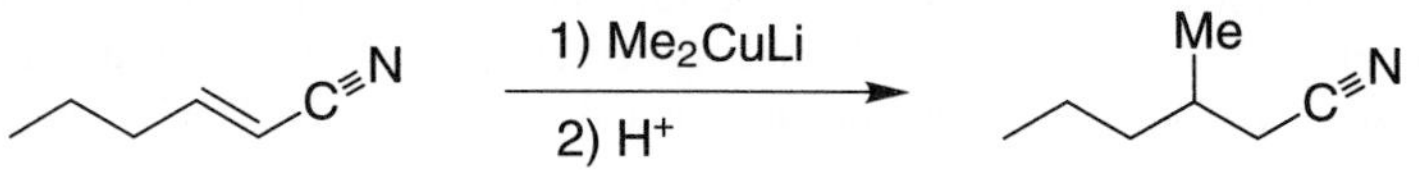

1) Me₂CuLi

2) H⁺

In order to do the next set of problems, you will need to review the lists of Michael donors and acceptors.

EXERCISE 7.77. Consider the following reagents:

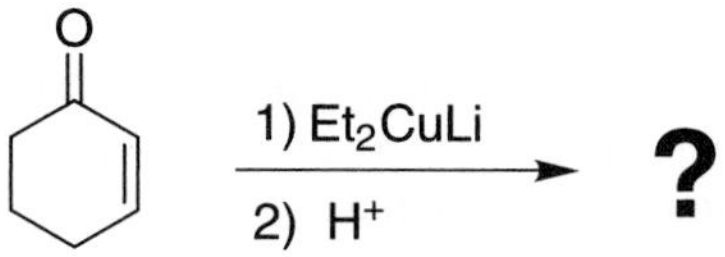

1) Et₂CuLi

2) H⁺

?

Do you expect these reagents to give a clean Michael reaction? (or do you expect a mess of products?) If you expect a Michael reaction, then draw the product that you expect.

Answer: In order to determine if we get a Michael reaction, we need to look for a Michael donor and a Michael acceptor. The α,β-unsaturated ketone is a Michael acceptor, and the lithium dialkyl cuprate is a Michael donor. So we do expect the following Michael reaction:

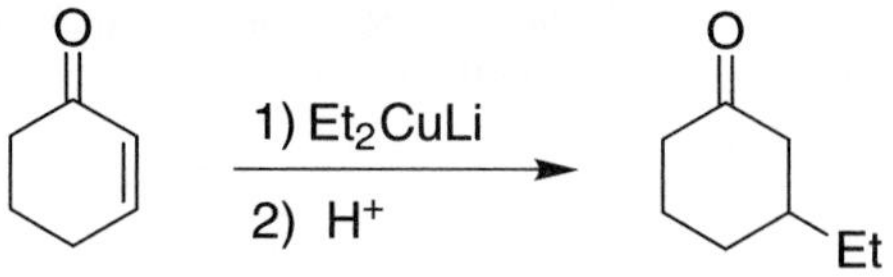

1) Et₂CuLi

2) H⁺

Et

Stare at this transformation for a few moments. It might be helpful to you in a synthesis problem one day. . . .

For each of the reactions below, determine whether you expect a clean Michael reaction. If so, then draw the product you expect. If you do not expect a clean Michael reaction, then you do not need to predict the products.

7.78.

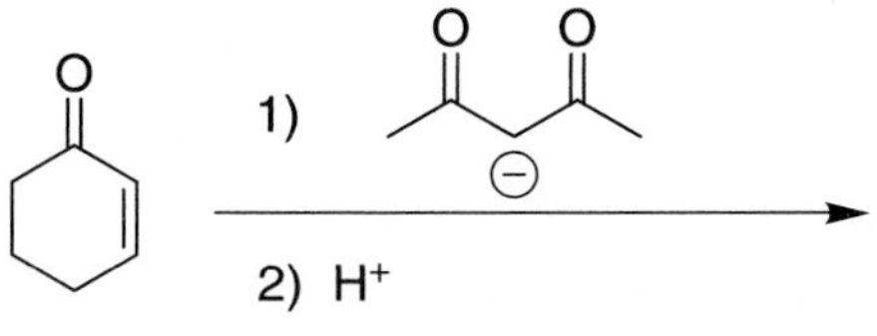

1)

2) H⁺

7.79.

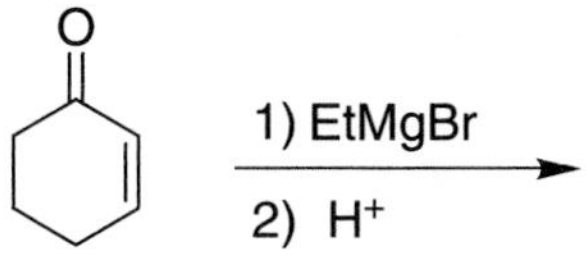

1) EtMgBr

2) H⁺

7.80.

1) Me₂CuLi

2) H⁺

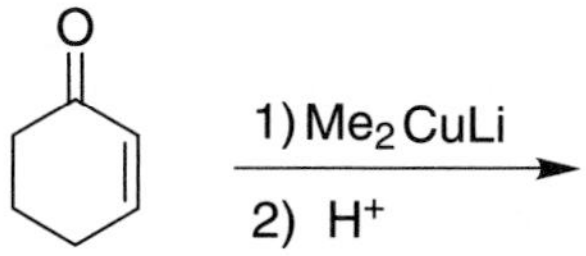

There is one more Michael donor that requires special mention. Enamines are very special Michael donors because they provide us with a useful synthetic strategy.

When we learned about ketones and aldehydes (Chapter 5), we saw that you can prepare an enamine by reacting a ketone with a secondary amine, under the following conditions:

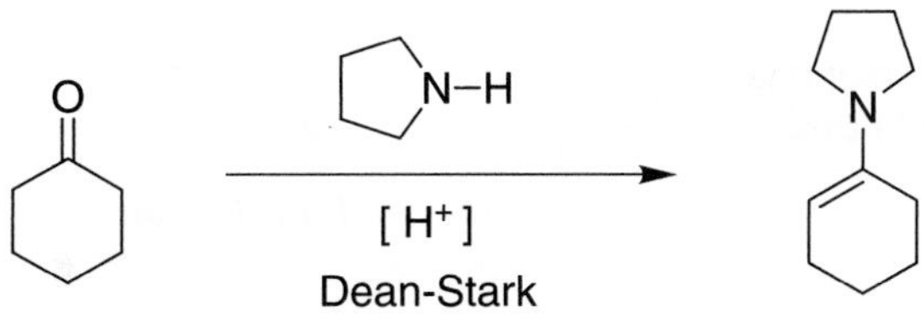

N–H

[H⁺]
Dean-Stark

This was our way of preparing an enamine. To understand how an enamine can function as a Michael donor, let's take a close look at the resonance structures:

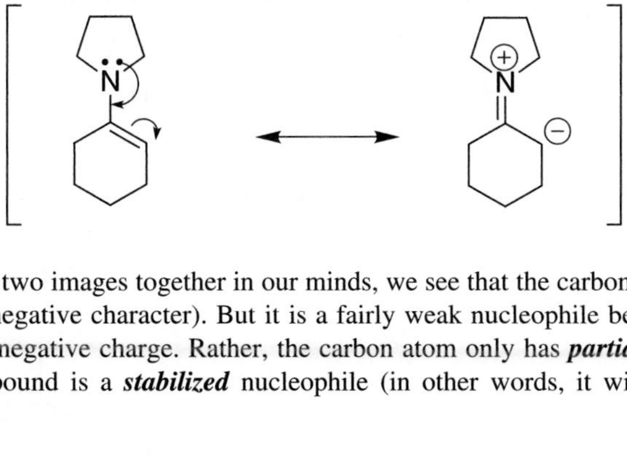

When we meld these two images together in our minds, we see that the carbon atom is nucleophilic (it has some partial negative character). But it is a fairly weak nucleophile because the compound does not have a *full* negative charge. Rather, the carbon atom only has *partial* negative character. Therefore, this compound is a *stabilized* nucleophile (in other words, it will be selective in its reactivity). This means we must add it to our list of Michael donors. If we use an enamine as a nucleophile to attack a Michael acceptor, we get a reaction like this:

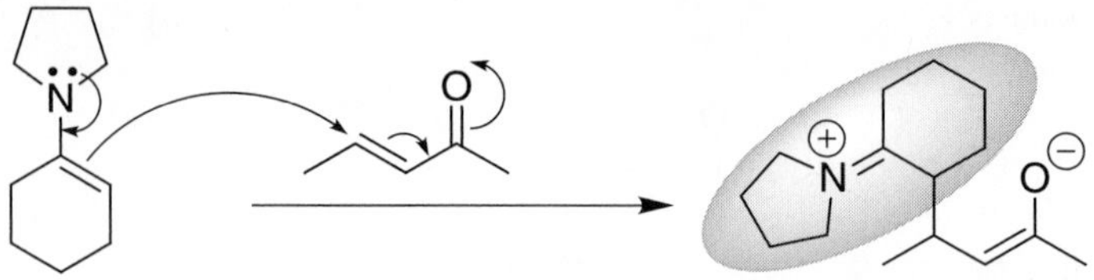

And then, the resulting iminium ion can be pulled off by adding water under acidic conditions (and under these conditions, the enolate gets protonated to form an enol, which tautomerizes to form a ketone):

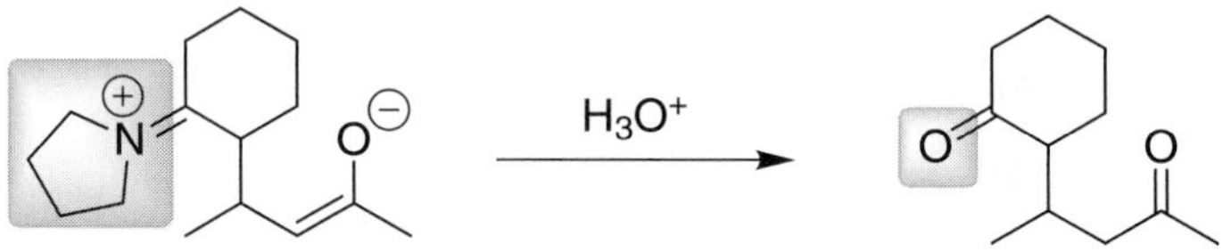

But why is this so important? Why am I singling out this one Michael donor, and why are we learning about enamines again? To understand the usefulness of enamines here, let's imagine that we wanted to do the following transformation:

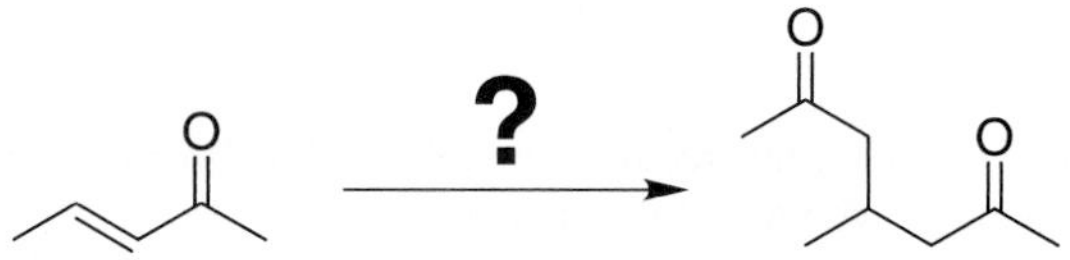

You decide that it should be simple. Your plan is to use a nucleophile that can attack the α,β-unsaturated ketone in a 1,4 addition:

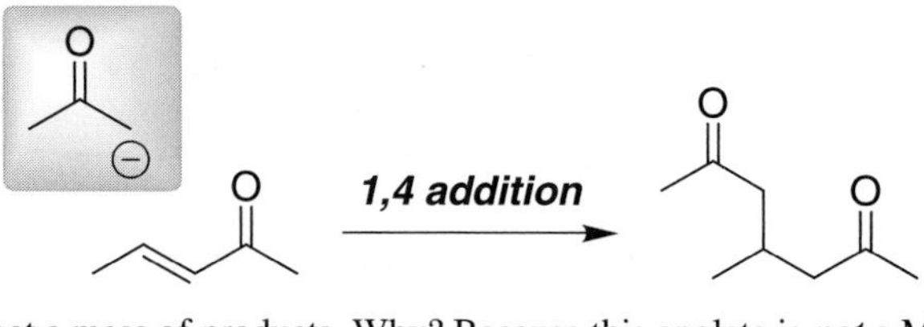

This is the nucelophile that you would need

1,4 addition

But when you try to do this, you get a mess of products. Why? Because this enolate is *not* a Michael donor, and therefore, it will not cleanly attack in a 1,4-addition. So, how do you get around this problem? This is where our enamine comes in handy.

Instead of using the enolate of acetone, suppose we convert the ketone into an enamine:

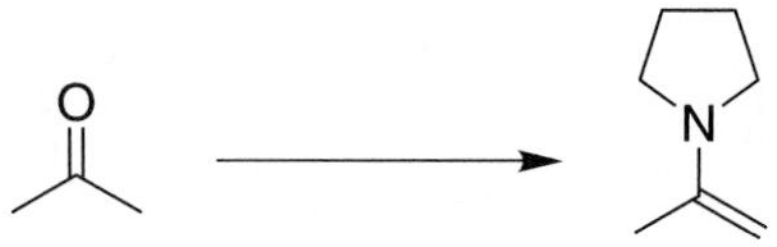

This enamine *is* a Michael donor, and it *will* attack cleanly in a 1,4-addition:

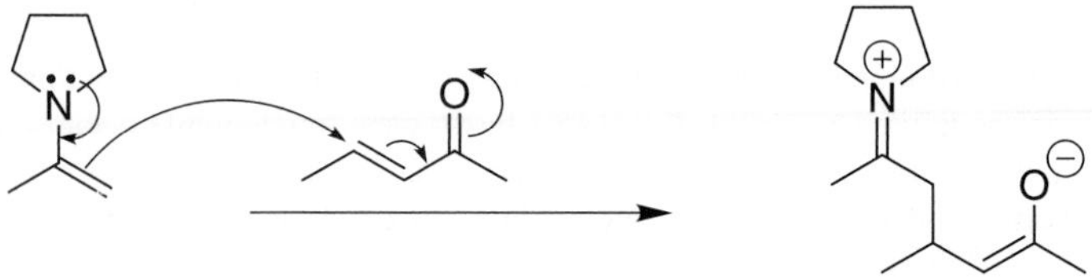

Finally, we use H_3O^+ to pull off the imminium group and protonate the enolate (which then turns into a ketone):

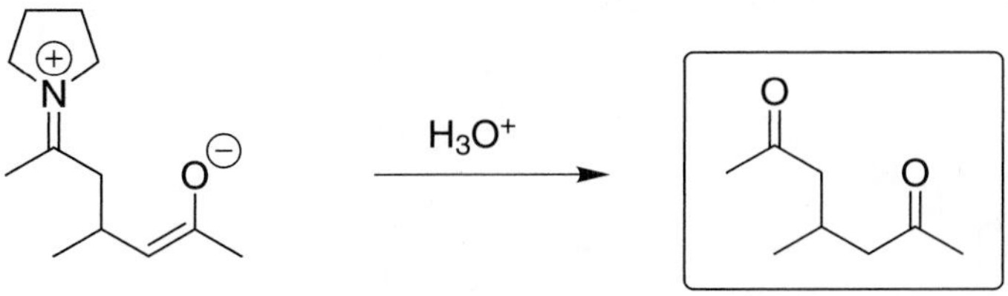

In the end, we made the product that we wanted. This synthetic strategy is called the *Stork enamine synthesis*, and it can come in handy when you are solving synthesis problems. Whenever you are trying to propose a synthesis, and you decide that you need an enolate to do a 1,4-addition, you will have a problem. Regular enolates are not stable enough to be Michael donors. But you can convert a ketone into an enamine, which *is* stable enough to be a Michael donor. Then, you can rip off the enamine in the end. The enamine serves as a way of temporarily modifying the reactivity of the enolate so that we can achieve the desired result. It is very clever when you really think about it.

EXERCISE 7.81. Propose a plausible synthesis for the following transformation:

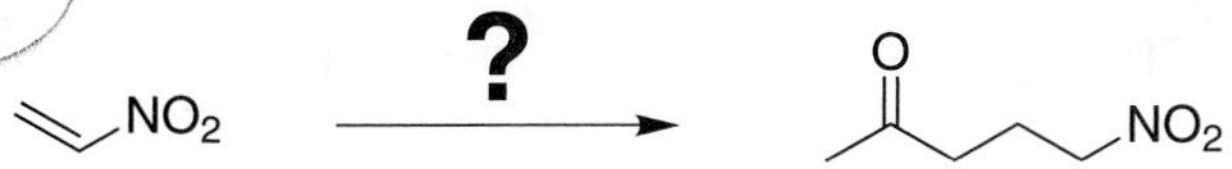

Answer: When we inspect this problem, we see that we need to add the following fragment:

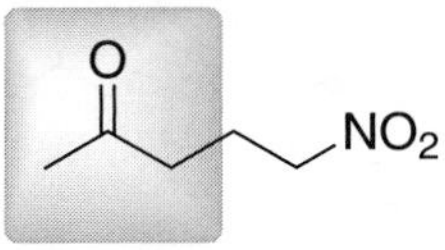

At the same time, we need to get rid of the double bond that was in the starting material. We can accomplish both at the same time by doing a 1,4 addition with the appropriate nucleophile. When we look carefully to see what nucleophile we would need, we realize that we will need to use the following:

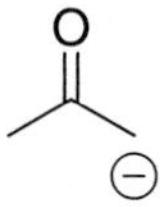

This enolate is *not* stable enough to be a Michael donor, so we realize that we need to use a Stork enamine synthesis:

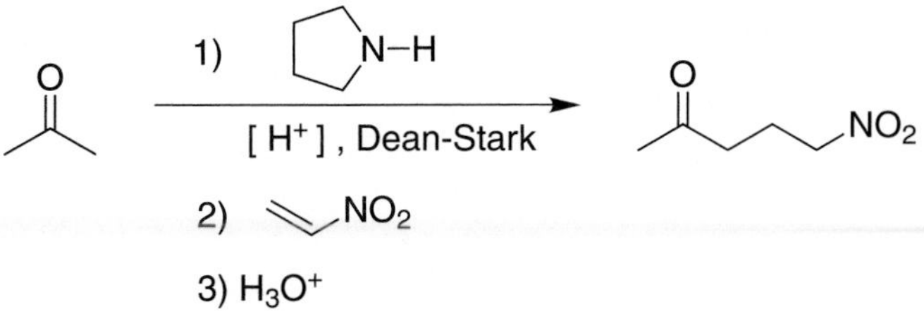

Propose a synthesis for each of the following transformations. In some cases, you will need to use a Stork enamine synthesis, but in other cases, it will not be necessary. Analyze each problem carefully to see if a Stork enamine synthesis is necessary.

7.82.

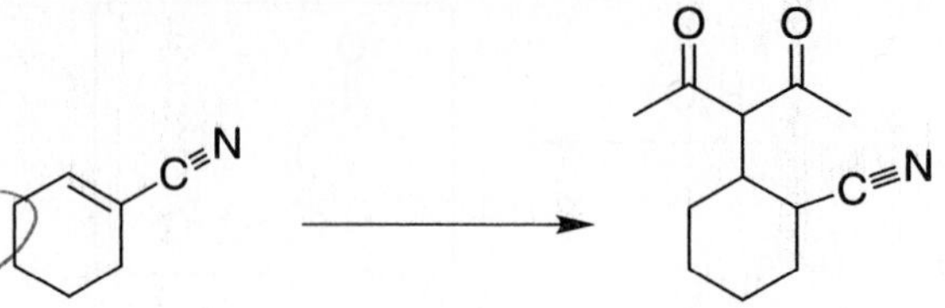

7.83.

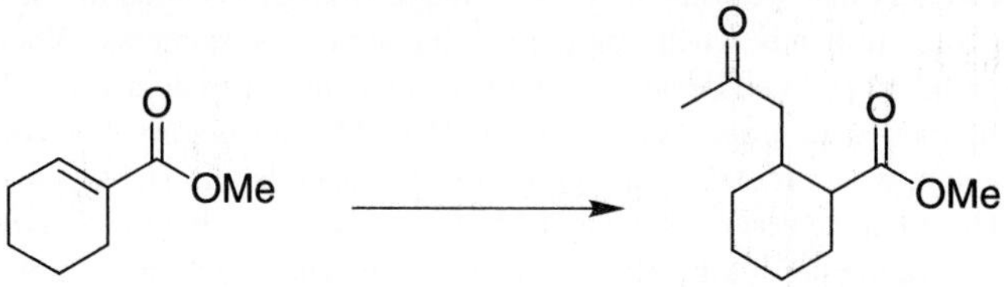

7.84.

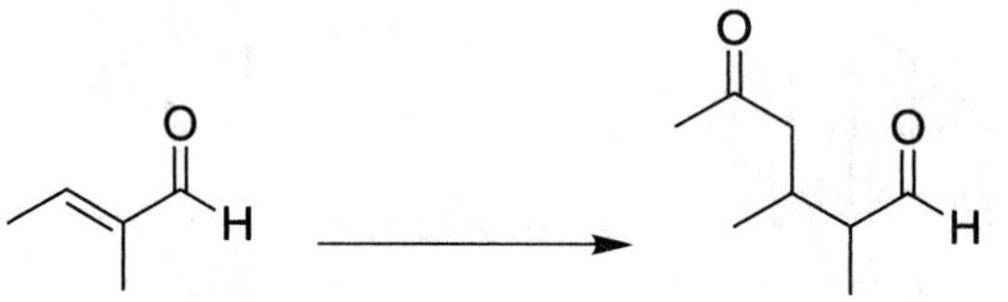

7.85.

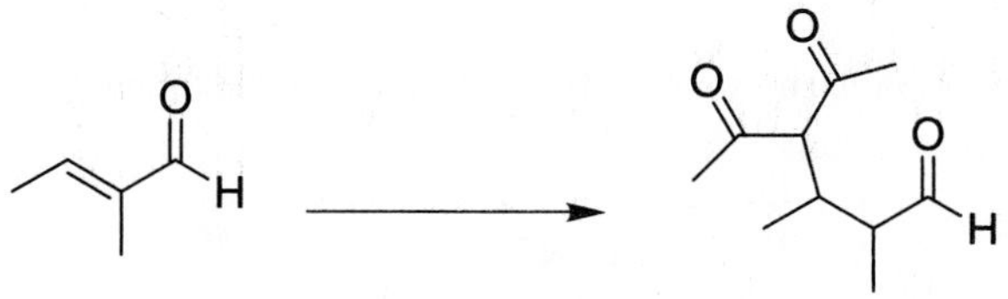

CHAPTER 8

AMINES

8.1 NUCLEOPHILICITY AND BASICITY OF AMINES AND AMIDES

Amines are classified based on the number of alkyl groups attached to the central nitrogen atom:

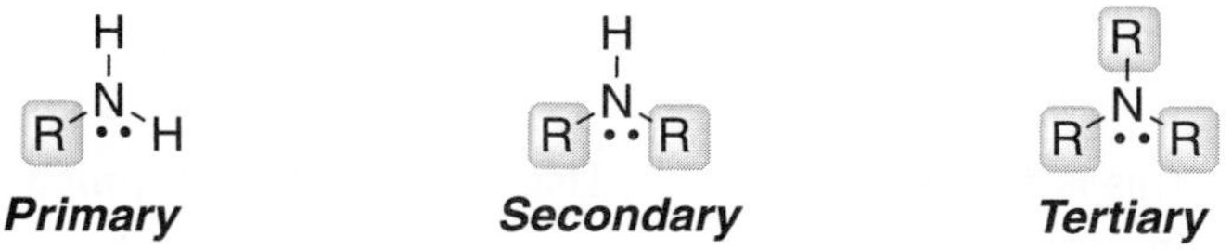

Primary   **Secondary**   **Tertiary**

The reactivity of all amines comes from the presence of a lone pair on the nitrogen atom. All amines have this lone pair:

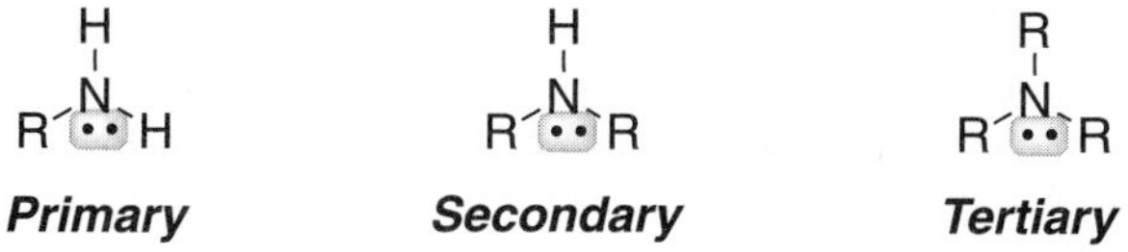

Primary   **Secondary**   **Tertiary**

This lone pair can function as a nucleophile (attacking an electrophile):

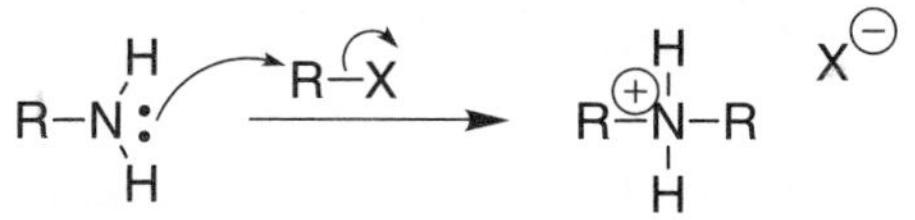

or it can function as a base (grabbing a proton):

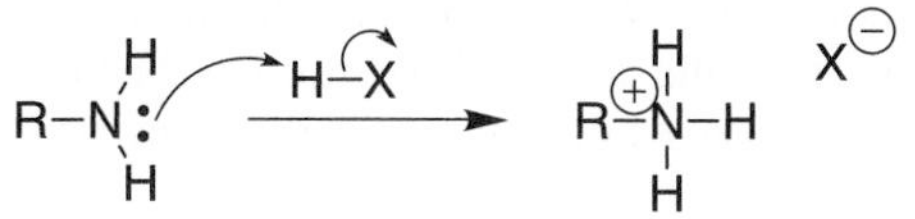

By focusing on this lone pair, we can understand why amines are good nucleophiles **and** good bases. When an amine participates in a reaction, the first step will always be one of these two possibilities: either the lone pair will grab a proton or the lone pair will attack an electrophile.

Of course, if we had a negative charge on the nitrogen, that would be even better. To get a negative charge on the nitrogen atom, we just have to deprotonate the amine. Tertiary amines don't have a proton for us to take off, but secondary and primary amines can be deprotonated to give the following types of compounds:

These compounds are stronger nucleophiles and stronger bases than neutral amines are. We call these compounds *amides*. Here are two examples of amides that we have seen so far in this course:

Lithium diisopropyl *amide*
(LDA)

Sodium *amide*

The term *amide* is actually a terrible name because we have already used this term to describe a type of carboxylic acid derivative:

This is an *amide*

This is also called an *amide*

Perhaps chemists could have given different names to these types of compounds (so that it would be less confusing to students). But historically, chemists have used the term *amide* to refer to both of these types of compounds—and old habits die hard. Since we are probably not going to convince chemists around the world to change their terminology, we will just have to get used to the terminology that is in use. Don't let this confuse you.

Now let's get back to neutral amines (without a charge). Some amines are actually less nucleophilic and less basic than regular amines. For example, compare the two amines:

An *alkyl* amine

An *aryl* amine

The first amine is called an *alkyl* amine because the nitrogen atom is connected to an alkyl group. The second compound is called an *aryl* amine because the nitrogen is connected to an aromatic ring. Aryl amines are less nucleophilic and less basic because the lone pair is delocalized into the aromatic ring. We can see this when we draw the resonance structures:

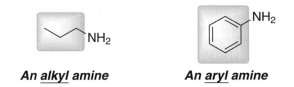

Since the lone pair is delocalized, it will be less available for use as a nucleophile or as a base. That does not mean that an aryl amine can't attack something. In fact, we will soon see a reaction where an aryl amine *is* used as a nucleophile. It *can* function as a nucleophile—but it is just *less* nucleophilic than an alkyl amine.

Now that we have had an introduction to amines, let's focus on a few ways to make amines.

8.2 PREPARATION OF AMINES THROUGH S$_N$2 REACTIONS

Suppose you wanted to make the following primary amine:

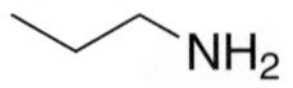

It might be tempting to suggest the following synthesis:

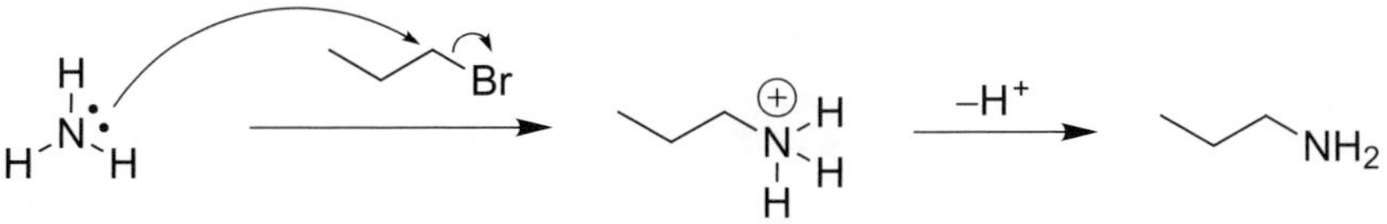

This is just an S$_N$2 reaction, followed by a deprotonation. This approach does work, BUT it is difficult to get the reaction to stop after mono-alkylation. The product is also a nucleophile, and it competes with NH$_3$ for the alkyl halide. It is an even better nucleophile than NH$_3$ because the alkyl group is electron donating. Therefore, it will attack again:

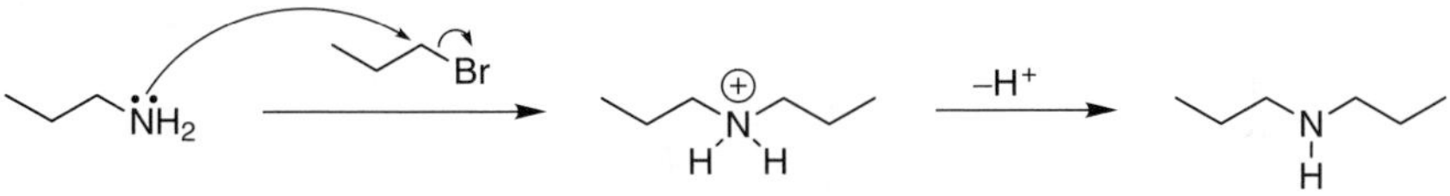

And then it attacks again:

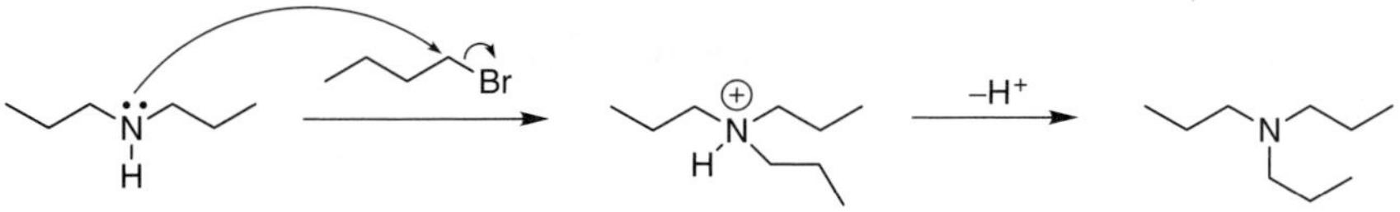

And then one last time:

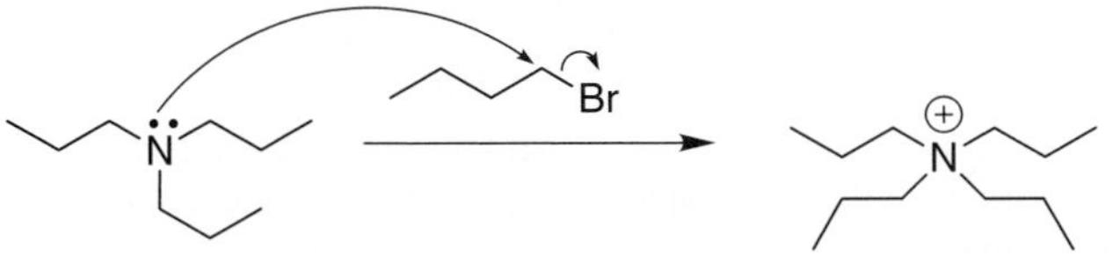

This time, there is no proton to take off

The final product has four alkyl groups (which is referred to as **quaternary**), and the nitrogen has a positive charge (which is referred to as an **ammonium** ion). So, we call this a **quaternary ammonium salt**. Because this compound does not have a lone pair, it is neither nucleophilic nor basic.

If our intention is to make a quaternary ammonium salt, then the synthesis above is a good approach. But what if we want to make a primary amine? We cannot simply alkylate ammonia:

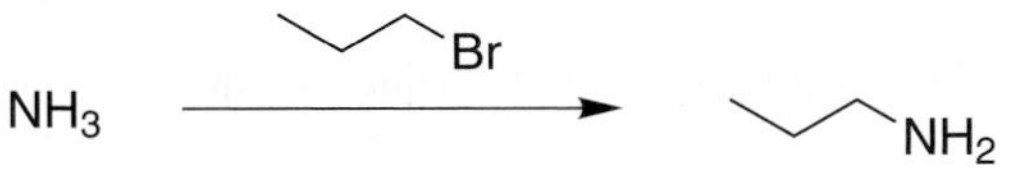

because it is too difficult to stop the reaction at this stage. The alkyl halide will react again with our primary amine. Even if we try to use only one mole of alkyl halide and one mole of ammonia, we will still get a mess of products. We will get some polyalkylated products, and we will get some ammonia that could not find any alkyl halide to react with. So what do we do?

To get around this problem, we use a clever trick. We use a starting amine that already has two dummy groups on it:

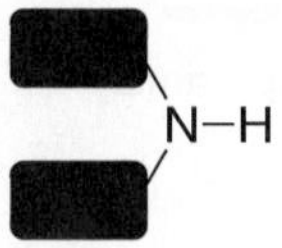

And we choose our dummy groups so that they are easily removable after we alkylate. So, we first alkylate, and then we pull the dummy groups off:

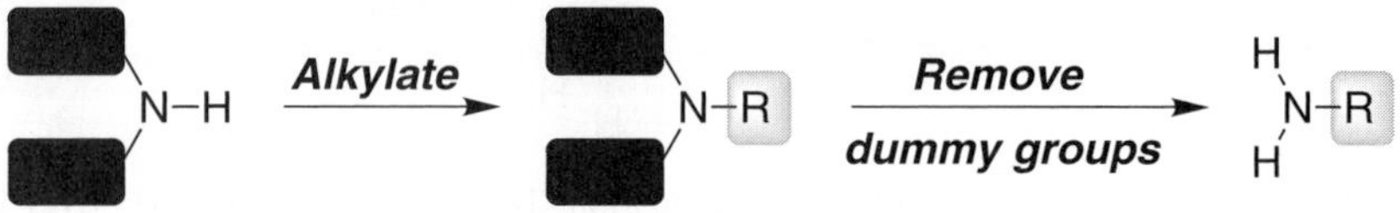

This strategy is called the Gabriel synthesis. It is a very good way of making primary alkyl halides, so let's explore this strategy in more detail.

We start with the following compound, called phthalimide:

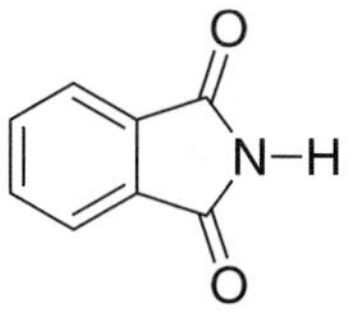

We use a base (KOH) to pull off the proton, and we get the following anion:

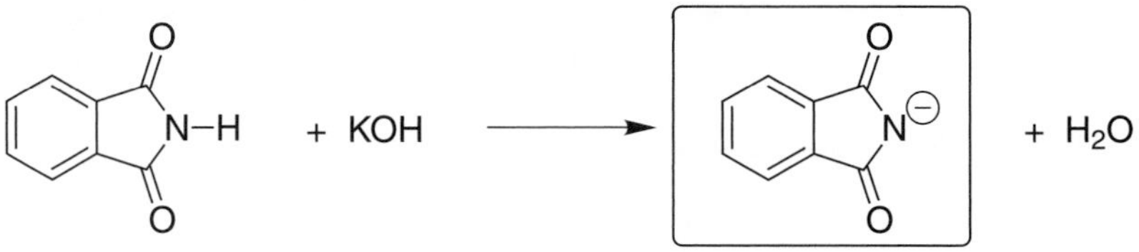

Notice that this negative charge is very stabilized (delocalized) by resonance. This is similar to the Michael donors that we saw in the end of the last chapter. It is a stabilized nucleophile. We use this nucleophile to attack an alkyl halide:

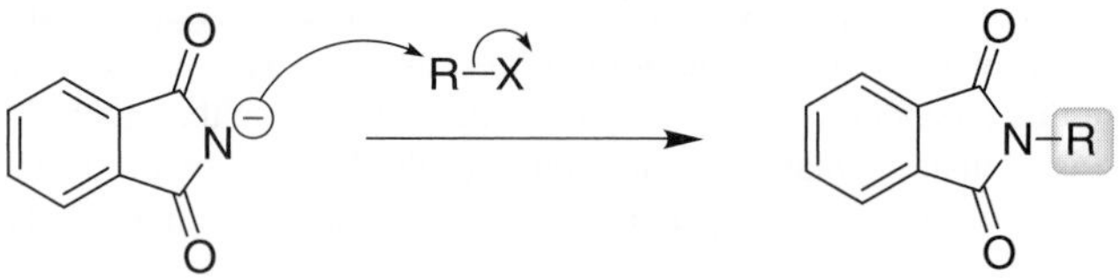

And then, we pull off the dummy groups using hydrazine, and that gives us our product:

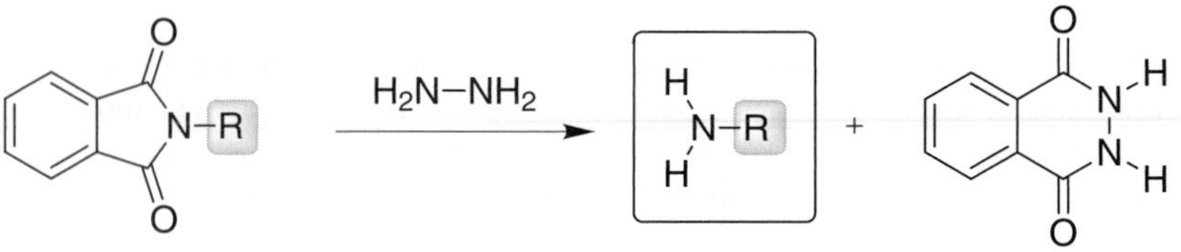

To understand how hydrazine pulls off the dummy groups, let's take a close look at the mechanism. If you remember what we covered in Chapter 6 (carboxylic acid derivatives), then you should find the steps of this mechanism to be very familiar:

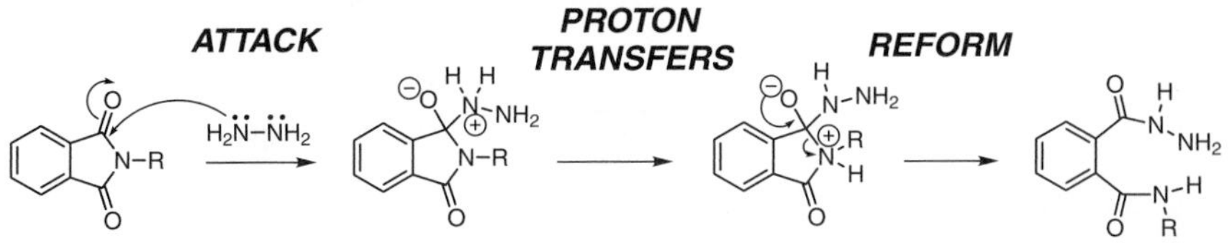

ATTACK · **PROTON TRANSFERS** · **REFORM**

These three steps should seem familiar (if they don't, you may want to review the chapter on carboxylic acid derivatives). Then we get the same three steps again:

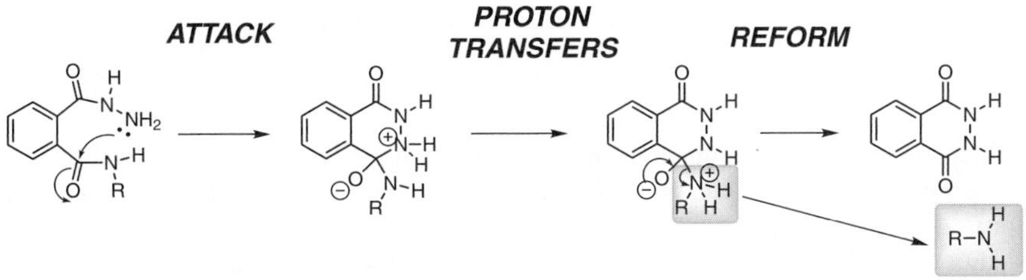

ATTACK · **PROTON TRANSFERS** · **REFORM**

The Gabriel synthesis can be summarized like this:

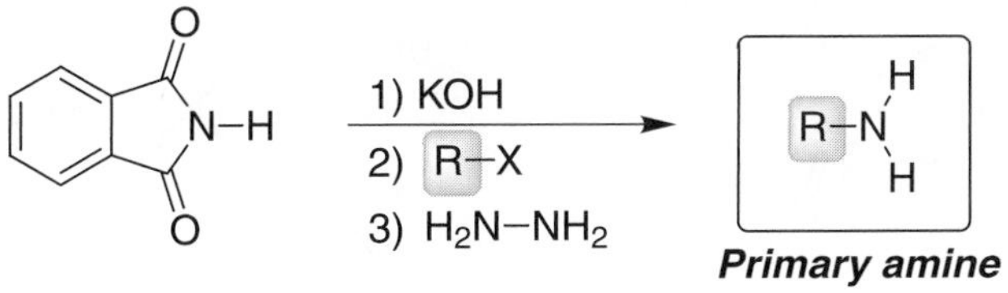

1) KOH
2) R-X
3) H$_2$N-NH$_2$

Primary amine

It is very useful for making primary amines from primary alkyl halides:

EXERCISE 8.1. Identify how would use a Gabriel synthesis to prepare the following amine:

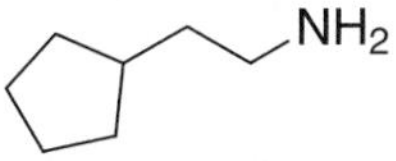

Answer: In order to do a Gabriel synthesis, we need to determine just one thing. We must decide what alkyl halide to use. And that is easy. We just draw a halogen instead of the NH$_2$, like this:

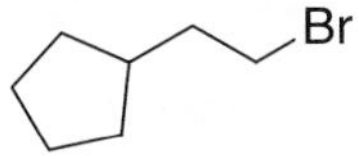

Now that we know what alkyl halide to use, we are ready to propose our synthesis:

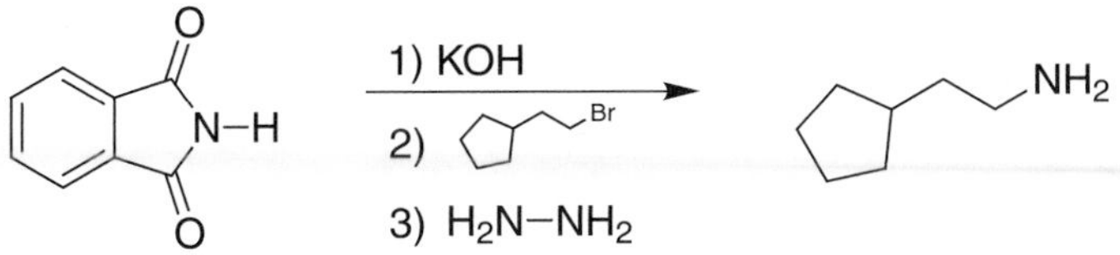

1) KOH
2)
3) H$_2$N-NH$_2$

Identify how you would use a Gabriel synthesis to prepare each of the following compounds.

8.2. ⟶

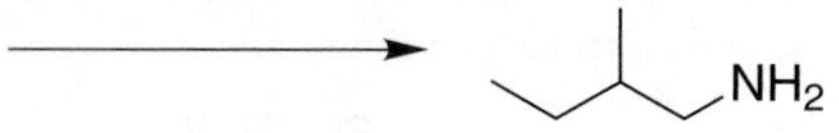

8.3. ⟶

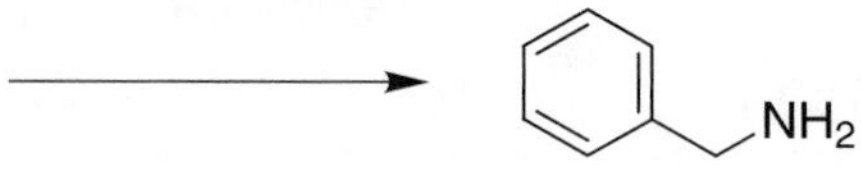

8.4. ⟶

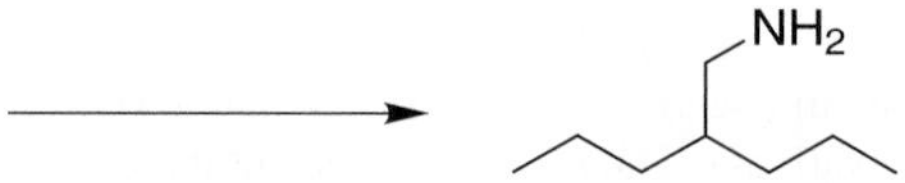

8.5. ⟶

The Gabriel synthesis has its limitations, however. Since it relies on an S_N2 reaction, it will work best with *primary* alkyl halides. It is not so great for secondary alkyl halides, and it will not work at all for tertiary alkyl halides.

In addition, it will not work for aryl halides because you cannot do an S_N2 reaction on an aryl halide:

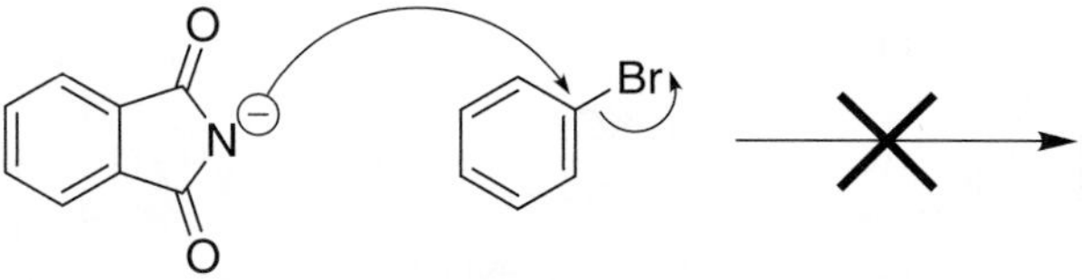

EXERCISE 8.6. Is it possible to prepare the following compound with a Gabriel synthesis?

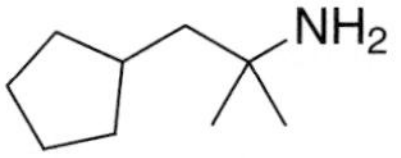

Answer: If we draw the alkyl halide, we would need,

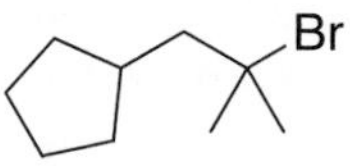

we find that it is a tertiary alkyl halide, so we *cannot* use a Gabriel synthesis to make our product.

For each of the following compounds, identify whether or not the compound could be made with a Gabriel synthesis:

8.7.

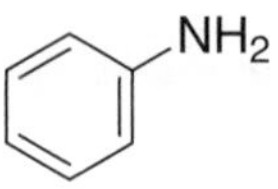

8.8.

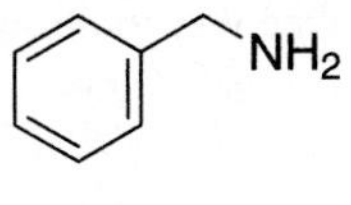

8.9.

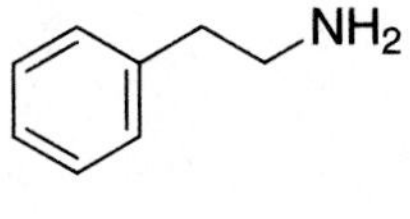

8.10.

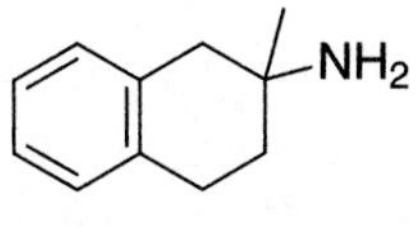

8.3 PREPARATION OF AMINES THROUGH REDUCTIVE AMINATION

In the previous section, we learned how to make amines via an S_N2 reaction. That method was best for making primary amines. In this section, we will learn a way to make secondary amines using a two-step synthesis (where the first step is a reaction that we have already seen). When we learned about ketones and aldehydes, we saw how to make *imines*:

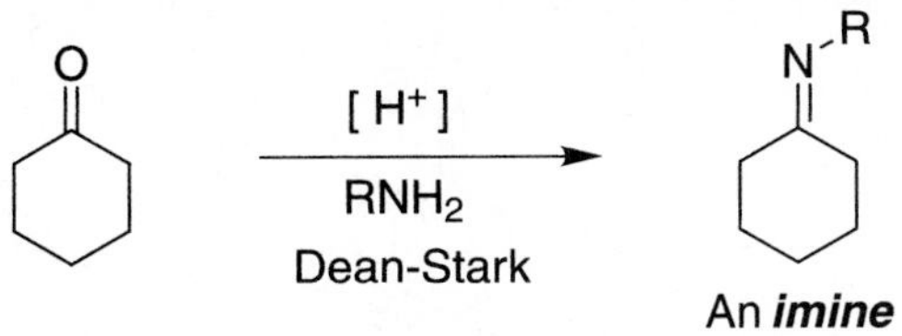

$$[H^+] \quad RNH_2 \quad Dean\text{-}Stark$$

An *imine*

We saw that a primary amine can react with a ketone (under acidic conditions) to give an imine. Now, we will put this reaction to use in making amines.

When we form an imine, we are forming the essential C-N bond that you need in order to have an amine:

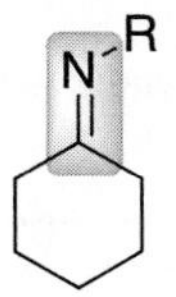

But the oxidation state is not correct. In order to get an amine, we need to do the following conversion:

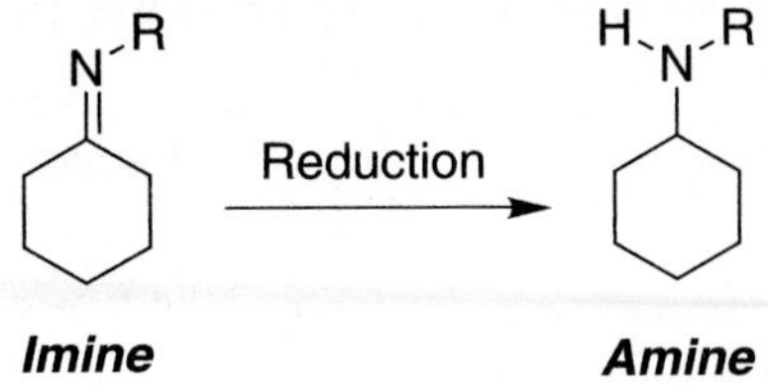

Reduction

Imine   *Amine*

In order to convert an *imine* into *amine*, we will need to do a reduction. One way to do so is to reduce the imine just like we reduce a ketone, using LAH:

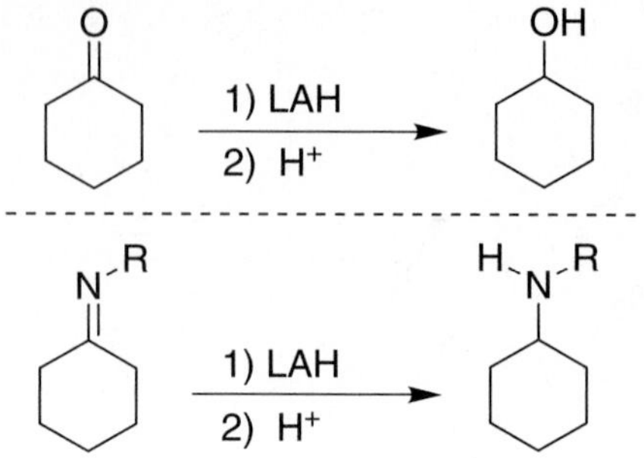

Alternatively, we can just hydrogenate the C=N double bond (using a catalyst):

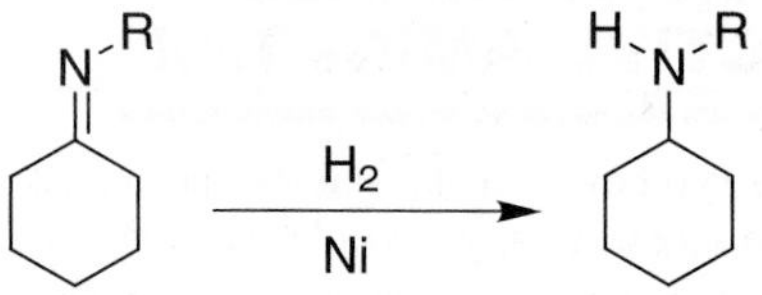

There are many other ways to reduce an imine as well. But the bottom line is that we now have a two-step synthesis for preparing amines:

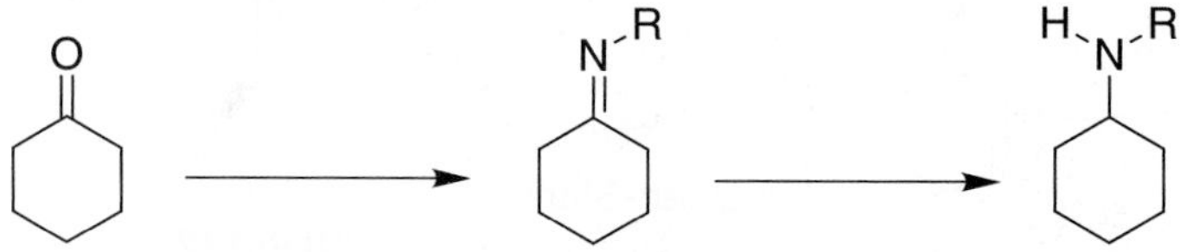

This reaction is called ***reductive amination*** because we are forming an amine (a process called *amination*) via a *reduction*. Students often have trouble pronouncing the word *amination* because our mouths tend to say the word animation instead. Try to say reductive amination ten times real fast, and you will see what I am talking about.

EXERCISE 8.11. Suggest an efficient synthesis for the following transformation:

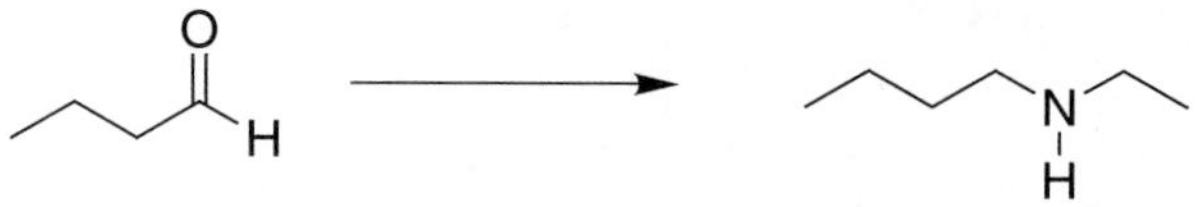

Answer: This problem is asking us to convert an aldehyde into an amine. This should alert us to the possibility of a reductive amination. If we used a reductive amination, our strategy would go like this:

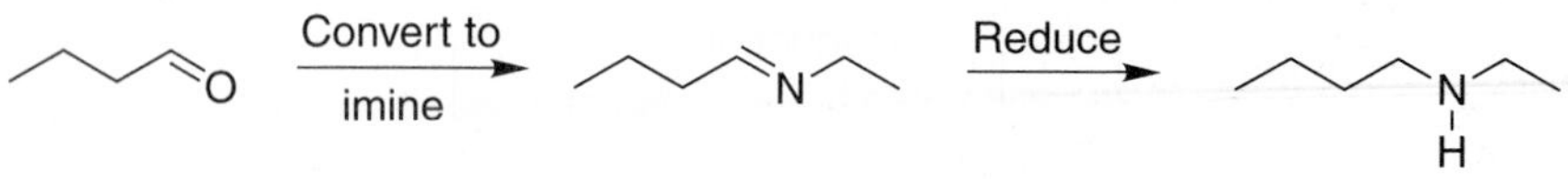

So our synthesis will look like this:

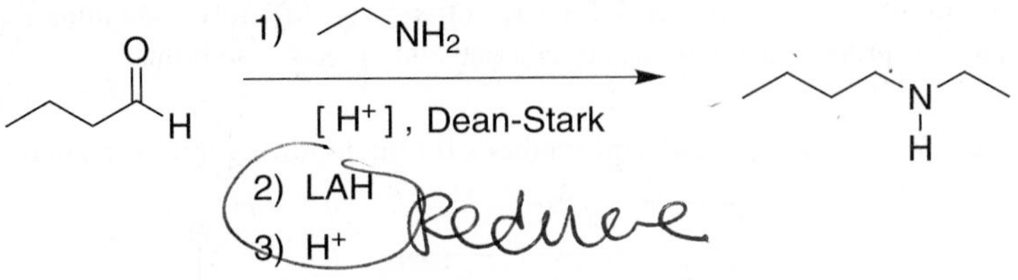

1) $\diagup\diagdown$ NH$_2$

[H$^+$], Dean-Stark

2) LAH Reduce

3) H$^+$

Suggest an efficient synthesis for each of the following transformations.

8.12.

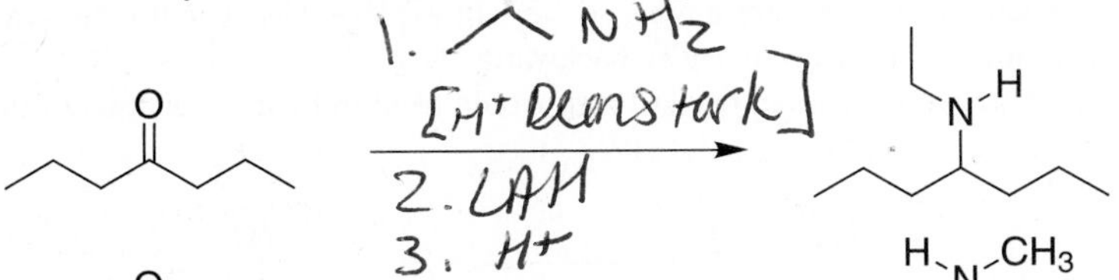

1. $\diagup\diagdown$ NH$_2$

[H$^+$ Dean-Stark]

2. LAH

3. H$^+$

8.13.

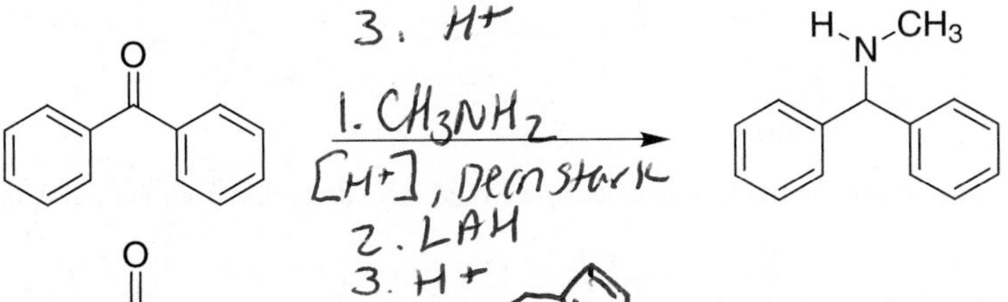

1. CH$_3$NH$_2$

[H$^+$], Dean-Stark

2. LAH

3. H$^+$

8.14.

1. NH$_2$

2. LAH

3. H$^+$

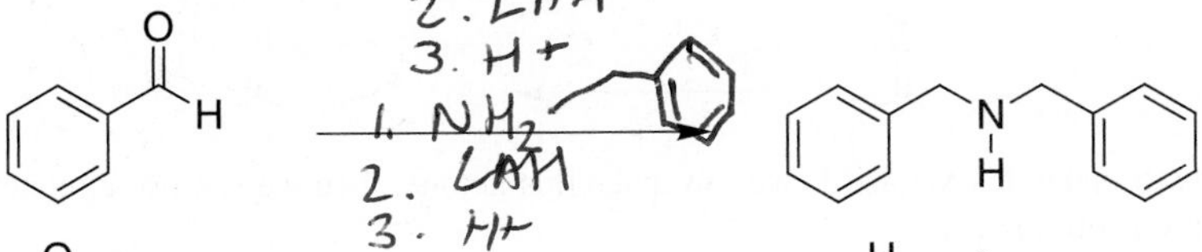

8.15.

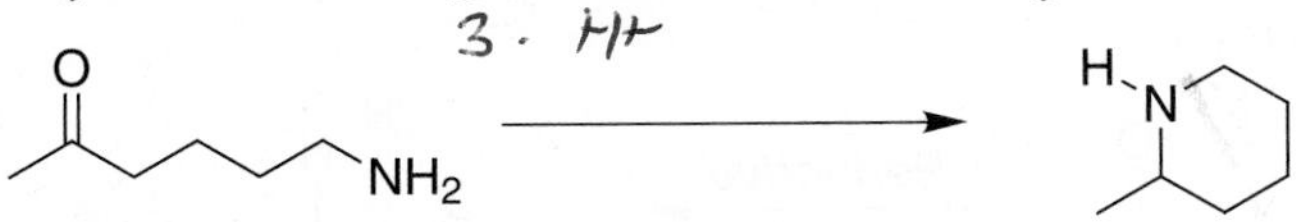

8.16.

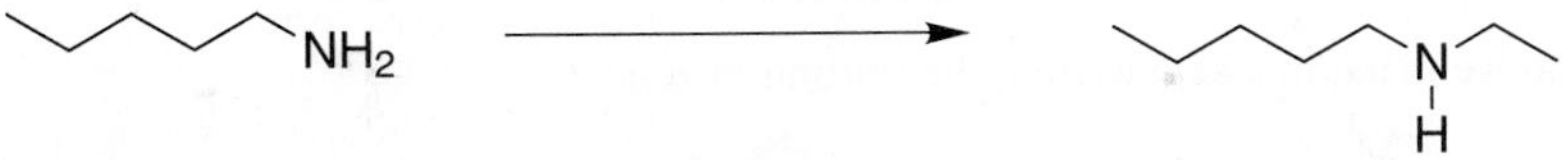

Reductive aminations are very useful because the starting material is a ketone or aldehyde. We have seen many ways to make ketones. This gives us a way to make amines from a variety of compounds:

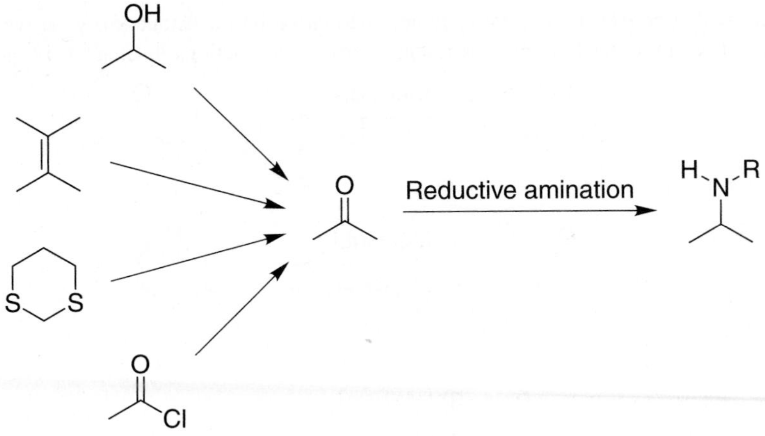

Reductive amination

All of these reactions are from Chapter 5. Students often have difficulty combining reactions from different chapters to propose a synthesis, so let's get some practice with this:

EXERCISE 8.17. Suggest an efficient synthesis for the following transformation:

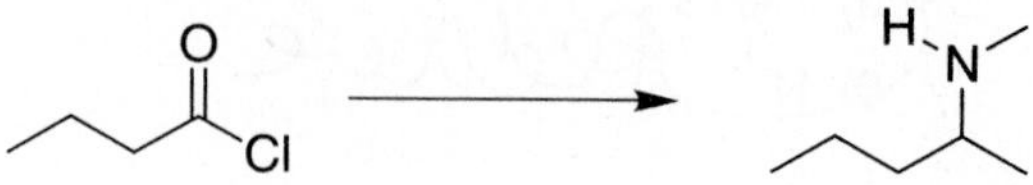
O
Cl
H
N

Answer: Our product is a secondary amine, so we will explore if we can do this synthesis using a reductive amination. Let's work our way backwards.

If we did use a reductive amination, our last step would need to be the reduction of the following imine:

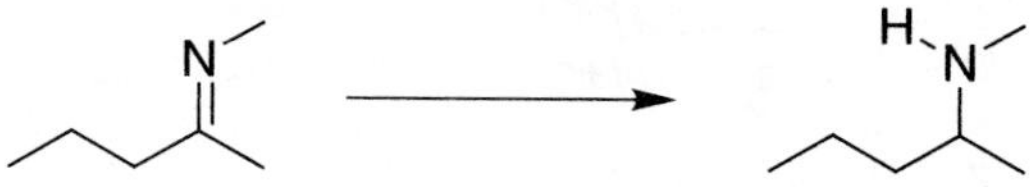
N
H
N

Imine

So, we would need to make the imine above, and we would have done that from the following ketone:

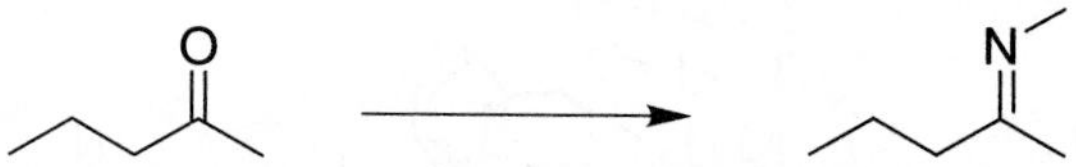
O
N

So, our goal is to make this ketone. If we can make this ketone, then we can use a reductive amination to form our product:

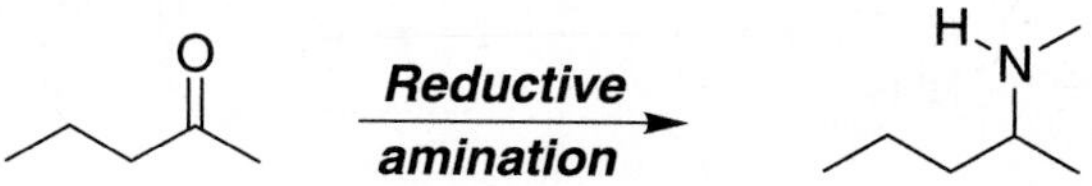
O
Reductive
amination
H
N

But how do we make this ketone from the starting material?

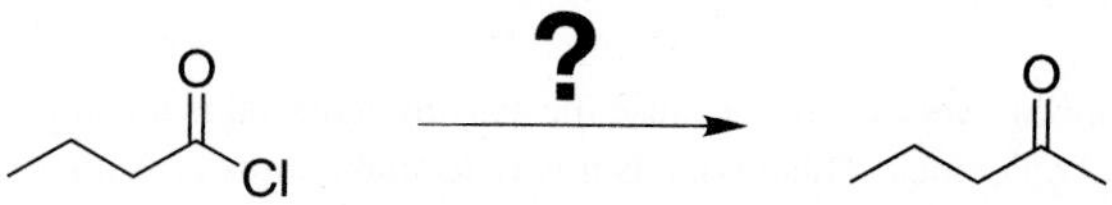
O
Cl
?
O

This would involve converting a carboxylic acid derivative into a ketone. So what we need is a crossover reaction (think back to our discussion of crossover reactions in Chapter 6):

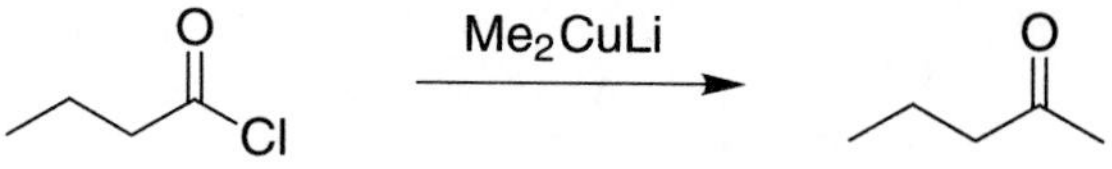
O
Cl
Me2CuLi
O

Therefore, our overall synthesis goes like this:

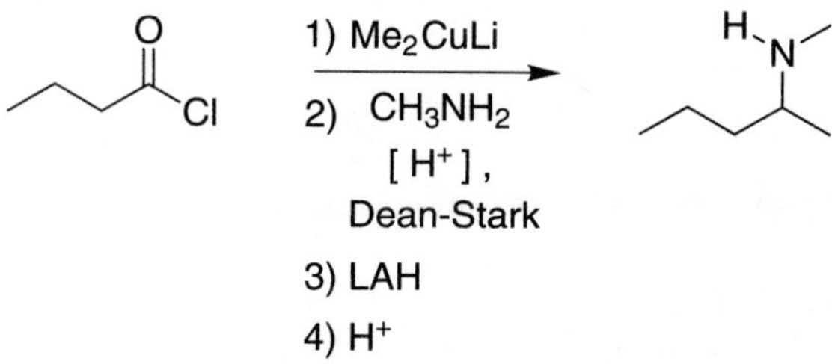
O
Cl
1) Me2CuLi
2) CH3NH2
[H+],
Dean-Stark
3) LAH
4) H+
H
N

Suggest an efficient synthesis for each of the following transformations. In each case, you should work your way backwards. Start by asking what ketone or aldehyde you would need in order to make the desired product via a reductive amination. Then, ask yourself how you could make that ketone from the starting material.

8.18.

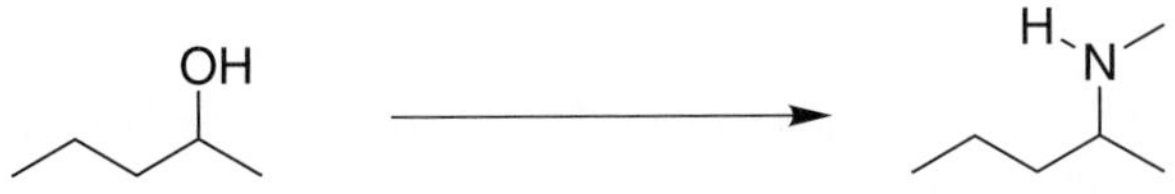

8.19.

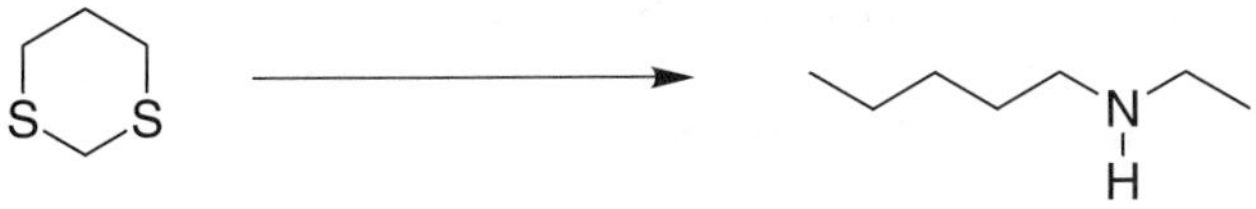

8.20.

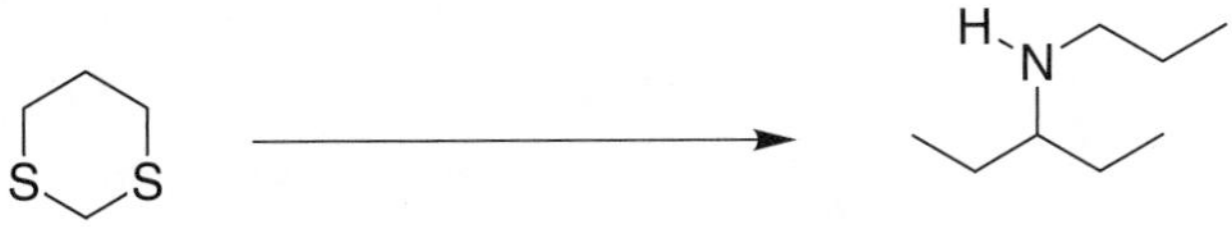

8.21.

8.22.

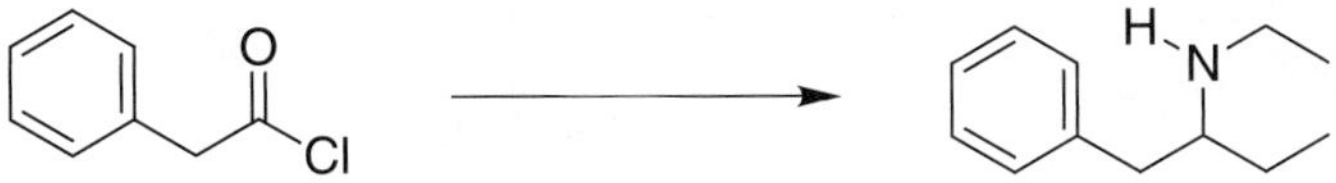

8.4 PREPARATION OF AMINES FROM AMIDES

We have seen how to prepare amines via an S_N2 reaction (the Gabriel synthesis), and we have seen how to prepare amines via reductive amination. In this section, we will see how to prepare an amine from an amide (when I say *amide* here, I am referring to a carboxylic acid derivative):

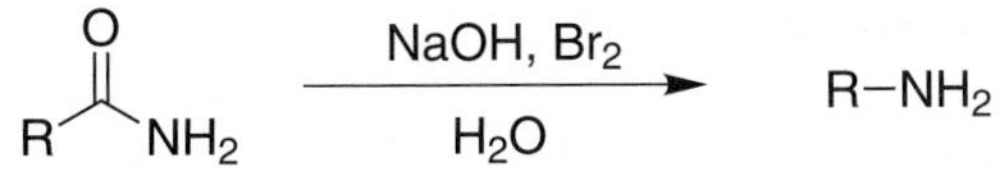

$$\underset{R}{\overset{O}{\|}}\underset{NH_2}{} \quad \xrightarrow[\text{H}_2\text{O}]{\text{NaOH, Br}_2} \quad R-NH_2$$

This reaction is called a Hoffman rearrangement, and it involves a very unique (and weird) step in the mechanism. Let's go through the mechanism step-by-step, so that we can see how it works.

In the presence of hydroxide, we can deprotonate the amide to give a negative charge that is stabilized by resonance:

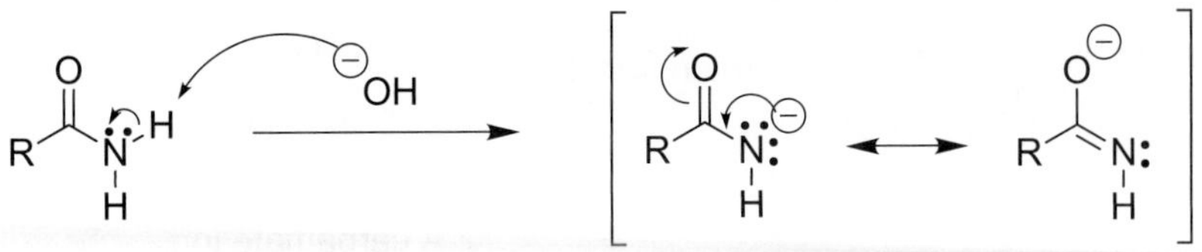

Stabilized by resonance

If you stare at this anion, you should realize how similar it is to an enolate.

The anion that we just formed can now function as a nucleophile to attack bromine:

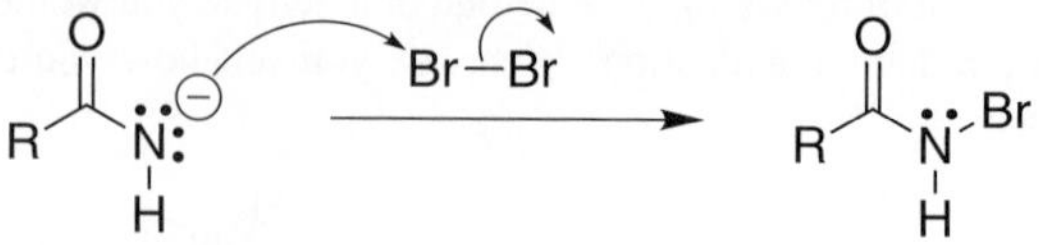

So far this is very similar to the haloform reaction (you might want to look at the first few steps of that reaction again, if you don't remember it).

Then, we deprotonate again:

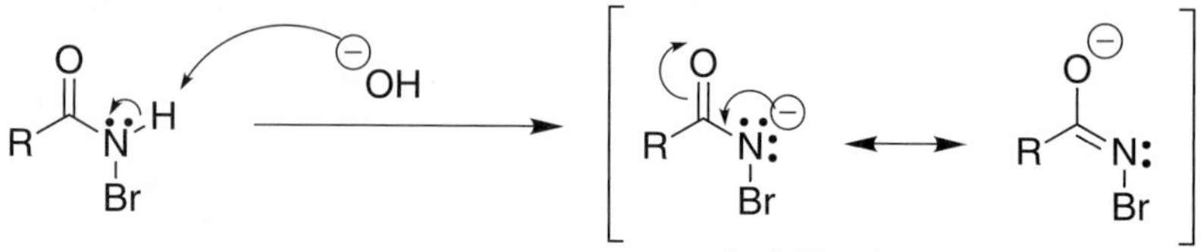

Stabilized by resonance

We might expect this anion to attack bromine again (just as we did a moment ago), but here is where something exceedingly strange happens:

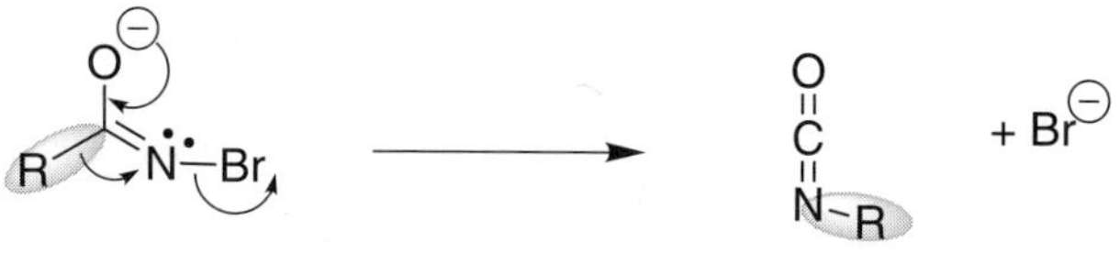

An isocyanate

We get a weird rearrangement to generate an isocyanate. The alkyl group (which is highlighted above) migrates and ends up being attached to the nearby nitrogen atom. This type of rearrangement is very unique. It is unlikely that you will see another reaction involving this type of rearrangement (there are similar reactions like this, but they are typically not covered in a first-year organic chemistry course). So, you probably don't have to worry about how you will know when to do this type of rearrangement in other situations. You should be OK if you just remember how to do it in this specific reaction.

The rest of the mechanism makes perfect sense. The isocyanate gets attacked by hydroxide to give a resonance-stabilized intermediate.

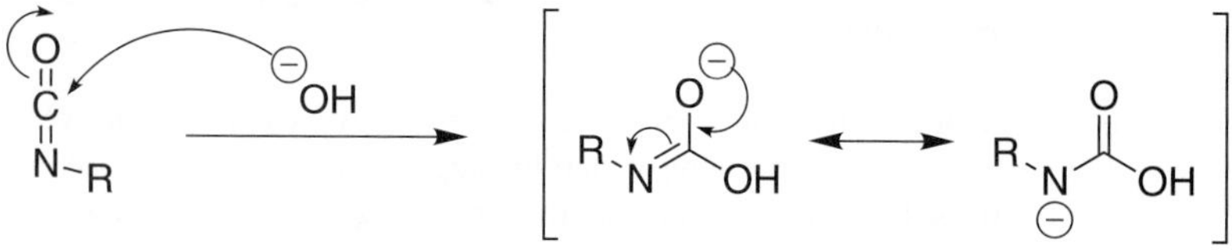

Proton transfers generate a carbamate ion

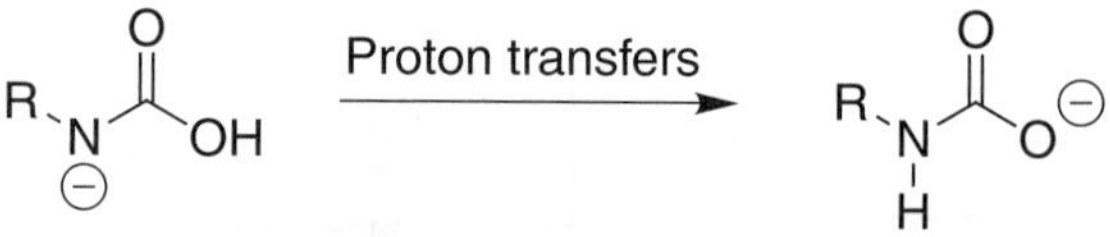

A carbamate ion

which then loses CO_2 and grabs a proton to form our product.

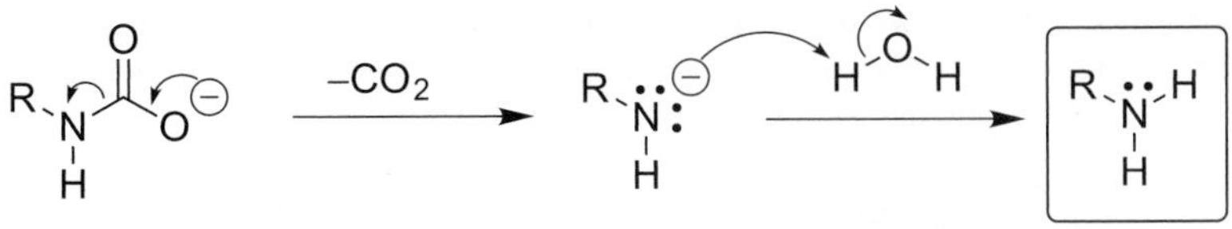

This mechanism is long and hard. It is probably one of the hardest mechanisms that you have seen (or will see) in this course, so don't get discouraged by its difficulty. Just go over it several times and realize that there is one step that is very unique (the rearrangement step).

In the end, the net result is to blast away the carbonyl:

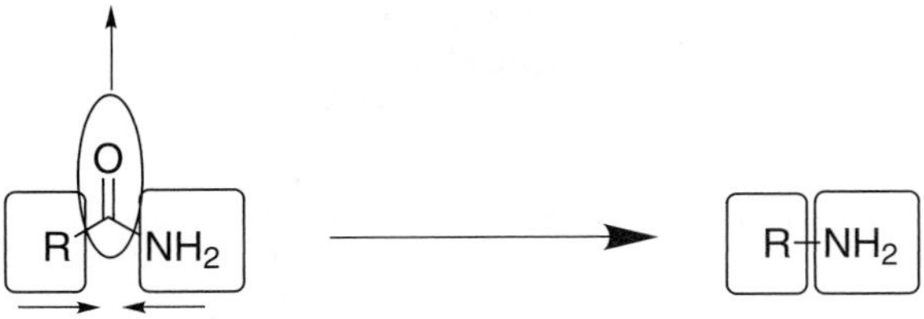

This method for preparing amines can be very useful because it gives us a way to make an amine from any carboxylic acid derivative:

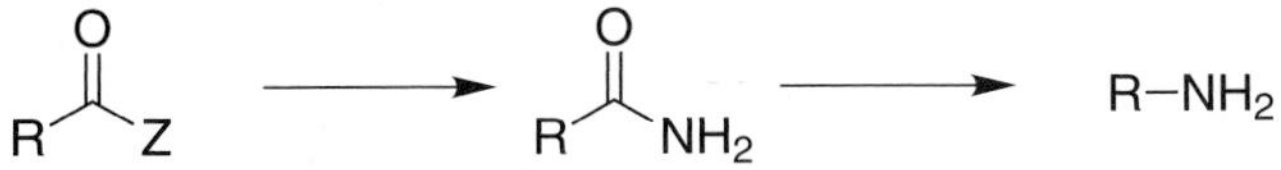

Just keep in mind that this approach will involve the *loss* of one carbon atom.

Let's see an example:

EXERCISE 8.23. Propose an efficient synthesis for the following transformation:

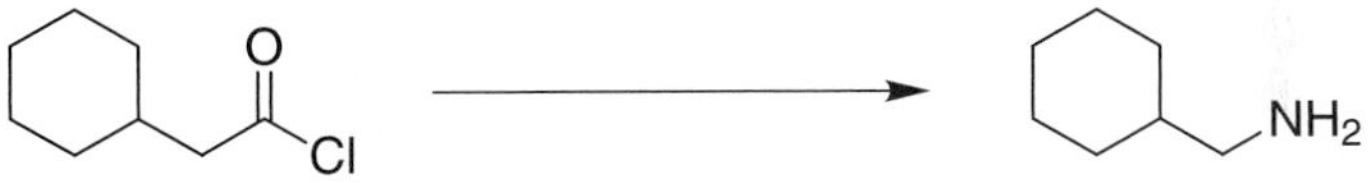

Answer: The starting material is an acyl halide, and the product is an amine. There is one clue that should direct us toward a Hoffman rearrangement. Notice that we are losing one carbon atom:

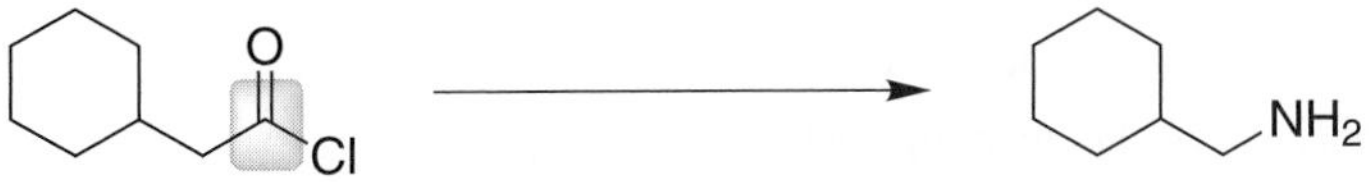

This should focus our attention on the possibility of a Hoffman rearrangement.

Now let's work our way backwards. In order to use a Hoffman rearrangement, we will need the following amide:

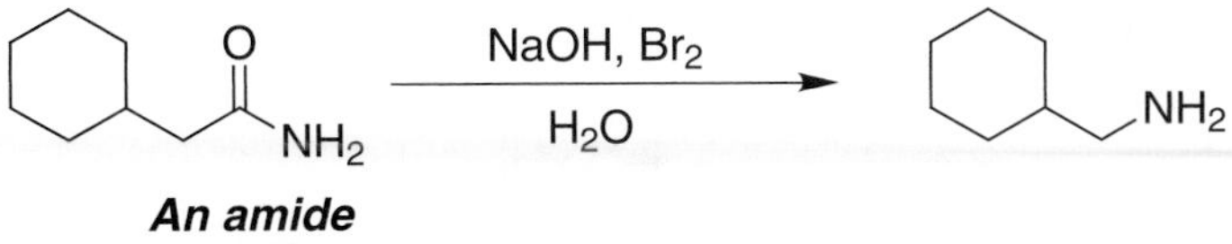

An amide

Working our way backwards, we then ask how to make this amide from the starting material. We have seen a way to do this. In Chapter 6, we learned about the interconversions between carboxylic acid derivatives, and we saw the following reaction:

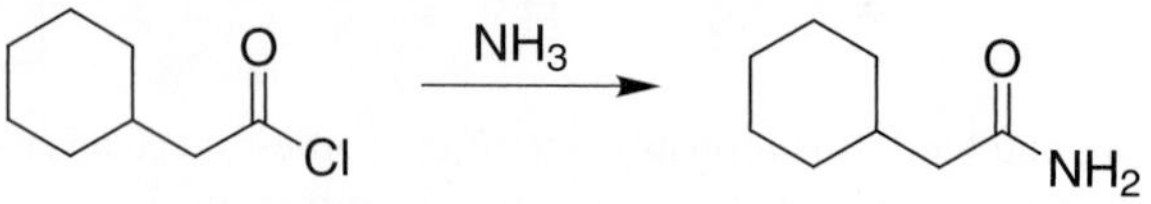

So now we have our synthesis. It goes like this:

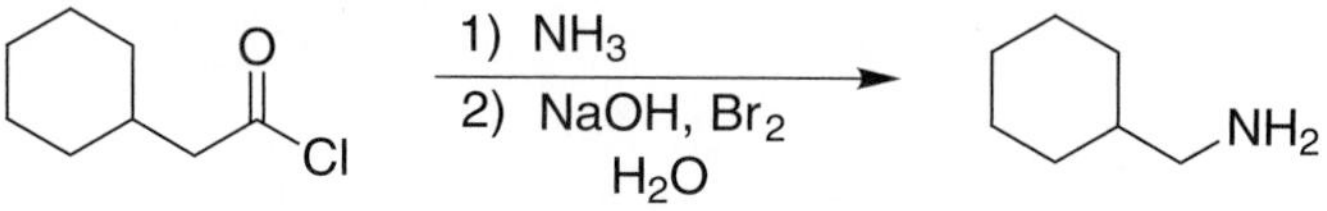

Propose an efficient synthesis for each of the following transformations:

8.24.

8.25.

8.26.

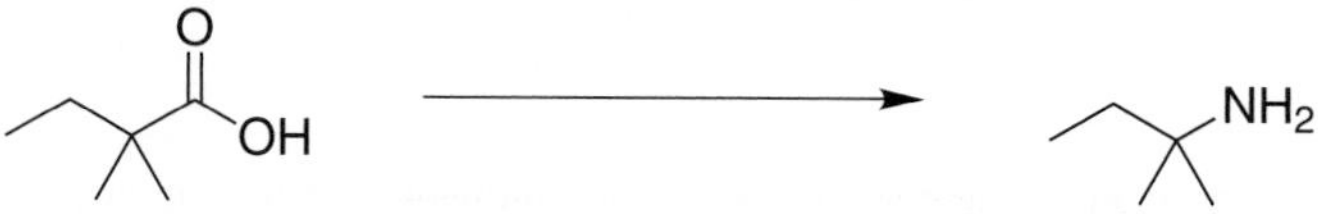

8.27.

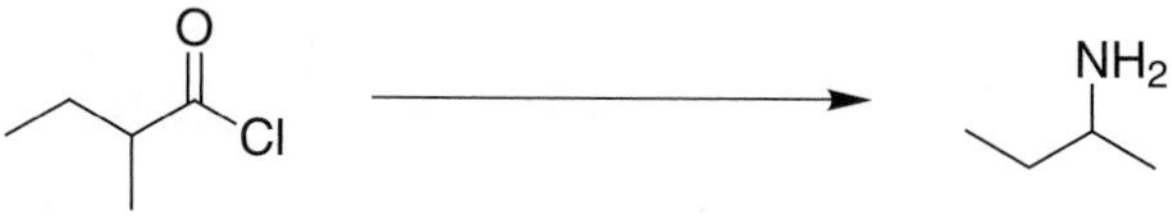

So far in this chapter, we have seen three ways to make amines:

1. From alkyl halides (Gabriel synthesis)
2. From ketones (reductive amination)
3. From amides (Hoffman rearrangement)

Now let's do some problems where we have to choose which method to use. If you do not remember the reagents for all three methods, then you should go back and review all three methods.
   Propose an efficient synthesis for each of the following transformations:

8.28.

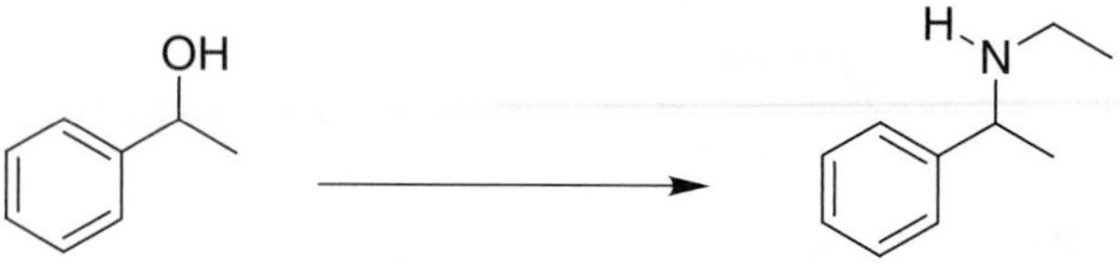

8.29.

8.30.

8.31.

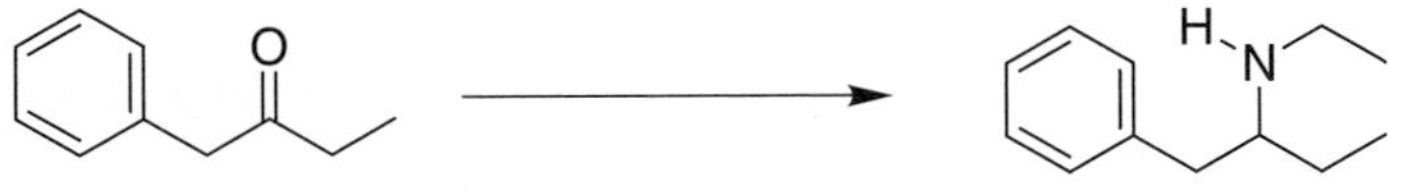

8.32.

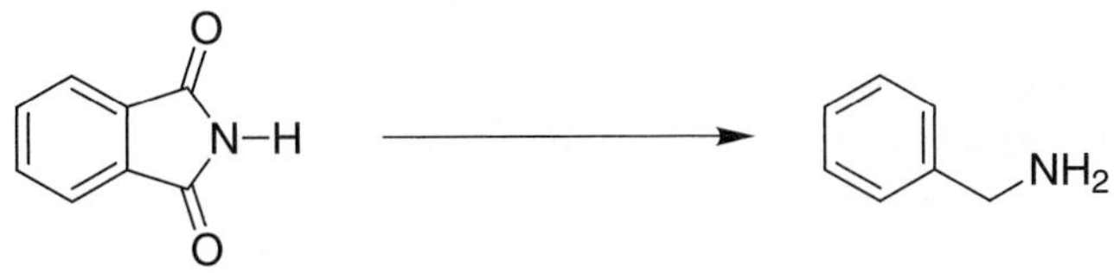

8.33.

There are yet other ways to make amines. We only covered three of the most common and most useful methods. You should look through your textbook and lecture notes to see if you are responsible for any other methods for preparing amines.

8.5 ACYLATION OF AMINES

So far in this chapter, we have focused on ways of *making* amines. For the rest of the chapter, we will shift our focus, and we will now start learning what reactions you can do with amines.

We will begin our survey with a reaction that we have already seen in a previous chapter. When we learned about carboxylic acid derivatives (Chapter 6), we saw that you can convert an acyl halide into an amide. For example:

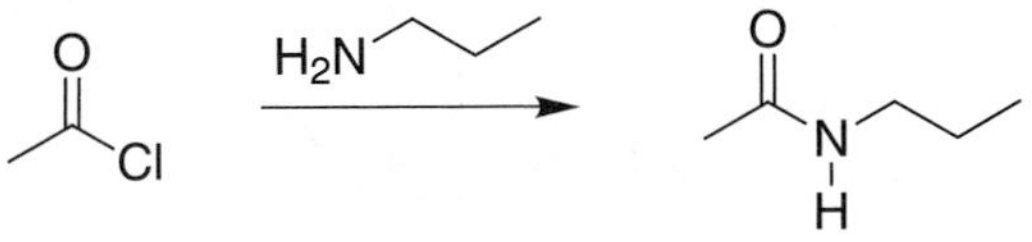

When we draw it this way (with the amine over the reaction arrow), the focus is on what happens to the acyl halide (that it is converted into an amide). But what if we choose to focus on the amine instead? In other words, let's rewrite the same exact reaction a bit differently. Let's put the acyl halide on top of the reaction arrow, as follows:

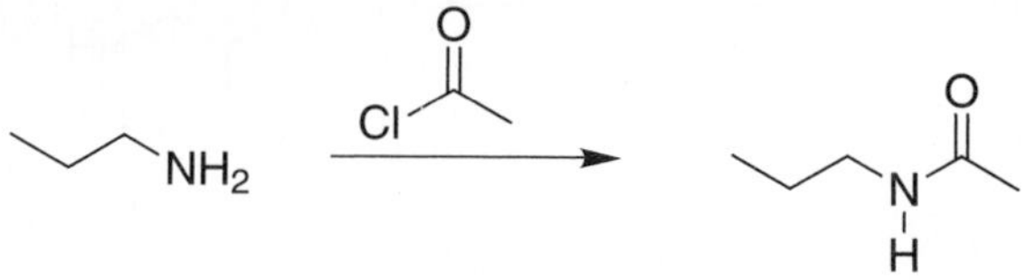

We have not changed the reaction at all. It is still the same exact reaction (an amine reacting with an acyl halide). But when we draw it like this, our attention focuses on converting the *amine* into an *amide*. The acyl halide is just the reagent that we use to accomplish this conversion.

Overall, we have placed an *acyl* group onto the amine:

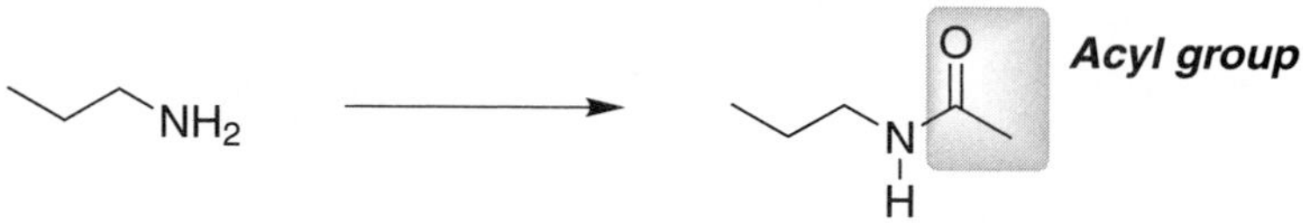

Therefore, we call this an *acylation* reaction.

Most primary and secondary amines can be acylated. Here is another example:

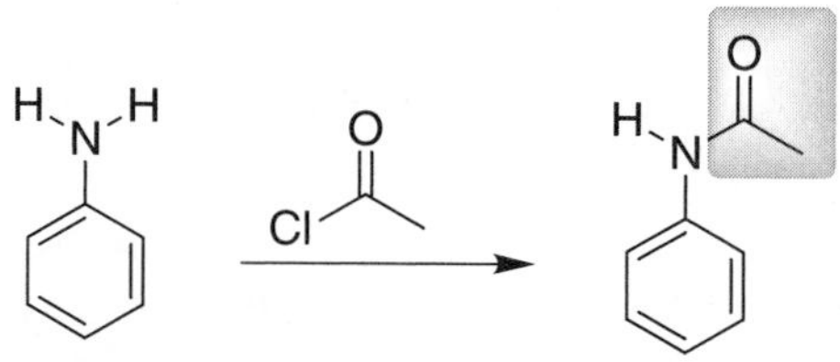

Now that we have seen how to acylate an amine, let's take a look at how to pull the acyl group back off. This reaction can also serve as a review for us. We saw this reaction in Chapter 6 on carboxylic acid derivatives as well. It is just the hydrolysis of an amide:

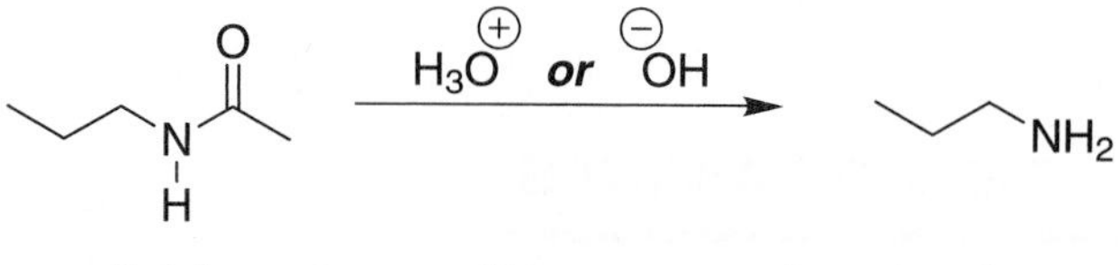

Notice that we have pulled the acyl group off to regenerate the amine. So we now know how to put an acyl group on and then take it back off:

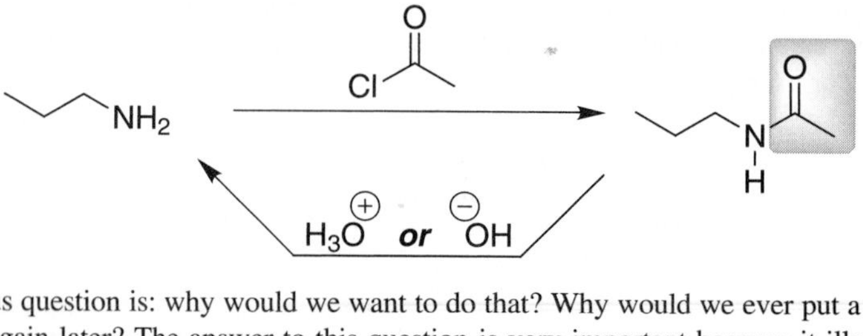

But the obvious question is: why would we want to do that? Why would we ever put a group on, just to take it off again later? The answer to this question is very important because it illustrates a common strategy that organic chemists use. Let's try to answer this question through a specific example.

Imagine that we want to do the following transformation:

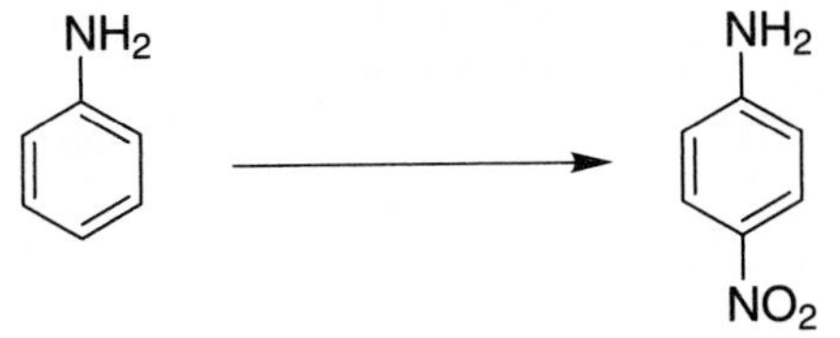

This seems easy to do. Do you remember how to put a nitro group onto a ring (we saw this in Chapter 3). We just used a mixture of nitric acid and sulfuric acid. The amino group is an activator, and so it will direct to the ortho and para positions, which is what we want here. So we propose the following:

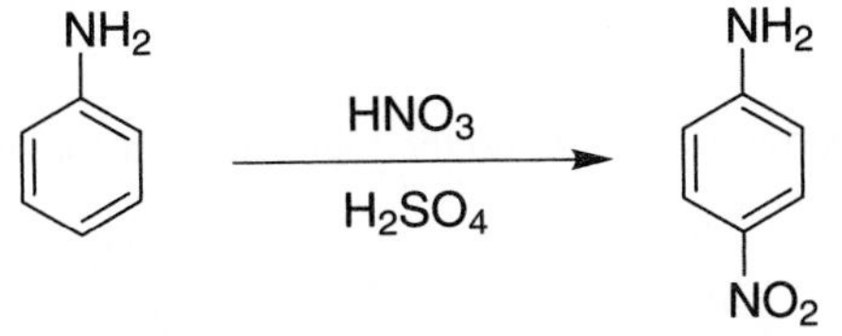

But when we try this, we find that it does not work. It is true that the amino group is a strong activator, but the problem is that it is **too strong** of an activator. You have a highly activated ring being exposed to a very strong oxidizing agent. The ring is so highly activated that the mixture of nitric acid and sulfuric acid will produce unwanted oxidation reactions. The ring will get oxidized, and we will destroy aromaticity, which is not a good thing if you just want to put a nitro group onto the ring. So how can we make our desired product?

The way to do it is actually very clever. We first acylate the amino group:

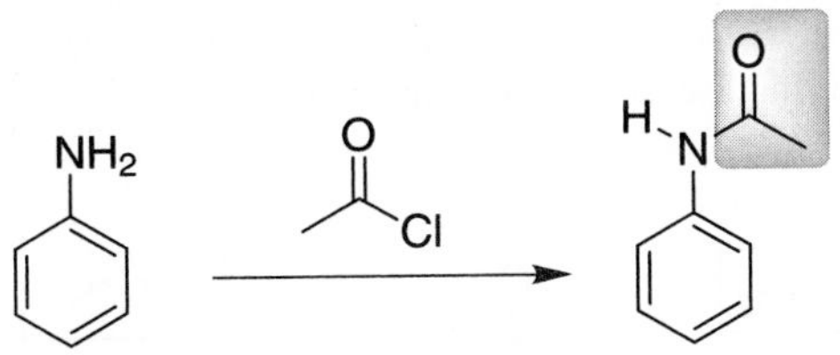

This converts the amino group (which is a **very strong** activator) into a **moderate activator**. Now that it is a moderate activator, we no longer get the unwanted oxidation reactions. So we can nitrate our ring without any problems:

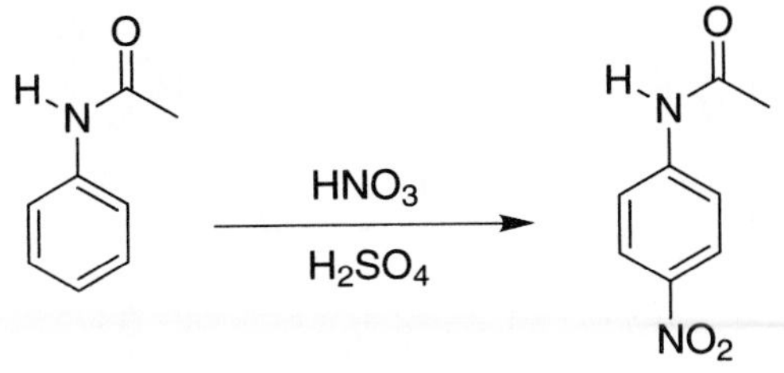

Then we pull off the acyl group to get our desired product:

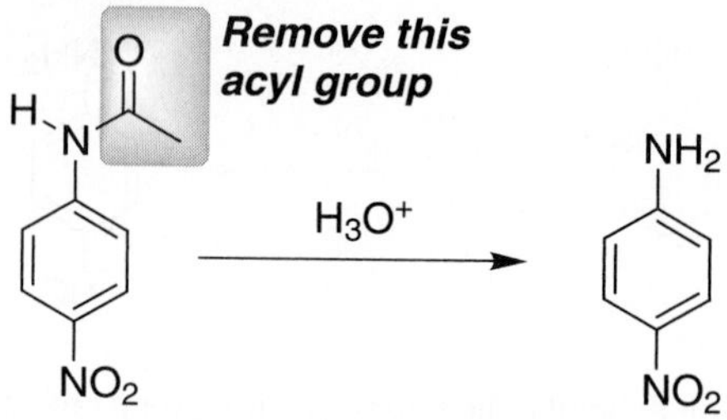

Think about what we just did. We used the acylation process as a way of *temporarily modifying* the electronics of the amino group, so that it would not ruin our reaction. Organic chemists use this strategy all of the time. This idea of temporarily modifying a functional group (and then converting it back later) is used in many other situations as well (not just for acylation of amines).

EXERCISE 8.34. Suppose we want to do the following transformation:

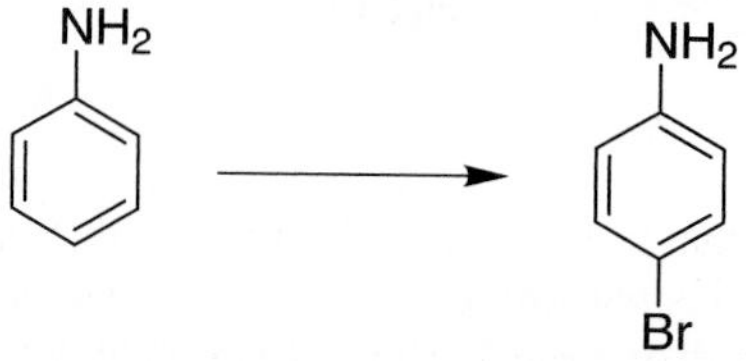

We try to do this transformation using Br_2, but we find that aniline is too reactive, and we get a mixture of mono-, di-, and tri-brominated products. What can we do to get our product without having to worry about polybromination?

Answer: The problem is the amino group: it is too strongly activating. To get around this obstacle, we use the strategy we have developed in this section. We temporarily acylate the amino group, and that makes the ring less activated (temporarily):

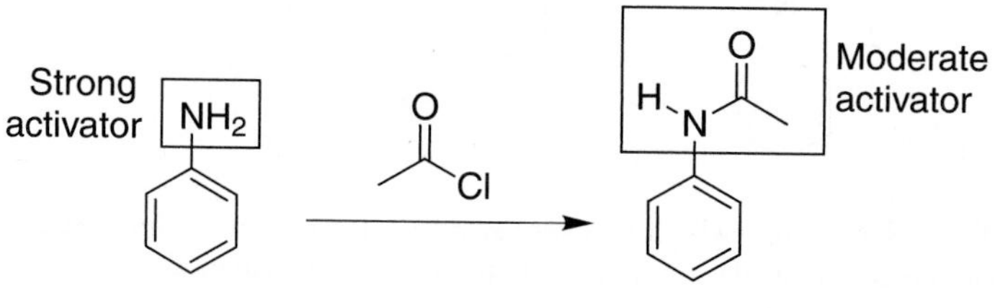

Now we are able to brominate:

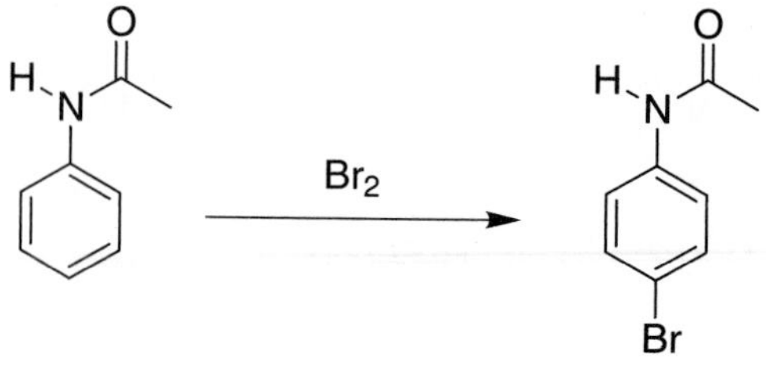

and that puts the bromine where we want it. Finally, we pull off the acyl group to get our product:

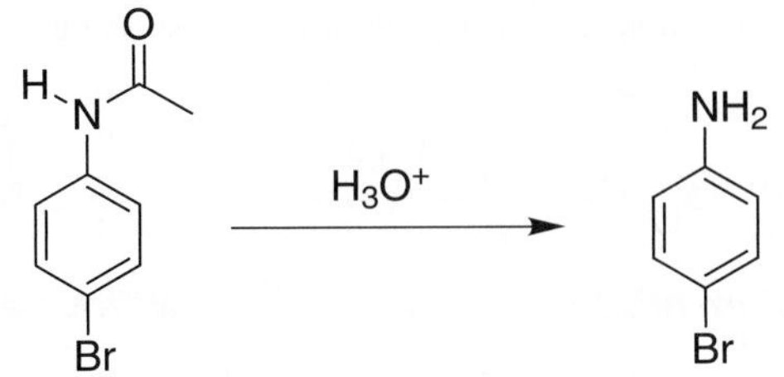

Propose an efficient synthesis for each of the following transformations:

8.35.

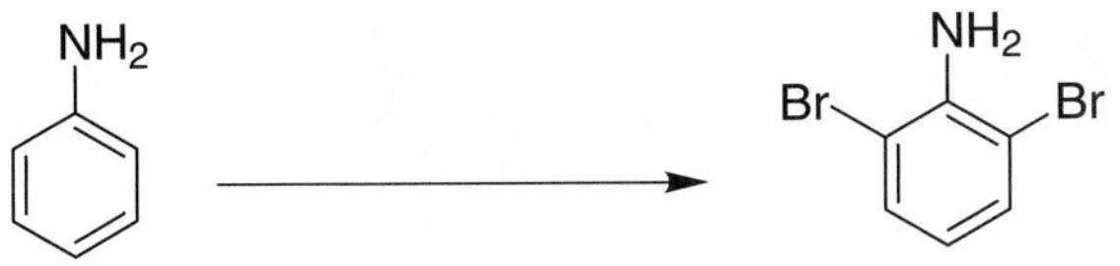

Hint: To solve this problem, you will need to use a blocking technique (sulfonation). If you don't remember how to do that, you can go back and look at Chapter 3 (section 3.8— Predicting and Exploiting Steric Effects).

8.36.

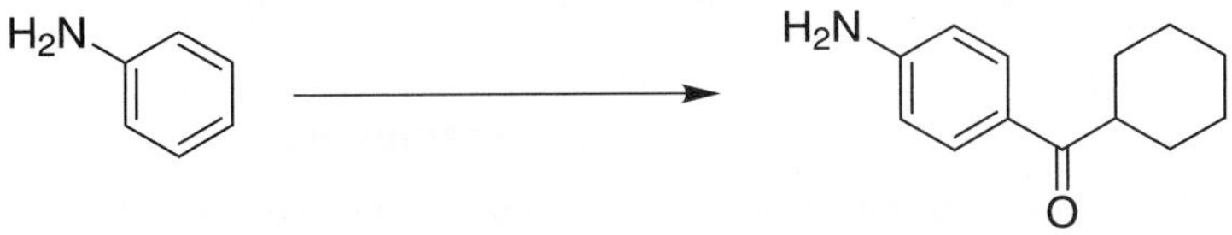

8.37.

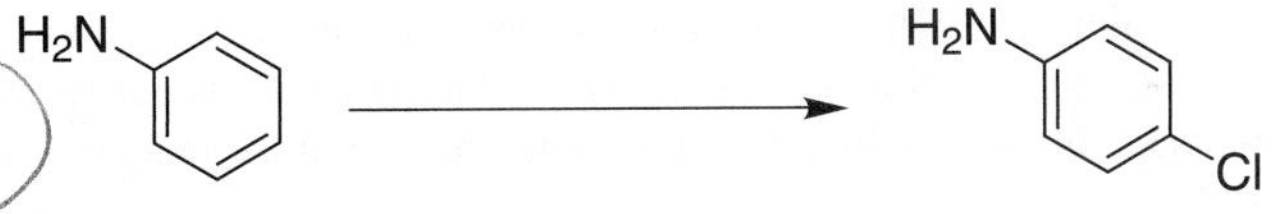

8.6 REACTIONS OF AMINES WITH NITROUS ACID

In this section, we will begin to explore the reactions that take place between amines and nitrous acid. Compare the structures of nit*rous* acid and nit*ric* acid:

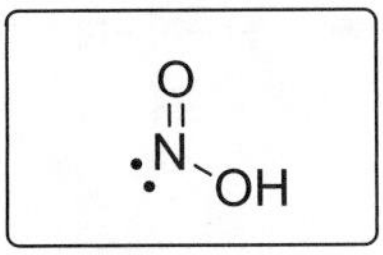

Nitrous acid 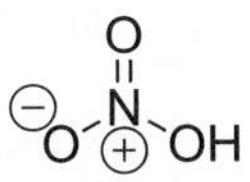   **Nitric acid**

When nitrous acid reacts with amines, the products are very useful. We will soon see that these products can be used in a large number of synthetic transformations. So let's make sure that you are comfortable with the reactions between amines and nitrous acid.

Let's start by looking at the source of nitrous acid. It turns out that nitrous acid is fairly unstable, and therefore, we cannot just purchase it. You won't find it stored in a bottle. Rather, we have to make nitrous acid in the reaction flask, and to do this, we use sodium nitrite ($NaNO_2$) and HCl:

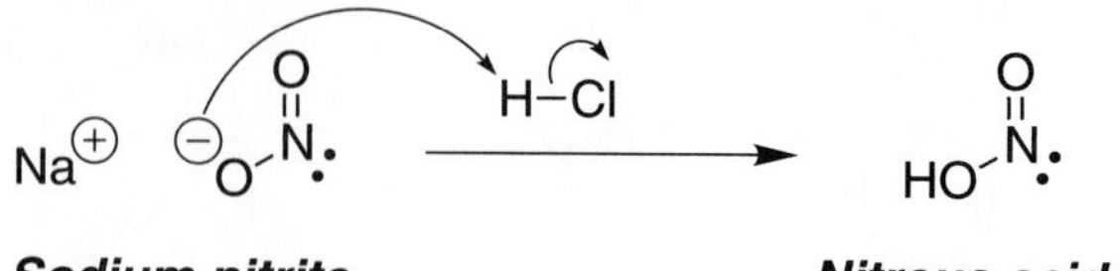

Sodium nitrite   **Nitrous acid**

Under these (acidic) conditions, nitrous acid gets protonated again, to produce a positively charged intermediate:

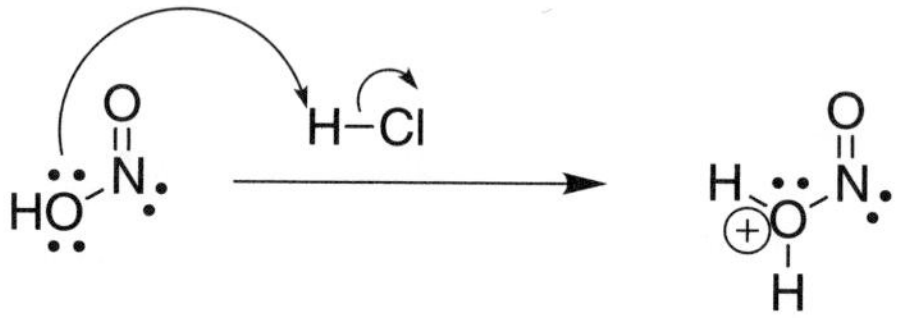

This intermediate can then lose water to give a highly reactive intermediate, called a nitrosonium ion:

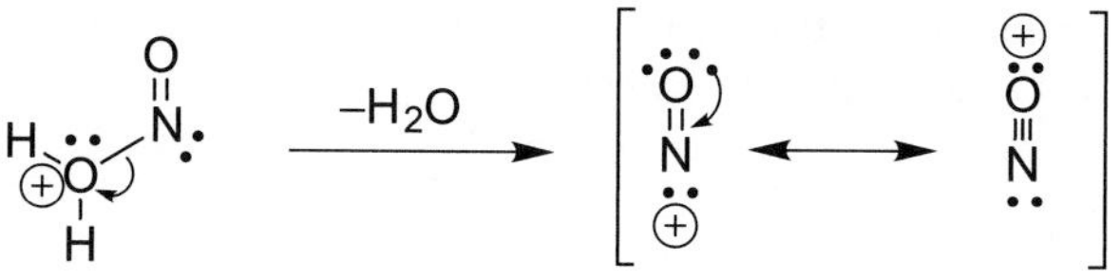

Nitrosonium ion

This intermediate (the nitrosonium ion) is the intermediate we need to focus on. Whenever we talk about an amine reacting with nitrous acid, we really mean to say that the amine is reacting with a nitrosonium ion (NO^+). You might notice the similarity between this intermediate and the NO_2^+ intermediate (that we used in nitration reactions). Do not confuse these two intermediates. NO^+ and NO_2^+ are different intermediates. In this section, we are only talking about the reactions of amines with the nitrosonium ion (NO^+).

Nitrosonium ions cannot be stored in a bottle. Instead, we must make them *in the presence of our amine*. That way, as soon as the nitrosonium ion is formed, it will immediately react with the amine before it has a chance to do anything else. This is called an *in situ* preparation.

So, now the question is: what happens when an amine reacts with a nitrosonium ion? Let's begin by looking at secondary amines (and then we will look at primary amines).

A secondary amine can attack a nitrosonium ion like this:

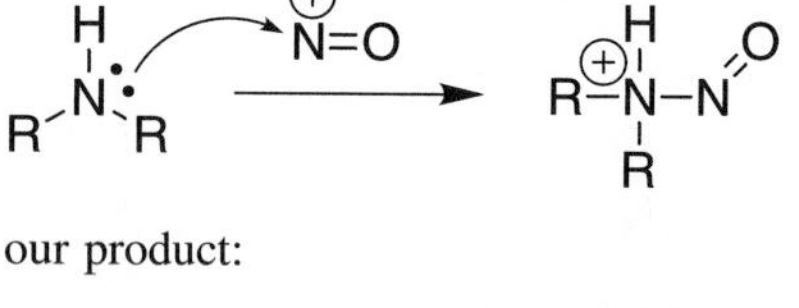

We then lose a proton to get our product:

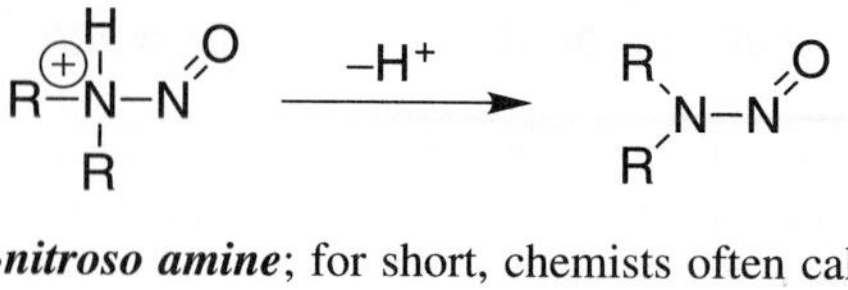

This product is called an *N-nitroso amine*; for short, chemists often call it a *nitrosamine*.

This reaction is not very useful to you in solving problems, but when a ***primary*** amine attacks a nitrosonium ion, we get a reaction that is extremely important. The amine attacks, to initially form a nitrosamine:

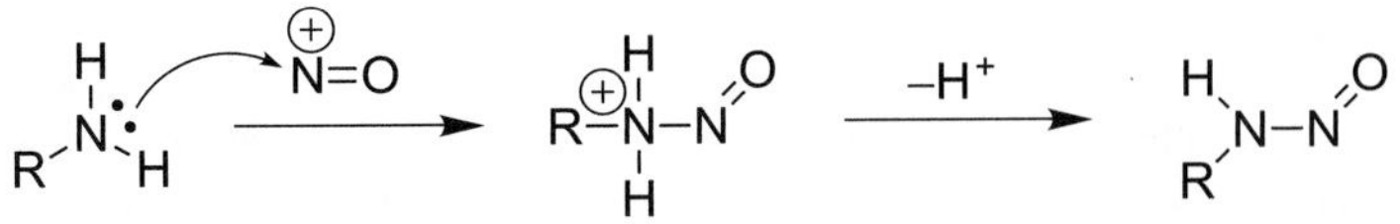

Since we used a primary amine to start with, we notice that we have one extra proton in our nitrosamine:

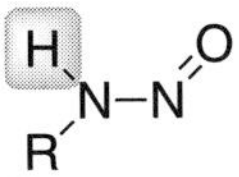

Because of this proton, the nitrosamine continues to react in the following way:

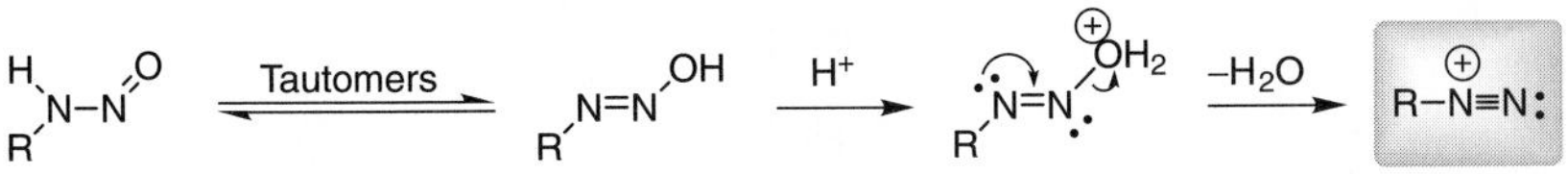

This product is called a ***diazonium*** ion. *Azo* means nitrogen, so ***diazo*** means two nitrogen atoms. And of course, ***onium*** means a positive charge. That is what we have here: two nitrogen atoms connected to each other and a positive charge—thus, the name ***diazonium***.

Primary ***alkyl*** amines will give ***alkyl*** diazonium salts, and primary ***aryl*** amines will give ***aryl*** diazonium salts:

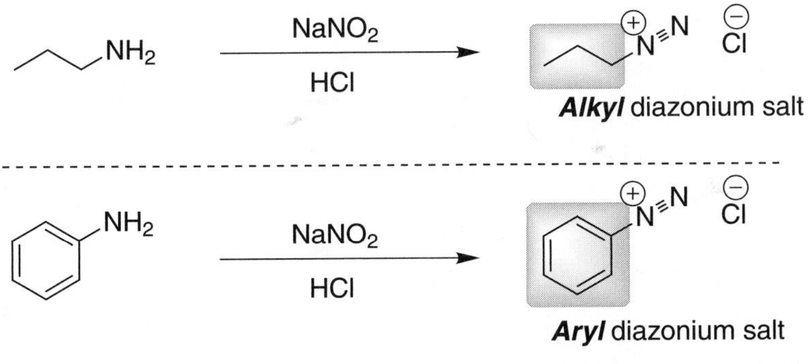

Alkyl diazonium salt

Aryl diazonium salt

Alkyl diazonium salts are not terribly useful. They are very explosive, and as a result, they are very dangerous to prepare. But ***aryl*** diazonium salts are much more stable, and they are incredibly useful, as we will see in the upcoming section. For now, let's just make sure that we know how to make diazonium salts:

EXERCISE 8.38. Predict the product of the following reaction:

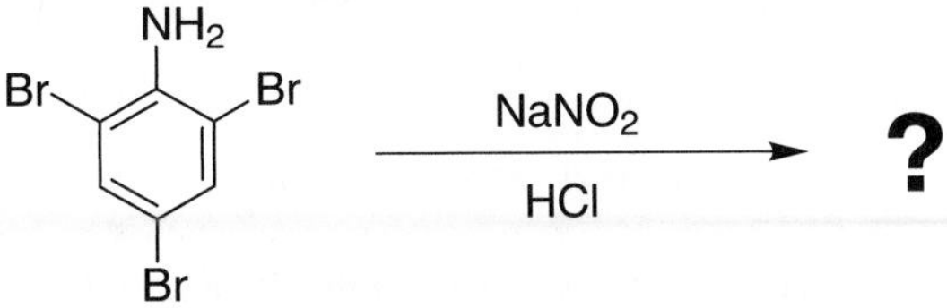

Answer: Our starting material is primary amine. These reagents (sodium nitrite and HCl) are used to form nitrous acid, which then forms a nitrosonium ion. Primary amines react with a nitrosonium ion to give a diazonium salt. So the product of this reaction is:

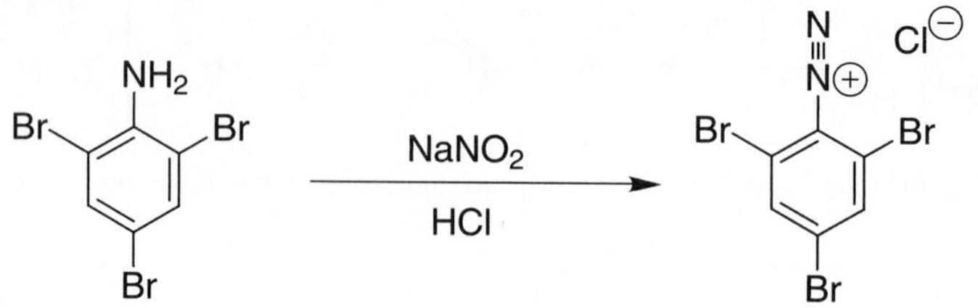

Predict the major product for each of the following reactions:

8.39.

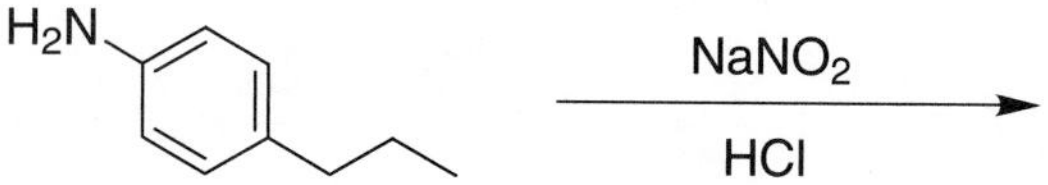

$$\text{(structure)} \xrightarrow[\text{HCl}]{\text{NaNO}_2}$$

8.40.

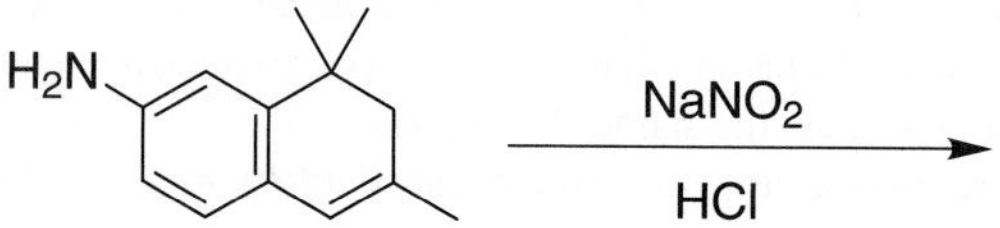

$$\text{(structure)} \xrightarrow[\text{HCl}]{\text{NaNO}_2}$$

8.41.

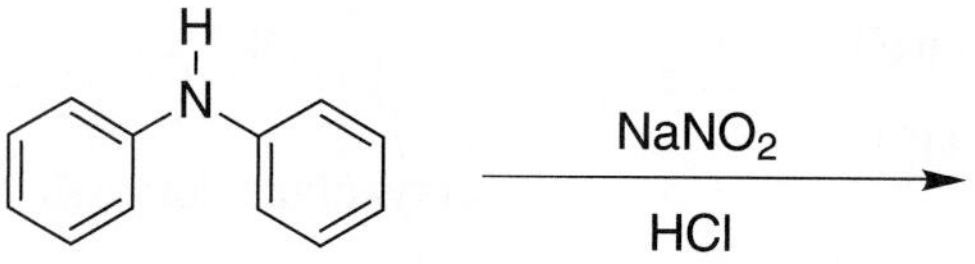

$$\text{(structure)} \xrightarrow[\text{HCl}]{\text{NaNO}_2}$$

8.42.

$$\text{(structure)} \xrightarrow[\text{HCl}]{\text{NaNO}_2}$$

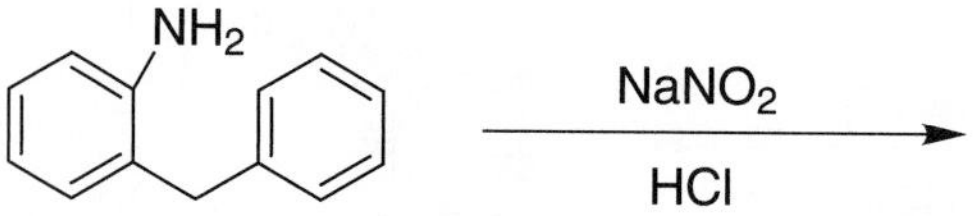

8.7 AROMATIC DIAZONIUM SALTS

In the previous section, we learned how to make aryl diazonium salts:

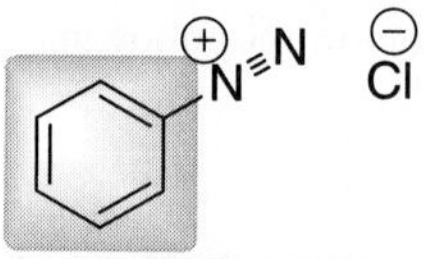

An *aryl* diazonium salt

Now we will learn what we can do with aryl diazonium salts. Here are a few reactions to start us off:

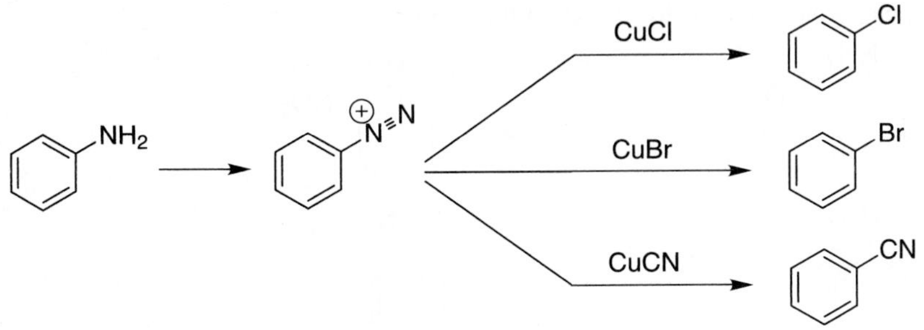

In all of these reactions, we are using copper salts as the reagents. These reactions are called *Sandmeyer reactions*. They are useful because they allow us to do transformations that we could not otherwise do with the chemistry that we learned in Chapters 3 and 4 (electrophilic and nucleophilic aromatic substitution).

These reactions allow us to convert an amino group into a halogen or a cyano group. As a case in point, earlier we did not see how to put a cyano group onto a ring; this is our first way to introduce a cyano group onto an aromatic ring. Let's get a bit of practice with some simple problems.

EXERCISE 8.43. What reagents would you use to do the following transformation:

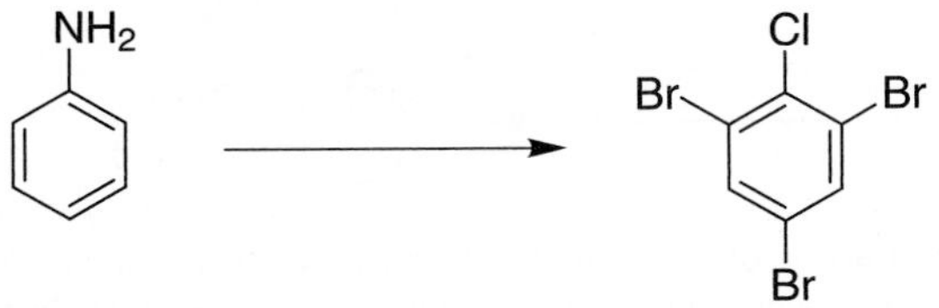

Answer: If we brominate aniline, we will find that the amino group is so activated that we will see a tribrominated product:

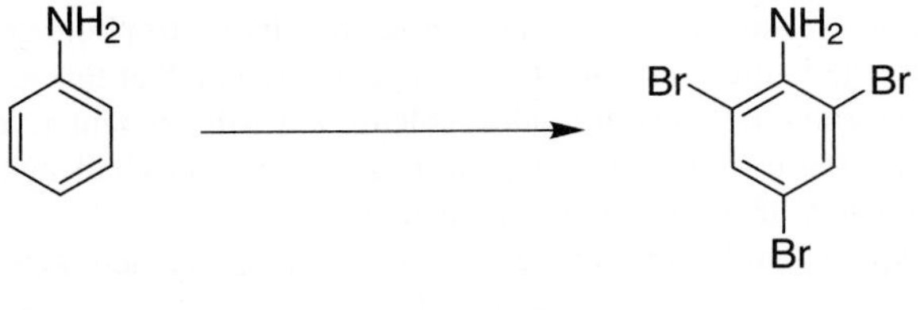

Then, we can convert the amino group into a chloro group. To do so, we make a diazonium salt, and then we use a Sandmeyer reaction:

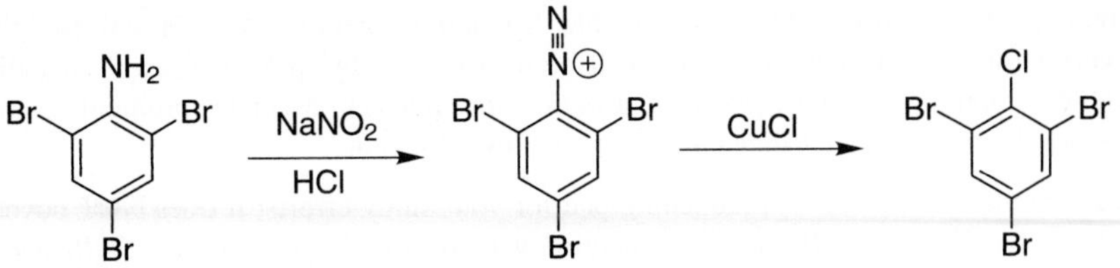

What reagents would you use to do each of the following transformations:

8.44.

8.45.

8.46.

8.47.

8.48.

So far, we have seen just a few of the many useful reactions that you can do with aryl diazonium salts. Some instructors will cover many of these reactions, whereas other instructors will cover just a few of these reactions. You should look through your lecture notes to see what you are responsible for. Your textbook will certainly show you at least two or three more reactions that you can do with aryl diazonium salts.

These reactions are very useful in synthesis problems. You will find that some problems will combine these reactions with the reactions that you learned in electrophilic aromatic substitution. These problems can range in difficulty, and they can get really tough at times. These types of synthesis problems are typically the most difficult problems you will encounter. You will find many, many such problems in your textbook. As you go through some of the challenging problems in your textbook, I will leave you with a bit of last-minute advice:

To master these types of problems, you need to do two important activities:

1. You must review the reactions and principles from the entire course. Go through your textbook and your lecture notes again and again and again. Make sure that you get to a point where you know all of the reactions cold (you should have command over all of the reactions).

2. You must do as many problems as possible. If you don't get practice, you will find that even a very strong grasp of the reactions will be insufficient. In order to truly master the art of problem solving, you must *practice, practice, practice*. Do as many problems as possible. You might even find that they can be fun, believe it or not.

This book was meant to serve as a launching pad for your study efforts; it does NOT cover everything that you need to know. Rather, my intention was to provide you with the *skills* and *understanding* that you need to study more efficiently. Good luck . . .

ANSWERS

CHAPTER 2

2.2. HÖ::ÖR   **2.3.** ⊖Ö::ÖR   **2.4.** [structure]   **2.5.** [structure]   **2.6.** [structure]

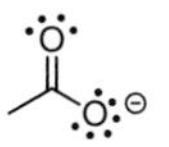

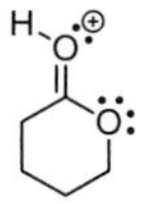

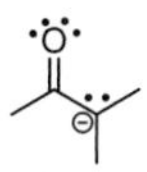

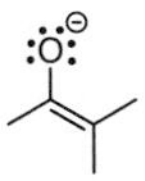

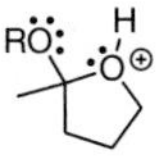

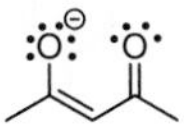

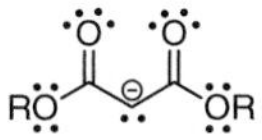

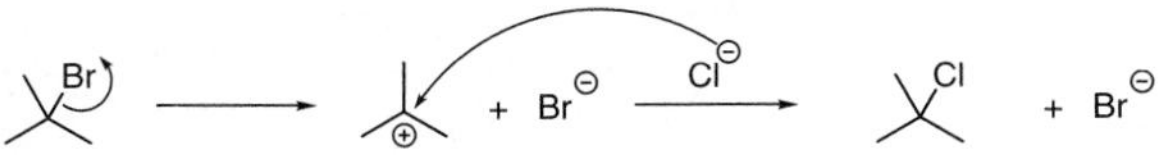

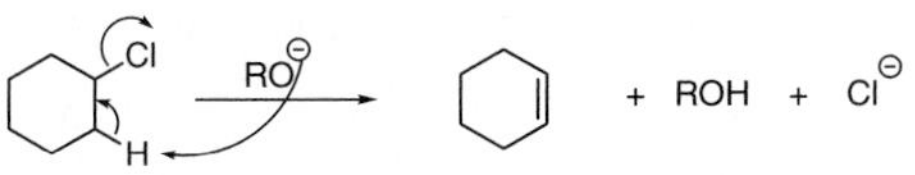

2.7. [structure]   **2.8.** [structure]   **2.9.** [structure]   **2.10.** [structure]

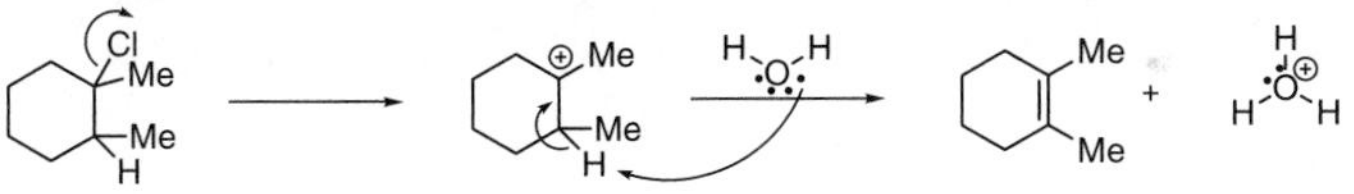

2.11. [reaction scheme]

2.12. [reaction scheme]

2.13. [reaction scheme]

2.14. [reaction scheme]

2.15. [reaction scheme]

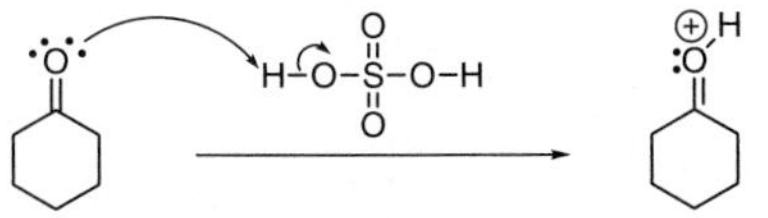

2.17. loss of a leaving group
2.18. proton transfer
2.19. rearrangement
2.20. nucleophile attacking an electrophile
2.21. proton transfer
2.22. loss of a leaving group
2.23. nucleophile attacking an electrophile
2.24. rearrangement
2.25. loss of a leaving group
2.26. nucleophile attacking an electrophile
2.28. proton transfer; nucleophile attacking an electrophile; proton transfer
2.29. nucleophile attacking an electrophile; proton transfer; proton transfer
2.30. proton transfer; nucleophile attacking an electrophile; loss of a leaving group

277

2.31. proton transfer; loss of a leaving group; nucleophile attacks an electrophile; proton transfer
2.32. proton transfer; nucleophile attacks an electrophile; proton transfer
2.33. nucleophile attacks an electrophile; loss of a leaving group; proton transfer

CHAPTER 3

3.2.

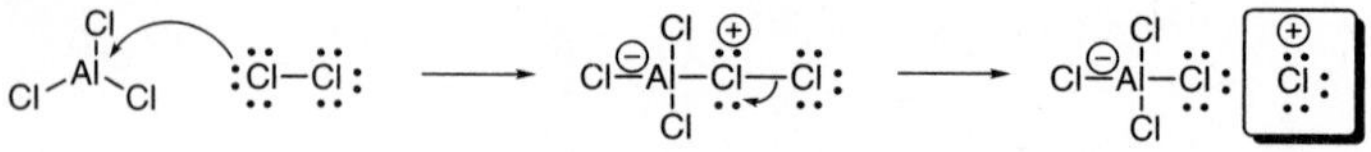

3.3.

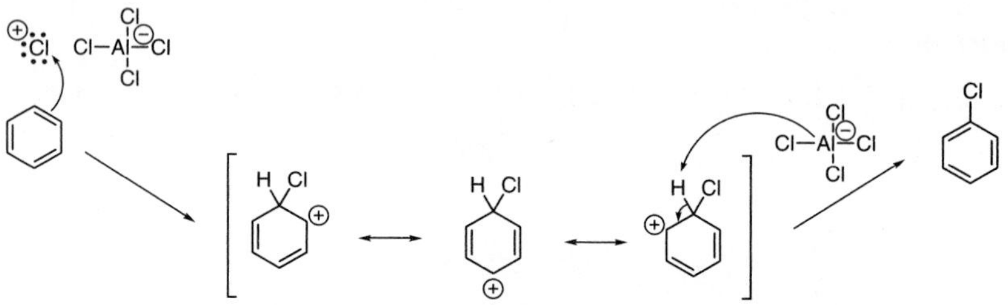

3.5.

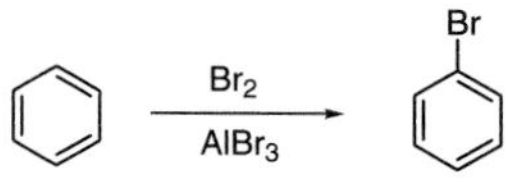

3.6.

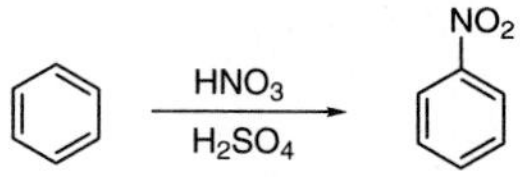

3.7.

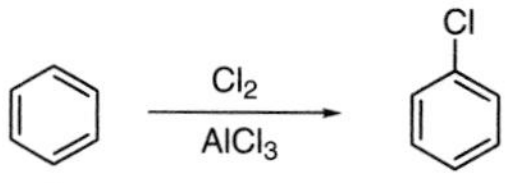

3.10.

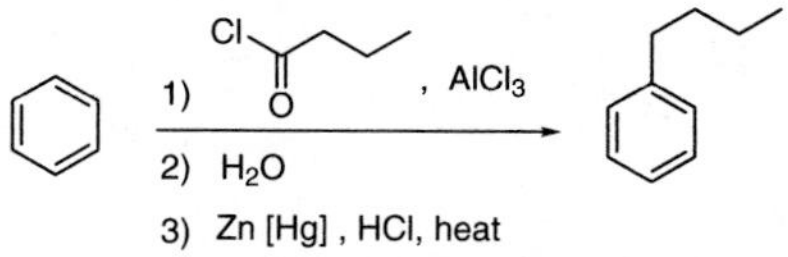

3.11.

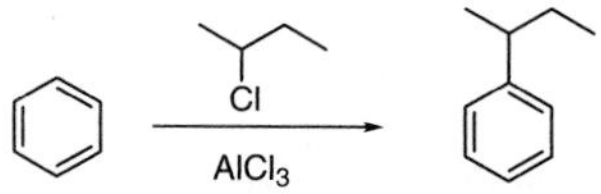

3.12.

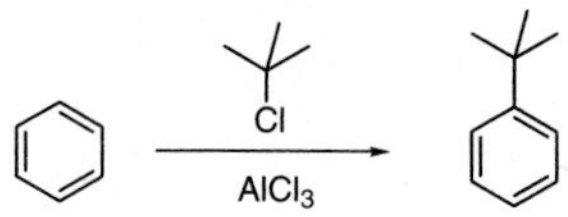

3.13.

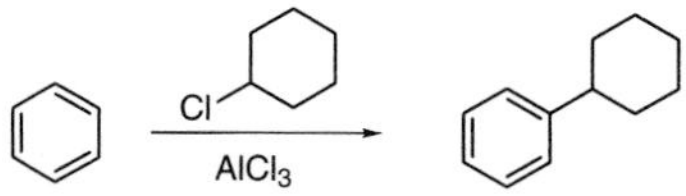

3.14.

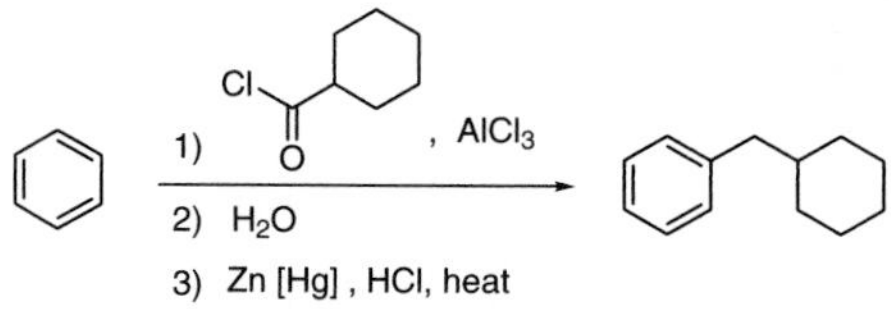

3.15.

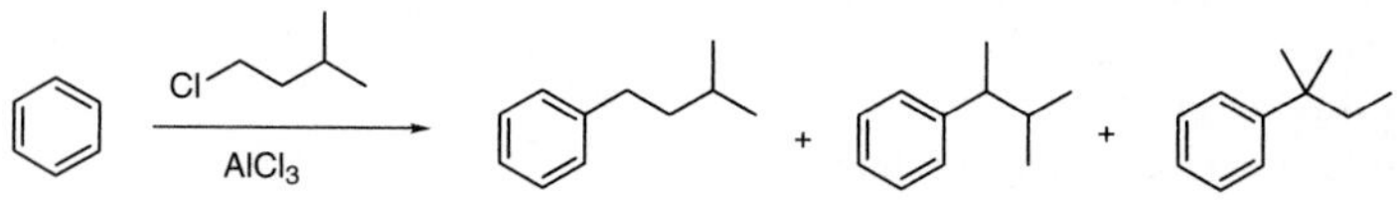

3.17.
Formation of acylium ion:

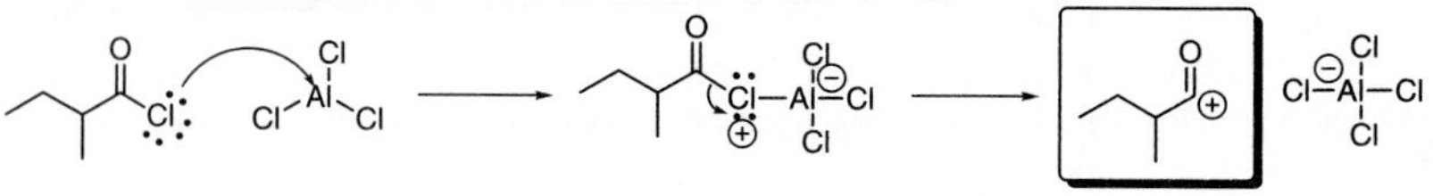

Electrophilic aromatic substitution:

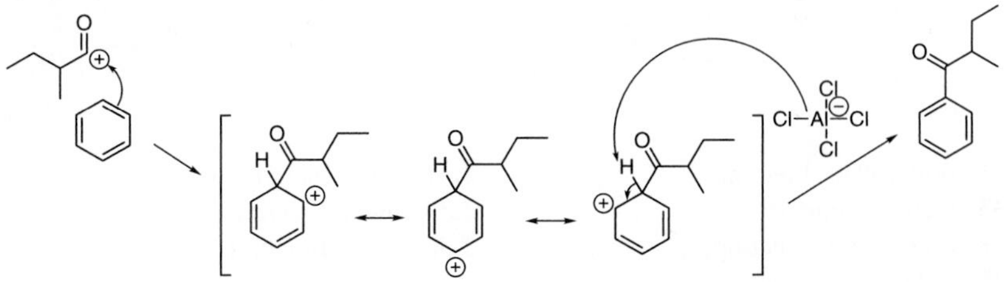

3.19.

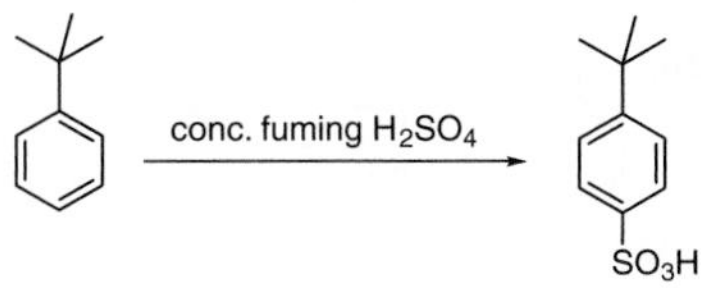

conc. fuming H₂SO₄

3.20

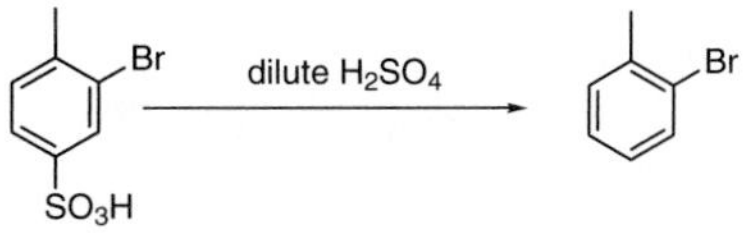

dilute H₂SO₄

3.21.

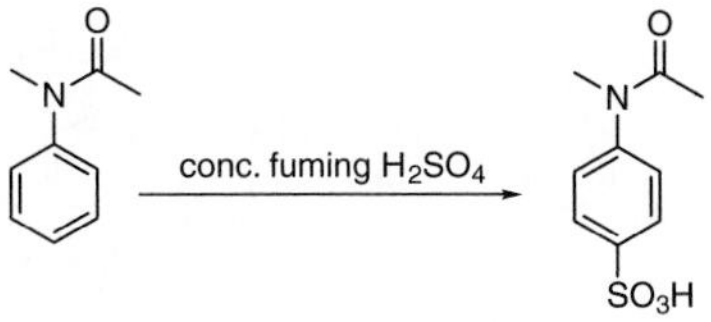

conc. fuming H₂SO₄

3.22.

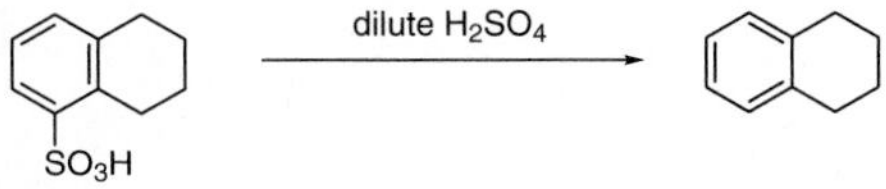

dilute H₂SO₄

3.23.

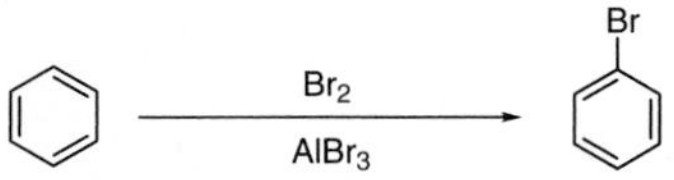

Br₂ / AlBr₃

3.24.

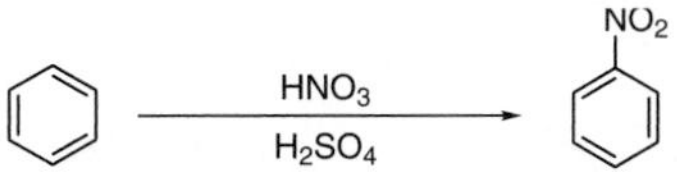

HNO₃ / H₂SO₄

3.25.

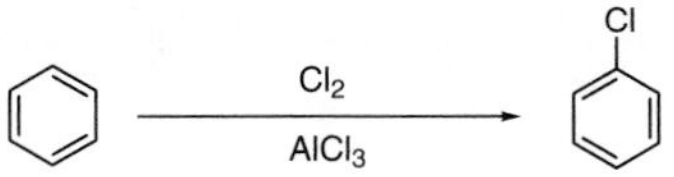

Cl₂ / AlCl₃

3.26.

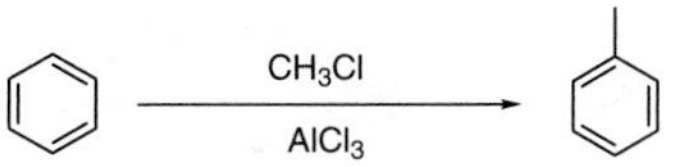

CH₃Cl / AlCl₃

3.27.

1) (O=)C-Cl , AlCl₃
2) H₂O
3) Zn [Hg] , HCl, heat

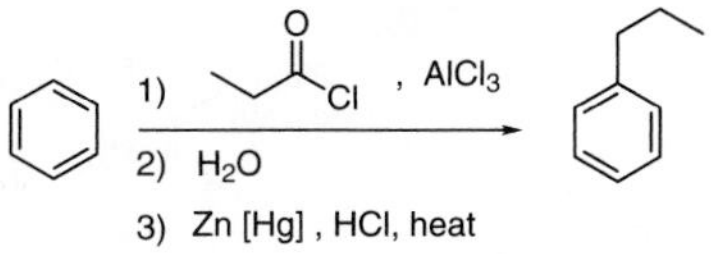

3.28.

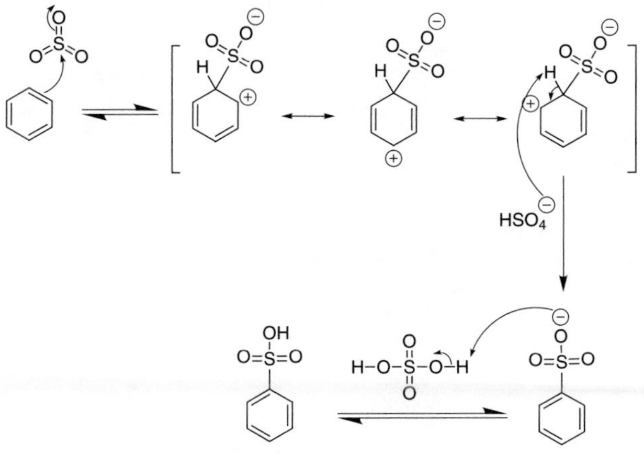

3.29.

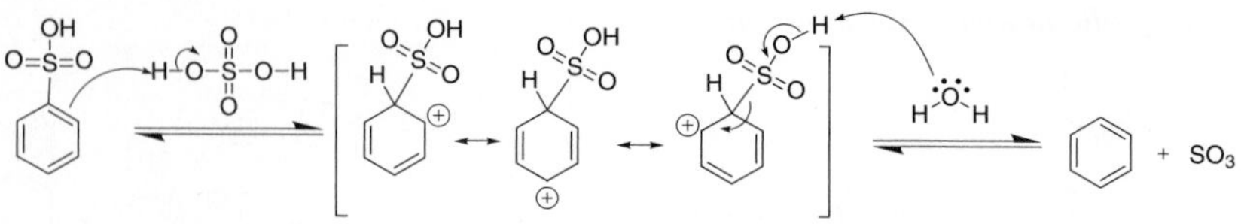

3.31. ortho, para directing   **3.32.** ortho, para directing
3.33. meta directing   **3.34.** meta directing
3.35. ortho, para directing   **3.36.** meta directing
3.37. ortho, para directing

3.39.

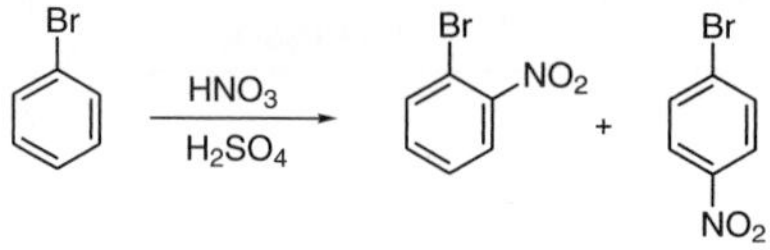

3.40.

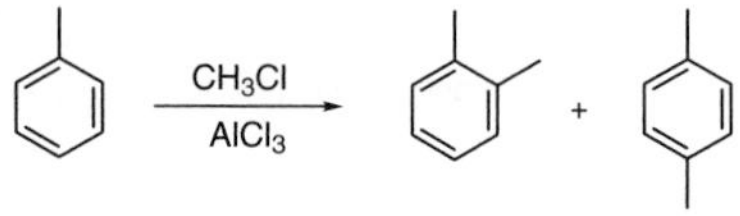

3.41.

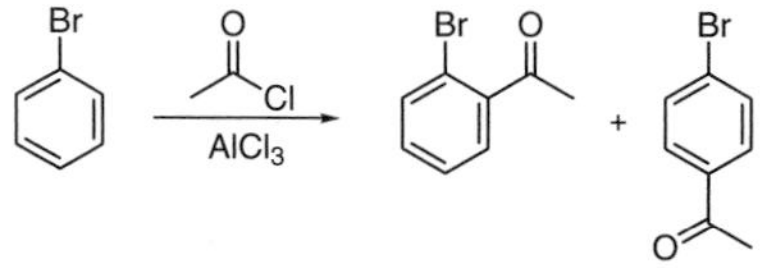

3.42.

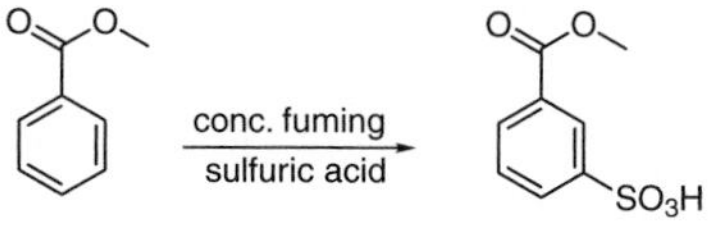

3.43.

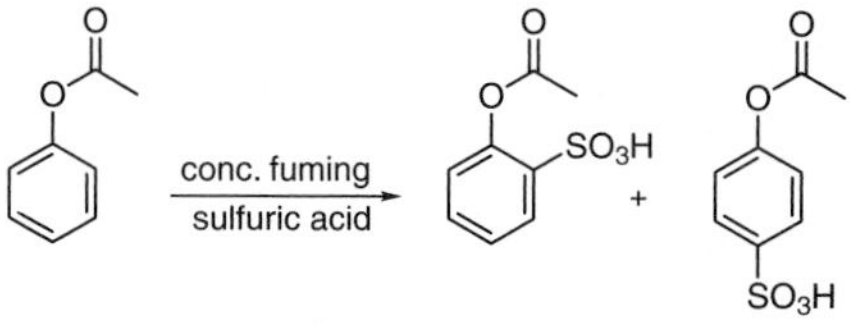

3.44.

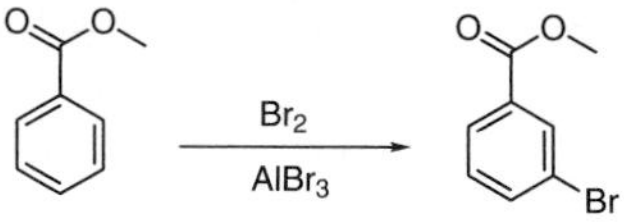

3.45.

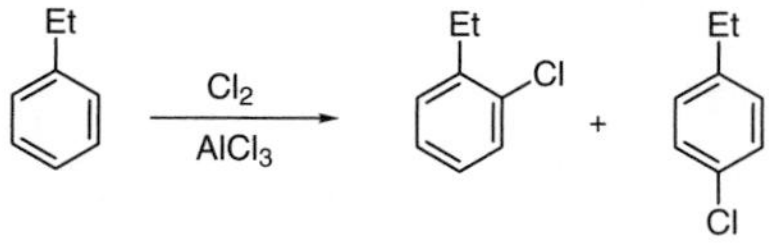

3.47. Strong Activator / Strong Deactivator

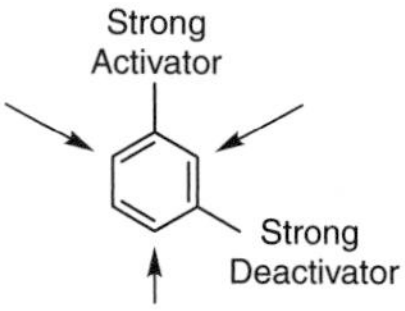

3.48. Weak Activator / Strong Deactivator

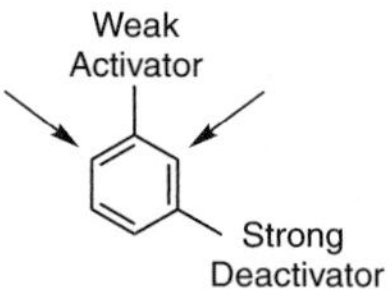

3.49. Strong Activator / Weak Activator

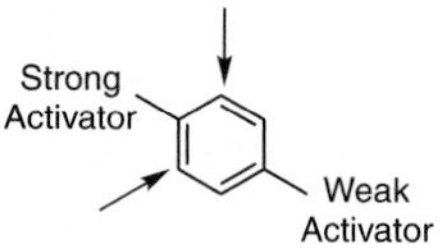

3.50. Strong Activator / Strong Deactivator

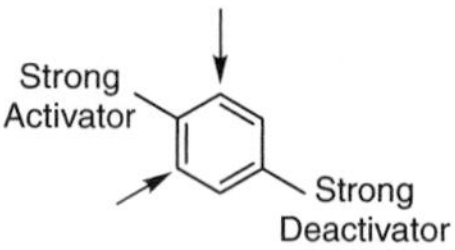

3.51. Strong Deactivator / Weak Deactivator

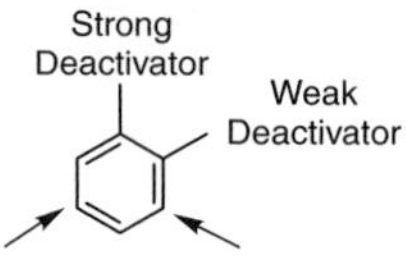

3.52. Strong Activator / Weak Activator

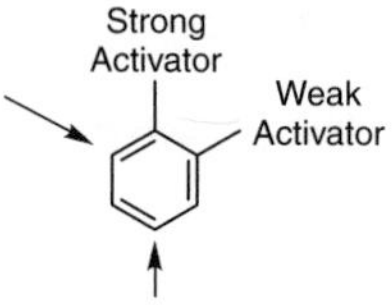

3.53. Weak Activator / Strong Deactivator

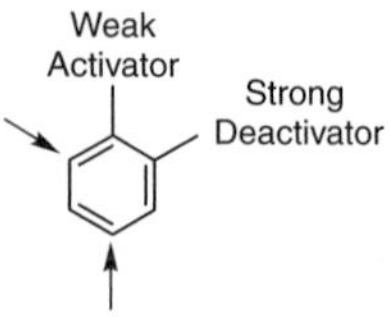

3.54. Strong Activator / Weak Activator

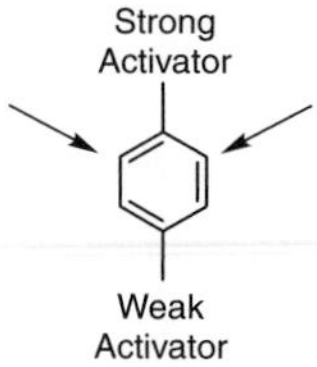

3.55. Strong Deactivator

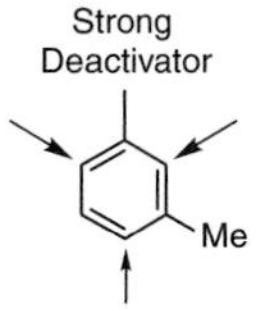

3.56. Strong Activator

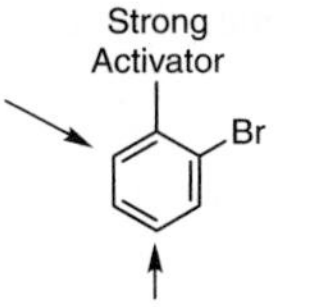

3.58.

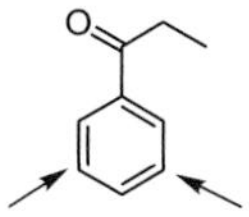

Moderate deactivator

3.59.

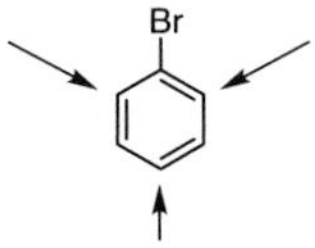

Weak activator

3.60.

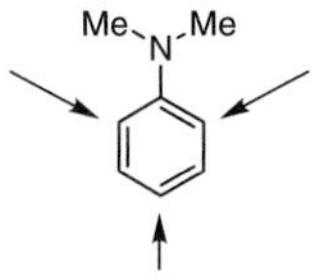

Strong activator

3.61.

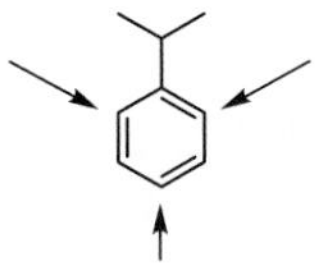

Weak activator

3.62.

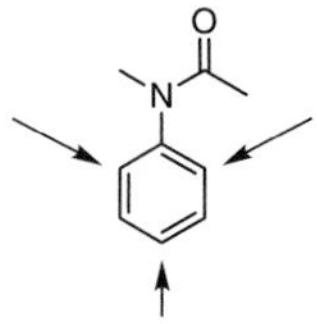

Moderate activator

3.63.

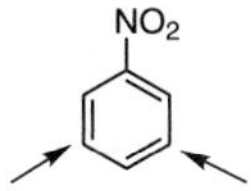

NO_2

Strong deactivator

3.64.

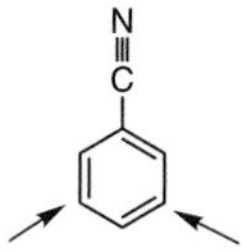

Moderate deactivator

3.65.

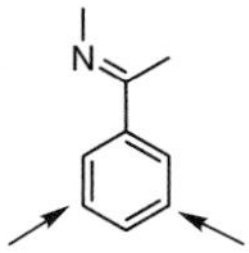

Moderate deactivator

3.66.

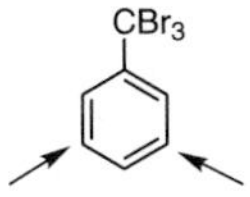

CBr_3

Strong deactivator

3.67.

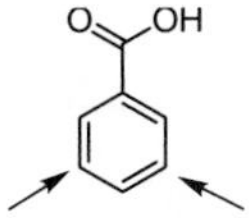

Moderate deactivator

3.68. The resonance structures show that the lone pair on the nitrogen atom is delocalized, and is spread throughout the ring, which strongly activates the ring. The effect is the same as if the lone pair was next to the ring.

3.70.

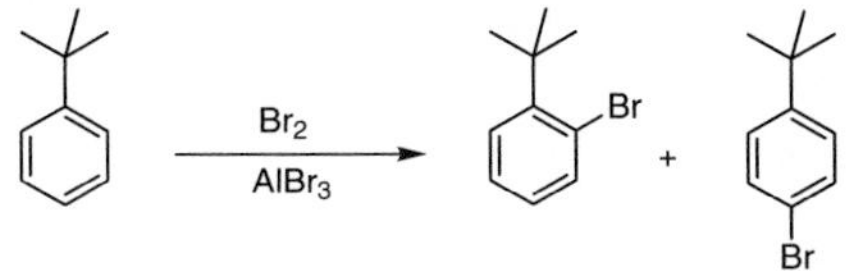

3.71.

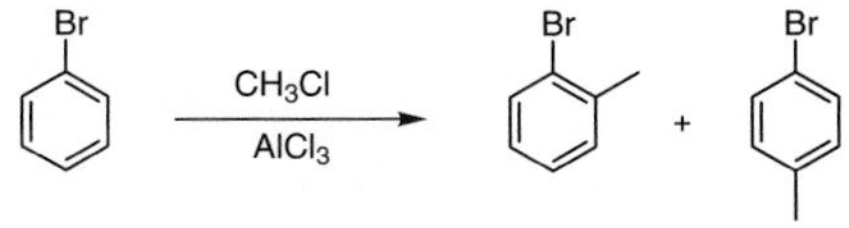

3.72.

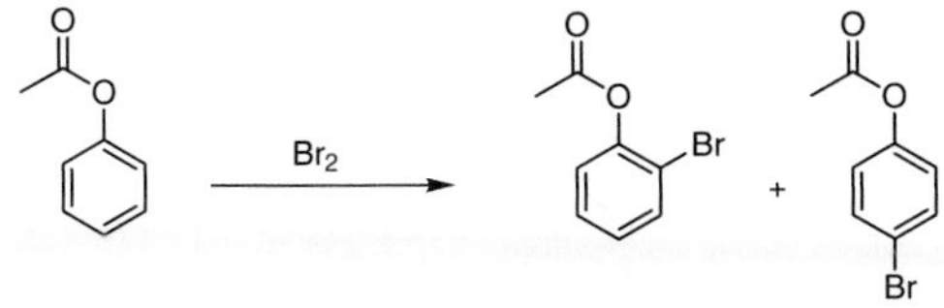

In this case, the ring is moderately activated toward bromination, so a Lewis acid is not necessary.

3.73.

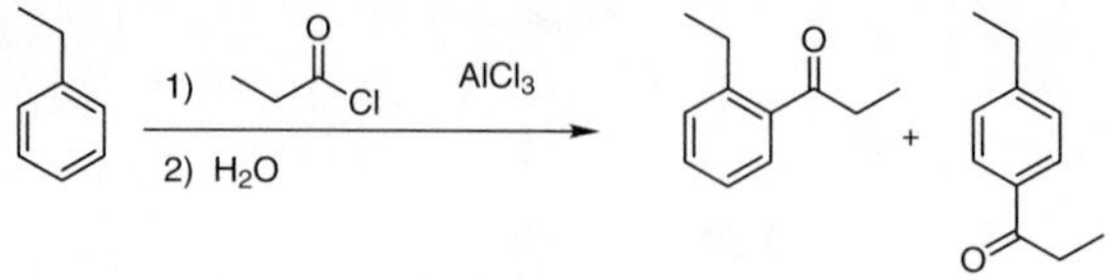

3.74.

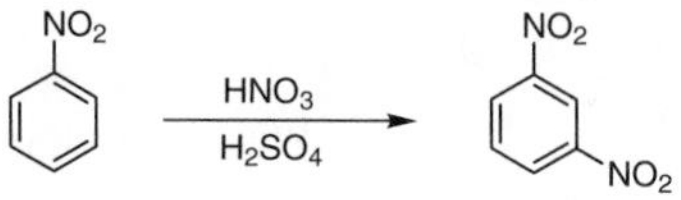

3.76.

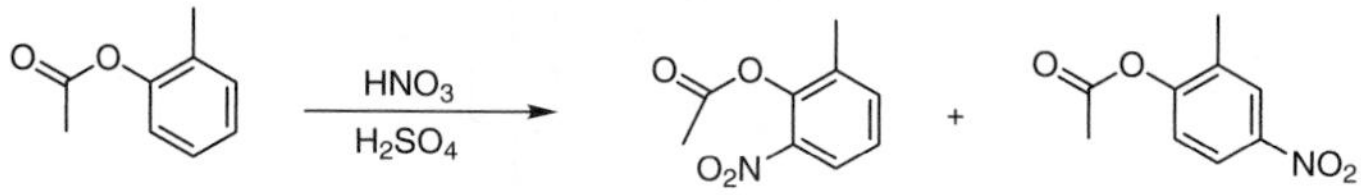

3.77.

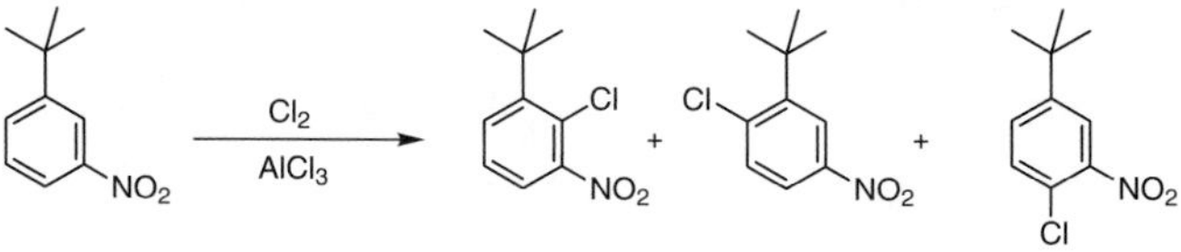

3.78.

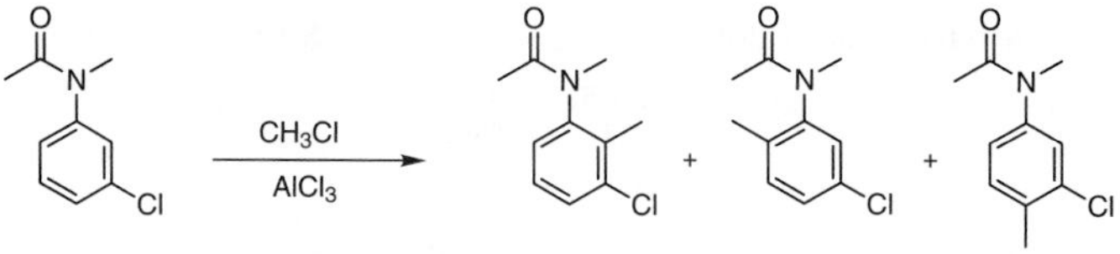

3.79.

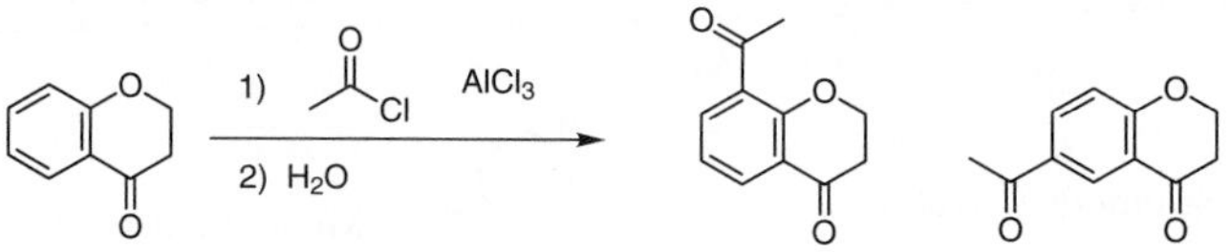

3.81.

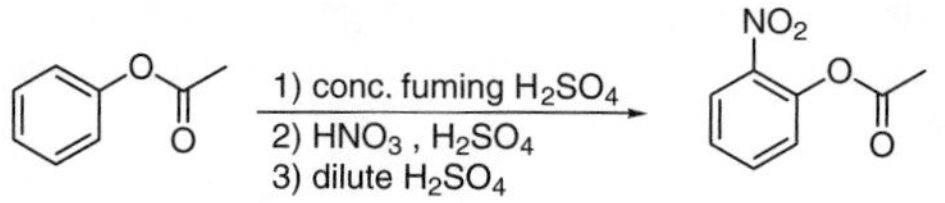

3.82.

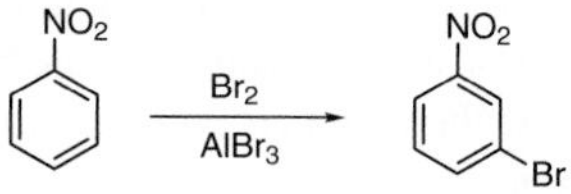

3.83.

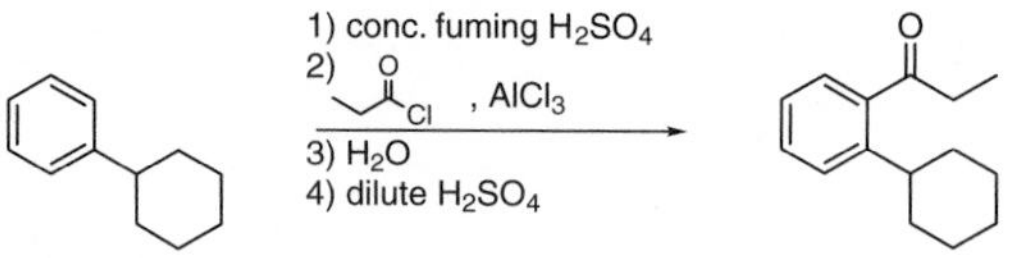

3.84.

3.85.

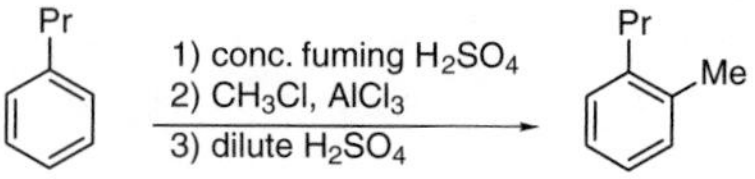

3.87.

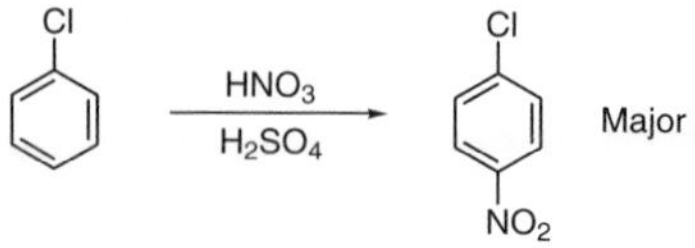

Major

3.88.

Major

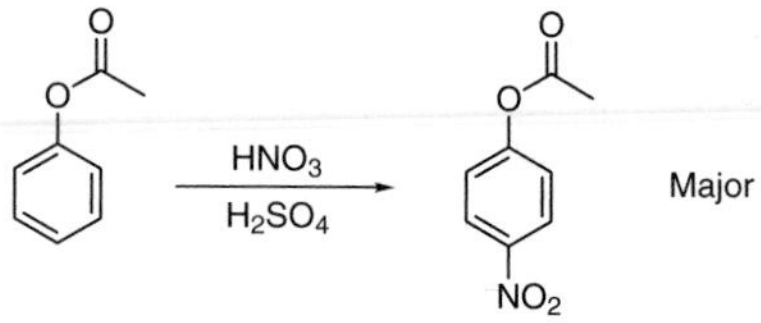

3.89.

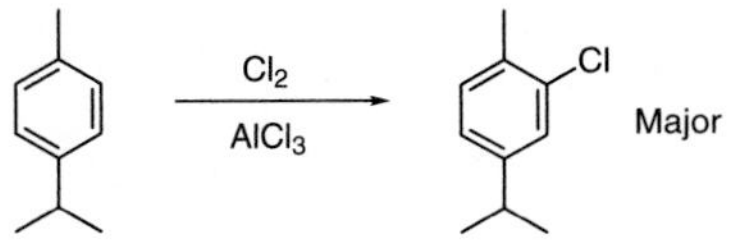

Cl₂ / AlCl₃ → Major

3.90.

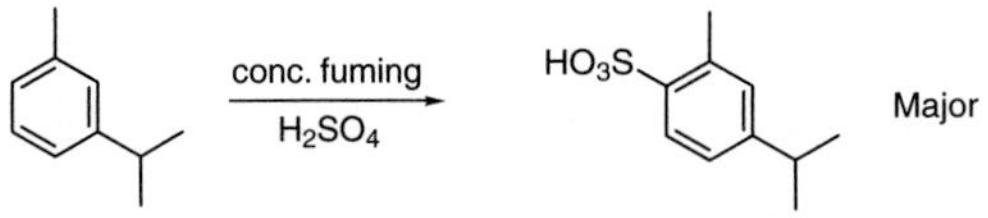

conc. fuming H₂SO₄ → Major

3.91.

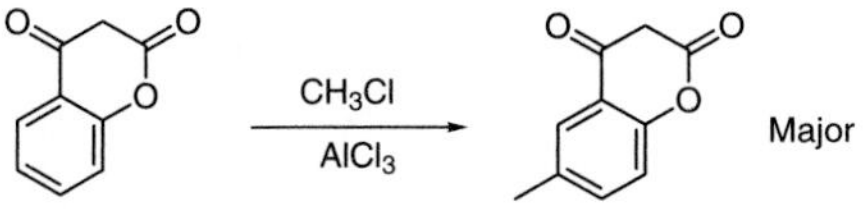

CH₃Cl / AlCl₃ → Major

3.92.

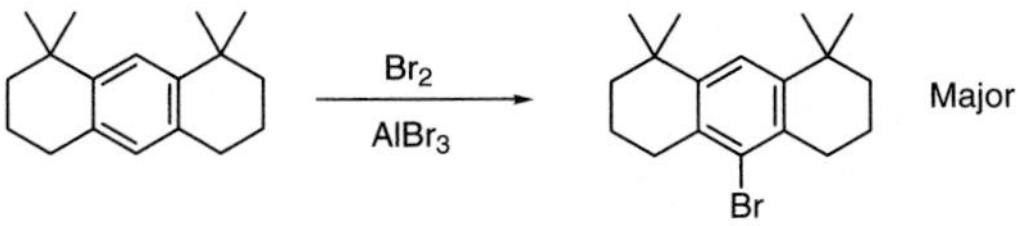

Br₂ / AlBr₃ → Major

3.94.

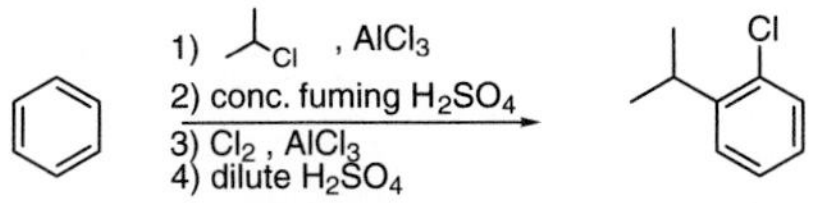

1) (isopropyl chloride), AlCl₃
2) conc. fuming H₂SO₄
3) Cl₂ , AlCl₃
4) dilute H₂SO₄

3.95.

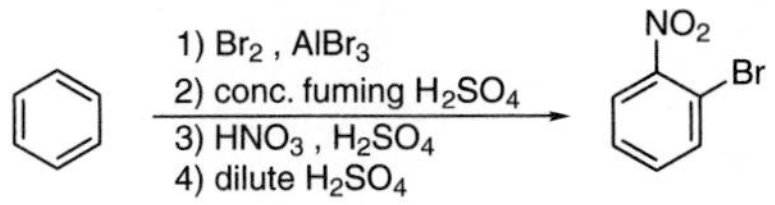

1) Br₂ , AlBr₃
2) conc. fuming H₂SO₄
3) HNO₃ , H₂SO₄
4) dilute H₂SO₄

3.96.

1) (tert-butyl chloride) , AlCl₃
2) conc. fuming H₂SO₄
3) CH₃Cl , AlCl₃
4) dilute H₂SO₄

3.97.

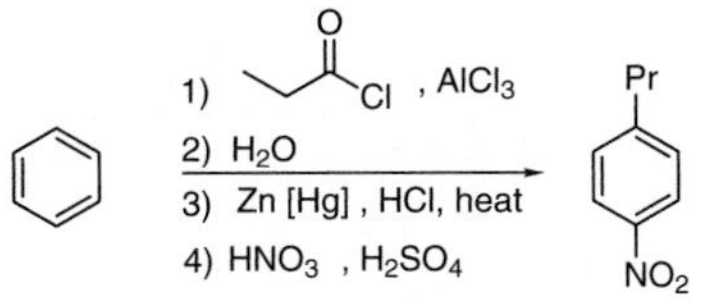

1) (propanoyl chloride), AlCl₃
2) H₂O
3) Zn [Hg] , HCl, heat
4) HNO₃ , H₂SO₄

3.98.

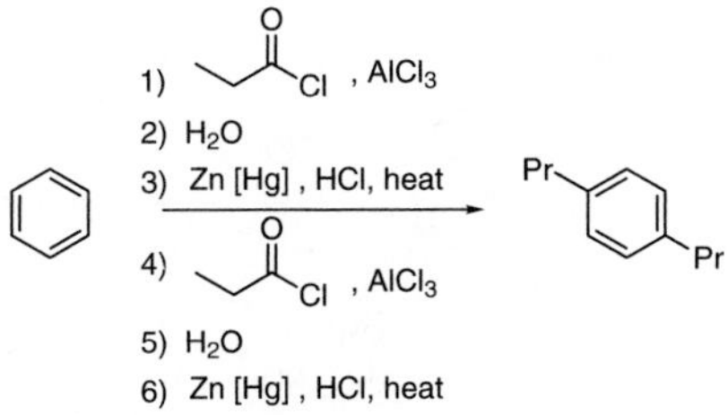

1) (propanoyl chloride) , AlCl₃
2) H₂O
3) Zn [Hg] , HCl, heat
4) (propanoyl chloride) , AlCl₃
5) H₂O
6) Zn [Hg] , HCl, heat

3.99.

1) HNO₃ , H₂SO₄
2) HNO₃ , H₂SO₄
3) EtCl, AlCl₃

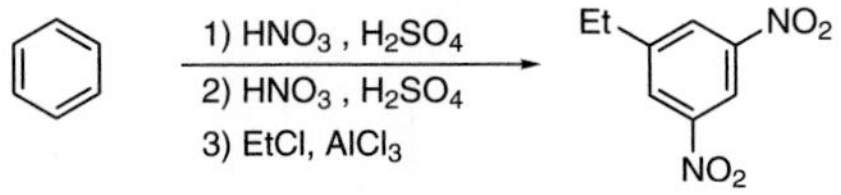

In Problem 3.99, we are putting on two nitro groups. The first nitro group deactivated the ring, and therefore, it is harder to put on the second nitro group. But we are able to put on the second nitro group by heating the reaction mixture.

3.100.

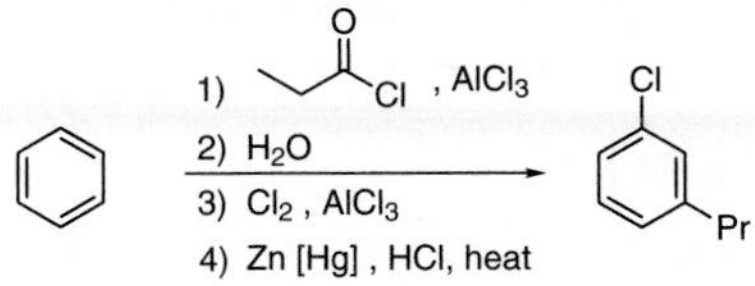

1) (propanoyl chloride) , AlCl₃
2) H₂O
3) Cl₂ , AlCl₃
4) Zn [Hg] , HCl, heat

3.101.

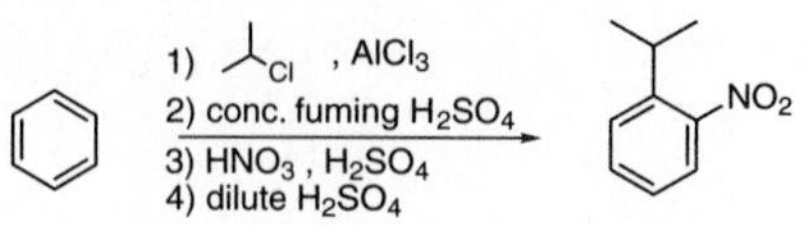

1) (isopropyl chloride) , AlCl$_3$
2) conc. fuming H$_2$SO$_4$
3) HNO$_3$, H$_2$SO$_4$
4) dilute H$_2$SO$_4$

3.102.

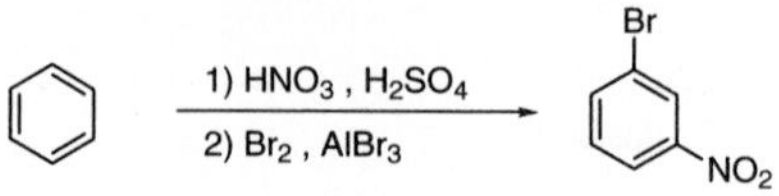

1) HNO$_3$, H$_2$SO$_4$
2) Br$_2$, AlBr$_3$

3.103.

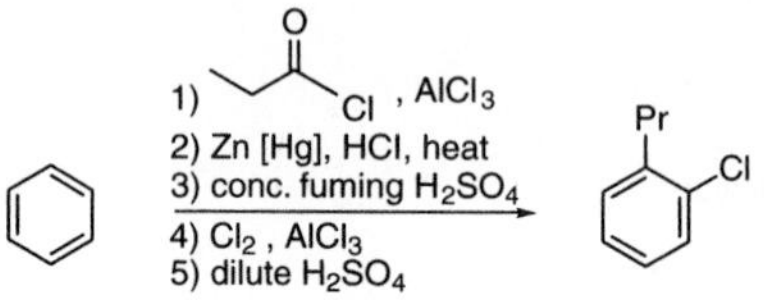

1) (propanoyl chloride) , AlCl$_3$
2) Zn [Hg], HCl, heat
3) conc. fuming H$_2$SO$_4$
4) Cl$_2$, AlCl$_3$
5) dilute H$_2$SO$_4$

3.104.

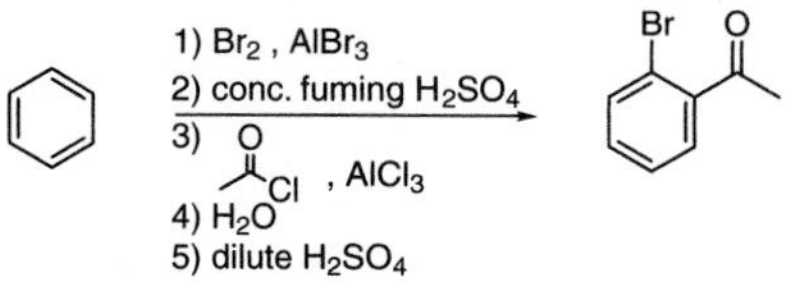

1) Br$_2$, AlBr$_3$
2) conc. fuming H$_2$SO$_4$
3) (acetyl chloride) , AlCl$_3$
4) H$_2$O
5) dilute H$_2$SO$_4$

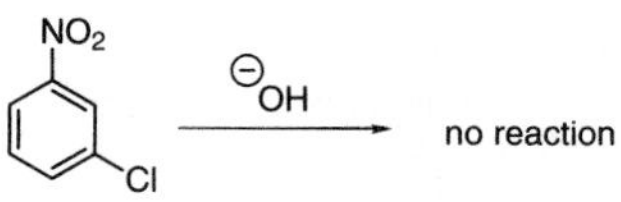

CHAPTER 4

4.2.

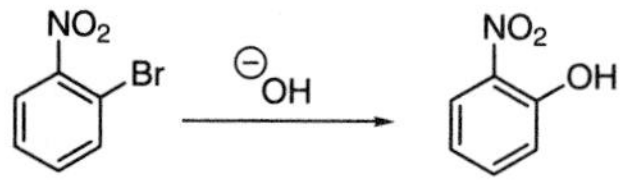

4.3. no reaction

4.4. no reaction

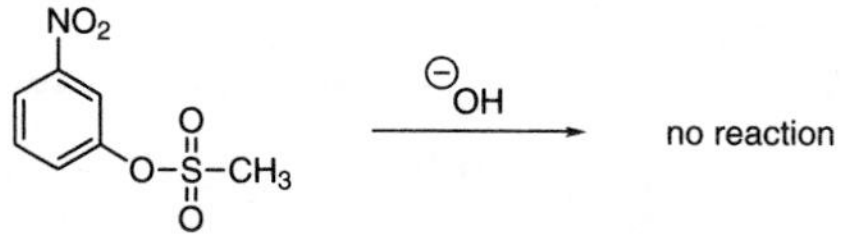

4.5.

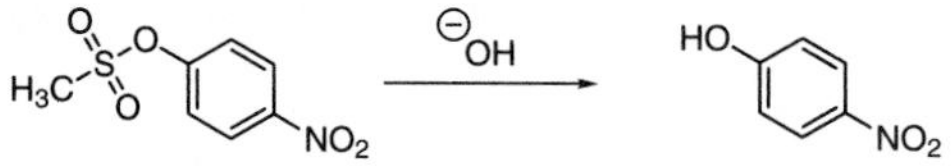

4.6. no reaction

4.7. no reaction

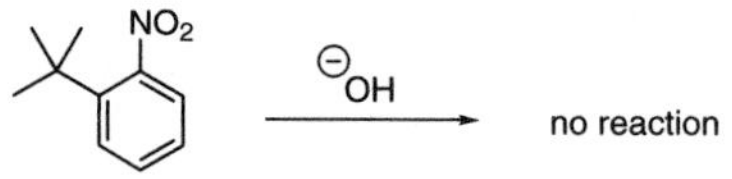

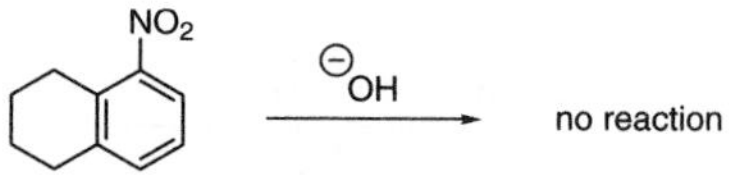

4.9.

MEISENHEIMER COMPLEX

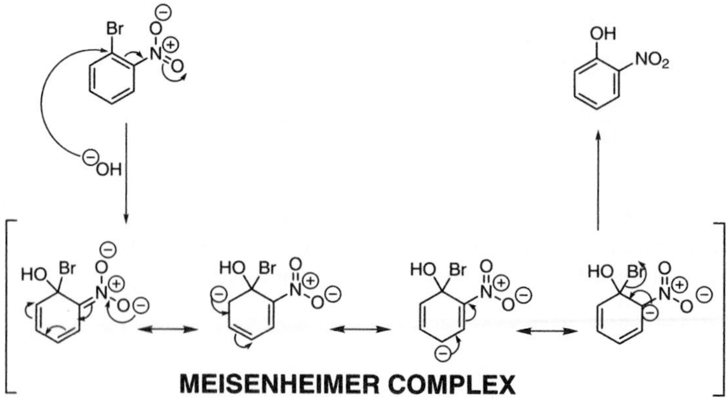

4.10.

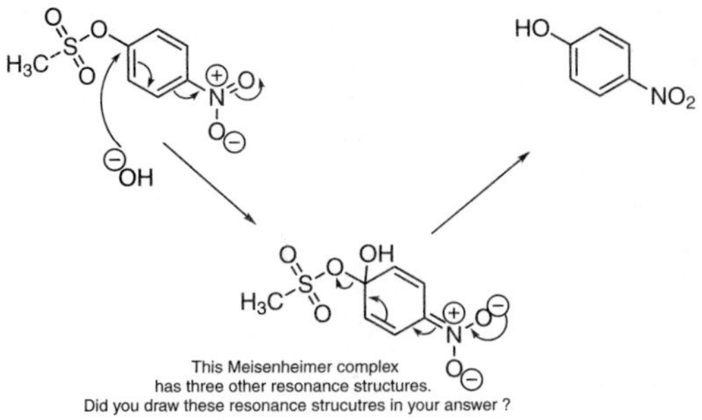

This Meisenheimer complex
has three other resonance structures.
Did you draw these resonance strucutres in your answer ?

4.11.

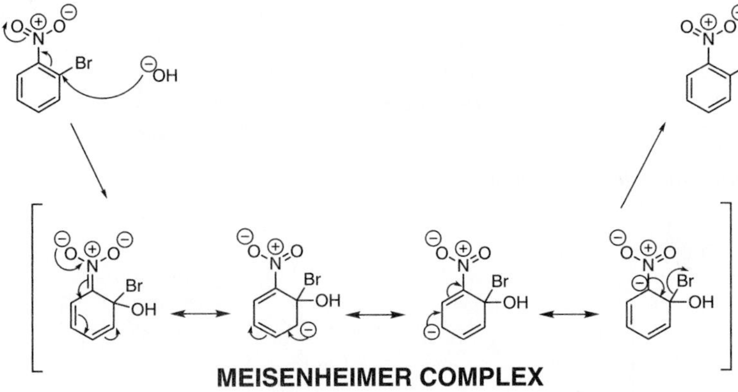

MEISENHEIMER COMPLEX

4.13.

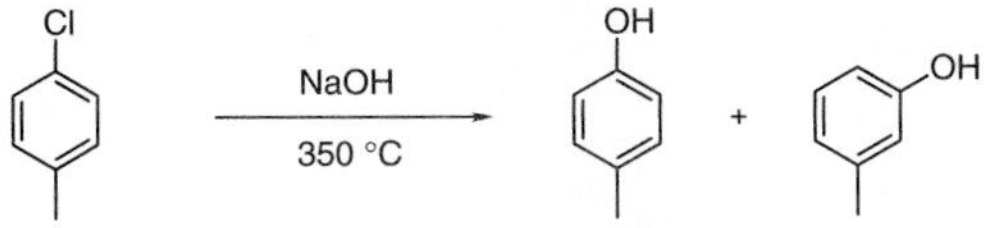

NaOH
350 °C

4.14.

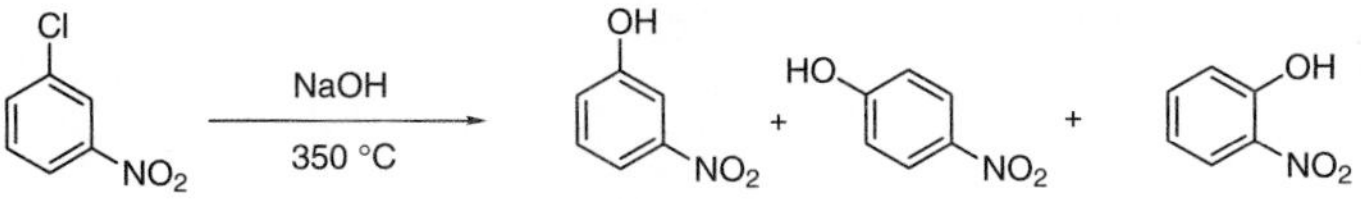

NaOH
350 °C

4.15.

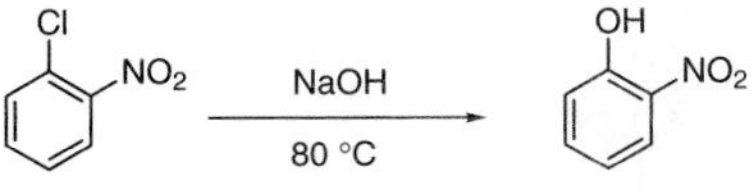

NaOH
80 °C

4.16.

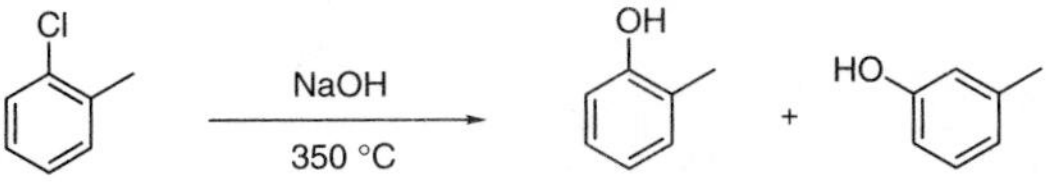

NaOH
350 °C

4.17.

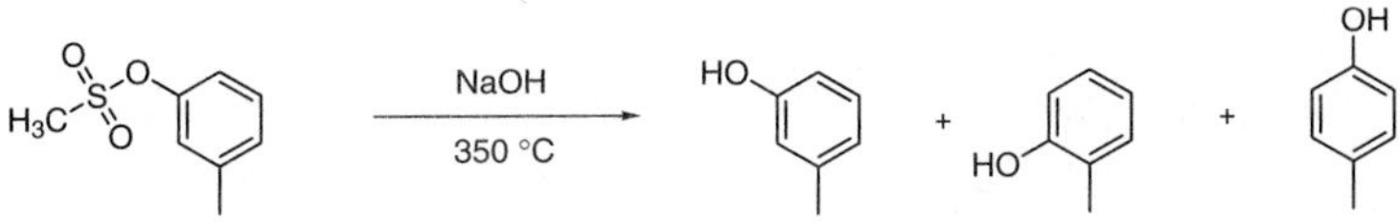

NaOH
350 °C

4.18.

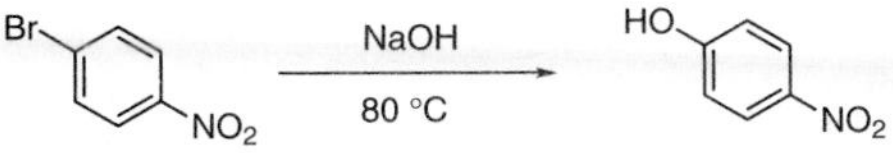

NaOH
80 °C

4.20.

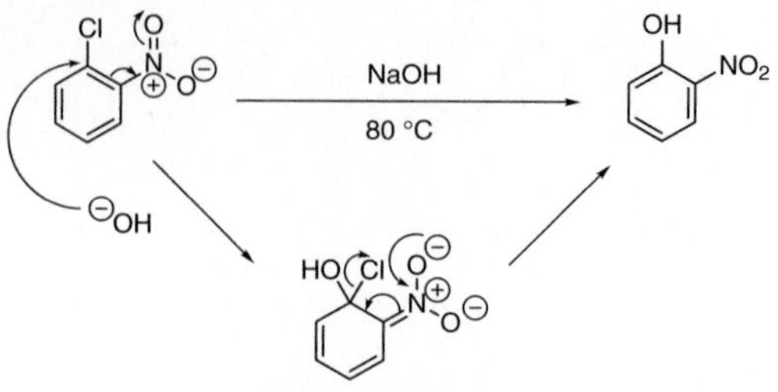

Meisenheimer Complex
(has other resonance structures)

4.21.

Formation of Cl⁺:

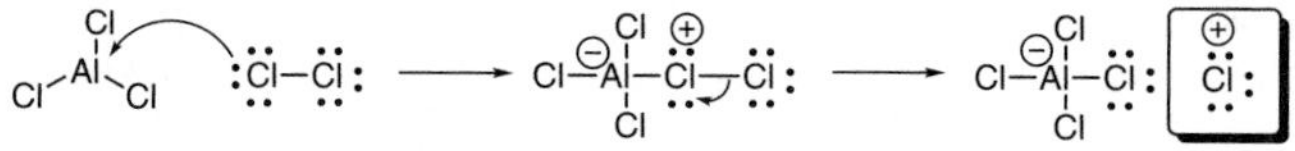

Electrophilic aromatic substitution

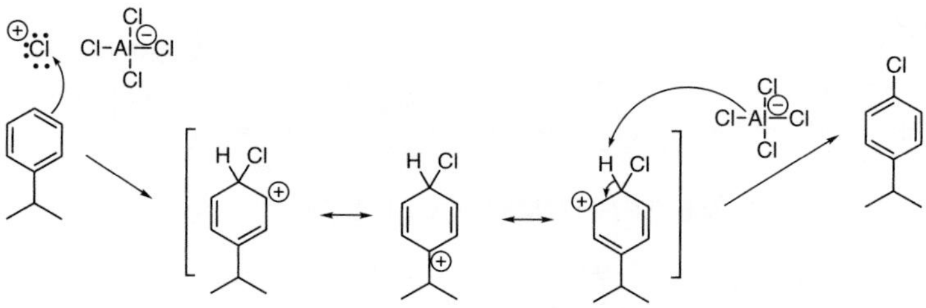

4.22.

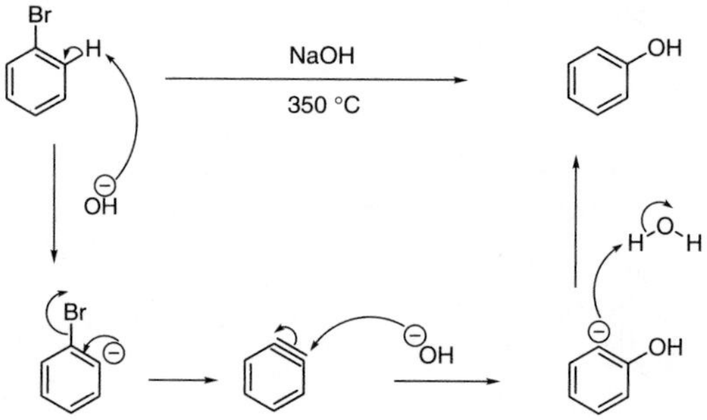

4.23.

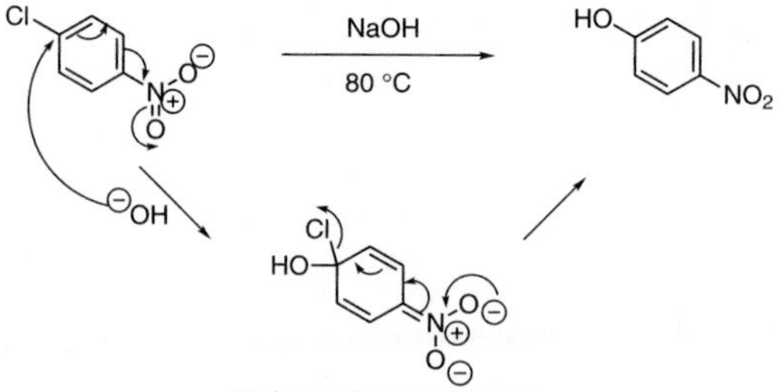

Meisenheimer Complex
(has other resonance structures)

4.24.

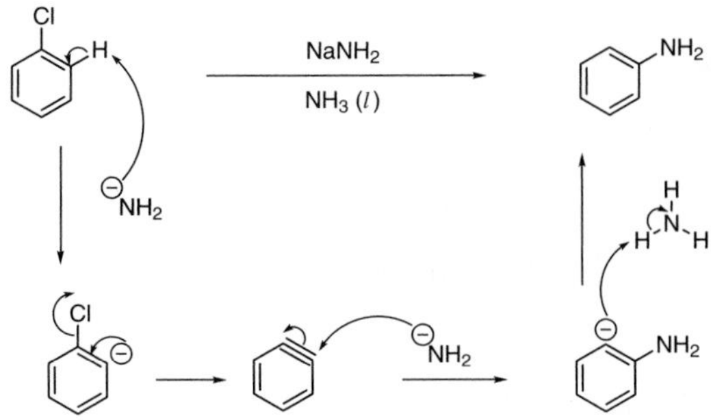

CHAPTER 5

5.2.

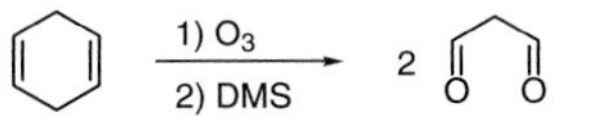

5.3.

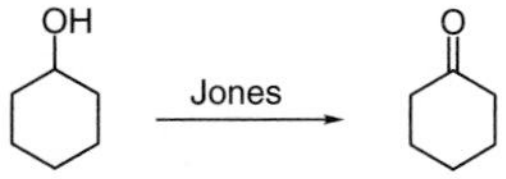

5.4.

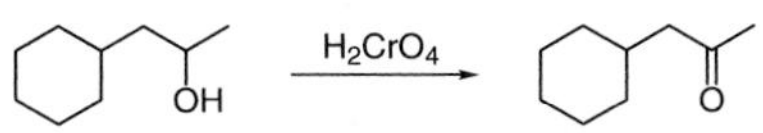

5.5.

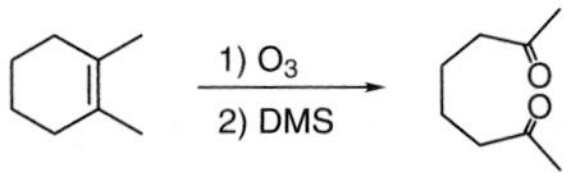

5.6.

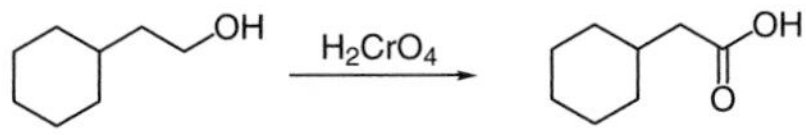

5.7.

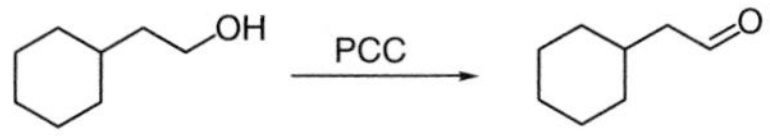

5.9.

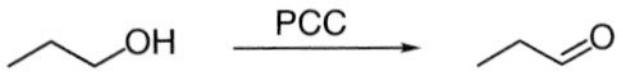

5.10.

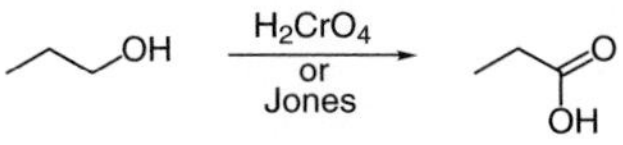

5.11.

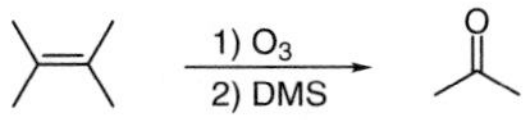

5.12.

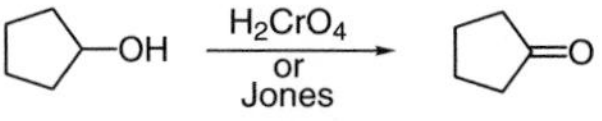

5.14.

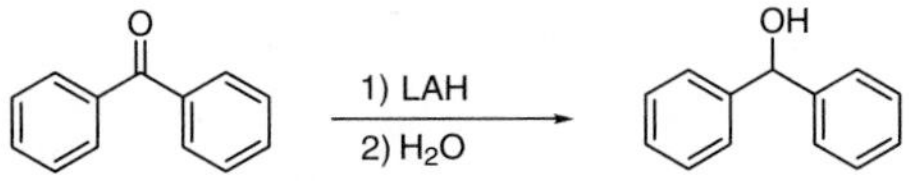

5.15.

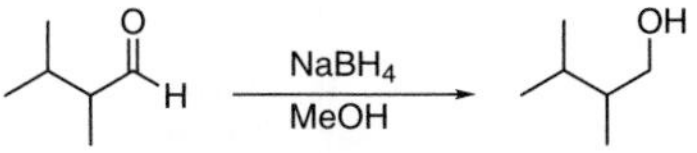

5.16.

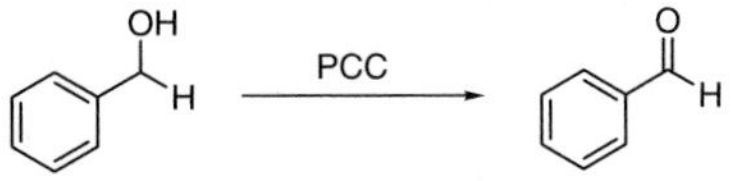

5.17.

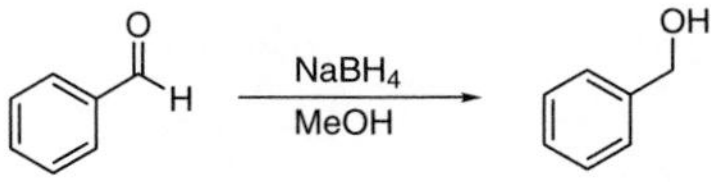

5.18.

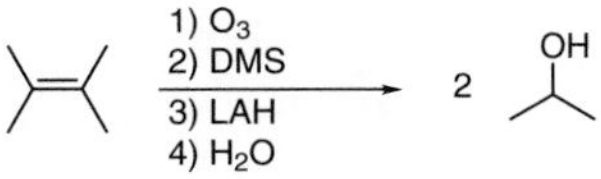

5.20.

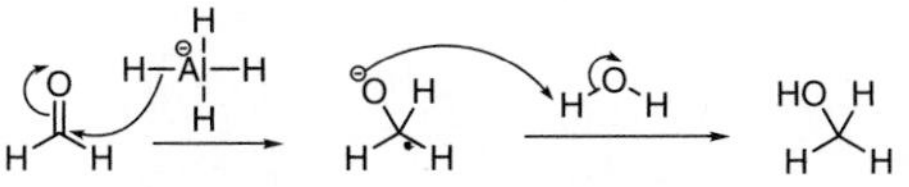

5.21.

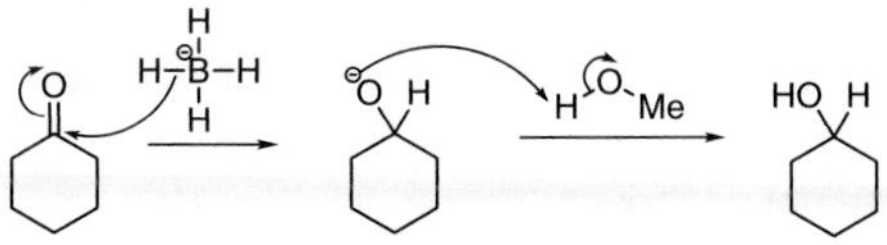

5.22.

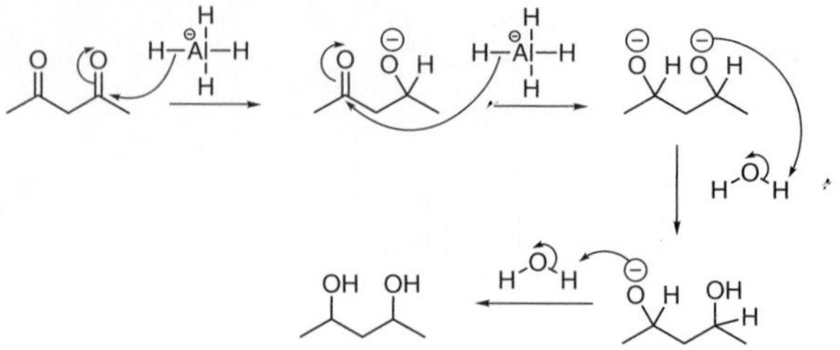

5.24.

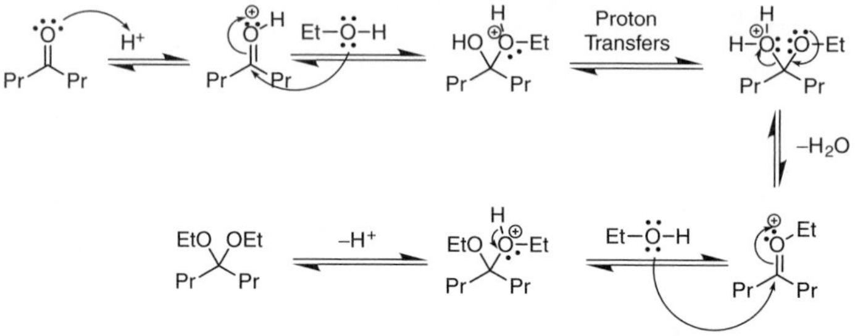

5.25.

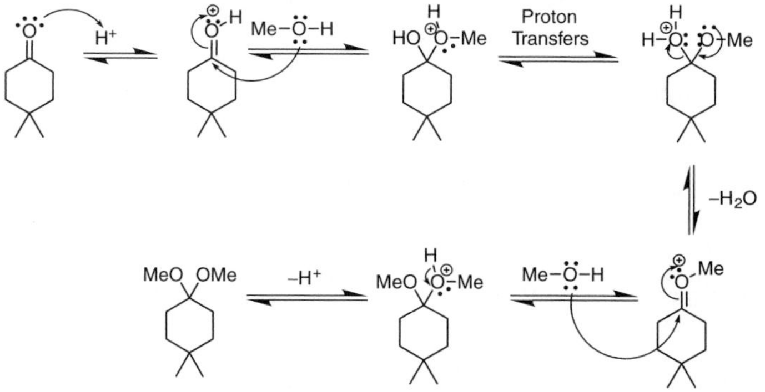

5.26.

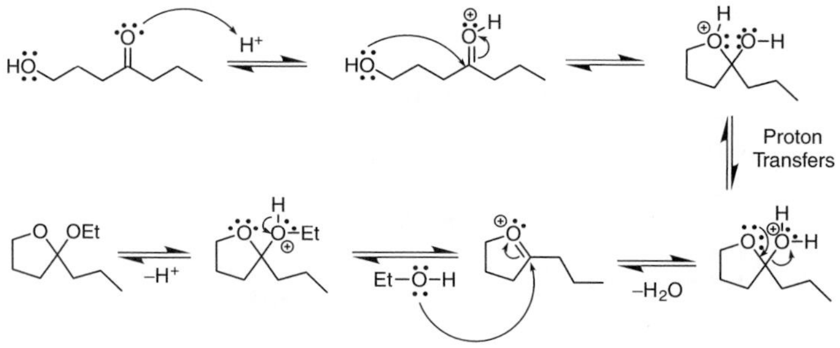

5.27.

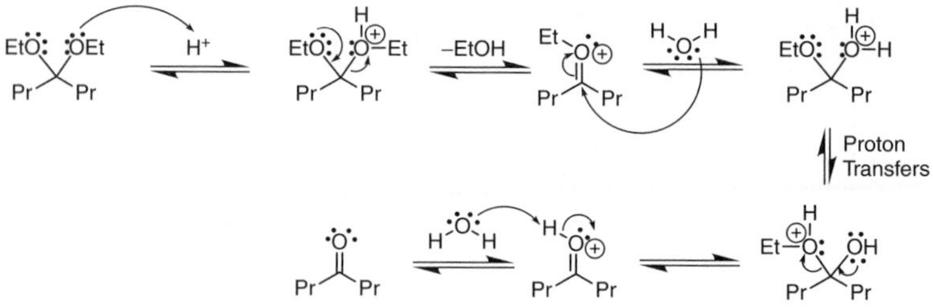

5.29.

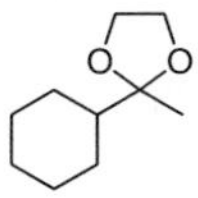

5.30.

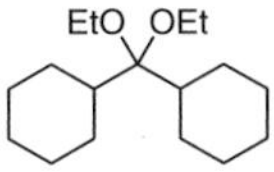

5.31.

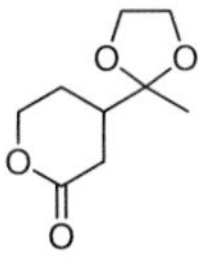

5.32.

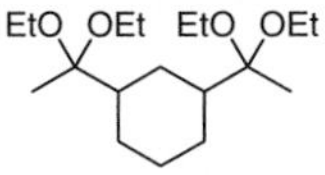

5.34.

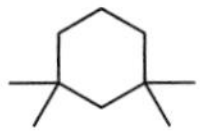

5.35.

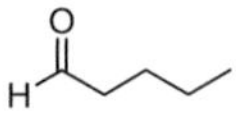

5.36.

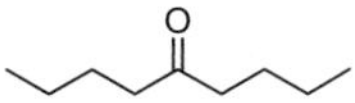

5.37.

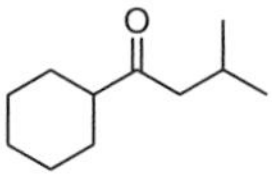

5.38.

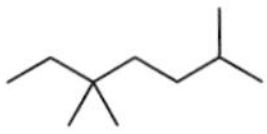

5.40.

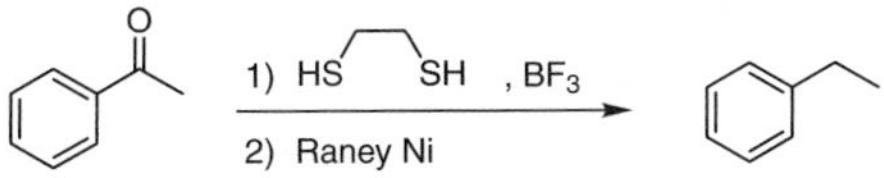

1) HS⌒SH , BF₃
2) Raney Ni

5.41.

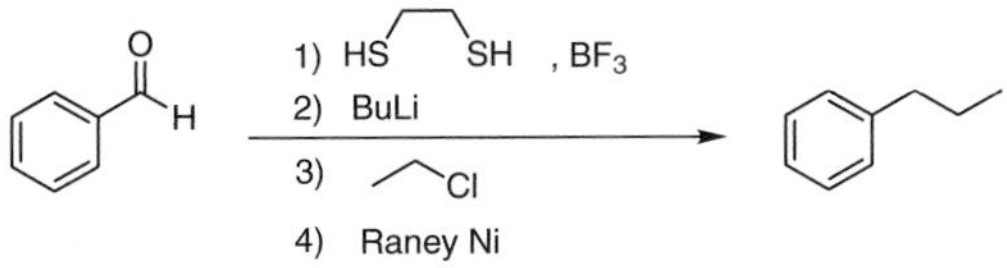

1) HS⌒SH , BF₃
2) BuLi
3) ⌒Cl
4) Raney Ni

5.42.

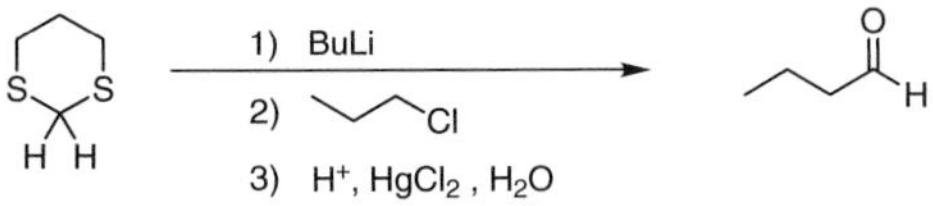

1) BuLi
2) ⌒Cl
3) H⁺, HgCl₂ , H₂O

5.43.

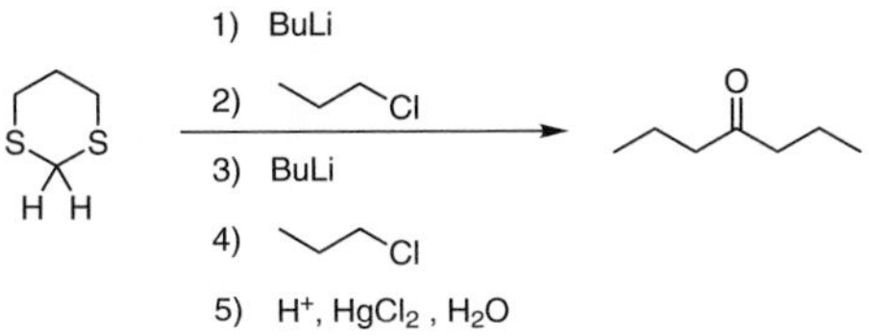

1) BuLi
2) ⌒Cl
3) BuLi
4) ⌒Cl
5) H⁺, HgCl₂ , H₂O

5.44.

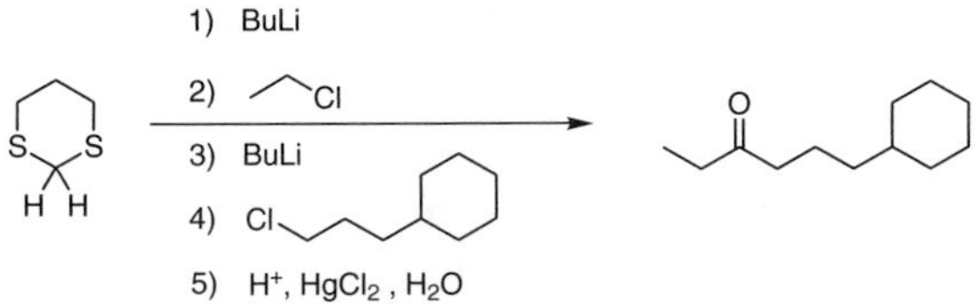

1) BuLi
2) ⌒Cl
3) BuLi
4) Cl⌒⌒
5) H⁺, HgCl₂ , H₂O

5.45.

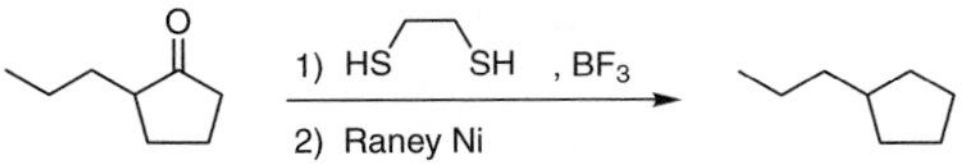

1) HS⌒SH , BF₃
2) Raney Ni

5.46.

1) HS⌒SH , BF₃
2) BuLi
3) Cl⌒
4) Raney Ni

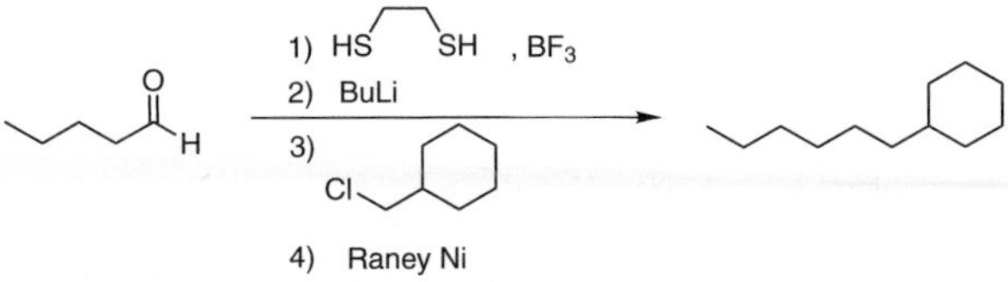

5.48.

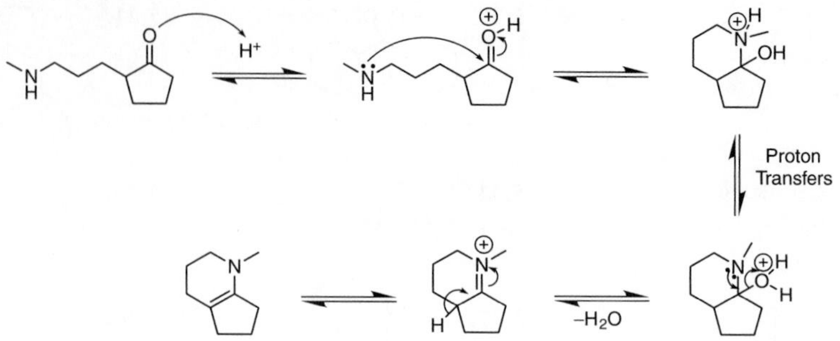

5.49.

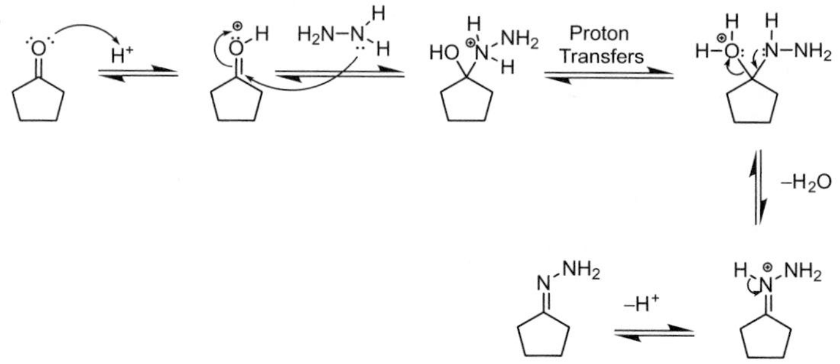

5.50.

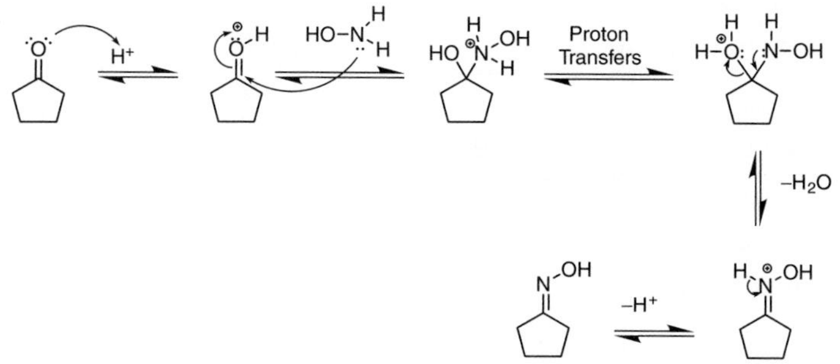

5.51.

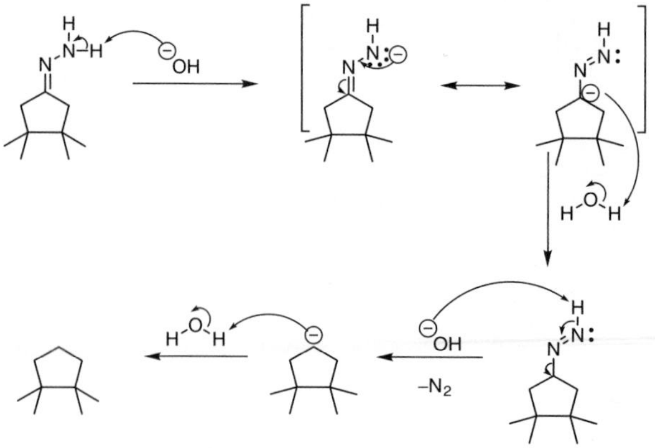

5.52.

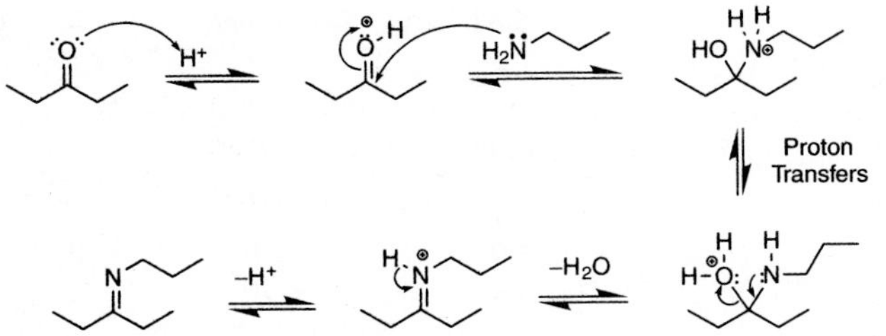

Proton Transfers

5.53.

Proton Transfers

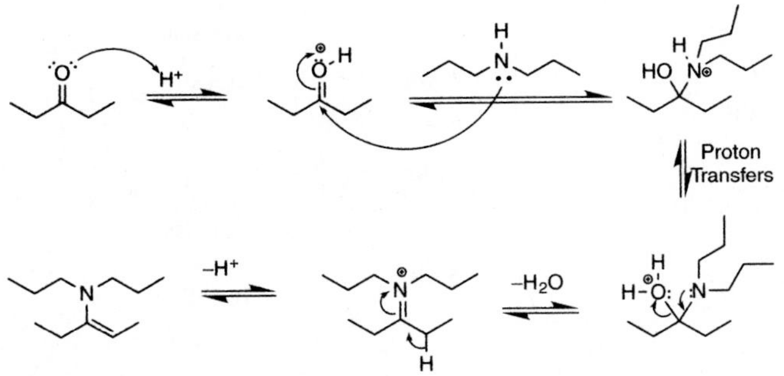

5.55. 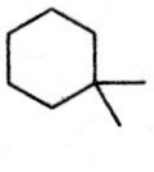 **5.56.** 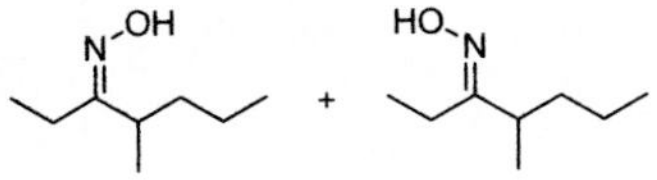 + **5.57.** 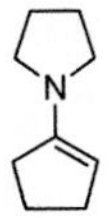 **5.58.**

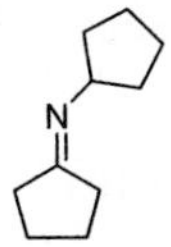

5.59. 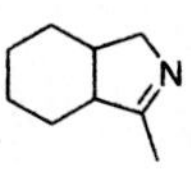 **5.60.** 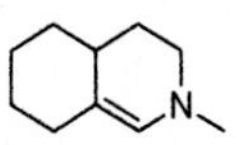 **5.62.** 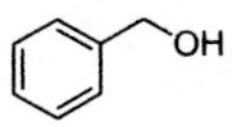 **5.63.** CH₃OH **5.64.**

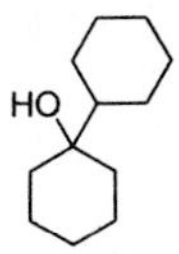

5.65. 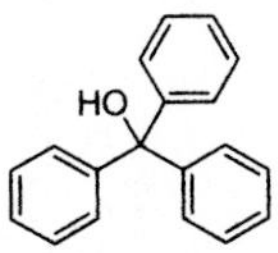 **5.67.** 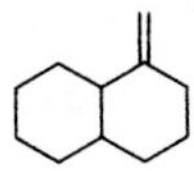 **5.68.**  **5.69.** 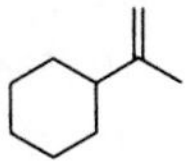 **5.71.**

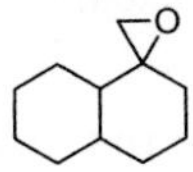

5.72. 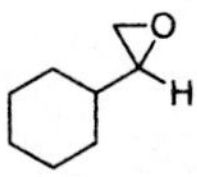 **5.73.** 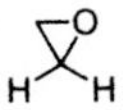 **5.75.** 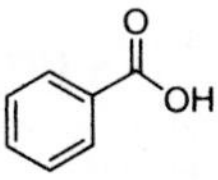 **5.76.** 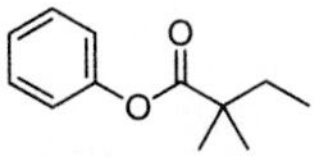 **5.77.**

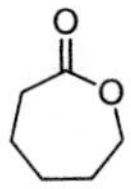

5.78. 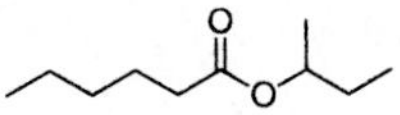 **5.79.**

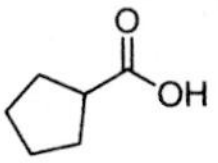

5.81. 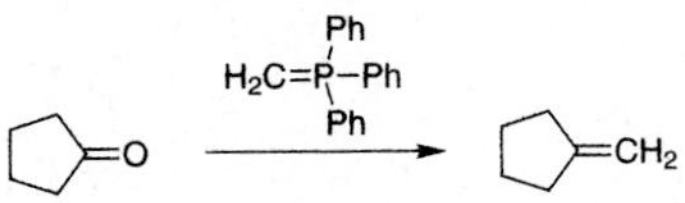 **5.82.**

5.83. 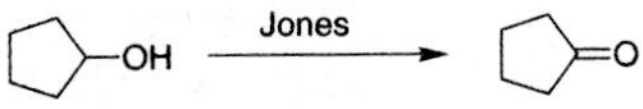 **5.84.**

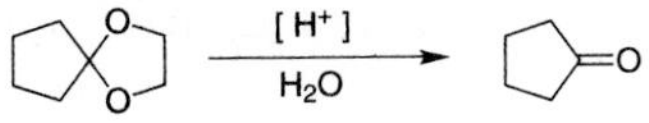

5.85. 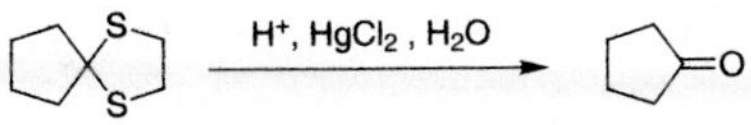 **5.86.**

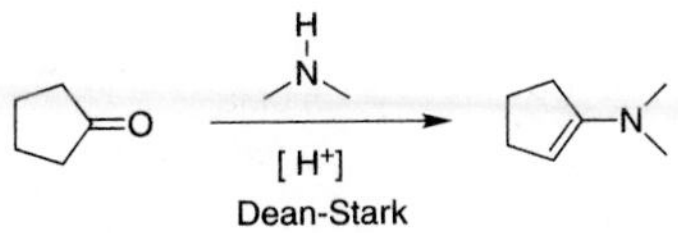

5.87.

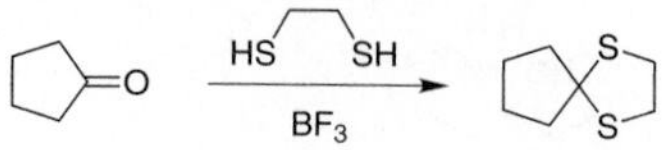

5.88.

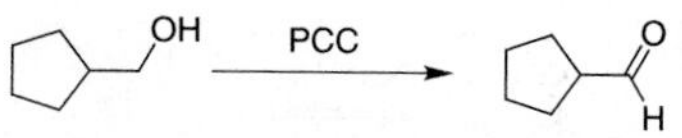

5.89.

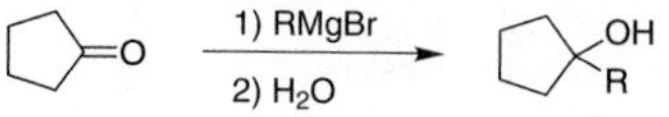

5.90.

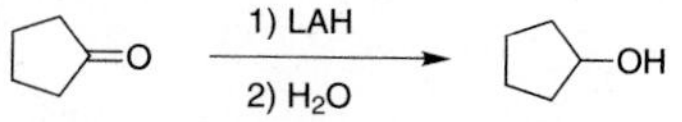

5.91.

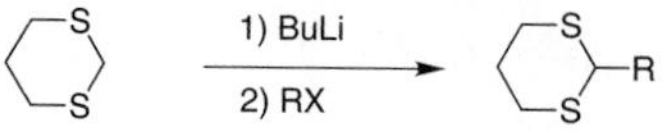

5.92.

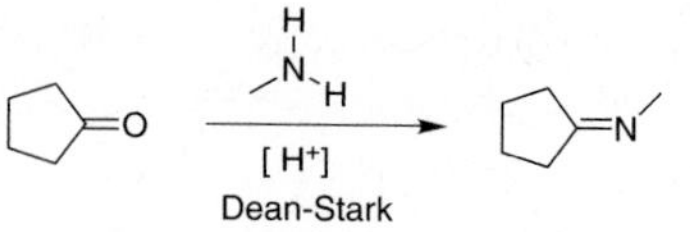

5.93.

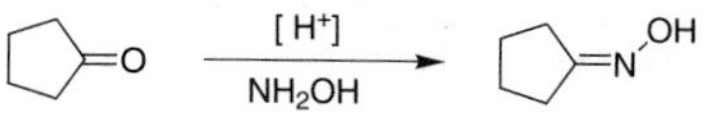

5.94.

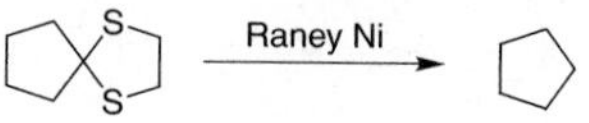

5.95.

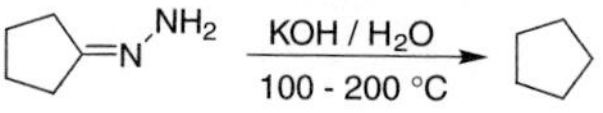

5.96.

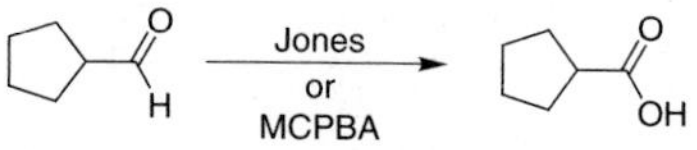

5.97.

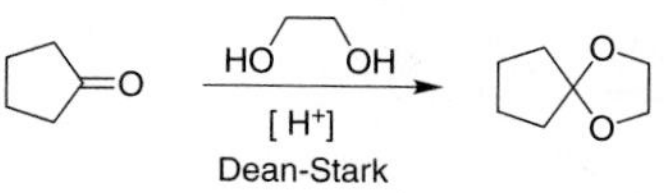

5.98.

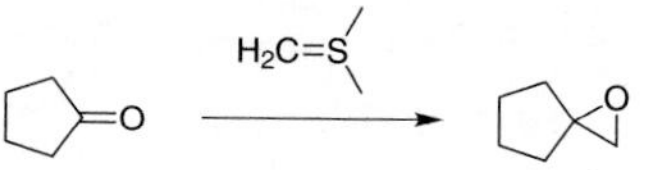

5.99.

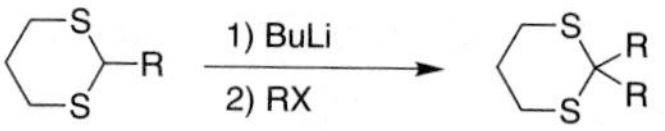

5.100.

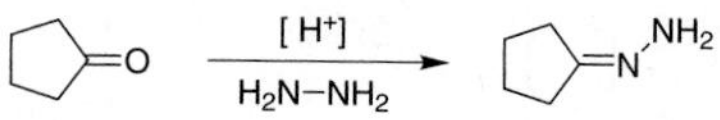

5.103.

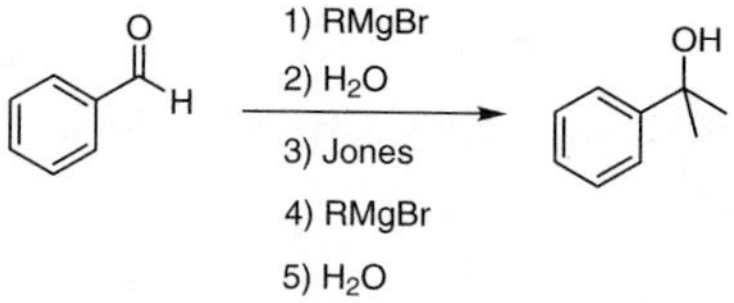

5.104.

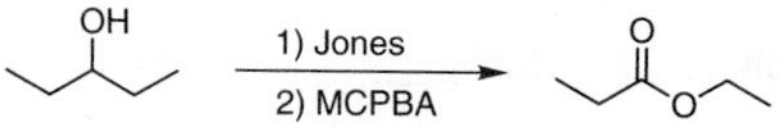

5.105.

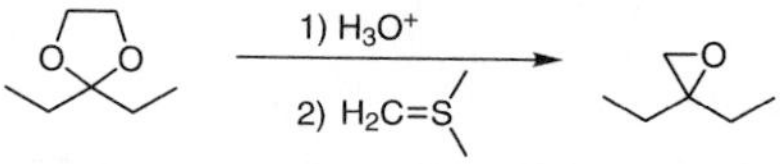

5.106.

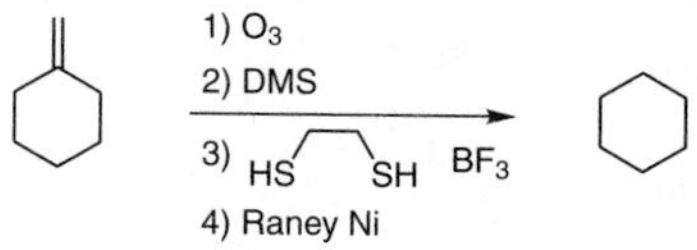

5.107.

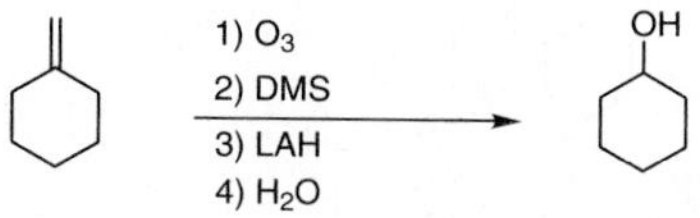

5.108.

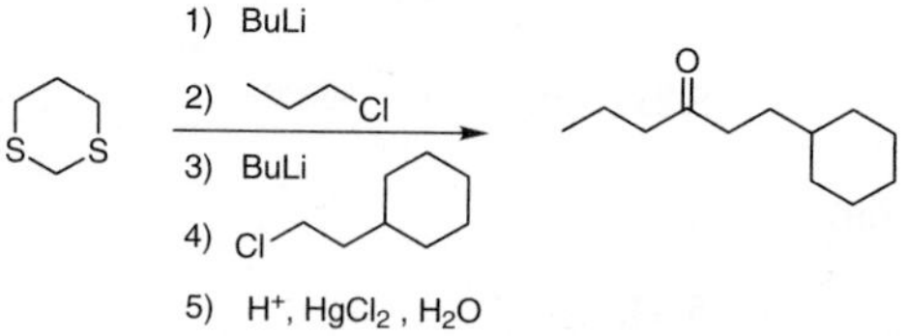

5.109.

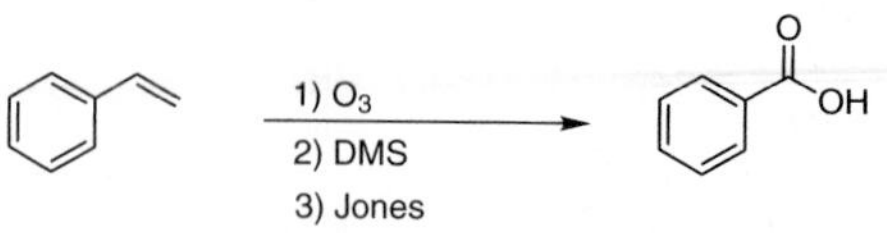

5.110.

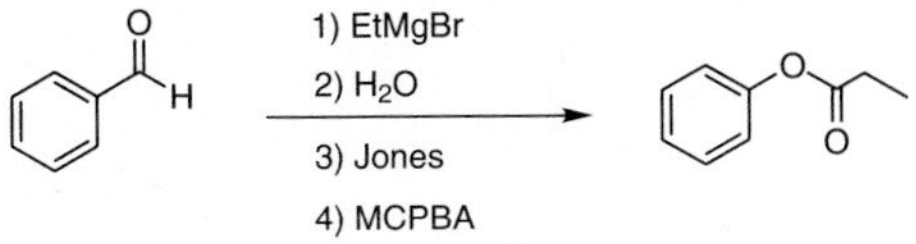

1) EtMgBr
2) H₂O
3) Jones
4) MCPBA

5.111.

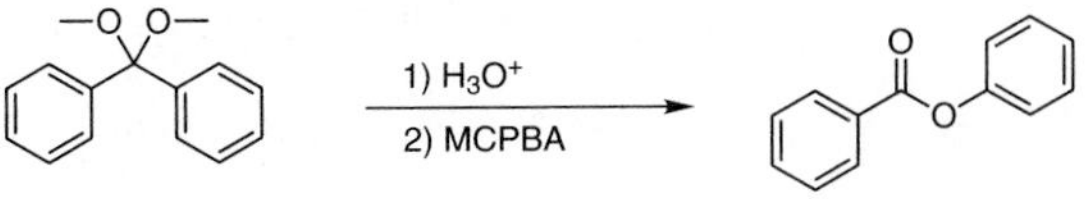

1) H₃O⁺
2) MCPBA

5.112.

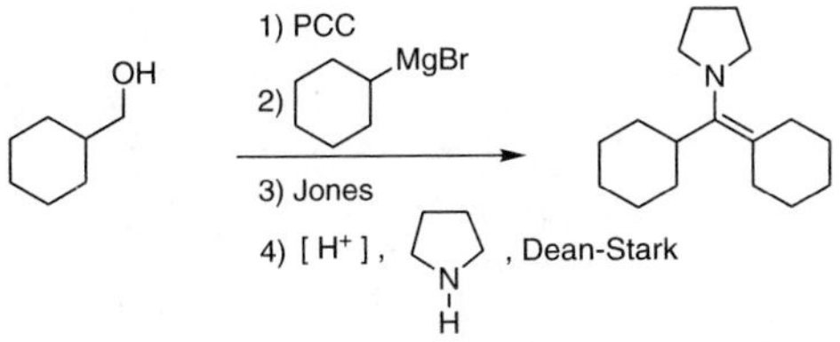

1) PCC
2) [cyclohexyl]MgBr
3) Jones
4) [H⁺], [pyrrolidine], Dean-Stark

5.113.

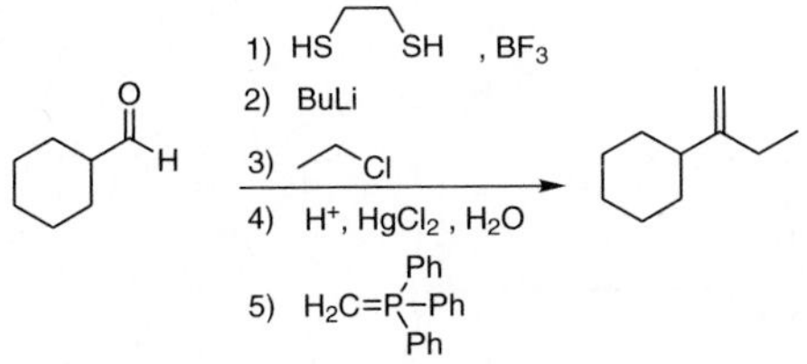

1) HS⌢SH , BF₃
2) BuLi
3) ⌒Cl
4) H⁺, HgCl₂ , H₂O
5) H₂C=P(Ph)(Ph)Ph

CHAPTER 6

6.2.

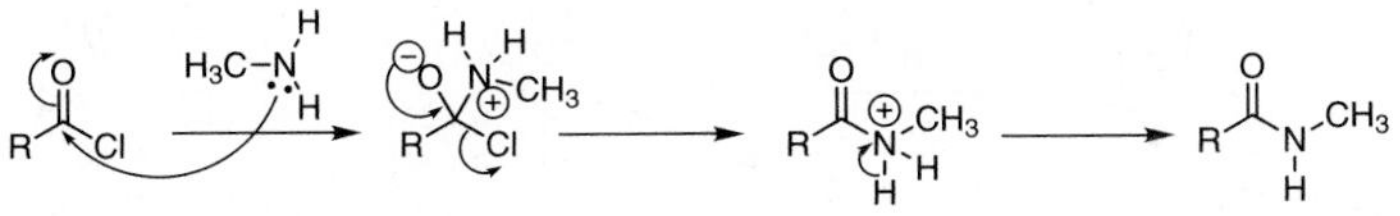

6.3.

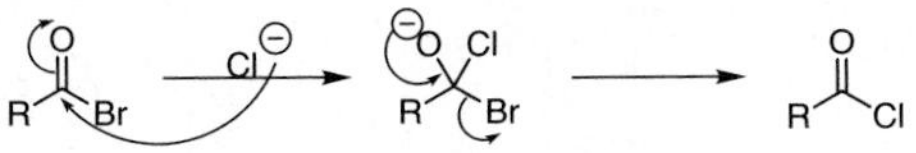

6.4.

Proton Transfers

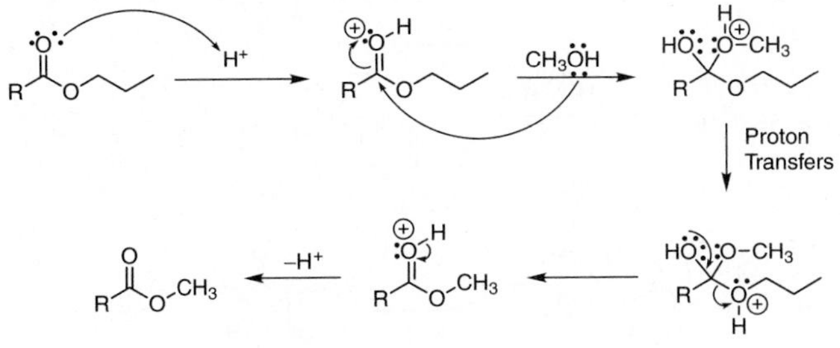

6.5.

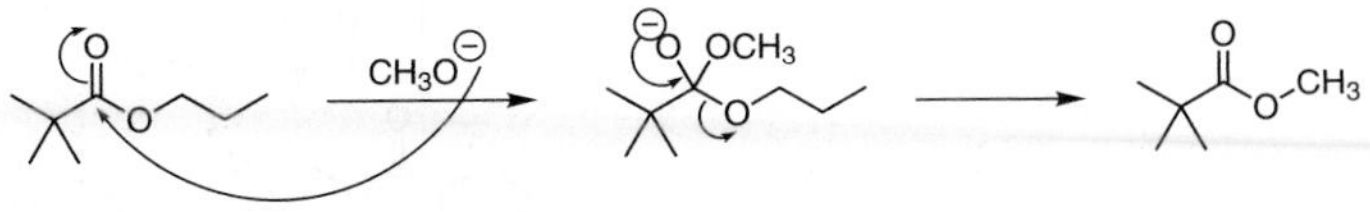

6.6.

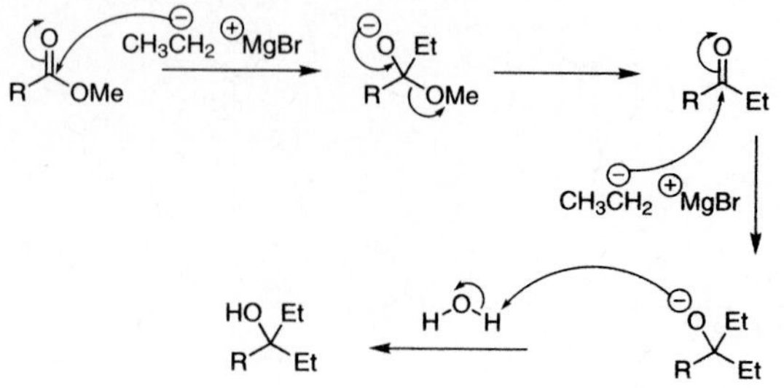

6.7.

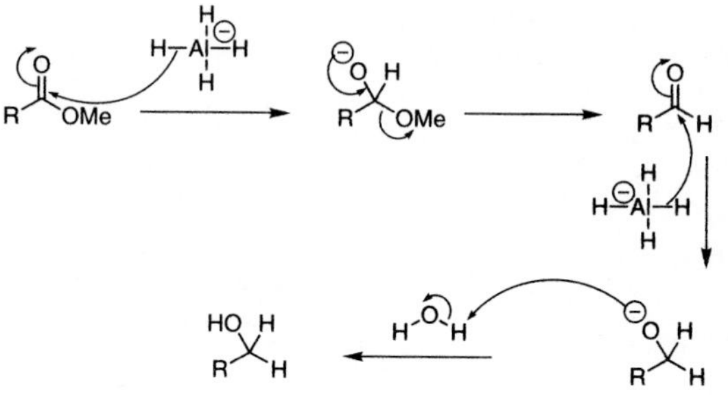

6.8.

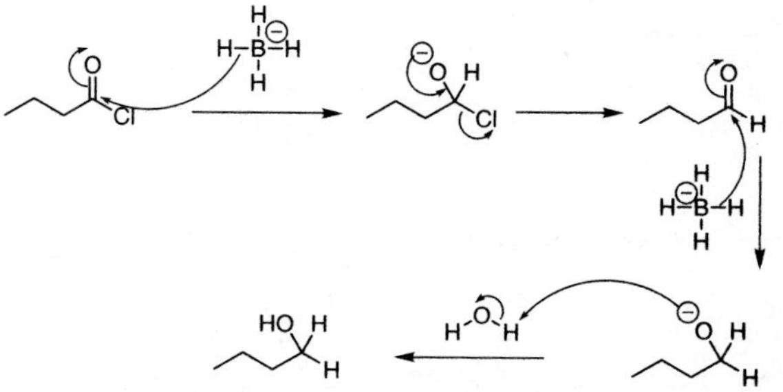

6.10.

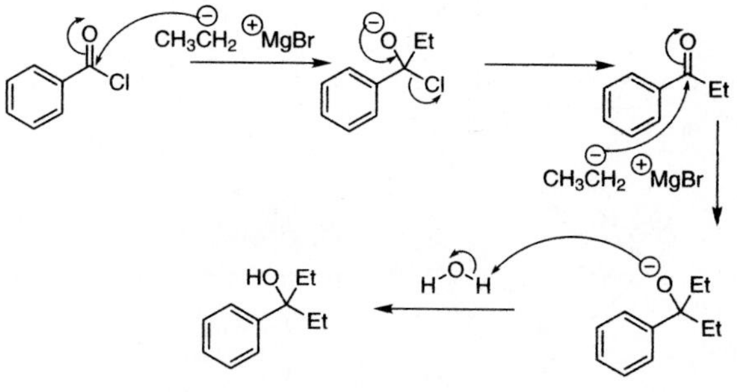

6.11.

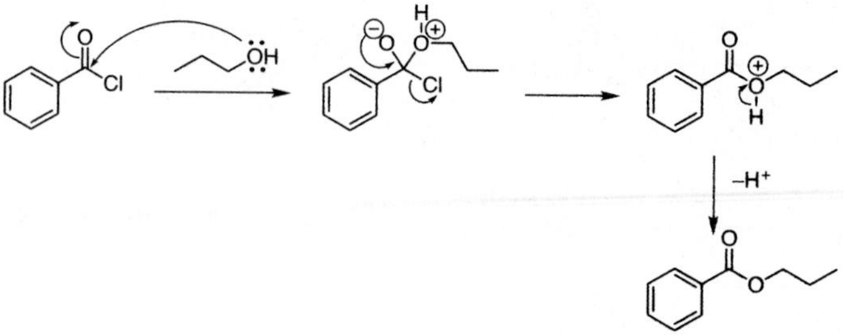

6.12.

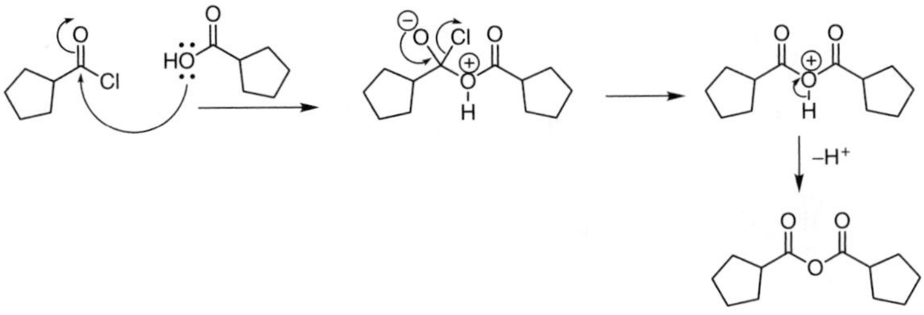
–H⁺

6.13.

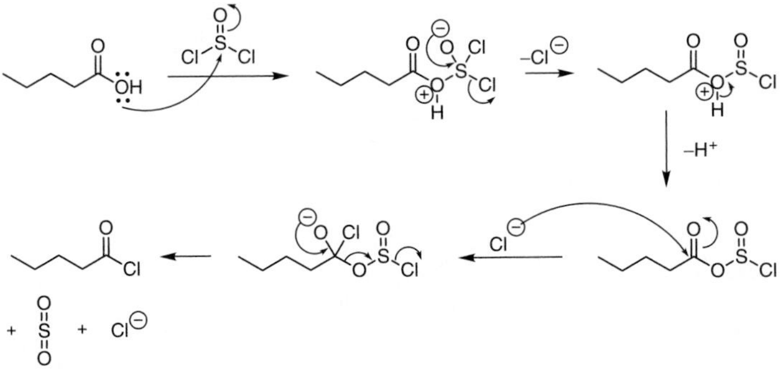
–Cl⊖
–H⁺

6.14.

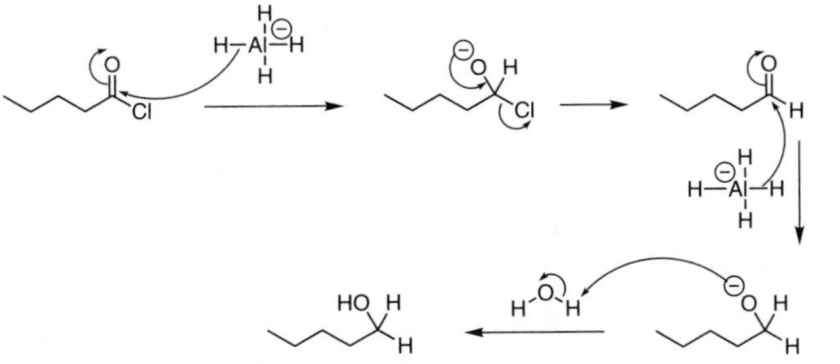

6.16.

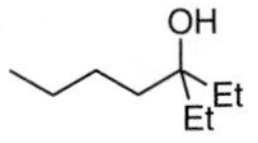

6.17.

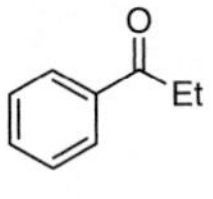

6.18.

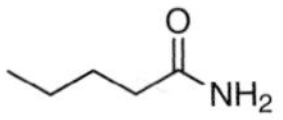

6.19.

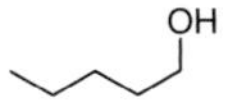

6.20.

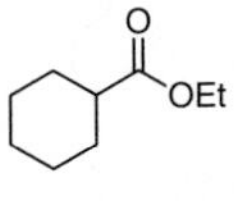

6.22.

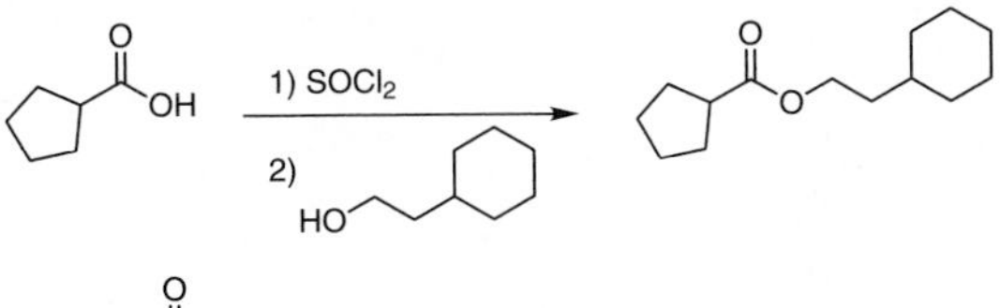
1) SOCl2
2)

6.23.

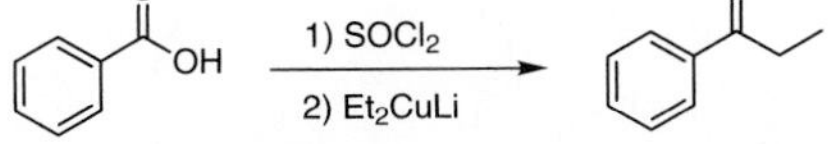
1) SOCl2
2) Et2CuLi

6.24.

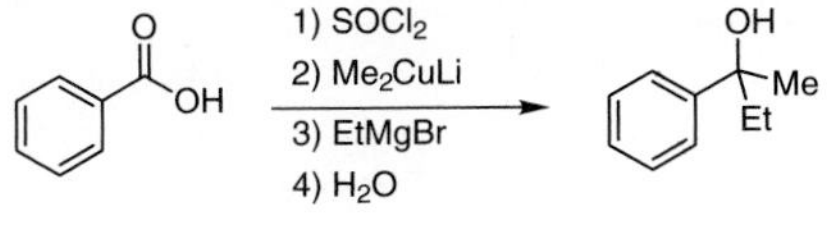
1) SOCl2
2) Me2CuLi
3) EtMgBr
4) H2O

6.25.

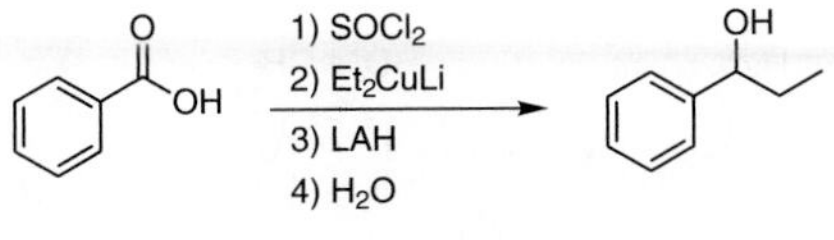
1) SOCl2
2) Et2CuLi
3) LAH
4) H2O

6.26.

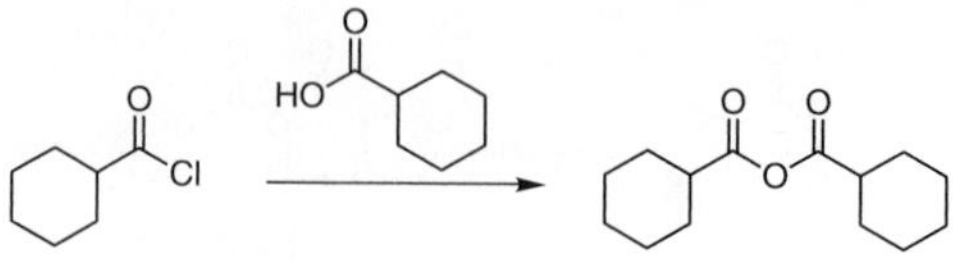

6.28.

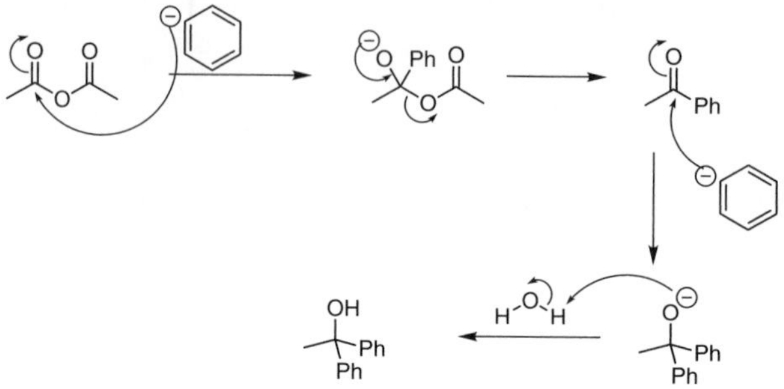

6.29.

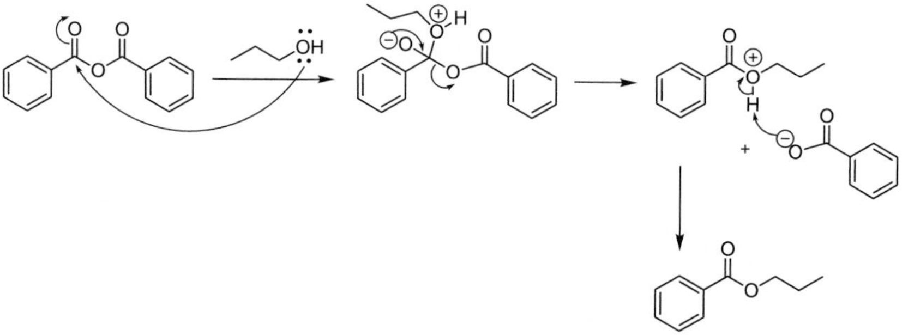

6.31.

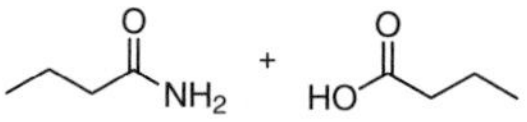

6.32.

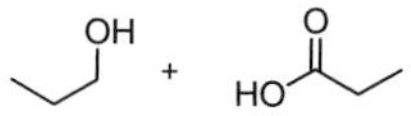

6.33.

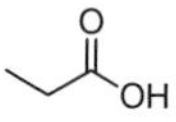

6.36.

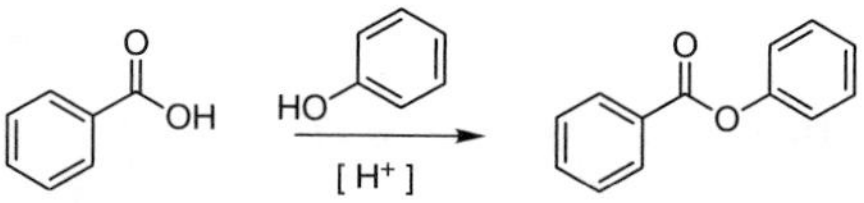

6.37.

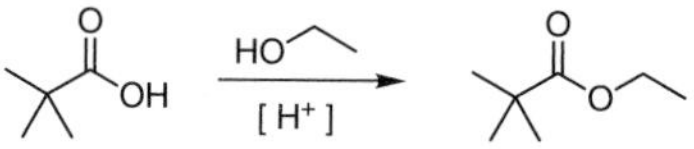

6.38.

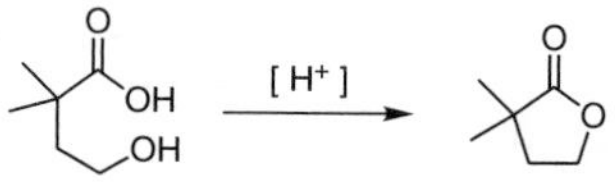

6.39.

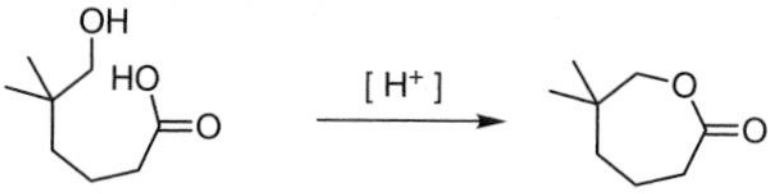

6.41.

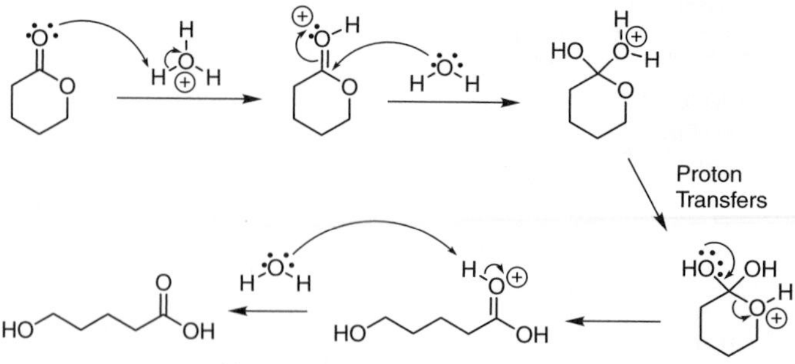

Proton Transfers

6.42.

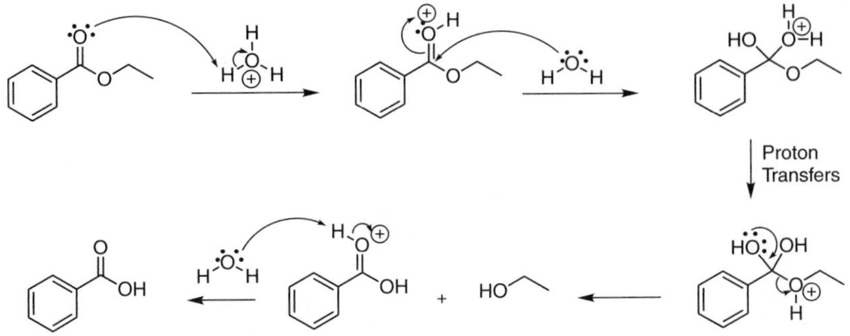

6.44.

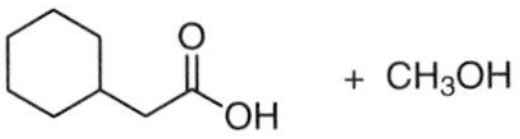

+ CH₃OH

6.45.

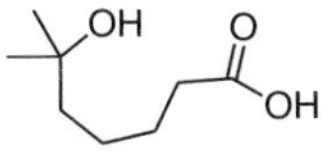

6.46.

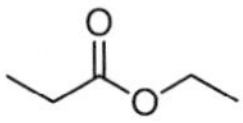

6.47.

+

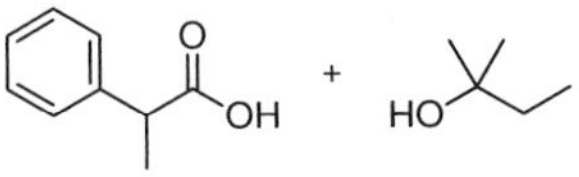

6.48.

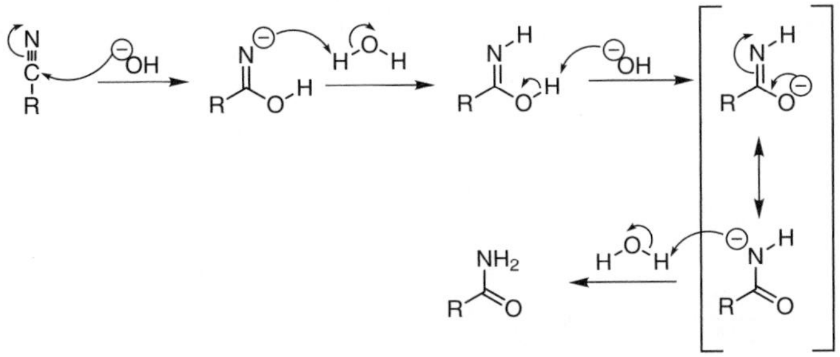

6.50.

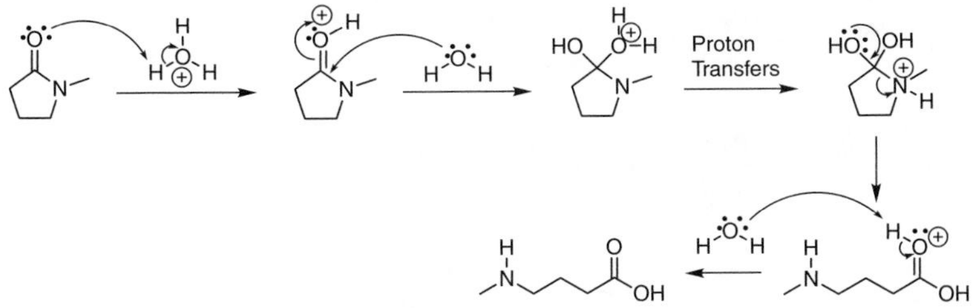

6.51.

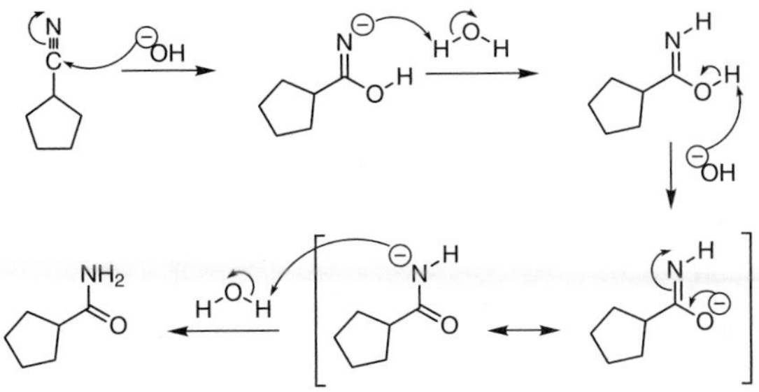

6.52.

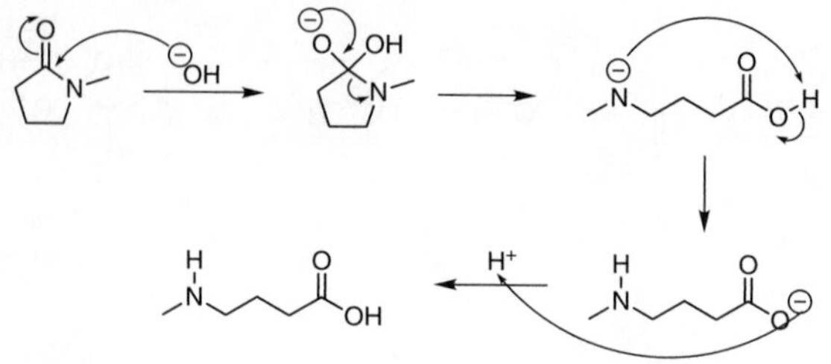

6.53.

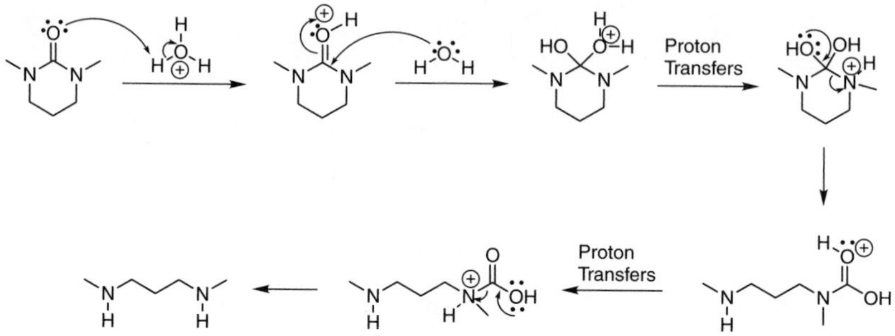

6.55.

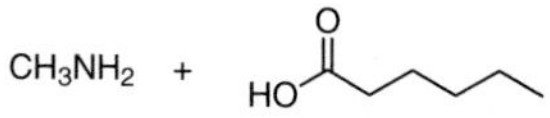

CH₃NH₂ +

6.56.

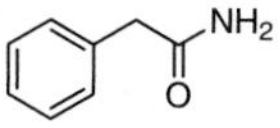

6.57.

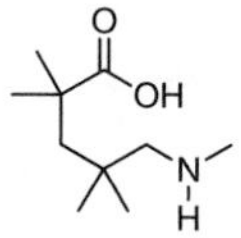

6.59.

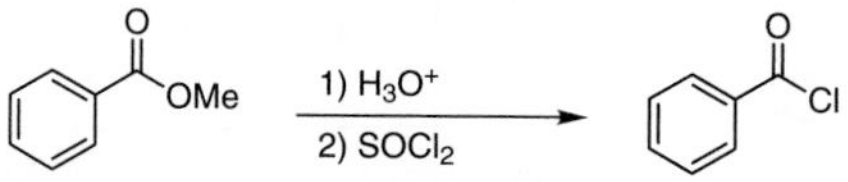

1) H₃O⁺

2) SOCl₂

6.60.

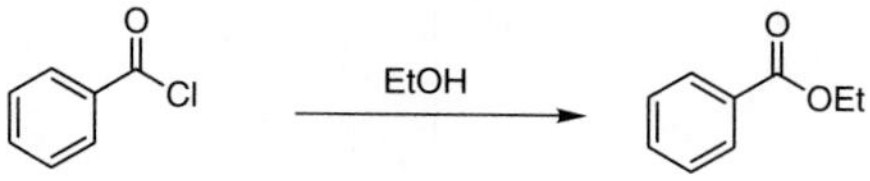

EtOH

6.61.

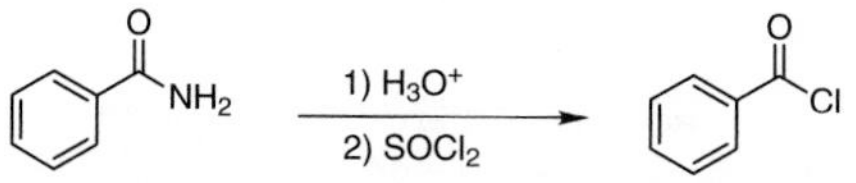

1) H₃O⁺

2) SOCl₂

6.62.

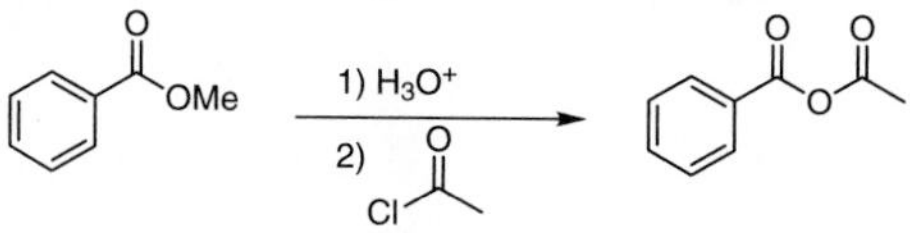

1) H₃O⁺

2)

6.63.

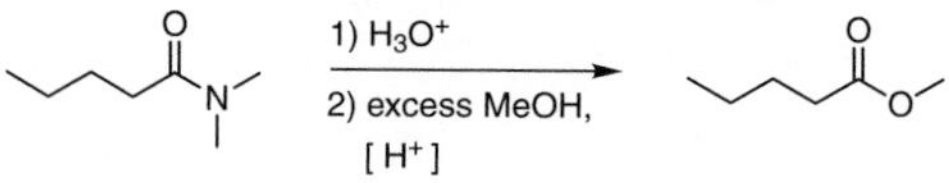

1) H₃O⁺

2) excess MeOH, [H⁺]

6.64.

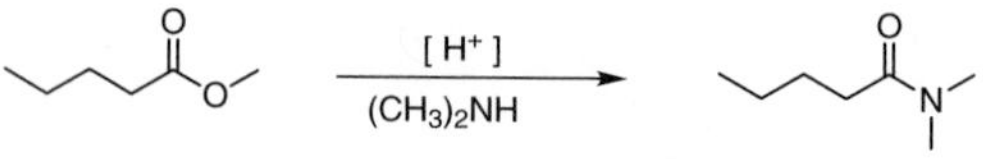

[H⁺]

(CH₃)₂NH

6.66.

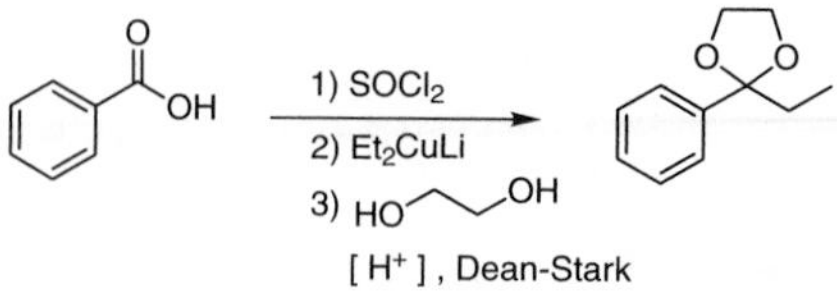

1) SOCl₂

2) Et₂CuLi

3) HO⌒OH

[H⁺], Dean-Stark

6.67.

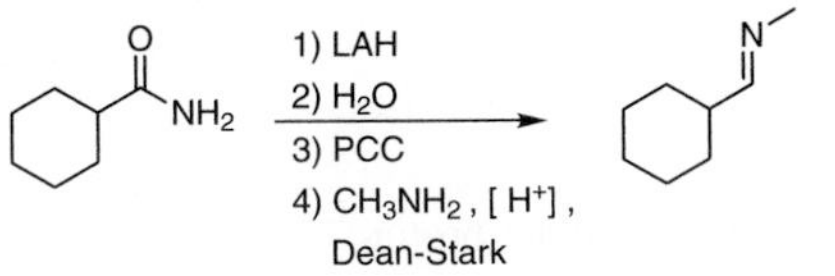

1) LAH
2) H₂O
3) PCC
4) CH₃NH₂ , [H⁺] , Dean-Stark

6.68.

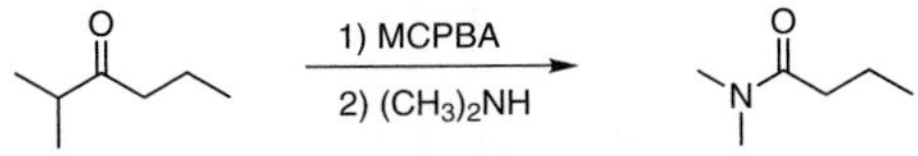

1) MCPBA
2) (CH₃)₂NH

6.69.

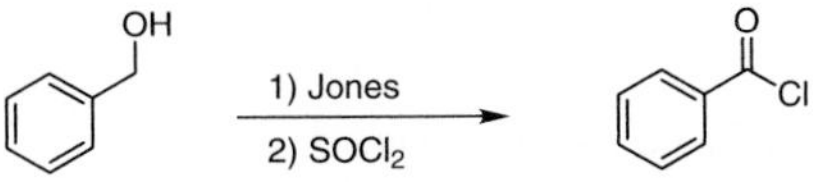

1) Jones
2) SOCl₂

6.70.

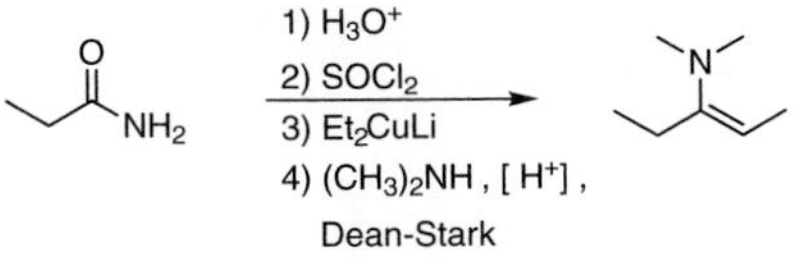

1) H₃O⁺
2) SOCl₂
3) Et₂CuLi
4) (CH₃)₂NH , [H⁺] , Dean-Stark

6.71.

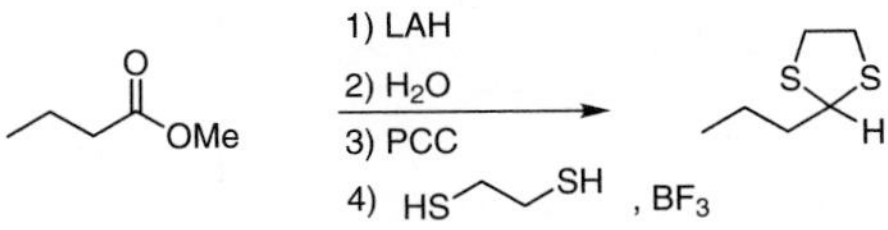

1) LAH
2) H₂O
3) PCC
4) HS⌣SH , BF₃

6.72.

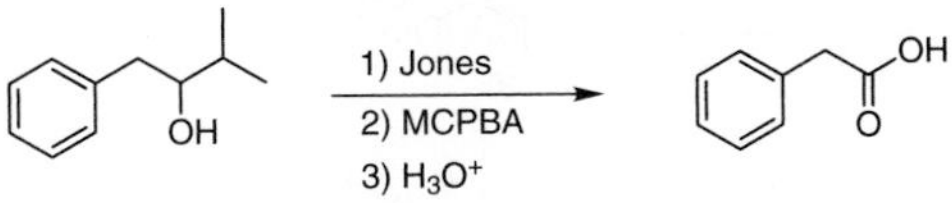

1) Jones
2) MCPBA
3) H₃O⁺

6.73.

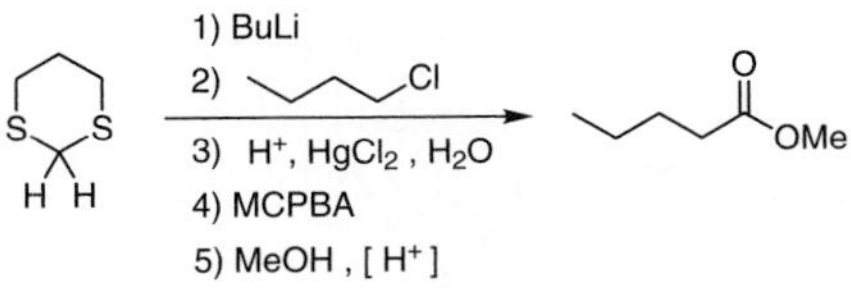

1) BuLi
2) ⌣⌣Cl
3) H⁺, HgCl₂ , H₂O
4) MCPBA
5) MeOH , [H⁺]

6.74.

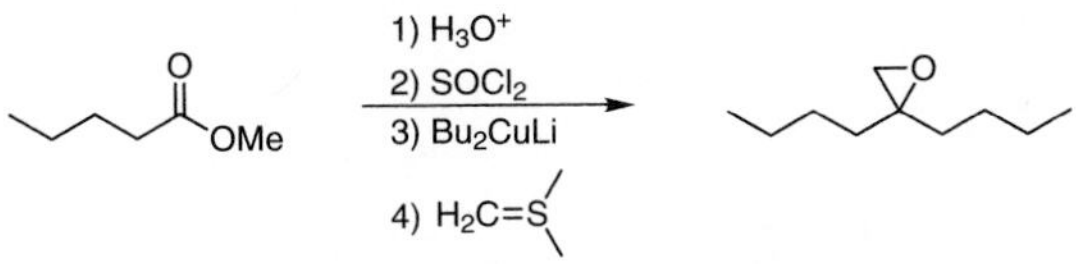

1) H₃O⁺
2) SOCl₂
3) Bu₂CuLi
4) H₂C=S⌣

6.75.

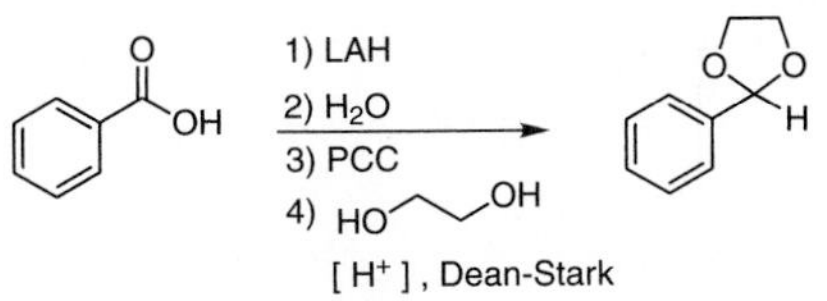

1) LAH
2) H₂O
3) PCC
4) HO⌣OH
[H⁺] , Dean-Stark

6.76.

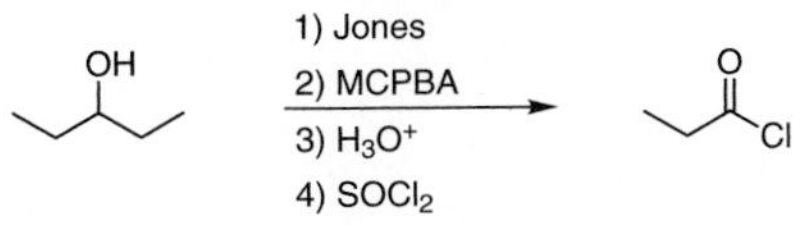

1) Jones
2) MCPBA
3) H₃O⁺
4) SOCl₂

6.77.

1) Jones
2) SOCl₂
3) (CH₃)₂NH

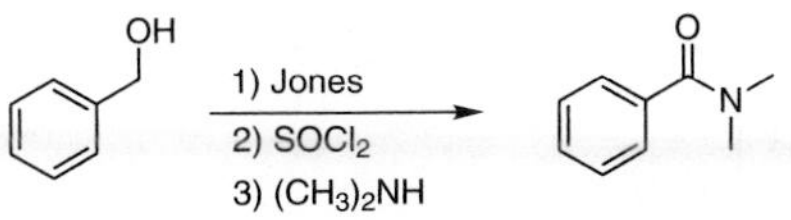

CHAPTER 7

7.2. one alpha proton:

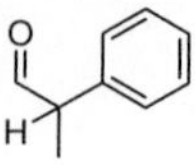

7.3. one alpha proton:

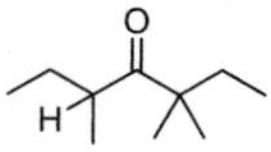

7.4. no alpha protons

7.5. four alpha protons:

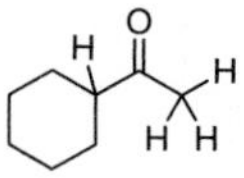

7.6. two alpha protons:

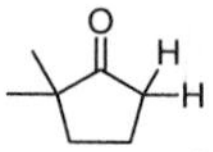

7.7. no alpha protons

7.9.

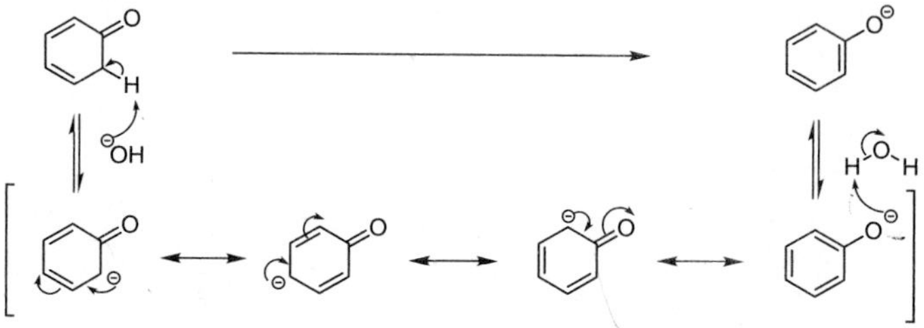

7.10.

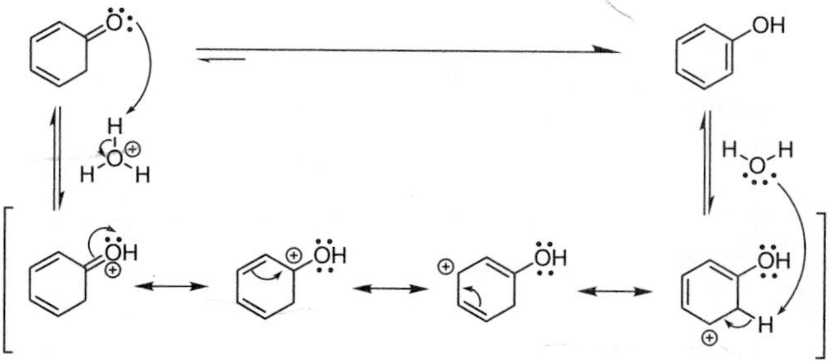

7.11.

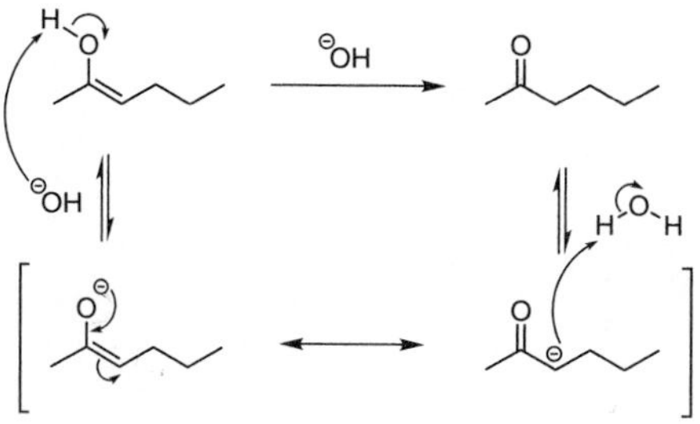

7.12.

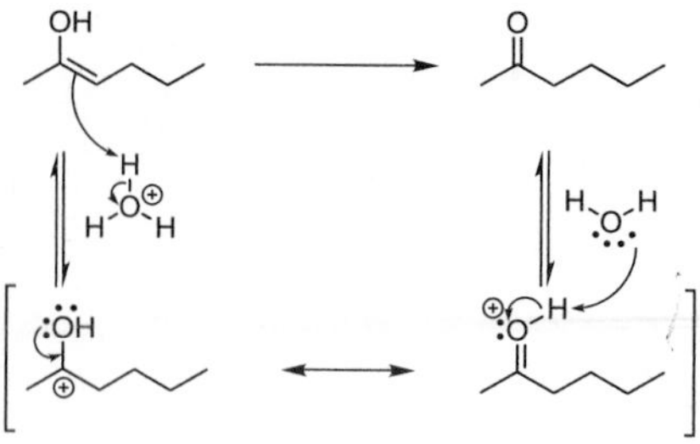

7.13.

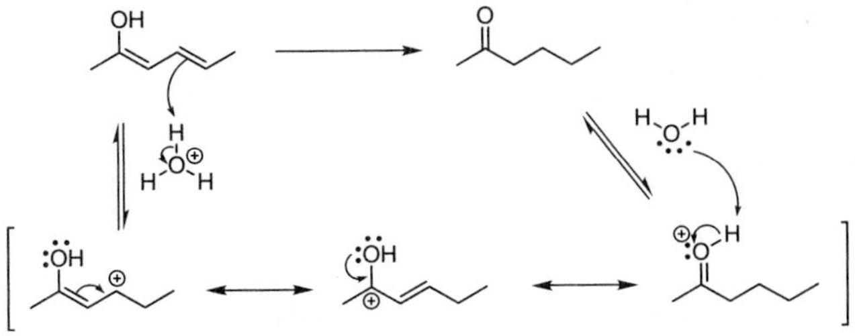

7.15.

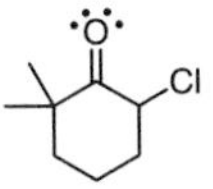

7.16.

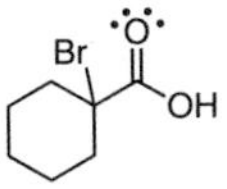

7.17.

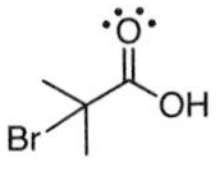

7.18.

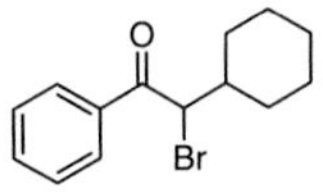

7.20.

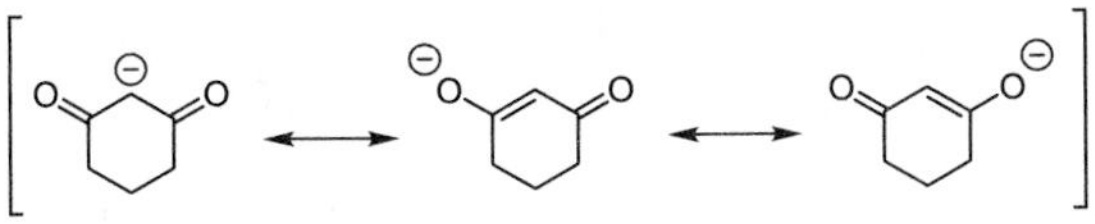

7.21.

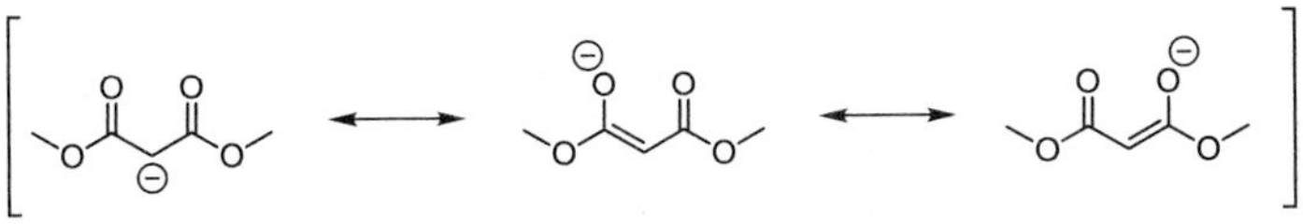

7.22.

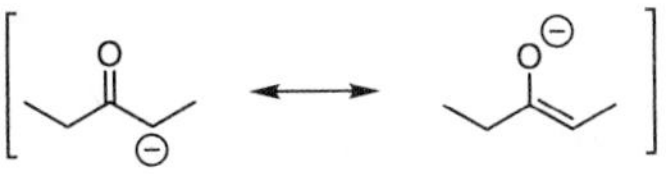

7.23.

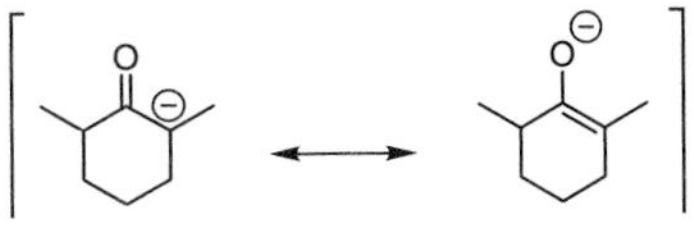

7.25.

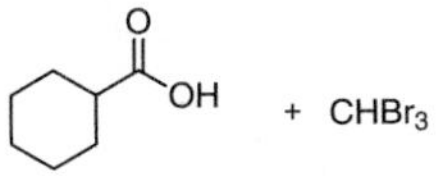
+ CHBr3

7.26.

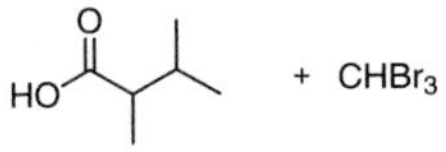
+ CHBr3

7.27.

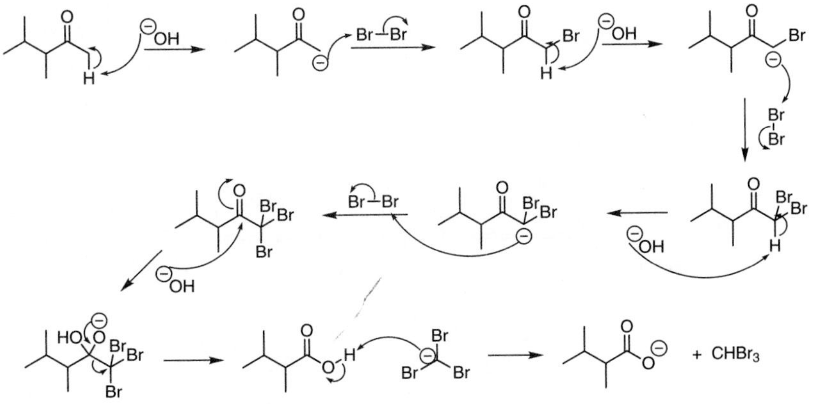
+ CHBr3

7.29.

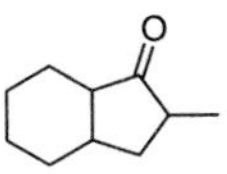

7.30.

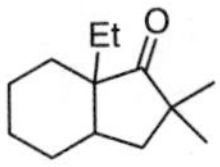

7.31.

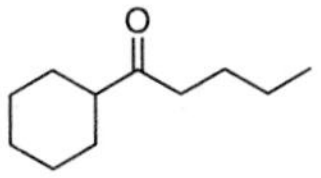

7.32.

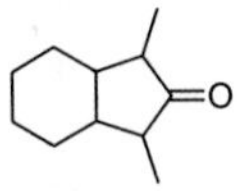

7.34.

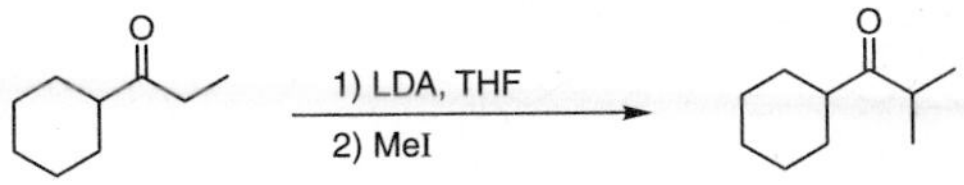
1) LDA, THF
2) MeI

7.35.

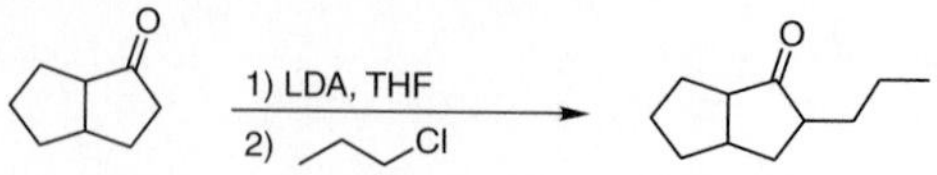

1) LDA, THF
2) ＣＩ

7.36.

1) LDA, THF
2) ＣＩ

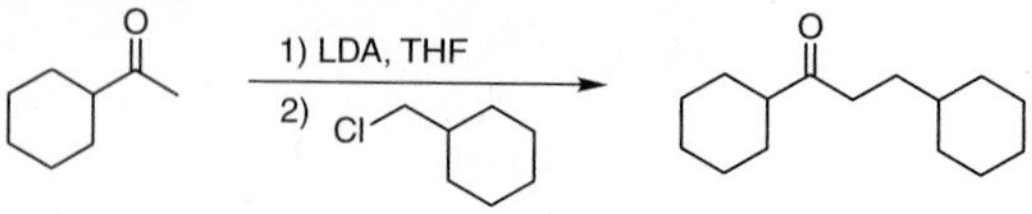

7.38.

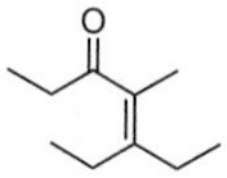

7.39.

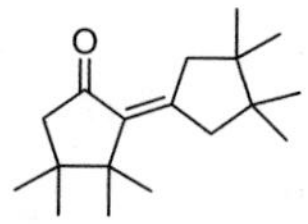

7.40.

+

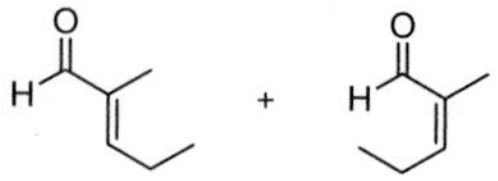

7.42.

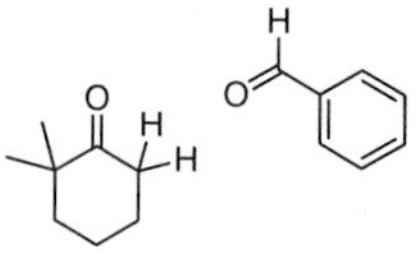

7.43.

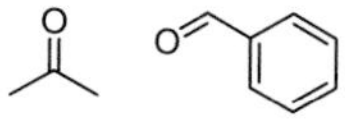

7.44.

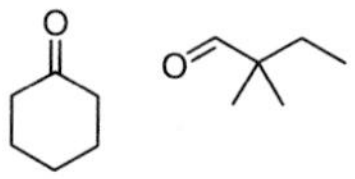

7.45.

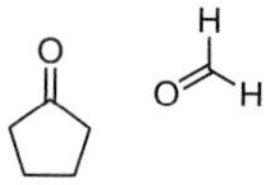

7.47.

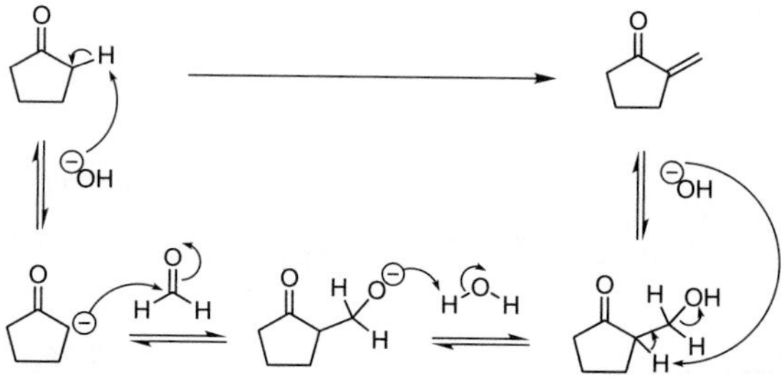

7.48.

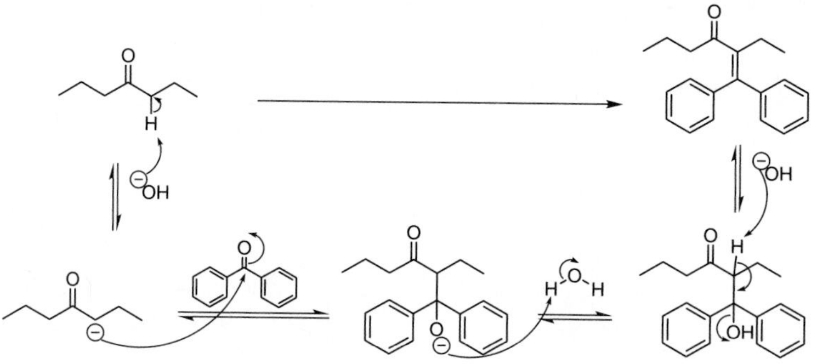

7.49.

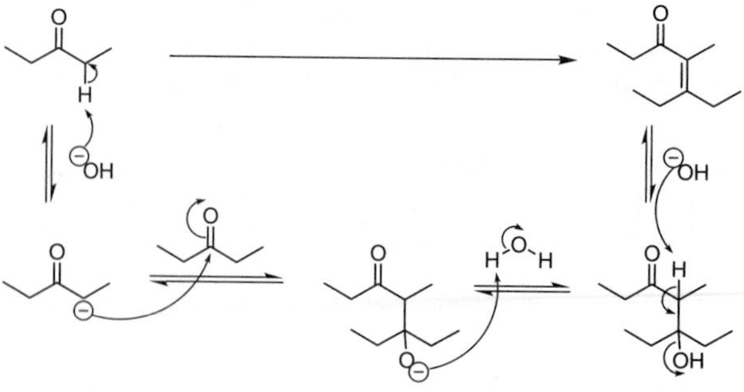

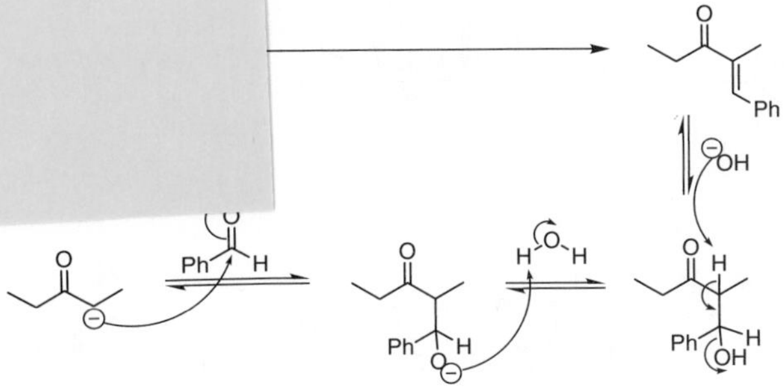

7.52.

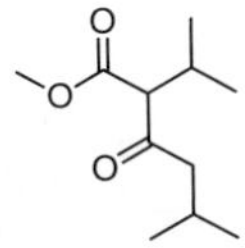

7.53.

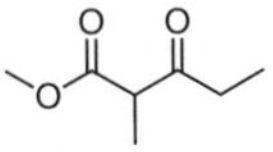

7.54.

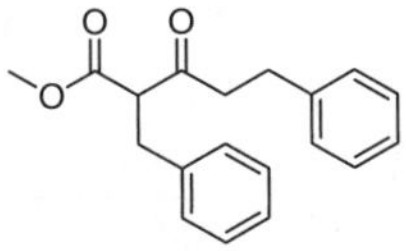

7.55.

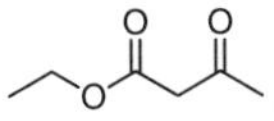

7.57.

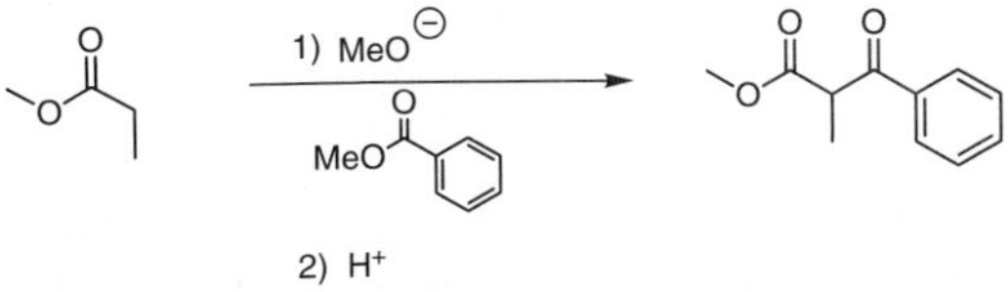

1) MeO⁻

MeO (benzoate)

2) H⁺

7.58.

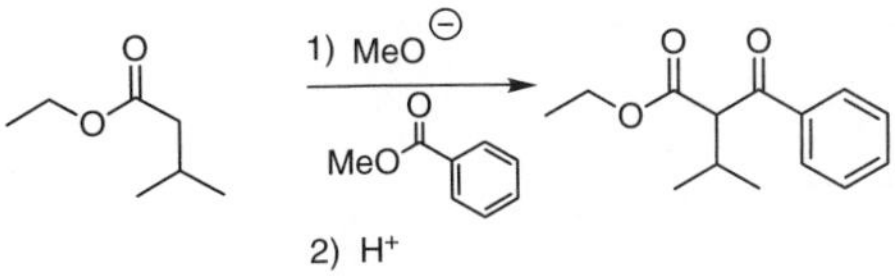

1) MeO⁻

MeO (benzoate)

2) H⁺

7.59.

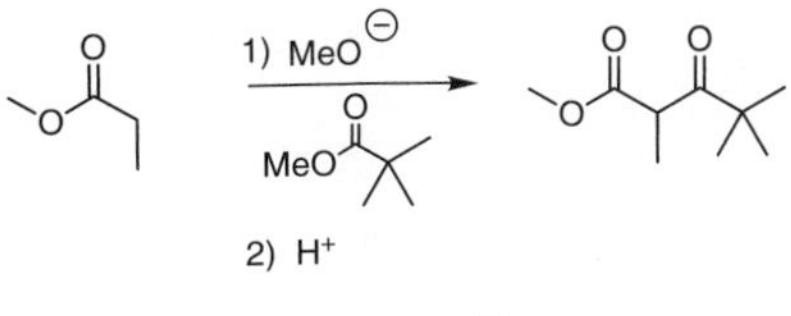

1) MeO⁻

MeO (pivalate)

2) H⁺

7.60.

1) MeO⁻

2) H⁺

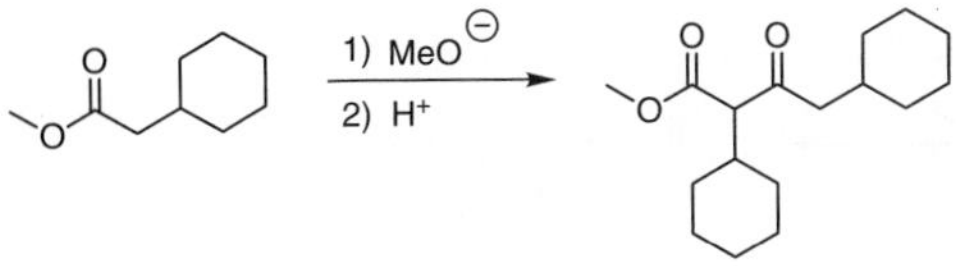

7.62.

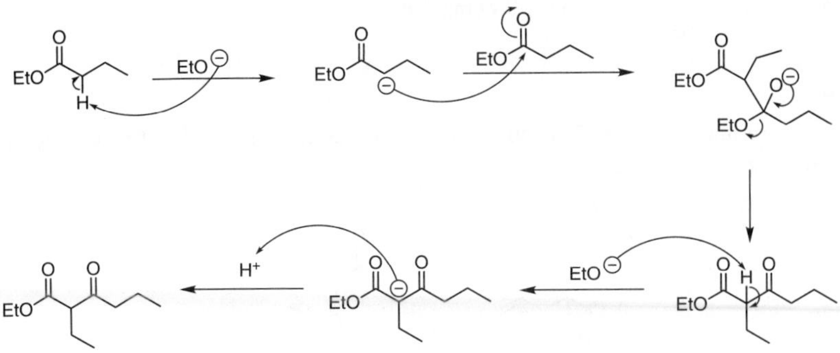

7.63.

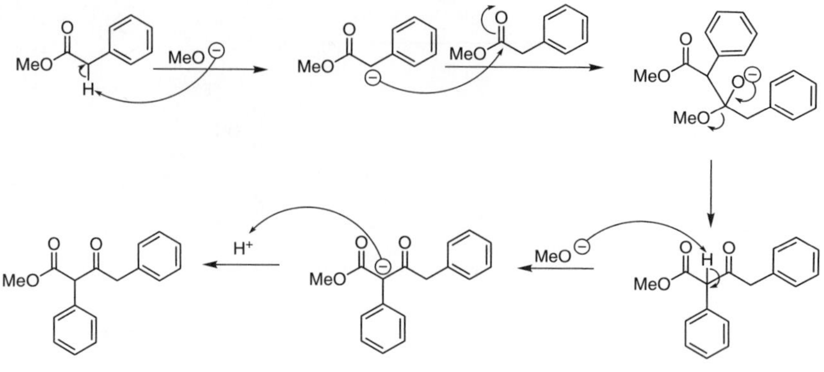

7.64.

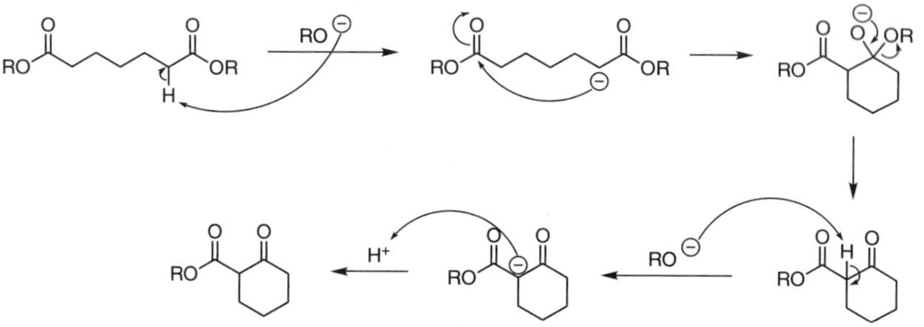

7.66.

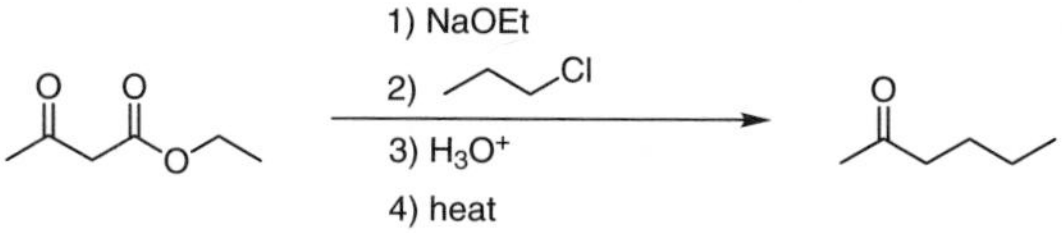

1) NaOEt
2) [structure with Cl]
3) H_3O^+
4) heat

7.67.

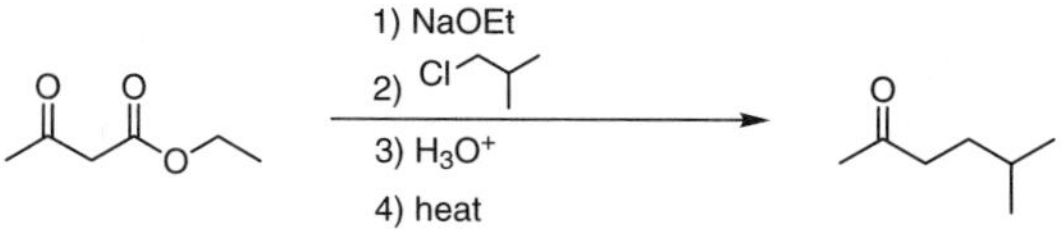

1) NaOEt
2) Cl [structure]
3) H_3O^+
4) heat

7.68.

1) NaOEt
2) Cl [benzyl structure]
3) H_3O^+
4) heat

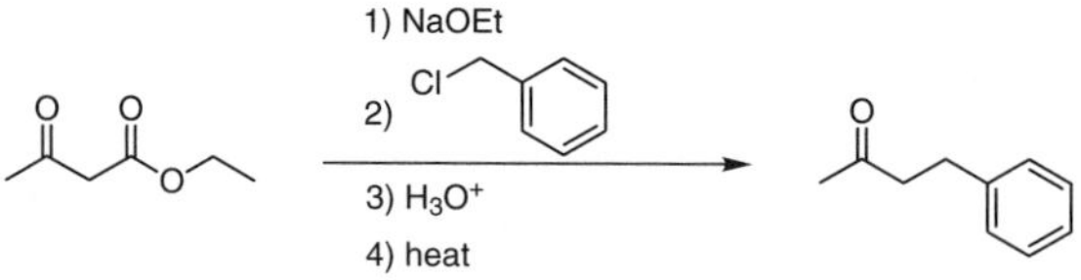

7.69. You would need to use the following halide:

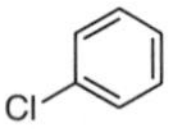

which will not undergo an S_N2 reaction (the carbon atom connected to the leaving group is sp^2 hybridized).

7.70.

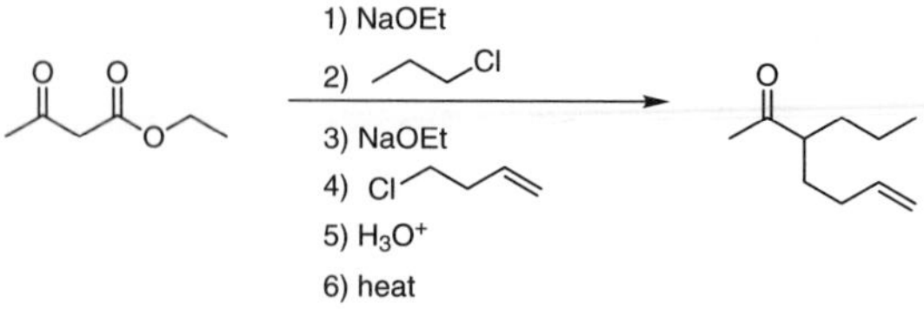

1) NaOEt
2) [structure with Cl]
3) NaOEt
4) Cl [structure]
5) H_3O^+
6) heat

7.71.

1) NaOEt
2) [CH2CH2CH3]Cl
3) NaOEt
4) CH3Cl
5) H3O+
6) heat

7.72.

1) NaOEt
2) [CH2CH2CH3]Cl
3) H3O+
4) heat

7.74.

1) NaOEt
2) [CH2CH2CH3]Cl
3) H3O+
4) heat

7.75.

1) NaOEt
2) [CH2CH2CH3]Cl
3) NaOEt
4) CH3Cl
5) H3O+
6) heat

7.76.

1) NaOEt
2) [benzyl]Cl
3) NaOEt
4) EtCl
5) H3O+
6) heat

7.78.

7.79. Will not give a clean Michael reaction. A Grignard reagent is not a good Michael donor. It is too reactive.

7.80.

7.82.

1)
2) [H+]

7.83.

1) [pyrrolidine] N–H
[H+], Dean-Stark
2)
3) H3O+

7.84.

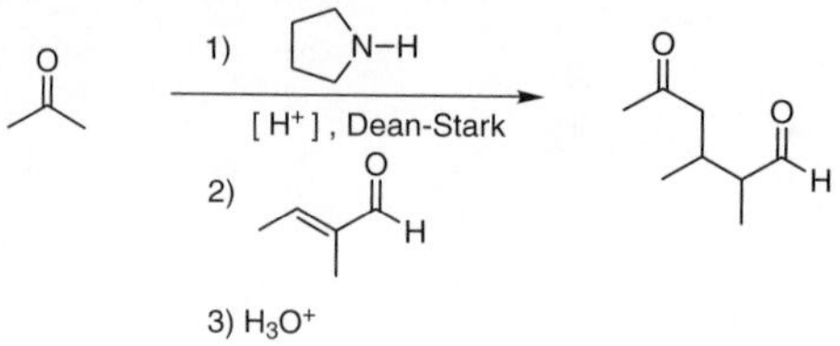

1) pyrrolidine (N–H)

[H⁺], Dean-Stark

2) (but-2-enal, methacrolein)

3) H_3O^+

7.85.

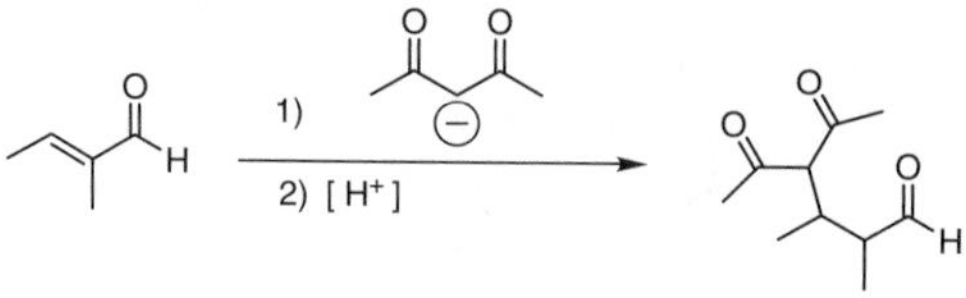

1) (pentane-2,4-dione anion)

2) [H⁺]

CHAPTER 8

8.2.

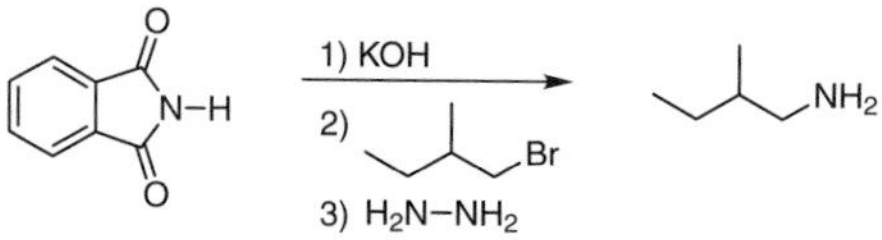

(phthalimide, N–H)

1) KOH

2) (1-bromo-2-methylbutane)

3) H_2N-NH_2

→ (2-methylbutylamine, NH₂)

8.3.

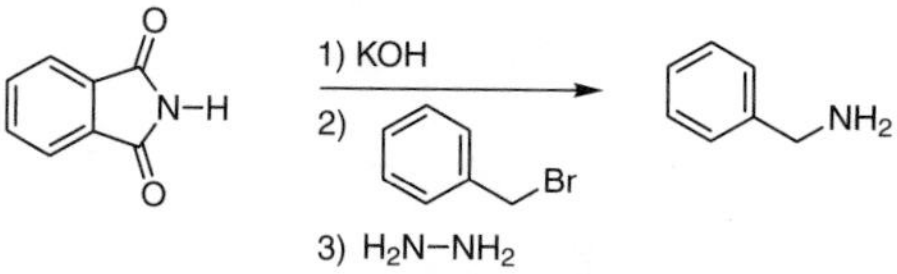

(phthalimide, N–H)

1) KOH

2) (benzyl bromide, Br)

3) H_2N-NH_2

→ (benzylamine, NH₂)

8.4.

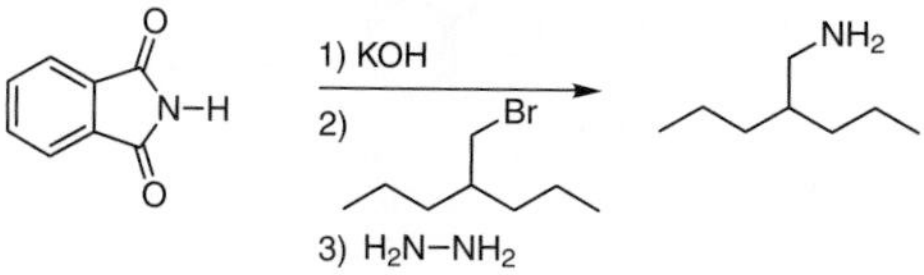

(phthalimide, N–H)

1) KOH

2) (Br compound)

3) H_2N-NH_2

→ (NH₂ product)

8.5.

(phthalimide, N–H)

1) KOH

2) (Br compound)

3) H_2N-NH_2

→ (H_2N product)

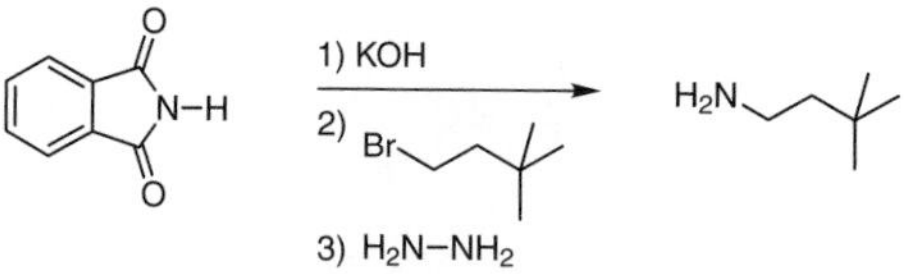

8.7. No   **8.8.** Yes   **8.9.** Yes   **8.10.** No

8.12.

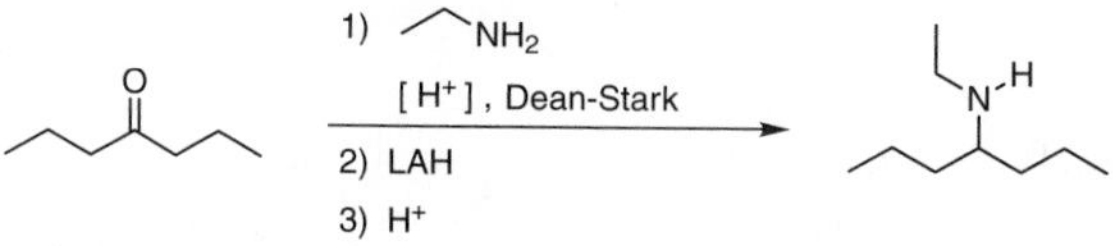

1) (NH_2 amine)

[H⁺], Dean-Stark

2) LAH

3) H^+

→ (N–H amine product)

8.13.

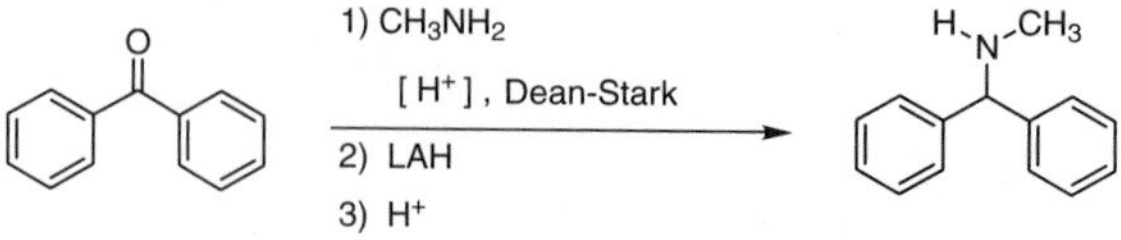

(benzophenone)

1) CH_3NH_2

[H⁺], Dean-Stark

2) LAH

3) H^+

→ ($H-N-CH_3$ product)

8.14.

(benzaldehyde)

1) (H_2N benzylamine)

[H⁺], Dean-Stark

2) LAH

3) H^+

→ (N–H dibenzylamine product)

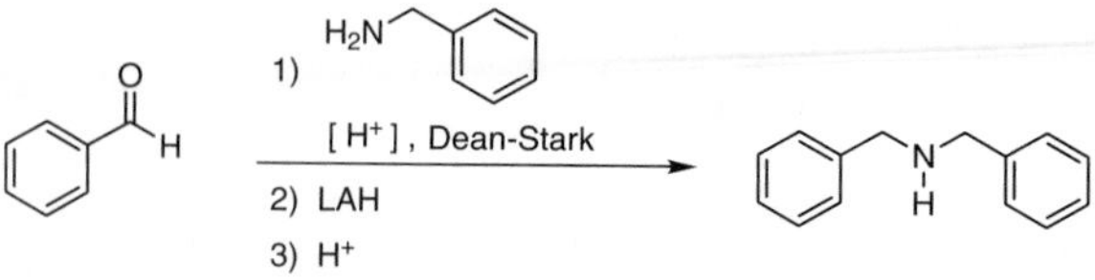

8.15.

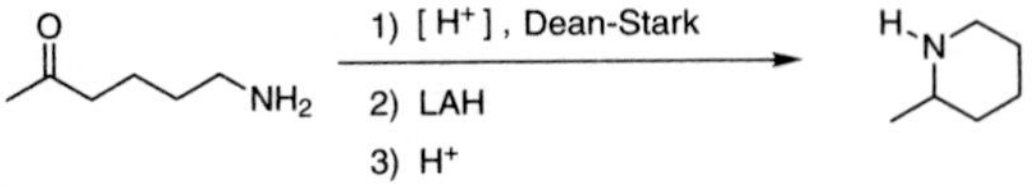

1) [H⁺], Dean-Stark

2) LAH

3) H⁺

8.16.

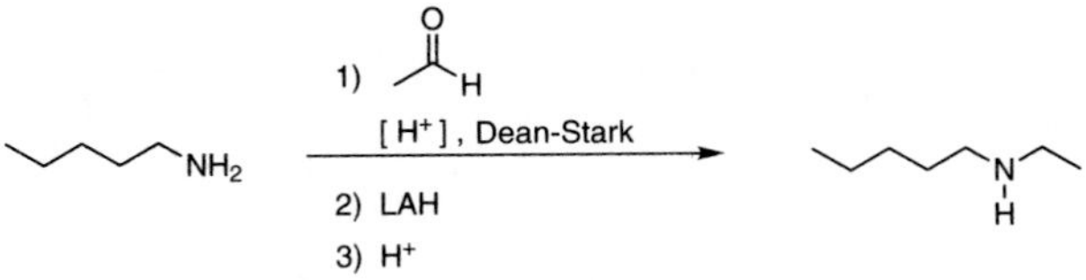

1) acetaldehyde

[H⁺], Dean-Stark

2) LAH

3) H⁺

8.18.

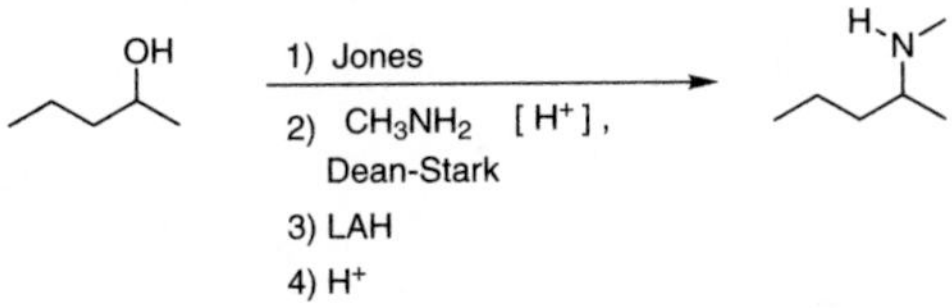

1) Jones

2) CH₃NH₂ [H⁺], Dean-Stark

3) LAH

4) H⁺

8.19.

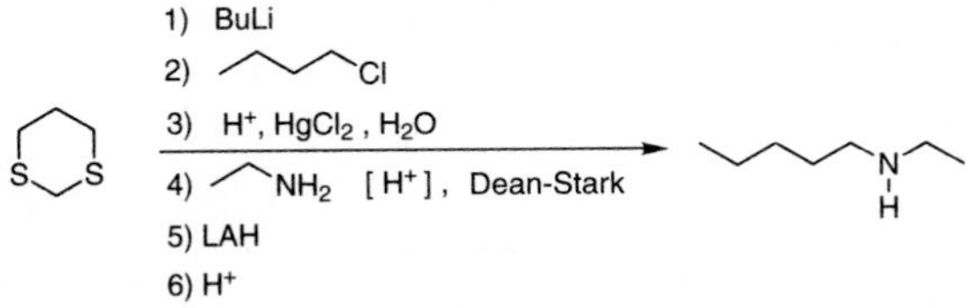

1) BuLi

2) propyl chloride

3) H⁺, HgCl₂, H₂O

4) NH₂ [H⁺], Dean-Stark

5) LAH

6) H⁺

8.20.

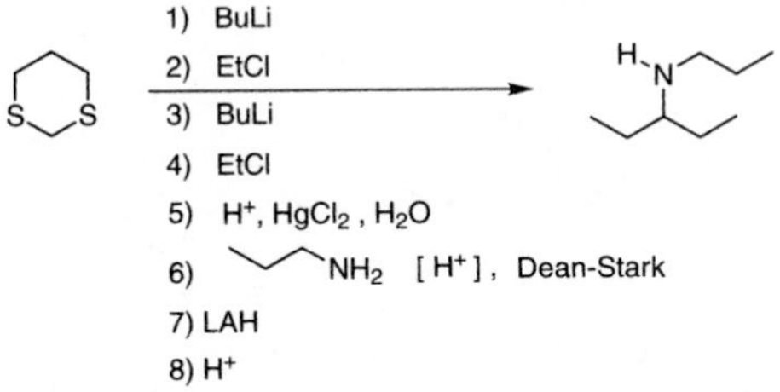

1) BuLi

2) EtCl

3) BuLi

4) EtCl

5) H⁺, HgCl₂, H₂O

6) NH₂ [H⁺], Dean-Stark

7) LAH

8) H⁺

8.21.

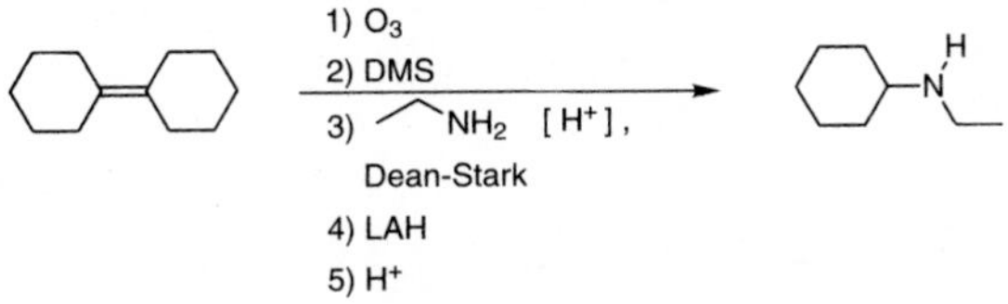

1) O₃

2) DMS

3) NH₂ [H⁺], Dean-Stark

4) LAH

5) H⁺

8.22.

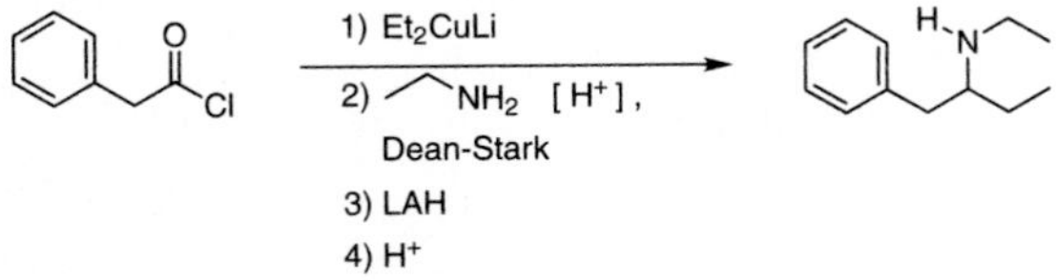

1) Et₂CuLi

2) NH₂ [H⁺], Dean-Stark

3) LAH

4) H⁺

8.24.

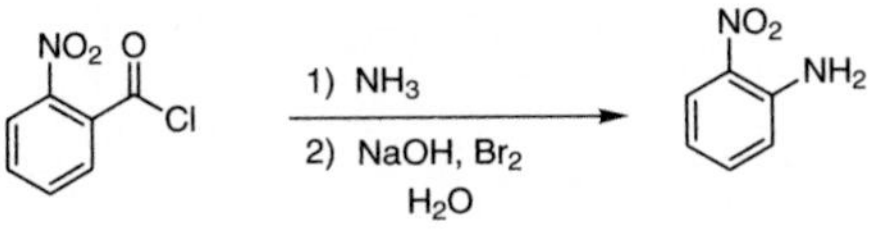

1) NH₃

2) NaOH, Br₂
   H₂O

8.25.

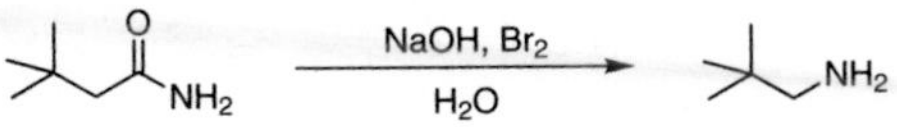

NaOH, Br₂
H₂O

8.26.

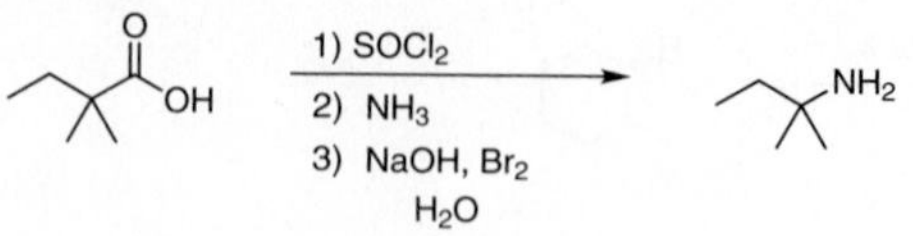
OH
1) $SOCl_2$
2) NH_3
3) NaOH, Br_2
H_2O
NH_2

8.27.

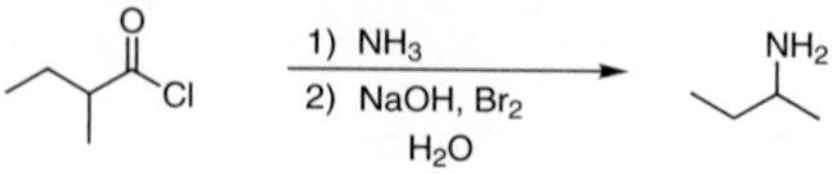
Cl
1) NH_3
2) NaOH, Br_2
H_2O
NH_2

8.28.

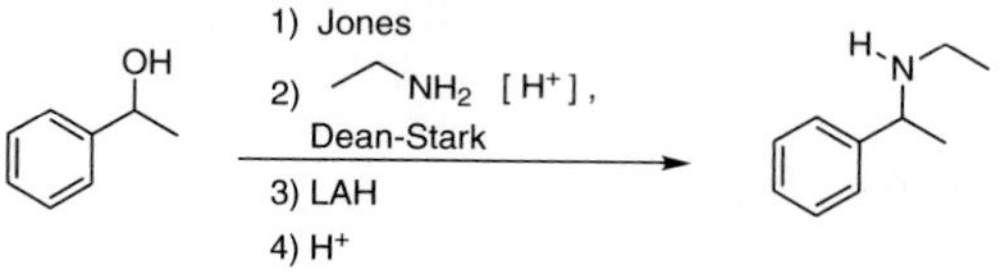
OH
1) Jones
2) NH_2 [H^+],
Dean-Stark
3) LAH
4) H^+
H
N

8.29.

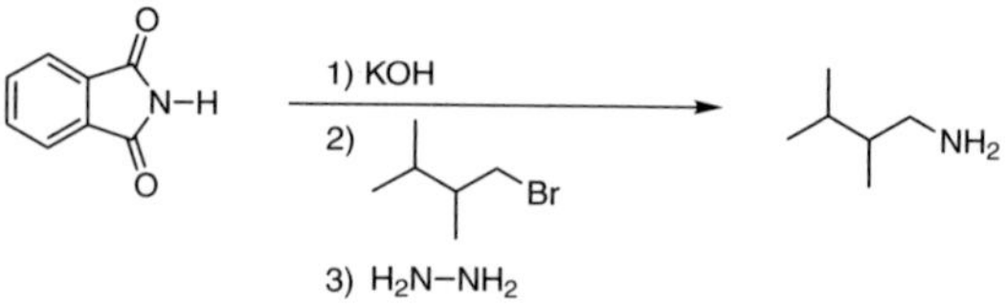
N–H
1) KOH
2)
Br
3) $H_2N–NH_2$
NH_2

8.30.

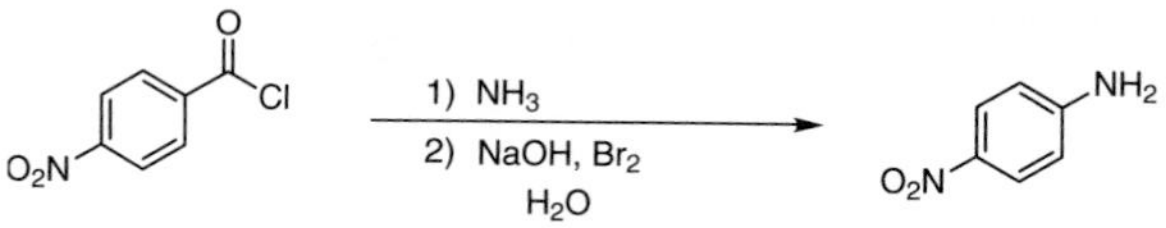
O_2N
Cl
1) NH_3
2) NaOH, Br_2
H_2O
O_2N
NH_2

8.31.

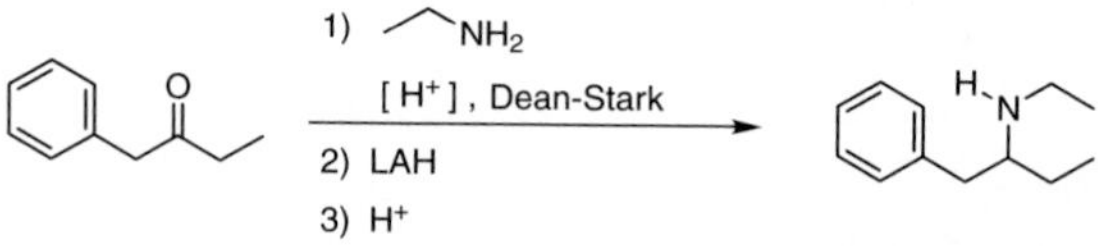
1) NH_2
[H^+], Dean-Stark
2) LAH
3) H^+
H
N

8.32.

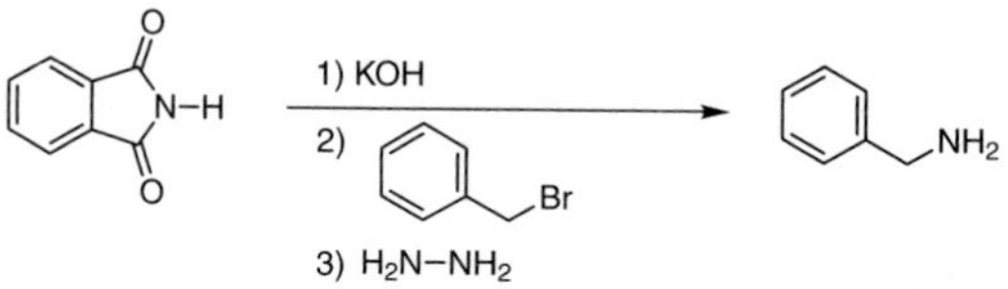
N–H
1) KOH
2)
Br
3) $H_2N–NH_2$
NH_2

8.33.

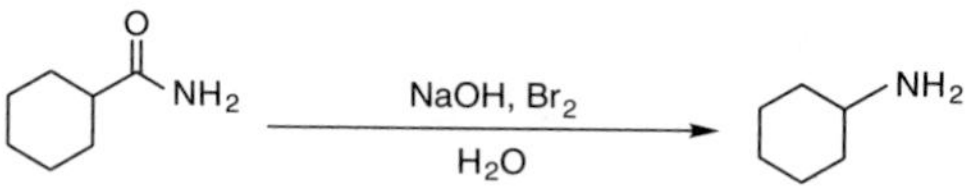
NH_2
NaOH, Br_2
H_2O
NH_2

8.35.

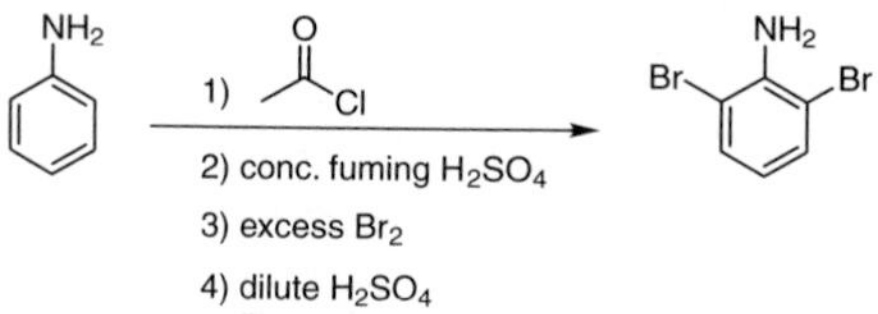
NH_2
1)
Cl
2) conc. fuming H_2SO_4
3) excess Br_2
4) dilute H_2SO_4
NH_2
Br
Br

8.36.

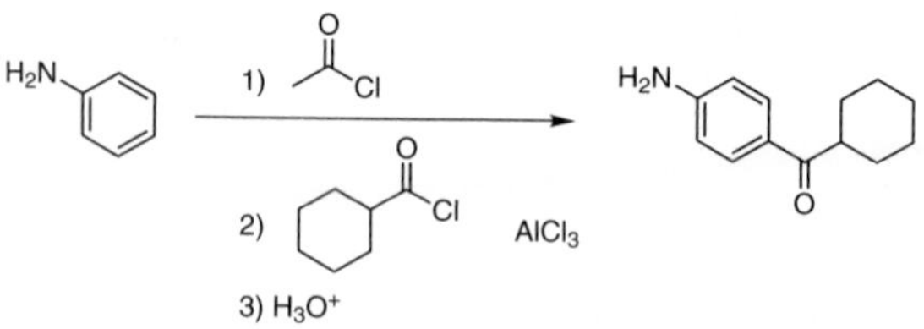
H_2N
1)
Cl
2)
Cl
$AlCl_3$
3) H_3O^+
H_2N
O

8.37.

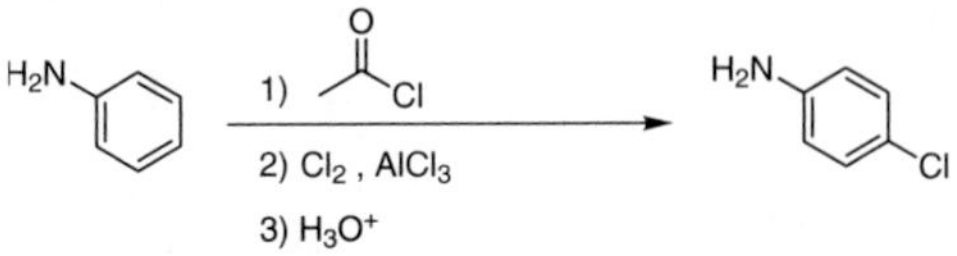
H2N
1) O Cl
2) Cl2 , AlCl3
3) H3O+
H2N
Cl

8.39.

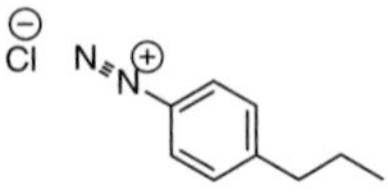
Cl
N
N

8.40.

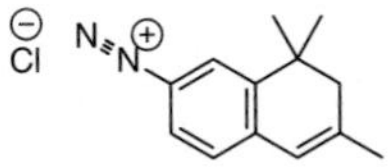
Cl
N
N

8.41.

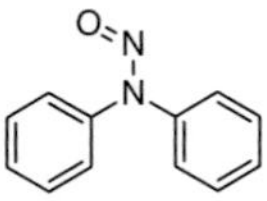
O
N
N

8.42.

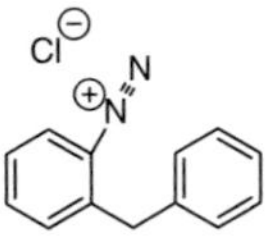
Cl
N
N

8.44.

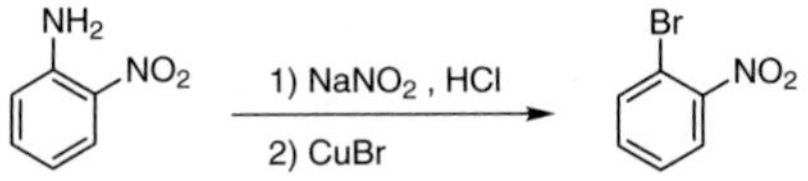
NH2
NO2
1) NaNO2 , HCl
2) CuBr
Br
NO2

8.45.

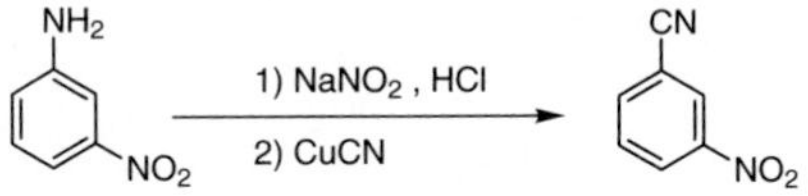
NH2
NO2
1) NaNO2 , HCl
2) CuCN
CN
NO2

8.46.

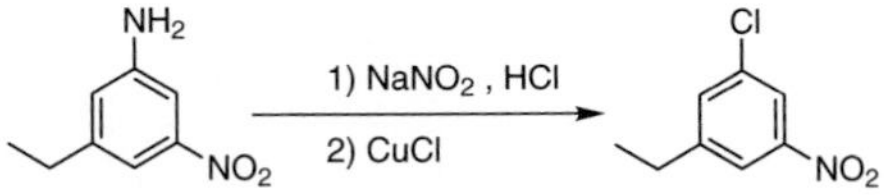
NH2
NO2
1) NaNO2 , HCl
2) CuCl
Cl
NO2

8.47.

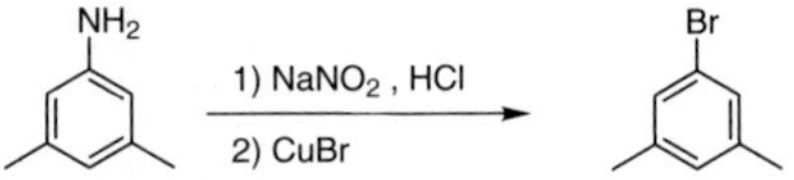
NH2
1) NaNO2 , HCl
2) CuBr
Br

8.48.

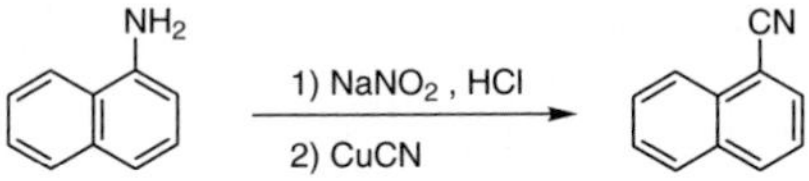
NH2
1) NaNO2 , HCl
2) CuCN
CN

INDEX

311